Light Filaments

Related titles on electromagnetic waves:

Dielectric Resonators, 2nd Edition Kajfez and Guillon
Electronic Applications of the Smith Chart Smith
Fiber Optic Technology Jha
Filtering in the Time and Frequency Domains Blinchikoff and Zverev
HF Filter Design and Computer Simulation Rhea
HF Radio Systems and Circuits Sabin
Microwave Field-Effect Transistors: Theory, design and application, 3rd Edition Pengelly
Microwave Semiconductor Engineering White
Microwave Transmission Line Impedance Data Gunston
Optical Fibers and RF: A natural combination Romeiser
Oscillator Design and Computer Simulation Rhea
Radio-Electronic Transmission Fundamentals, 2nd Edition Griffith, Jr
RF and Microwave Modeling and Measurement Techniques for Field Effect Transistors Gao
RF Power Amplifiers Albulet
Small Signal Microwave Amplifier Design Grosch
Small Signal Microwave Amplifier Design: Solutions Grosch
2008+ Solved Problems in Electromagnetics Nasar
Antennas: Fundamentals, design, measurement, 3rd Edition Blake and Long
Designing Electronic Systems for EMC Duff
Electromagnetic Measurements in the Near Field, 2nd Edition Bienkowski and Trzaska
Fundamentals of Electromagnetics with MATLAB®, 2nd Edition Lonngren *et al.*
Fundamentals of Wave Phenomena, 2nd Edition Hirose and Lonngren
Integral Equation Methods for Electromagnetics Volakis and Sertel
Introduction to Adaptive Arrays, 2nd Edition Monzingo *et al.*
Microstrip and Printed Antenna Design, 2nd Edition Bancroft
Numerical Methods for Engineering: An introduction using MATLAB® and computational electromagnetics Warnick
Return of the Ether Deutsch
The Finite Difference Time Domain Method for Electromagnetics: With MATLAB® simulations Elsherbeni and Demir
Theory of Edge Diffraction in Electromagnetics Ufimtsev
Scattering of Wedges and Cones with Impedance Boundary Conditions Lyalinov and Zhu
Circuit Modeling for Electromagnetic Compatibility Darney
The Wiener–Hopf Method in Electromagnetics Daniele and Zich
Microwave and RF Design: A systems approach, 2nd Edition Steer
Spectrum and Network Measurements, 2nd Edition Witte
EMI Troubleshooting Cookbook for Product Designers Andre and Wyatt
Transmission Line Transformers Mack and Sevick
Electromagnetic Field Standards and Exposure Systems Grudzinski and Trzaska
Practical Communication Theory, 2nd Edition Adamy
Complex Space Source Theory of Spatially Localized Electromagnetic Waves Seshadri
Electromagnetic Compatibility Pocket Guide: Key EMC facts, equations and data Wyatt and Jost
Antenna Analysis and Design Using FEKO Electromagnetic Simulation Software Elsherbeni, Nayeri and Reddy
Scattering of Electromagnetic Waves by Obstacles Kristensson
Adjoint Sensitivity Analysis of High Frequency Structures with MATLAB® Bakr, Elsherbeni and Demir
Developments in Antenna Analysis and Synthesis Vol. 1 and Vol. 2 Mittra
Advances in Planar Filters Design Hong
Post-Processing Techniques in Antenna Measurement Castañer and Foged
Nano-Electromagnetic Communication at Terahertz and Optical Frequencies, Principles and applications Alomainy, Yang, Imran, Yao and Abbasi
Nanoantennas and Plasmonics: Modelling, design and fabrication Werner, Campbell and Kang
Electromagnetic Reverberation Chambers: Recent advances and innovative applications Andrieu
Radio Wave Propagation in Vehicular Environments Azpilicueta, Vargas-Rosales, Falcone and Alejos
Advances in Mathematical Methods for Electromagnetics Kobayashi and Smith
Emerging Evolutionary Algorithms for Antennas and Wireless Communications Goudos

Light Filaments

Structures, challenges and applications

Edited by
Jean-Claude Diels, Martin C. Richardson and
Ladan Arissian

The Institution of Engineering and Technology

Published by SciTech Publishing, an imprint of The Institution of Engineering and Technology, London, United Kingdom

The Institution of Engineering and Technology is registered as a Charity in England & Wales (no. 211014) and Scotland (no. SC038698).

First published 2021

The Institution of Engineering and Technology
Michael Faraday House
Six Hills Way, Stevenage
Herts, SG1 2AY, United Kingdom

www.theiet.org

British Library Cataloguing in Publication Data
A catalogue record for this product is available from the British Library

ISBN 978-1-78561-240-4 (hardback)
ISBN 978-1-78561-241-1 (PDF)

Typeset in India by MPS Limited
Printed in the UK by CPI Group (UK) Ltd, Croydon

Contents

About the editors

Jean-Claude Diels is a professor at the Department of Physics and Astronomy at the University of New Mexico, USA. He co-authored the book, *Lasers: The Power and Precision of Light*, celebrating the 50th anniversary of the laser, and the graduate Textbook: "Ultrashort Laser Pulse Phenomena." He received the 2006 Engineering Excellence Award of the Optical Society of America. His lab pioneered the field of filamentation in air.

Martin C. Richardson is a professor at the College of Optics and Photonics at the University of Central Florida. He has published over 400 scientific articles in professional scientific journals and presented numerous invited and plenary talks. He holds more than 20 patents and has chaired many international conferences, including IQEC, ICHSP, and several SPIE meetings. He is a former Associate Editor of JQE, a recipient of the Schardin Medal, and a Fellow of OSA.

Ladan Arissian holds the appointment of physicist at the National Institute of Standards and Technology (NIST) in Boulder, CO, USA. She previously held the position of a Research Associate Professor at the University of New Mexico in Albuquerque and an Adjunct Professor at the University of Ottawa in Canada. Her areas of expertise are ultrafast optics, precision spectroscopy, and strong-field physics.

Introduction

Jean-Claude Diels[1]

I.1 Light filament, an old concept

With the advent of the laser in 1960, it seemed that a practical source of plane optical waves had been found. The ideal laser is in fact producing Gaussian beams which diffract as documented in the work of Kogelnik and Li [1] reproduced in all graduate textbooks of optics. The earlier report of the possibility of self-trapping came as early as 1962 in a paper of Askar'yan [2]. It is remarkable that his approach resurfaced more than 20 years later, with dielectric microspheres taking the place of electrons [3]. The concept of an intensity-dependent index of refraction resulting in "self-trapping" of a laser beam became popular though the papers of Chiao *et al.* [4] and Talanov [5,6]. Given a sufficiently high intensity, the index of refraction will increase proportionally to the intensity, as indicated by the dashed curve in Figure I.1(a). This spatial dependence of the index of refraction results in a lensing effect. The wave fronts will curve, and the rays, perpendicular to the wave fronts, will bend toward the axis.

There exists a critical *power* of self-trapping. Of the numerous definitions of self-trapping [7], we choose for simplicity to define the critical power P_{cr} as the power for which the phase factor on the axis of the Gaussian beam, $arctan(z/\rho_0)$, exactly compensates the nonlinear phase shift $2\pi n_2 I/\lambda$:

$$P_{cr} = \frac{\lambda_\ell^2}{4\pi n_0 \bar{n}_2}. \tag{I.1}$$

It was observed in the early 1960s that intense nanosecond pulses sent in transparent material left a fine rosary of broken glass. The concept of "filament" took shape when Akhmanov theorized that a Taylor expansion of the intensity-dependent nonlinear index could have a positive first-order term ($n_2 I$) and a negative second-order term ($-n_4 I^2$). As shown by Akhmanov [8], the higher order negative term prevents the collapse of the beam. As the intensity of the self-focused beam increases, the defocusing effect of the higher order term becomes dominant and will lead to the defocusing of the laser pulse. Subsequently, as the beam diameter expands, the self-focusing takes over (Figure I.1(b)) and another cycle of

[1]Center for High Technology Materials and Physics Department, University of New Mexico, Albuquerque, USA

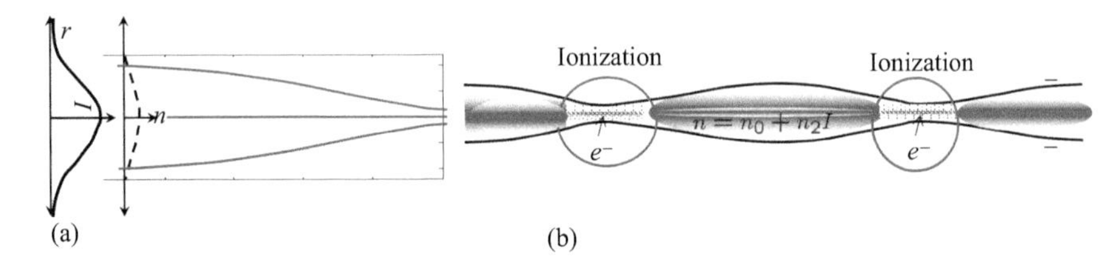

Figure I.1 (a) For high-energy laser pulses, the nonlinear index of refraction of the transparent medium is intensity dependent, which in the case of a Gaussian-like laser pulse is equivalent to a convex lens that leads to a self-focusing phenomenon. (b) Through focusing the intensity on the axis increases until a higher order nonlinearity takes over. In the case of filamentation in air, the higher order nonlinearity is multiphoton ionization resulting in a plasma. The density of the plasma increases until its defocusing effect becomes dominant, leading to defocusing. Subsequently, as the plasma density decreases due to defocusing, the self-focusing takes over and another cycle of focusing–defocusing will be established

focusing–defocusing will be established. This cycle could repeat itself as far as there is sufficient intensity for self-focusing resulting in the formation of a stable waveguide that can exist for distances exceeding the Rayleigh range [8,9]. This was the time of the cold war, and physicists did not fail to participate. For Shen (UC Berkeley), self-induced waveguides do not exist: it is always a "moving focus" [10,11].

In the moving focus model, as the power increases along the leading edge of the pulse, it will reach the critical power (Equation I.1), corresponding to focusing at infinity. As the power increases beyond that point, the focal point recedes toward the source. This spot moving toward the source at superluminal velocity (in the case of short pulses) creates the illusion of a filament [10–12]. The damage tracks in the form of a string of beads produced in transparent solids by intense nanosecond pulses fit the moving focus description. In the case of fs pulses, the moving focus model implies that the few mJ energy observed in a single filament would be stretched over distances of the order of a meter. With the observed diameter of the order of 100 μm, the intensity and energy density are orders of magnitude below the values needed to explain the plethora of nonlinear effects observed. In the other temporal limit, a moving focus cannot exist with cw beams. Stable waveguides have been observed in liquids at very low optical powers, exploiting the nonlinearity of microemulsions [3]. Whether moving focus or self-induced, waveguide does not only depend on the pulse duration. As detailed in Chapter 2, the description of subnanosecond UV filament definitely follows the pattern of self-induced waveguides.

I.2 Filamentation in air

For sufficiently high laser intensity, air itself is a nonlinear medium. As for filamentation in solids, the discovery of air filaments was through laser damage. Tiny "pits" were observed in mirrors placed in the path of an initially uniform high-power laser beam, at several meters from the source. This laser damage was attributed to a nonlinear propagation effect in air for femtosecond near-infrared (IR) [13] and ultraviolet [14] pulses. What was initially perceived as a minor annoyance (pits in a mirror) turned into a major research area, giving birth to a biannual international filament conference series. It appeared to offer an unexpected answer to a call for proposal from the Department of Defense starting with "Given that atmospheric effects impose a fundamental limitation on the propagation of light. . . ." In fact, atmospheric effects impose only a fundamental limitation on linear propagation of light. Early experiments showed that a filament sent through the heat of a candle was unaffected, while the same candle positioned under the pre-filamented macroscopic beam prevented the formation of a filament. Turbulence of the atmosphere is of a much larger scale than the filament transverse dimensions. Radiation at 800 nm is at the limit of the human eye sensitivity. However, mJ energy filamented near an IR pulse creates a spectacular rainbow of colors on a distant target (Figure I.2(a)).

It may sound strange to publish a book on filamentation in air, nearly three decades after the first observation. One would have expected the comprehensive report of Couairon and Mysyrowicz [15] to mark the conclusion of extensive research on this topic. This has not been the case, as new aspects of filament research have proliferated since. This book is also far from closing this area of

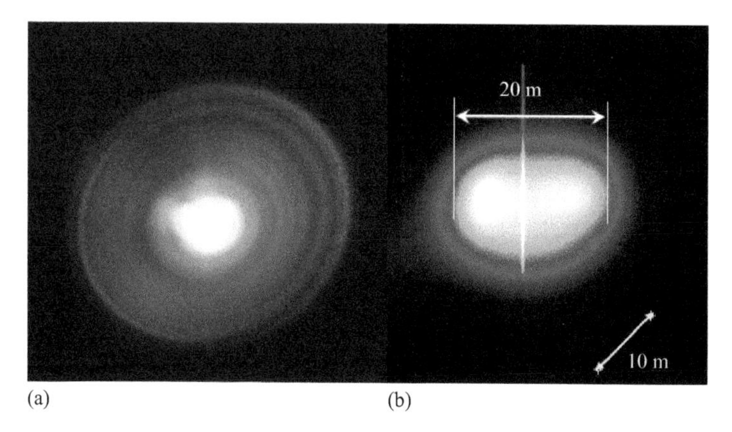

Figure I.2 (a) Typical pattern of conical emission of a 800-nm filament, as observed at 25 m from the laser source. (b) Profile of the white light conical emission of filaments observed on a cirrus cloud at 10-km altitude. The pattern has a diameter of 20 m. Pictures courtesy of J.P. Wolf and J. Kasparian, University of Geneva

research. With the development of high repetition rate sources, new aspects of waveguiding are emerging based on the hydrodynamic perturbation created by each filamenting pulse.

The ebb and flow of funding were often driven by a misinterpretation of research papers. A major rush for funding came after publication of the trace of a filament sent skyward [16],* and the recording of range-gated spectra of atmospheric constituent at up to 13 km [17]. Many casual readers have concluded that filaments propagated for at least 13 km (even though there was no statement to that effect in these reports). It is the white light (conical emission) that is projected by multiple filaments onto the sky, as shown in Figure I.2(b), which shows the white light produced by multiple filaments and projected on a cirrus cloud at 10-km altitude.

It has been established that a single filament at 800 nm carrying a few mJ of energy propagates only distances of the order of a few meters [18]. Pulses of the order of hundreds of mJ are known to break up into multiple filaments. Interaction between these filaments is a complex problem still being investigated (Chapter 5 of this book). Through fusion (concatenation of these filaments [19,20]), plasma trails have been generated over distances of the order of 100 m [21]. The title "Kilometer-range filamentation" [21] may have misled some superficial readers that km-long filaments had been produced. The paper demonstrated in fact a postponement of the start of filamentation by temporal focusing. A downchirped pulse is launched into the atmosphere, to be compressed by its positive dispersion, a technique proposed earlier for UV filaments [22].

I.3 It is more complex than a simple balance of focusing–defocusing

Laser filamentation is among the most fascinating topics in light and matter interaction where the optical pulse is modified due to propagation in plasma. Most often, in studies of light–matter interaction, the result of the interaction is a modification of matter. An electron, an ion, or a photon that is generated by ionization, or a photon generated by a high harmonic process, are the tangible measurements that allow us to reconstruct the scene of the interaction. After the laser pulse has been sapped by the interaction, it is often ignored. In filament studies, however, the generated plasma leaves its imprint on the optical pulse itself. A modified index of refraction due to a contribution of chemical bonds, free electrons, and ions creates a waveguide for the light to propagate beyond the Rayleigh range. As the laser pulse forms and modifies the plasma, it is, in turn, modified by the plasma.

The model of self-induced waveguide is an elegant oversimplification; the reality is more complex. The sheer number of physical phenomena associated with the filaments, such as conical emission, THz emission, harmonic generation, spatial

*Figure 12.1 of Chapter 12.

replenishment [23], self-healing [24], made this field rich of physics and potential applications, but considerably more complex than anticipated. To add confusion to the interpretation, there is an obscure curtain of the order of a meter that separates the prepared initial condition from the observed filament. Substantial temporal [25,26] and spatial [27,28] reshaping of the initial pulse profile takes place in amplitude and phase, such that the pulse ultimately reaching a self-focus of the size of the filament is very different from the macroscopic beam launched in the atmosphere. In addition, we have shown that the state of polarization of the beam is changed during the process of self-focusing that precedes filamentation [29,30].

Filament is a manifestation of beam reshaping due to nonlinear absorption in the preparation (self-focusing) phase of the filament, resulting in "axicon" self-focusing, which is known to result in non diffracting "Bessel beams" [31]. Another model that has been proposed is based on the measurement of the power dependence of the nonlinear index indicating a sign reversal at the intensities typical of 800-nm filaments [32]. At the heart of the difficulties and confusion in the understanding of IR filaments is that the intensities in the filament core reach tens of TW/cm^2. As a result of the high peak intensity and short pulse duration, a plethora of nonlinear effects, such as self-phase modulation, four-wave mixing, self-steepening and pulse compression, third harmonic generation, THz-generation, and Raman processes to cite a few, occur.

Beyond the fs, ps, and ns, there is a much slower time constant that has generated recent interest: the shock wave generated by a filament subsists for a fraction of ms, opening the prospects of generating a permanent acoustic waveguide with high repetition rate filaments [33]. At the single optical period timescale, long-wavelength single filaments also offer a promise of creating km-long self-guided channels [34].

I.4 Chapters organization

This book is organized from general overviews to more specialized topics. The first few chapters provide a comparison of filamentation in different wavelength ranges. The bulk of the research over the last decades has concentrated on the Ti:sapphire laser systems. Chapter 1, by Aurélien Houard and André Mysyrowicz of the Laboratoire d'Optique Appliquée (LOA), presents a broad picture of the state of the art of filamentation at 800 nm, with applications to laser-induced discharges, improving the speed of trains, supersonic drag reduction, plasma antenna, coherent THz emission, and lasing in air.

Chapter 2 by Ali Rastegari, Alejandro Aceves and Jean-Claude Diels from the University of New Mexico, (A.R. and J.-C. D.) and Southern Methodist University (A. A.) introduces filamentation at UV wavelengths (<300 nm). The higher energy of the UV filament is advantageous for a shadowgraphy study of the shock wave generated by filaments. A similar shadowgraphy study leads to the application of UV filaments to high-resolution (isotopic) laser-induced breakdown spectroscopy of solids at atmospheric pressure. Another application where UV filaments have an edge is laser-induced discharges.

The next two chapters relate to filamentation at the other end of the spectrum. Chapter 3 by S. Ya. Tochitsky and C. Joshi of the University of California, Los Angeles (UCLA) presents filament research at the longest wavelength end of the spectrum. As detailed in Chapter 2, the longer the wavelength, the shorter the pulse duration and the higher the peak power required. The UCLA group met this challenge by using CO_2 laser amplifiers at a combination of pressure† and power broadening, such as to reach several Joules in 3 ps. Self-guiding in air was demonstrated with this multiterawatts 10-μm pulse. Numerical modeling indicates that a high-power long-wavelength IR laser holds the promise of forming a single-centimeter-diameter channel km-scale atmospheric propagation.

Considerable progress has also been made in the development of mid-IR sources that may become the workhorse of a new generation of filaments. These developments are presented in Chapter 4 by an international team led by Jens Biegert from ICFO – Institut de Ciencies Fotoniques, The Barcelona Institute of Science and Technology.

The previous chapters were dealing mostly with single filaments at a single wavelength. High-energy beams typically produce multiple filaments. It is desirable to have some control over these filaments and understand their interactions. Chapter 5 by Shermineh Rostami of NanoSpective Inc. and CREOL (University of Central Florida) in Orlando presents an overview of filament-to-filament interactions.

Consideration of other nonlinear processes (stimulated Raman scattering, 3:1 resonance‡ for two-color filaments) and some detailed description of numerical approaches used to study such complex dynamics is the topic of Chapter 6 by Alejandro Aceves, Alexander Sukhinin, and Edward Downes from Southern Methodist University (SMU).

The confined real estate of laboratories does not generally give experimentalists the luxury to study filamentation with collimated beams. Focused beams are often used to shorten the distance needed for filaments to form. This brings up a fundamental question: does a focused beam still form a filament? This question is answered in detail in Chapter 7 on "Filamentation of femtosecond laser pulses under tight focusing" by Andrey Ionin, Daria Mokrousova, and Leonid Seleznev from the Lebedev Institute in Moscow.

The initial condition for creating a filament is not limited to sending a collimated or convergent Gaussian beam. With proper beam shaping, can one generate curved filaments? This intriguing possibility, and many others, is addressed in Chapter 8 on "Linear and nonlinear exotic light wave packets, physics and applications" by Dimitris G. Papazoglou and Stelios Tzortzakis of the University of Crete, Heraklion.

It was mentioned earlier in this introduction that filamentation involves a number of physical phenomena, too numerous to be incorporated on a simple analytical description. However, a comprehensive numerical simulation has been developed and is the topic of Chapter 9 by Arnaud Couairon from the "Institut

†All rotational lines of P branch are merged in a broad band at pressures >10 atm.

‡3:1 resonance refers to the ratio of UV/IR frequencies.

Polytechnique de Paris," Christoph Heyl from DESY (Hamburg), and Cord Arnold from Lund University.

The Kerr effect not only plays a central role in the formation of filaments but also in the generation of the white light conical emission. In the pre-filamentation formation stage, the Kerr effect can be exploited to amplify a seed pulse. This phenomenon called Kerr instability amplification is the topic of Chapter 10 by M. Nesrallah, T. J. Hammond, A. Almalki, G. Bart, C. R. McDonald, T. Brabec, and G. Vampa from the University of Ottawa.

Equally important as the Kerr effect is the role of the defocusing plasma, which is the object of Chapter 11 on "Plasma studies in filament" by Xin Lu, Tingting Xi, and Jie Zhang of the Chinese Academy of Sciences.

This overview of filaments is complemented by a chapter dedicated to the impact the filaments had and will continue to have on atmospheric science. Chapter 12 on "Filamentation for atmospheric remote sensing and control" by Jérome Kasparian and Jean-Pierre Wolf covers a whole range of applications from light detection and ranging, remote spectroscopy, whispering gallery modes in water droplets, laser-induced condensation, cloud research, and laser-induced lightning.

Last but not least, Chapter 13 by Martin Richardson et al. of CREOL, University of Central Florida, covers multiple filament engineering, and applications to ablation, RF emission, acoustic shock waves and high altitude propagation.

References

[1] H. W. Kogelnik and T. Li. Laser beams and resonators. *Appl. Opt.*, 5: 1550–1567, 1966.

[2] G. A. Askar'yan. Effects of gradients of a strong electromagnetic beam on electrons and atoms. *Sov. Phys. JETP*, 15: 1088–1090, 1962.

[3] E. Freysz, M. Afifi, A. Ducasse, B. Pouligny, and J. R. Lalanne. Giant optical non-linearities of critical micro-emulsions. *J. Phys. Lett.*, 46: 181–187, 1985.

[4] R. Y. Chiao, E. Garmire, and C. H. Townes. Self-trapping of optical beams. *Phys. Rev. Lett.*, 13: 479–482, 1964.

[5] V. I. Talanov. Self-focusing of electromagnetic waves in nonlinear media. *Radiophys*, 8: 257, 1964.

[6] V. I. Talanov. Self-focusing of waves in nonlinear media. *JETP Lett.*, 2: 138, 1965.

[7] J.-C. Diels and W. Rudolph. *Ultrashort Laser Pulse Phenomena*. Elsevier, ISBN 0-12-215492-4; second edition, Boston, MA, 2006.

[8] S. A. Akhmanov, A. P. Sukhorukov, and R. V. Khokhlov. Self-focusing and self-trapping of intense light beams in a nonlinear medium. *Sov. Phys JETP*, 23: 1025–1033, 1966.

[9] S. A. Akhmanov, A. P. Sukhorukov, and R. V. Khokhlov. Development of an optical waveguide in the propagation of light in a nonlinear medium. *Sov. Phys JETP*, 24: 198–201, 1967.

[10] Y. R. Shen and M. M. Loy. Theoretical interpretation of small-scale filaments of light originating from moving focal spots. *Phys. Rev. A*, 3: 2099–2105, 1971.

[11] Y. R. Shen. *The Principles of Nonlinear Optics*. John Wiley & Sons, New York, NY, 1984.

[12] A. Brodeur, C. Y. Chien, F. A. Ilkov, S. L. Chin, O. G. Kosareva, and V. P. Kandidov. Moving focus in the propagation of ultrashort laser pulses in air. *Opt. Lett.*, 22: 304–306, 1997.

[13] A. Braun, G. Korn, X. Liu, D. Du, J. Squier, and G. Mourou. Self-channeling of high-peak-power femtosecond laser pulses in air. *Opt. Lett.*, 20: 73–75, 1995.

[14] X. M. Zhao, P. Rambo, and J.-C. Diels. Filamentation of femtosecond UV pulses in air. In *QELS, 1995*, volume 16, page 178 (QThD2), Baltimore, MA, 1995. Optical Society of America.

[15] A. Couairon and A. Mysyrowicz. Femtosecond filamentation in transparent media. *Phys. Rep.*, 441: 49–189, 2007.

[16] L. Woeste, S. Wedeking, J. Wille, *et al.* Femtosecond atmospheric lamp. *Laser Optoelektron.*, 29: 51–53, 1997.

[17] P. Rairoux, H. Schillinger, S. Niedermeier, *et al.* Remote sensing of the atmosphere using ultrashort laser pulses. *Appl. Phys. B*, 71: 573–580, 2000.

[18] M. Kolesik, D. Mirell, J. C. Diels, and J. V. Moloney. On the higher-order Kerr effect in femtosecond filaments. *Opt. Lett.*, 35: 3685–3687, 2010.

[19] S. Tzortzakis, L. Bergé, A. Couairon, M. A. Franco, B. S. Prade, and A. Mysyrowicz. Breakup and fusion of self-guided femtosecond light pulses in air. *Phys. Rev. Lett.*, 86: 5470–5473, 2001.

[20] S. Tzortzakis, G. Méchain, G. Patalano, M. A. Franco, B. S. Prade, and A. Mysyrowicz. Concatenation of plasma filaments created in air by femtosecond infrared laser pulses. *Appl. Phys. B*, 76: 609–612, 2003.

[21] M. Durand, A. Houard, B. Prade, *et al.* Kilometer range filamentation. *Opt. Express*, 21: 26836–26845, 2013.

[22] X. M. Zhao, J.-C. Diels, C. Y. Wang, and J. Elizondo. Femtosecond ultraviolet laser pulse induced electrical discharges in gases. *IEEE J. Quantum Electron.*, 31: 599–612, 1995.

[23] M. Mlejnek, E. M. Wright, and J. V. Moloney. Dynamic spatial replenishment of femtosecond pulse propagating in air. *Opt. Lett.*, 23: 382–384, 1998.

[24] M. Kolesik and J. V. Moloney. Self-healing fs light filaments. *Opt. Lett.*, 29: 590–592, 2004.

[25] A. C. Bernstein, T. S. Luk, T. R. Nelson, J.-C. Diels, and S. Cameron. Observation of multiple pulse-splitting of ultrashort pulses in air. In *QELS 2001, QFC6*, pages 262–263, Baltimore, MY, 2001. Optical Society of America.

[26] A. C. Bernstein, T. S. Luk, T. R. Nelson, A. McPherson, J.-C. Diels, and S. M. Cameron. Asymmetric ultra-short pulse splitting measured in air using frog. *Appl. Phys. B*, B75 (1): 119–122, 2002.

[27] K. D. Moll, G. Fibich, and A. L. Gaeta. Self-similar optical wave collapse: Observation of the Townes profile. *Phys. Rev. Lett.*, 90: 203902, 2003.

[28] T. D. Grow, A. A. Ishaaya, L. T. Vuong, A. L. Gaeta, R. Gavish, and G. Fibich. Collapse dynamics of super-Gaussian beams. *Opt. Express*, 14: 5468–5475, 2006.

[29] L. Arissian, D. Mirell, J. Yeak, S. Rostami, and J.-C. Diels. Effect of light polarization on plasma distribution and filament formation. In *Proceedings of the 12th International Conference on Multiphoton Processes*, Sapporo (Japan), 2011.

[30] D. Close, C. Giuliano, R. Hellwarth, L. Hess, F. McClung, and W. Wagner. Evolution of circularly polarized light in a Kerr medium. *IEEE J. Quantum Electron.*, QE-2: 553–557, 1966.

[31] A. Dubietis, E. Gaisauskas, G. Tamosauskas, and P. Di Trapani. Light filaments without self-channeling. *Phys. Rev. Lett.*, 92: 253903-1–253903-5, 2004.

[32] V. Loriot, E. Hertz, O. Fauchet, and B. Lavorel. Measurement of high order Kerr refractive index of major air components. *Opt. Express*, 17: 13439–13434, 2009.

[33] Y.-H. Cheng, J. K. Wahlstrand, N. Jhajj, and H. M. Milchberg. The effect of long timescale gas dynamics on femtosecond filamentation. *Opt. Express*, 21: 4740, 2013.

[34] S. Tochitsky, E. Welch, and C. Joshir. Megafilament in air formed by self-guided terawatt long-wavelength infrared laser. *Nat. Photonics*, 13: 41–46, 2018.

Chapter 1

Femtosecond laser filamentation and applications

Aurélien Houard and André Mysyrowicz

Femtosecond filamentation is a spectacular phenomenon whereby a short laser pulse propagates non-linearly through a transparent medium allowing high peak light intensities to be transferred over distances far exceeding the beam Rayleigh length, as if diffraction were suppressed. In this chapter, after a brief description of some key features of filamentation, we put the accent upon some potential applications.

1.1 Basic concepts

The filamentation process is initiated by the optical Kerr effect, responsible for a light intensity-dependent index of refraction. A positive self-induced change of refractive index, even if minute, is cumulative upon propagation and leads to beam self-focusing. Self-focusing overcomes diffraction provided the initial power of the pulse exceeds a critical value P_{cr} [1]. In air, P_{cr} is a few GW for a pulse at 800 nm of 50-fs duration corresponding to a few mJ of pulse energy (in a transparent solid such as fused silica, a few μJ are sufficient). Upon further propagation, the peak intensity becomes high enough to ionize air molecules in a multiphoton process. Beam collapse is then arrested by high field ionization and the defocusing effect of the ensuing plasma. A complex behaviour emerges, where diffraction, group velocity dispersion, optical Kerr effect, absorption due to ionization, space-time focusing, pulse self-steepening, plasma defocusing and plasma reabsorption all contribute. A filament core surrounded and replenished by a reservoir of photons bath is emerging, as shown in Figure 1.1 showing the calculated propagation of a laser pulse centred at 800 nm with initial power $P{\sim}P_{cr}$.

One can distinguish three stages of propagation. The first stage occurs in the absence of ionization and corresponds to the self-focusing regime. The second stage begins with the onset of ionization. It leads first to a continuous plasma column followed by quasiperiodic cycles of focusing/defocusing due to

Laboratoire d'Optique Appliquée, ENSTA Paris, CNRS, Ecole Polytechnique, Institut Polytechnique de Paris, Palaiseau, France

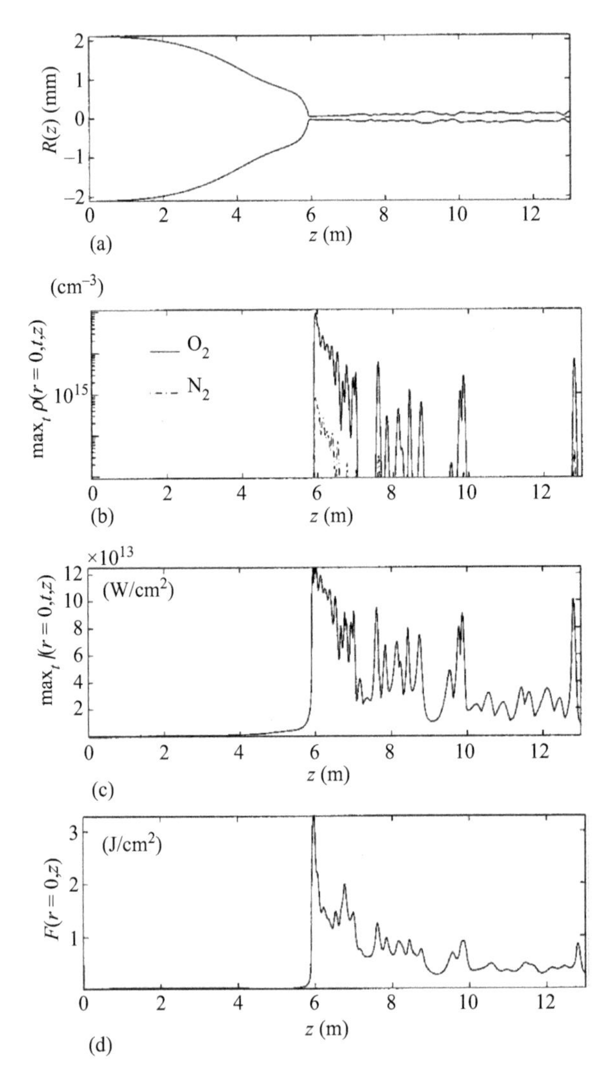

Figure 1.1 (a) Mean radius, (b) peak electron density, (c) intensity, (d) fluence of the filament core as a function of distance for a collimated input pulse at 800 nm with 5-mJ energy and 50-fs duration [2]

competition between the optical Kerr effect and defocusing, and then by more intermittent focusing/defocusing cycle as the pulse power progressively decreases. Finally, in a third stage, not shown in the figure, the pulse power falls below P_{cr}. Bright light channels, stabilized by a competition between self-focusing and diffraction, propagate with little loss over long distances with a beam diameter increasing slowly [3]. The theoretical description of the filamentary propagation requires extensive simulations. Because of space-time coupling, there are no

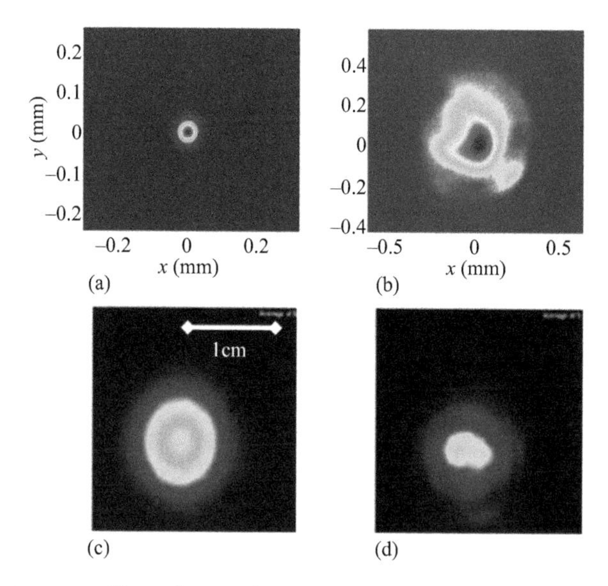

Figure 1.2 Beamprofile in fused silica: after filamentation (a) and without filamentation (b) [9]. Beam profile measured in air 19 m from an fs laser at 800 nm: after filamentation (c) and without filament (d) [8]

analytical solutions. Several reviews on the theory of filamentation are accessible [4–6]. In this chapter, we briefly put the accent on some characteristic features of filaments before describing some possible applications.

1.2 Self-actions during filamentation

1.2.1 Filamentation is characterized by multiple self-actions

Beam self-stabilization: Following beam collapse, the diameter of the beam becomes stable, as seen on a large scale $l \gg d$, where d is average filament diameter. A microscopic inspection reveals a more complex behaviour, with the beam diameter slightly varying with distance due to the dynamic competition between Kerr self-focusing and plasma defocusing, as shown in Figure 1.1.

Pulse self-cleaning: The onset of ionization is accompanied by an improvement of the beam spatial profile, as seen in Figure 1.2 [7,8]. At high incident peak powers $P \gg P_{cr}$, the beam splits into a multifilament pattern. Here again, self-cleaning of the multifilament pattern has been observed, as shown in Figure 1.3.

Self-frequency broadening: The rapid variation of the refractive index during the pulse and the ionization process leads to an important spectral broadening of the pulse. In air, the broadband continuum generated by an fs pulse at 800 nm covers the entire visible part of the spectrum [11].

Self-shortening of the pulse duration: Complex interactions occurring during beam collapse lead to successive pulse splitting and shortening, as shown in

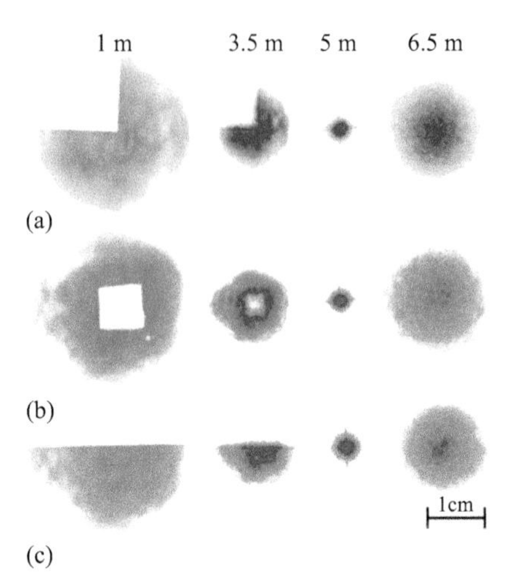

Figure 1.3 *Burningpatterns on photographic plates showing the beam profile of a 700-fs laser pulse with incident power* P~*102* P_{cr}. *The strongly distorted initial profile of the focused beam, recorded at different distances from the laser, exhibits near-complete recovery of circular profile after focus (from [10]). A mask reduces the input laser energy by 25% in (a), 10% in (b) and 50% in (c)*

Figure 1.4 [12]. Extraction of the pulse at the right distance can yield pulses of duration close to the single cycle limit, as further discussed in Section 1.3.

Group velocity dispersion: Group velocity dispersion plays an important role in filamentation. This can be understood by considering the role of the time-dependent change of refractive index during the pulse (see Figure 1.5). In the normal dispersion region, the ascending (descending) part of the pulse generates instantaneous frequencies on the redder (bluer) side of the pulse spectrum. Such frequencies separate from the pulse peak upon propagation if the system in the normal dispersion region. By contrast, in the anomalous dispersion, the generated frequencies are sent back to the peak of the pulse, leading to pulse stabilization. As an example, Figure 1.6 compares the simulated propagation of a femtosecond laser pulses in the normal and anomalous dispersion region of fused silica. With a laser pulse at 800 nm (corresponding to the normal dispersion region), multiple pulse splitting and shortening is apparent. By contrast, a laser pulse of the same incident peak intensity exhibits a smooth profile in the anomalous dispersion region. A self-guided pulse akin to a lossy spatio-temporal soliton emerges when $P{\sim}P_{\mathrm{cr}}$ [9].

In recent years, several groups have started investigating short laser pulse propagation at mid IR wavelengths, where group velocity dispersion in air is low or even negative. The mechanism preventing beam collapse is different from the case of visible or near IR pulses. We refer to [13] and Voronin and Zheltikov [14] for work in this domain.

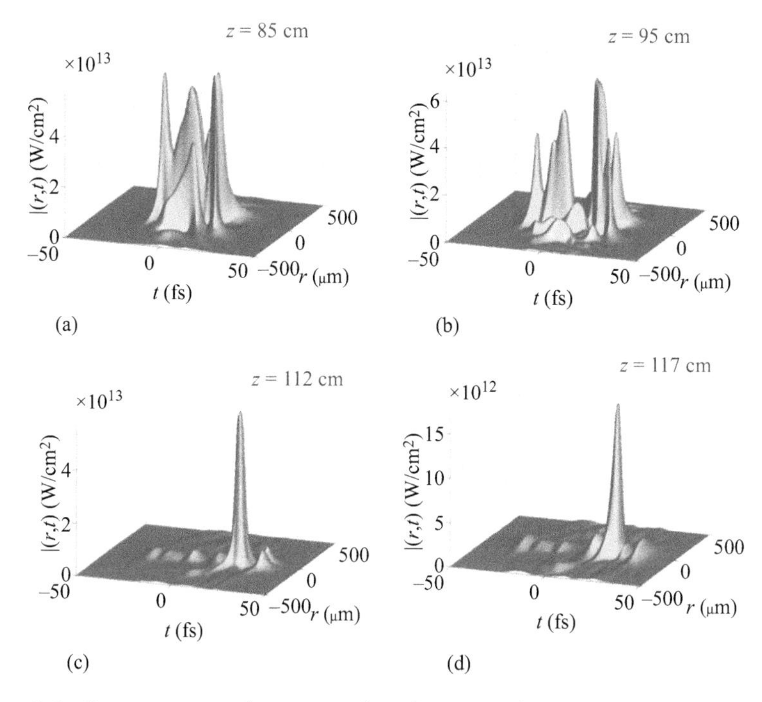

Figure 1.4 Spatio-temporal intensity distribution at (a) 85 cm, (b) 95 cm, (c) 112 cm, and (d) 117 cm after the focusing lens [12]

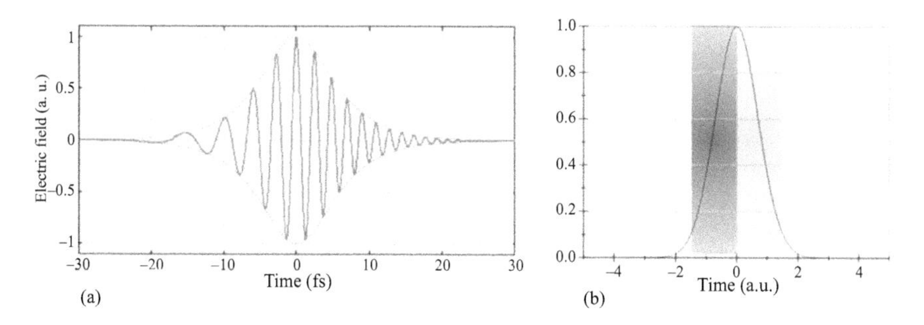

Figure 1.5 Self-phase modulation shifts the spectrum in the front and the back of the pulse to the red and blue, respectively

Formation of low-density channels in air: Following beam collapse in air, part of the pulse energy is deposited in a long channel through Raman rotational excitation, ionization and inverse Bremsstrahlung [15]. This fast energy deposition is responsible for a sudden heating of air and the formation of a radially expanding cylindrical shockwave, leaving behind a central low-density channel with the same geometry as the filament, as shown in Figure 1.7 [16]. This under-dense channel then diffusively decays over a millisecond timescale, as shown in Figure 1.8.

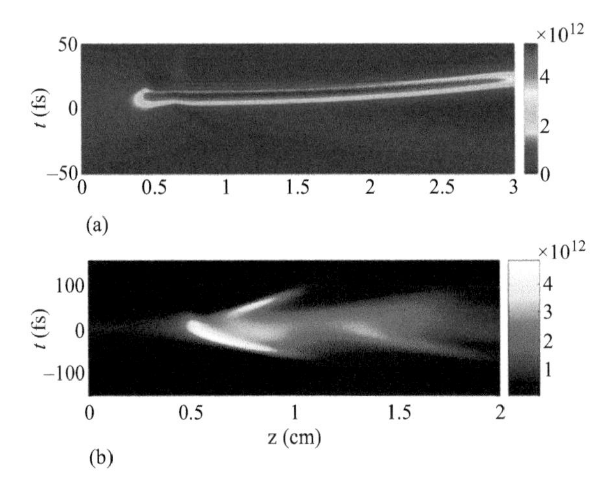

Figure 1.6 Numerical calculation of the pulse evolution during filamentation in silica for a femtosecond pulse with a central wavelength at 1.6 μm (a) and at 800 nm (b) [9]

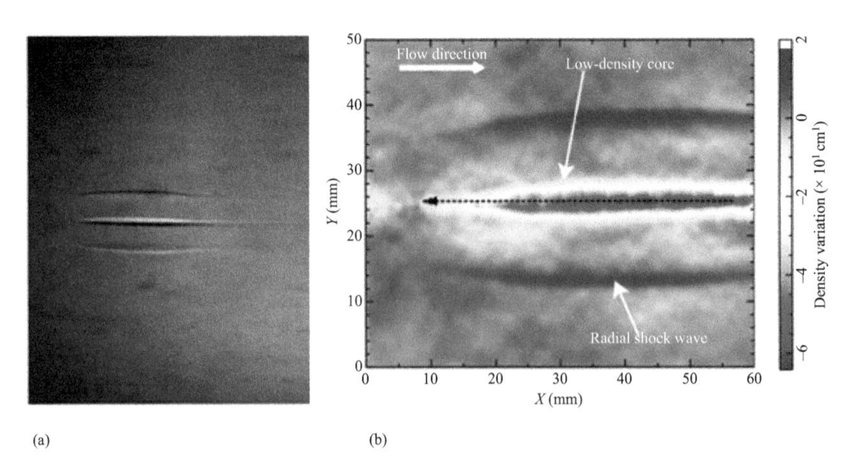

Figure 1.7 Image a low-density channel observed 6 μs after the filament formation in air: (a) Schlieren image showing the radial shockwave; (b) Density map obtained by interferometry [17]

The dynamics of the expanding shock wave and under-dense channel formation in air has been studied by means of transverse interferometry (see Figure 1.8) [17,18]. In the case of relatively tight focusing (*f*/35), channels with a density decrease of more than 60% are observed, lasting for more than 90 ms.

Such long-lived hydrodynamic structures have been used to create remote virtual waveguides in air [19,20]. By placing several filaments in a square or circle,

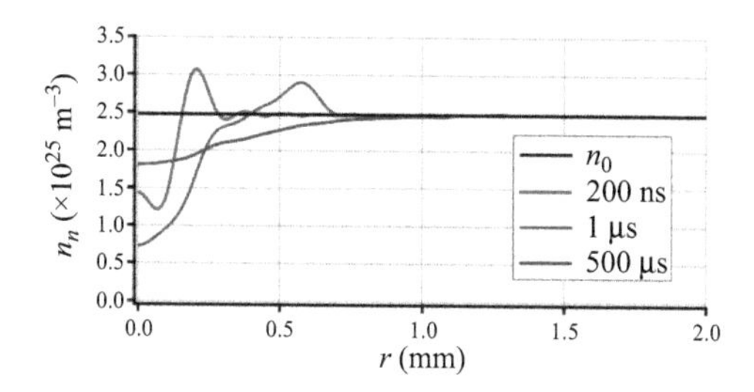

Figure 1.8 Measured evolution of air density after the formation of a filament [18]

one obtains an increased air density region in the centre of the configuration, the over-density of which can act as a long-lasting waveguide for lasers at other wavelengths, as demonstrated by the Milchberg group at the University of Maryland [19].

Filamentation-induced low-density channels in air also play an important role in the triggering and guiding of electric discharges [21,22] and for aerodynamic control [17], as discussed next.

1.3 Applications

1.3.1 Spatio-temporal femtosecond laser pulse improvement

For many ultrafast applications, it is important to start with a short laser pulse of high quality. Of particular importance is a smooth spatial profile and the absence of pulse precursor. One can exploit the self-cleaning action of filaments to improve the quality of a femtosecond laser pulse. The scheme is shown in Figure 1.9(a). An fs pulse with strongly distorted beam profile and with significant pre-pulse energy is incident on two counterpropagating filaments. The retroreflected pulse displays considerable improvement in the beam profile. Also, pre-pulse energy is removed (see Figure 1.9(b) and (c)) [23].

1.3.2 Temporal pulse improvement

There are strong incentives to obtain laser pulses with duration close to the fundamental limit of one optical cycle. Such short pulses are essential for XUV (extreme ultraviolet) attosecond pulse generation, for ultrafast streaking physics and may reduce the costs to reach the ultra-relativistic intensity regime, since the cost of amplifiers scales with pulse energy. One can exploit the property of filaments to reduce the duration of the propagating pulse close to the single cycle limit. Figure 1.10 shows the experimental set-up used by U. Keller group in Zurich to demonstrate this effect [24]. Filamentation occurs in two successive Argon cells, with optimized pressure. Figure 1.11 shows the pulse duration measured after each cell [25].

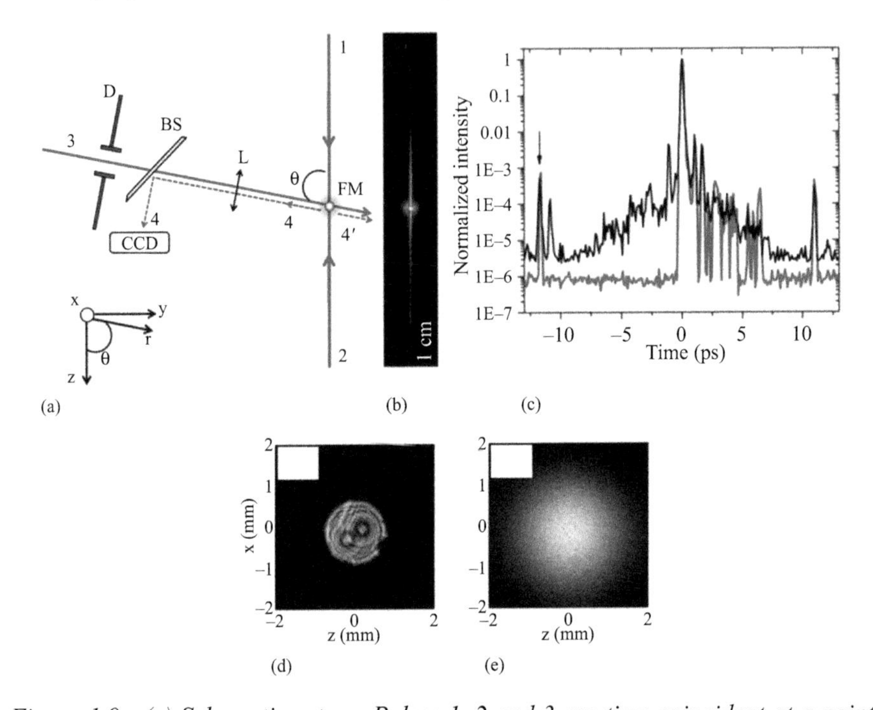

Figure 1.9 *(a) Schematic set-up. Pulses 1, 2 and 3 are time coincident at a point of intersection. D, diaphragm; BS, beam splitter; L, lens; CCD, charge-coupled device camera; FM, filament mirror; (b) photograph of the plasma luminescence from the two counterpropagating filaments forming the mirror. (c) Temporal contrast of the probe (in black) and of the reflected beam (red) measured with a third-order cross-correlator. Fluence profile of a probe femtosecond pulse with strongly distorted spatial profile before (d) and after retroreflection on the filament mirror (e)*

1.3.3 Triggering and guiding of electric discharges

Plasma filaments can be particularly useful for the remote manipulation of high voltage discharges. They can trigger and guide megavolt discharges in air over several metres [26,27], carry high DC currents with reduced losses [28] or deviate arcs from their natural path [22]. These properties are of great interest for applications such as the laser lightning rod [29,30], virtual plasma antennas for radio-frequency transmission [31] or high-voltage switch [32].

The physical mechanism is the following. As discussed in the previous part, laser energy deposition in the filament leads to the appearance of a hot, low-density channel at the centre of the filament path. In addition, a fraction of the free electrons produced by the laser pulse does not recombine on the parent ions but becomes attached to neutral oxygen molecules. These loosely bound electrons can be easily released by current heating, leading to a decrease of the leader inception

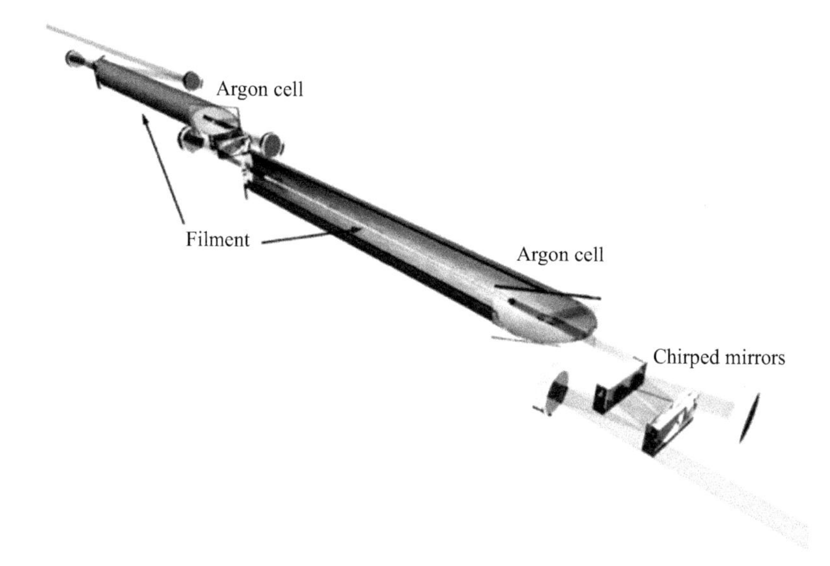

Figure 1.10 Experimental set-up for the shortening of fs pulses via double filamentation in Argon [24]

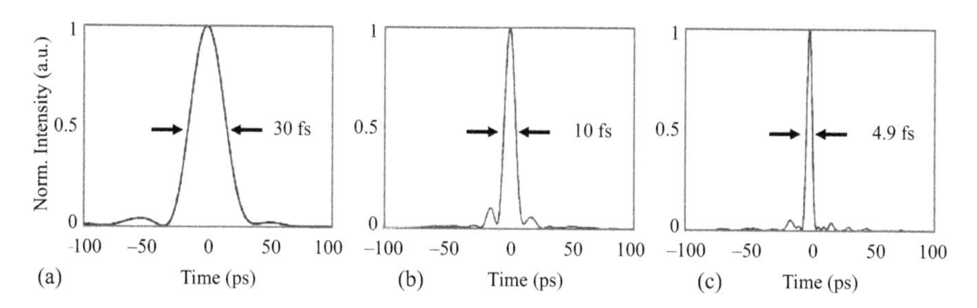

Figure 1.11 Pulse duration at (a) input of first filament stage, (b) after first filament stage, (c) after second filament stage (from [25])

voltage [33]. This offers a privileged path for electric discharges [21,34]. A filament decreases the breakdown voltage in a gas by more than 30% and guides the discharge over the perfectly straight path defined by the laser; see Figure 1.12. Recently, a European-funded project has been started with the aim to exploit this effect for implementing an active lightning rod. For more details, see [30].

Fast high-voltage triggers: Classical high voltage triggering switches based on electronics and avalanche breakdown present a jitter (defined as the standard deviation of the switching time delay) in the sub-microsecond range. This jitter is a severe limitation for multistage high-voltage generator and for applications where high-voltage pulses have to be precisely synchronized [32].

Since laser filaments decrease the breakdown voltage of a spark gap in a very fast and reproducible way, they also reduce significantly the jitter of an air switch, which is on the order of 0.2 ns, as shown in Figure 1.13.

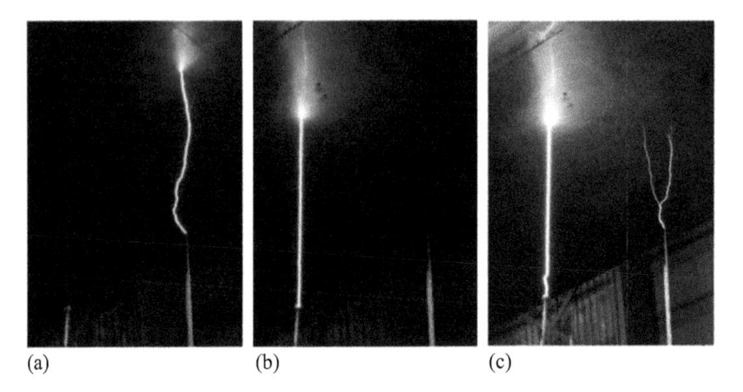

Figure 1.12 Diversion action of a filament on electric discharges. (a) A spontaneous discharge always originates at the sharp grounded electrode on the right. (b and c) With a filament, the guided discharge originates always at a rounded ground electrode located further apart from the HV electrode plate (from [22])

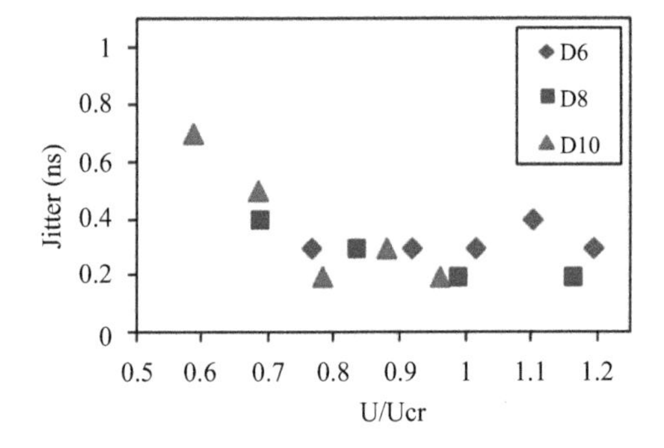

Figure 1.13 Jitter of the switch as a function of normalized voltage for inter-electrode gaps of 6, 8 and 10 mm [32]

In addition, the long and uniform plasma channels formed by the filament offer the possibility to trigger simultaneously several air spark gaps with an excellent synchronization. This effect has been used to build a multistage Marx generator triggered in atmospheric air by a single filament. The generator is composed of six stages connected by five gap switches similar to the one presented in the previous section. Every stage includes six capacitors of 2 nF connected in parallel par metal plates and charged up to 27 kV (slightly below self-breakdown). The two electrodes on each gap switch are pierced on the axis with a hole of 3–4.5 mm in diameter and separated by 11 mm. When a beam of multiple filaments is formed inside the generator, all stages of the generator are connected simultaneously and a voltage pulse of 160 kV is generated at the output of the generator. This voltage pulse has a rise time of a few nanoseconds and a sub-nanosecond jitter, as shown in Figure 1.14 [35].

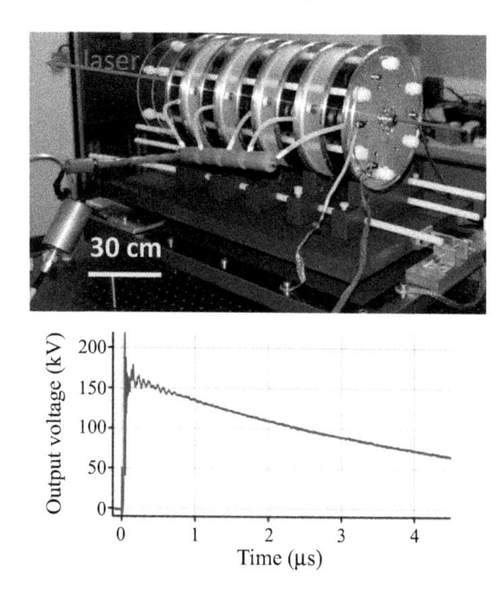

Figure 1.14 High-voltage Marx generator fully triggered by a single filament. Voltage rise time of 5 ns for a voltage pulse of 160 kV

1.3.4 Improving transport

1.3.4.1 Improving the speed of trains

The contact between pantograph and power line sets a limit to the speed of fast trains, because of the mechanical friction and the generation of unwanted oscillations in the power line at high speeds. Experiments performed at the test site of TGV trains in France have shown that this difficulty could in principle be overcome by connecting the current collecting pantograph to the power supply line via cm long electric discharges induced by filamentation [28]. A contactless connection proved efficient for both DC and AC power supply, transmitting an electric power in the MW range with little power loss, amounting to 1%–2% (see Figure 1.15).

1.3.4.2 Improving supersonic flights

When a vehicle moves in atmosphere at supersonic velocity, a shock wave is generated, leading to a considerable increase of the drag experienced by the vehicle and of its fuel consumption. A common strategy to reduce the drag is to give a slender shape to the front of the vehicle, in the form of a spike. Mechanical instabilities and manoeuvrability constraints put a limit to this approach. Several authors have suggested that depositing an immaterial plasma spike could circumvent these limitations.

Experimental tests performed in the supersonic wind tunnel of ONERA in Meudon, France with a mobile TW laser showed that filamentation led to a significant reduction of the drag experienced by a blunt test model [17]. The filament induced by an fs laser pulse emerging from the nose of the test model led to a low-density channel that perturbed the conical wave, with a transient reduction by 50%

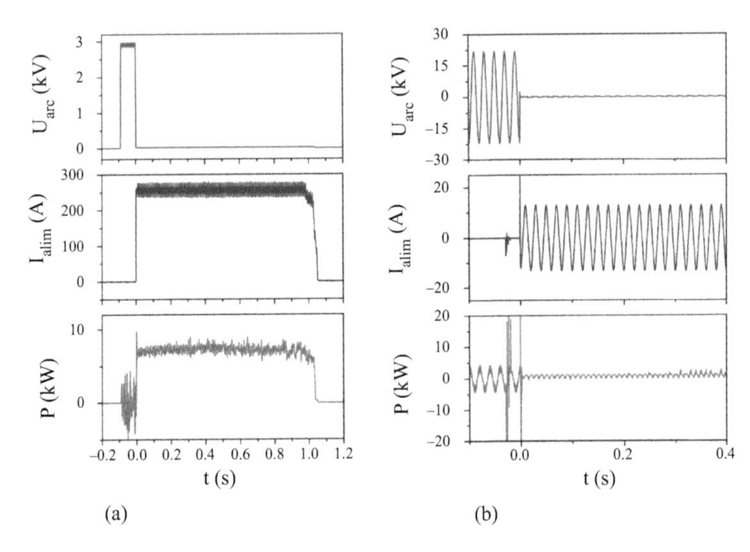

Figure 1.15 Filament-induced power supply: (a) DC power coupling, (b) AC power coupling. The arrival of the laser pulse at t = 0 leads to a rapid drop of the voltage of the charged electrode and to a surge of the current. The lower trace shows the power dissipated in the discharge

of the drag experienced by the test model at Mach 3 (Figure 1.16). It was also shown that weightless manoeuvrability can be achieved by a slight displacement of the beam direction from the module axis.

1.3.5 Virtual plasma antenna

There are incentives to replace conventional antennas by virtual antennas where plasma replaces metal. Such antennas could be broadband, easily reconfigurable, easily deactivated, stealthy and would lead to a reduction of co-site interference between multiple adjacent antennas [36]. Most designs so far use low-pressure plasmas confined inside solid dielectric vessels. Recently, a new concept based on filament-guided electric discharges has been demonstrated [31,37]. This virtual plasma antenna is tuneable in a large frequency range (100 MHz–1 GHz).

Figure 1.17(a) shows a proof of principle implementation. A metre-scale conductive plasma column is formed in air by guiding an electric discharge generated by a compact Tesla coil (output voltage of 350 kV) using an ultrashort laser pulse (700 fs, 300 mJ @ 800 nm) undergoing filamentation. Radio frequency (RF) power is then injected in the plasma by means of an inductive coupler. Radio emission was detected with a remote patch antenna of 100 MHz–1 GHz bandwidth (Figure 1.17(b)).

In the first demonstration experiment of Brelet *et al.*, the duration of the emission was limited by the lifetime of the discharge to 100 ns. Techniques to

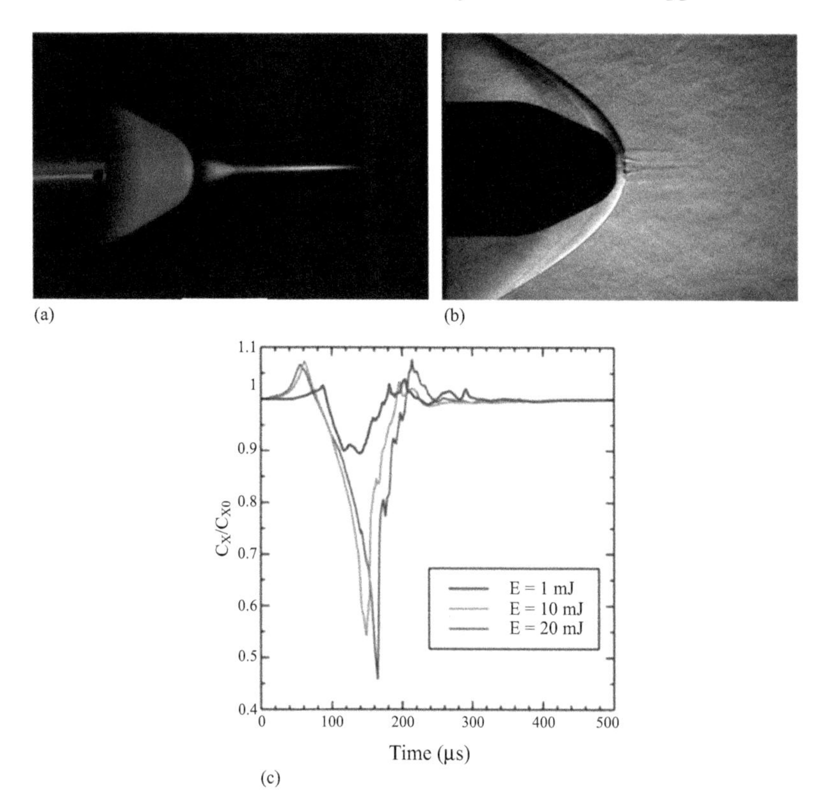

Figure 1.16 (a) Still photography of the filament produced in front of the blunt body. (b) Schlieren image showing the shockwave at Mach 3 and its perturbation by the filament in the presence of the Mach 3 airflow [17]. (c) Normalized drag signal measured on the model after the formation of the filament for three different laser energies

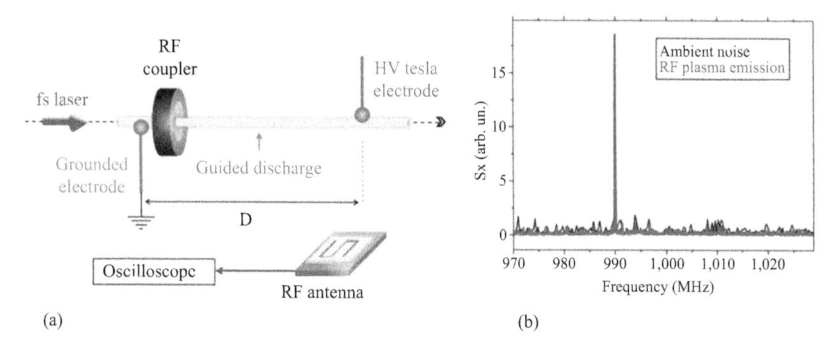

Figure 1.17 (a) Experimental set-up used for RF coupling in the plasma. (b) Emitted RF signal strength with (red) or without (black) the plasma column when the coupler is excited at 990 MHz

increase the lifetime of the plasma antenna have been since developed. The idea is to inject a long-lived current from a secondary source in the guided discharge. An increase of the plasma lifetime to the millisecond regime has been demonstrated [38,39].

1.3.6 Remote production of secondary radiation sources

The remote production of filaments in air leads to the production of coherent and directed secondary radiation sources, ranging from THz to UV.

1.3.6.1 Filament-induced coherent terahertz radiation in air

Because of the Lorentz force, the plasma created during filamentation in air is left with a longitudinal oscillation, heavily damped by collisions. This moving plasma column acts as a progressive antenna and gives rise to a forward oriented conical emission of short bursts of THz radiation, as shown in Figure 1.18 [40].

Although the conversion efficiency is low, on the order of 10^{-9}, this THz source has several advantages. Since filaments can be deposited at km distance from the laser source [41], it leads to the possibility to remotely create a short THz pulse close to a distant target, avoiding the problem of THz wave attenuation by air humidity. Also, since there is no material other than air for THz production, there is no risk of damage limiting the initial laser power. The THz intensity is then expected to increase by a factor $N \sim P/P_{cr}$. This was confirmed with a laser of 2-TW peak power (producing around 20 filaments), where an increase of THz signal by a factor of 20 as compared to a single filament was observed at a distance of 20 m from the laser (see Figure 1.19) [42]. Even more interestingly, it was shown that it is possible to interfere constructively the THz emission from separate filaments [43]. With proper organization of filaments in the form of a phased array, directionality and an increase of the THz intensity scaling like N^2 is expected.

1.3.6.2 Filament-induced coherent UV emission in air

The plasma column created in air during the filamentation of an fs laser pulse at 800 nm can act as an amplifying medium for UV radiation. Stimulated amplification of an injected seed at 337, 391 and 428 nm has been observed from nitrogen molecules following filamentation with a femtosecond pump laser pulse at 800 nm [44,45]. Figure 1.20 shows the energy level structure of N_2 and N_2+ and the corresponding laser transitions. Amplification at 337 nm takes place if the pump pulse is circularly polarized but disappears with linearly polarized pump light, while a reverse polarization dependence is observed for the two other wavelengths. While the origin of amplification of lines at 391 and 428 nm is still controversial, the excitation mechanism for the 337-nm emission is well understood: the energy distribution of free electrons generated by the circularly polarized pump pulse at 800 nm has an initial peak at $2U_p$, where $U_p = e^2 I/2c\varepsilon_0 m\omega^2$ is the free electron ponderomotive energy. For a laser peak intensity of 1.5×10^{14} W/cm^2, this corresponds to 14 eV, sufficient to excite neutral molecules from ground level $X^1\Sigma^+{}_g$ of the neutral molecule to excited level $C^3\Pi^+{}_u$ by multiple electron-neutral

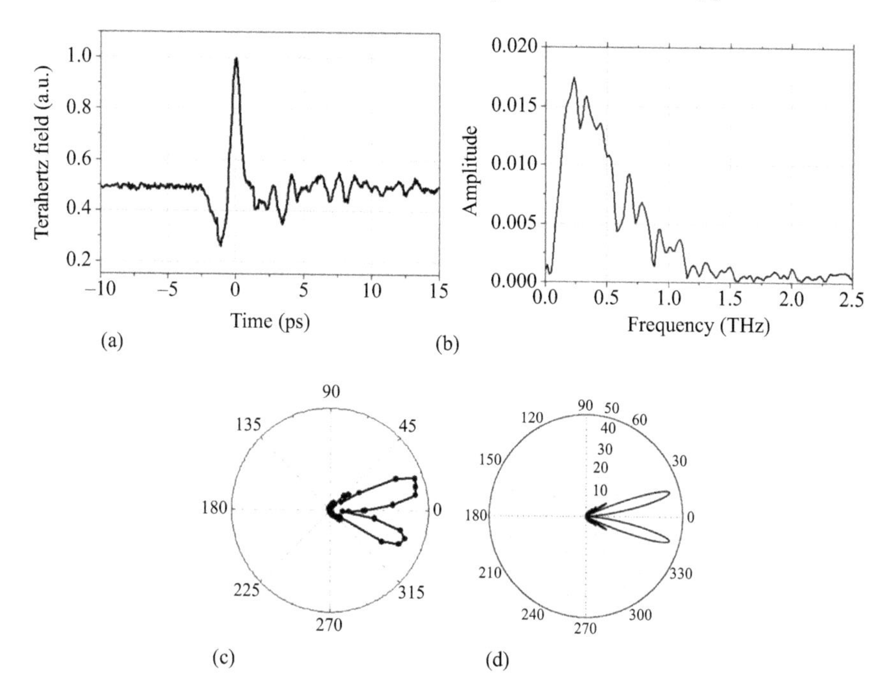

Figure 1.18 *(a) Temporal shape of the THz field emitted by a single filament measured by electro-optic sampling in ZnTe crystal and (b) corresponding spectrum obtained by Fourier transform. (c) Angular distribution of the THz emission measured at 0.1 THz and (d) calculated using the transition Cerenkov model of [40]*

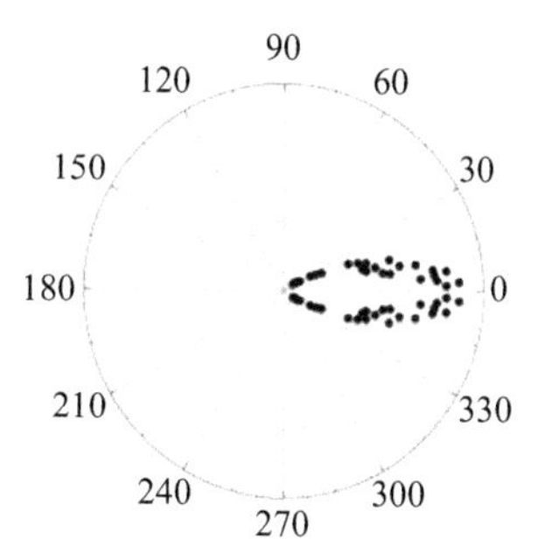

Figure 1.19 *Angular distribution of the THz from filaments generated in air by a terawatt laser beam [42]*

molecule collisions, producing a transient inversion of population between levels $C^3\Pi^+{}_u$ and $B^3\Pi_g{}^+$ [46]. A dream of atmosphere scientists is to have a lasing signal in air propagating backwards, towards the pump source, since it would considerably increase the signal detection efficiency. Interestingly, the 337-nm

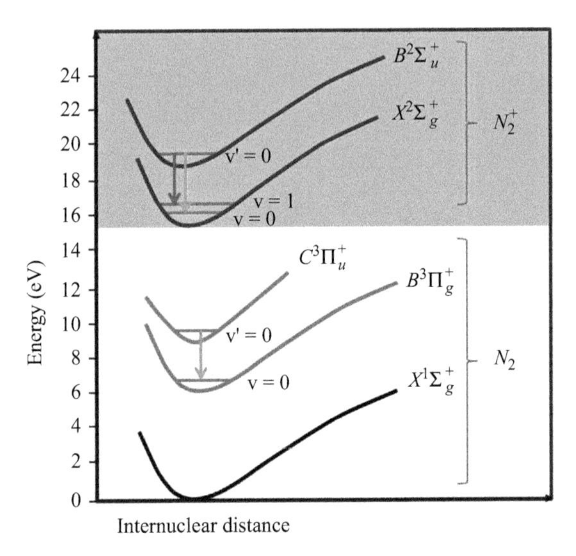

Figure 1.20 Energy-level structure of N_2 and N_2+ and the corresponding laser transitions

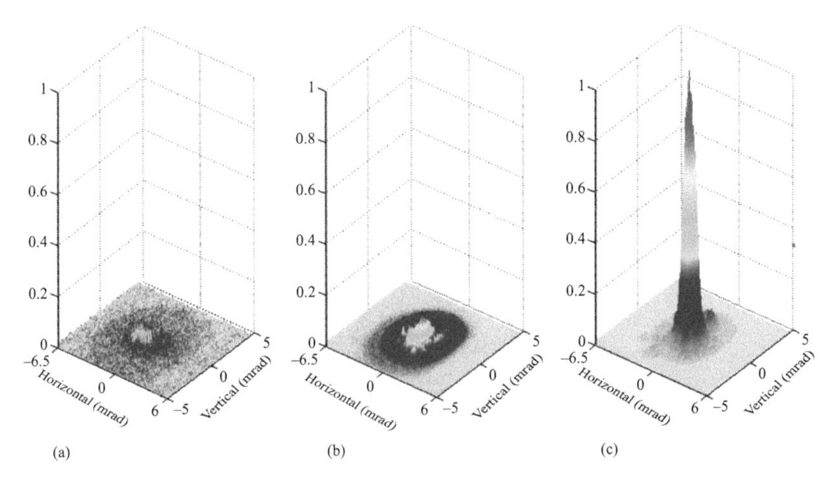

Figure 1.21 Spatial profile of the backward ASE (a), the seed pulse (b) and the backward amplified 337-nm radiation (c) in nitrogen [46]

amplified emission has also been observed in the backward directions with respect to the pump pulse direction, as shown in Figure 1.21. Unfortunately, oxygen acts as a quenching mechanism, preventing so far the observation of significant backward stimulated emission in air with oxygen concentration exceeding 10% [47].

1.4 Conclusion

In this chapter, we have briefly described several properties of filaments and their potential use in applications. The list of presented items is far from being exhaustive. Many applications can be seen as futuristic. Their viability will depend on the development of reliable femtosecond laser sources.

Acknowledgment

This work has been partially supported by DGA (Délégation Générale à l'Armement), France.

References

[1] Marburger J. H. 'Self-focusing: theory'. *Progress in Quantum Electronics* 1975; 4: 35–110.

[2] Couairon A., Tzortzakis S., Bergé L., Franco M., Prade B. and Mysyrowicz A. 'Infrared femtosecond light filaments in air: simulations and experiments'. *Journal of the Optical Society of America B* 2002; 19: 1117–1131.

[3] Mechain G, Couairon A, Andre Y.-B. *et al.* 'Long-range self-channeling of infrared laser pulses in air: a new propagation regime without ionization'. *Applied Physics B* 2004; 79: 379–382.

[4] Couairon A. and Mysyrowicz A. 'Femtosecond filamentation in transparent media'. *Physics Reports* 2007; 441(2): 47–189.

[5] Chin S. L., Hosseini S. A., Liu W. *et al.* 'The propagation of powerful femtosecond laser pulses in optical media: physics, applications, and new challenges.' *Canadian Journal of Physics* 2005; 83(9): 863–905.

[6] Kolesik M. and Moloney J. V. 'Modeling and simulation techniques in extreme nonlinear optics of gaseous and condensed media'. *Reports on Progress in Physics* 2014; 77(1): 016401.

[7] Moll K. D., Gaeta A. L. and Fibich G. 'Self-similar optical wave collapse: observation of the Townes profile'. *Physical Review Letters* 2003; 90: 203902.

[8] Prade B., Franco M., Mysyrowicz A. *et al.* 'Spatial mode cleaning by femtosecond filamentation in air'. *Optics Letters* 2006; 31: 2601–2603.

[9] Durand M., Jarnac A., Houard A. *et al.* 'Self-guided propagation of ultrashort laser pulses in the anomalous dispersion region of transparent solids: a new regime of filamentation'. *Physical Review Letters* 2013; 110: 115003.

[10] Milian C., Jukna V., Couairon A. *et al.* "Laser beam self-symmetrization in air in the multifilamentation regime'. *Journal of Physics B* 2015; 48: 094013.

[11] Kasparian J., Sauerbrey R., Mondelain D. *et al.* 'Infrared extension of the supercontinuum generated by femtosecond terawatt laser pulses propagating in the atmosphere'. *Optics Letters* 2000; 25: 1397–1399.
[12] Couairon A., Biegert J., Hauri C. P. *et al.* 'Self-compression of ultra-short laser pulses down to one optical cycle by filamentation'. *Journal of Modern Optics* 2006; 53(1–2): 75–85.
[13] Panagiotopoulos P., Whalen P., Kolesik M. *et al.* 'Super high power mid-infrared femtosecond light bullet'. *Nature Photonics* 2015; 9: 543–548.
[14] Voronin A. A. and Zheltikov A. M. 'Long-wavelength infrared solitons in air'. *Optics Letters* 2017; 42: 3614–3617.
[15] Cheng Y.-H., Wahlstrand J. K., Jhajj N. and Milchberg H. M. 'The effect of long timescale gas dynamics on femtosecond filamentation'. *Optics Express* 2013; 21: 4740.
[16] Wahlstrand J. K., Jhajj N., Rosenthal E. W., Zahedpour S. and Milchberg H. M. 'Direct imaging of the acoustic waves generated by femtosecond filaments in air'. *Optics Letters* 2014; 39: 1290.
[17] Elias P.-Q., Severac N., Luyssen J.-M. *et al.* 'Improving supersonic flights with femtosecond laser filamentation'. *Science Advances* 2018; 4: *eaau5239*.
[18] Point G., Milián C., Couairon A., Mysyrowicz A. and Houard A. 'Generation of long-lived underdense channels using femtosecond filamentation in air'. *Journal of Physics B* 2015; 48: 094009.
[19] Jhajj N., Rosenthal E., Birnbaum R., Wahlstrand J. and Milchberg H. M. 'Demonstration of long-lived high power optical waveguides in air'. *Physical Review X* 2014; 4: 011027.
[20] Lahav O., Levi L., Orr I. *et al.* 'Long-lived waveguides and sound-wave generation by laser filamentation'. *Physical Review A* 2014; 90: 021801.
[21] Tzortzakis S., Prade B., Franco M., Mysyrowicz A., Hüller S. and Mora P. 'Femtosecond Laser-guided Electric Discharge in Air'. *Physical Review E* 2001; 64: 57401.
[22] Forestier B., Houard A., Revel I. *et al.* 'Triggering, guiding and deviation of long air spark discharges with femtosecond laser filament'. *AIP Advances* 2012; 2: 012151.
[23] Jarnac A., Durand M., Houard A. *et al.* 'Spatiotemporal cleaning of a femtosecond laser pulse through interaction with counterpropagating filaments in air'. *Physical Review A* 2014; 89: 023844.
[24] Hauri C. P., Kornelis W., Helbing F. W. *et al.* 'Generation of intense, carrier-envelope phase-locked few-cycle laser pulses through filamentation'. *Applied Physics B* 2004; 79: 673.
[25] Zaïr A., Guandalini A., Schapper F. *et al.* 'Spatio-temporal characterization of few-cycle pulses obtained by filamentation'. *Optics Express* 2007; 15: 5394.
[26] Pépin H., Comtois D., Vidal F. *et al.* 'Triggering and guiding high-voltage large-scale leader discharges with sub-joule ultrashort laser pulses'. *Physics of Plasmas* 2001; 8: 2532.

[27] Rodriguez M., Sauerbrey R., Wille H. *et al.* 'Triggering and guiding megavolt discharges by use of laser-induced ionized filaments'. *Optics Letters* 2002; 27: 772.

[28] Houard A., D'Amico C., Liu Y. *et al.* 'High current permanent discharges in air induced by femtosecond laser filamentation'. *Applied Physics Letters* 2007; 90: 171501.

[29] Zhao X. M., Diels J.-C., Braun A. *et al.* 'Use of self-trapped filaments in air to trigger lightning' in *Ultrafast Phenomena, Springer Series in Chemical Physics*. New York, NY: Springer-Verlag 1994; 60: 233.

[30] Produit T., Walch P., Herkommer C. *et al.* 'The laser lightning rod project'. *The European Physical Journal: Applied Physics* 2021; 92: 30501.

[31] Brelet Y., Houard A., Point G. *et al.* 'Radiofrequency plasma antenna generated by femtosecond laser filaments in air'. *Applied Physics Letters* 2012; 101: 264106.

[32] Arantchouk A., Houard A., Brelet Y. *et al.* 'A simple high-voltage high current spark gap with subnanosecond jitter triggered by femtosecond laser filamentation'. *Applied Physics Letters*. 2013; 102: 163502.

[33] Comtois D., Pepin H., Vidal F. *et al.* 'Triggering and guiding of an upward positive leader from a ground rod with an ultrashort laser pulse—II: modeling'. *IEEE Trans. on Plasma Science* 2003; 31: 387.

[34] Vidal F., Comtois D., Chien C.-Y. *et al.* 'Modeling the triggering of streamers in Air by ultrashort laser pulses'. *IEEE Transactions on Plasma Science* 2000; 28: 418.

[35] Arantchouk A., Point G., Brelet Y. *et al.* 'Compact 180-kV Marx generator triggered in atmospheric air by femtosecond laser filaments'. *Applied Physics Letters* 2014; 104: 103506.

[36] Borg G. G., Harris J. H., Martin N. M. *et al.* 'Plasmas as antennas: theory, experiment and applications'. *Physics Plasmas* 2000; 7: 2198.

[37] Théberge F., Gravel J.-F., Kieffer J.-C., Vidal F. and Châteauneuf M. 'Broadband and long lifetime plasma-antenna in air initiated by laser-guided discharge'. *Applied Physics Letters* 2017; 111: 073501.

[38] Arantchouk L., Honnorat B., Thouin E., Point G., Mysyrowicz A. and Houard A. 'Prolongation of the lifetime of guided discharges triggered in atmospheric air by femtosecond laser filaments up to 130 μs'. *Applied Physics Letters* 2016; 108: 173501.

[39] Théberge F., Daigle J. F., Kieffer J. C. *et al.* 'Laser-guided energetic discharges over large air gaps by electric-field enhanced plasma filaments'. *Scientific Report* 2017; 7: 40063.

[40] D'Amico C., Houard A., Franco M. *et al.* 'Conical forward THz emission from femtosecond laser filamentation in air'. *Physical Review Letters* 2007; 98: 235002.

[41] Durand M., Houard A., Prade B. *et al.* 'Kilometer range filamentation'. *Optics Express* 2012; 21: 26836.

[42] D'Amico C., Houard A., Akturk S. *et al.* 'Forward THz radiation emission by femtosecond filamentation in gases: theory and experiment'. *New Journal of Physics* 2008; 10: 013015.

[43] Mitryukovskiy S. I., Liu Y., Prade B., Houard A. and Mysyrowicz A. 'Coherent interaction between the THz radiation emitted by filaments in air'. *Laser Physics* 2014; 24: 094009.
[44] Liu Y., Ding P., Lambert G., Houard A., Tikhonchuk V. T. and Mysyrowicz A. 'Recollision-induced superradiance of ionized nitrogen molecule'. *Physical Review Letters* 2015; 115: 133203.
[45] Yao J., Jiang S., Chu W. *et al.* 'Population redistribution among multiple electronic states of molecular nitrogen ions in strong laser fields'. *Physical Review Letters* 2016; 116: 143007.
[46] Ding P., Oliva E., Houard A., Mysyrowicz A. and Liu Y. 'Lasing dynamics of neutral nitrogen molecules in femtosecond filaments'. *Physical Review A* 2016; 94: 043824.
[47] Gui J., Zhou D., Zhang X. *et al.* 'Quenching effect of O_2 on cavity-free lasing of N_2 pumped by femtosecond laser pulses'. *Acta Photonica Sinica* 2020; 49: 1149013.

Chapter 2
UV filaments

Ali Rastegari,[1,2] Alejandro Aceves,[3] and Jean-Claude Diels[1,2]

2.1 Introduction

Much of the research on filaments has been driven by the availability of short and intense pulses, such as the Ti:sapphire femtosecond laser systems. More efficient sources, such as ytterbium-doped YAG disk lasers capable of high repetition rate, have shifted the emphasis to even longer wavelengths. However, it is shown in this chapter that the longer the wavelength, the shorter the pulses required to form a filament. This is because the collisional ionization of the molecules by the photo-electrons accelerated in the field of the laser creates an "avalanche ionization" of the medium, most effective at longer wavelengths. The intensity within a filament being "clamped" at a specific value [1], operating at shorter wavelengths, makes it possible to create more energetic single filaments. Filaments of 300 mJ energy have been observed at 266 nm. This chapter starts with a discussion of the main qualitative differences between UV and mid-IR filaments: from multiphoton ionization in the UV to tunnel ionization in the near- to mid-IR. A general qualitative analysis of the properties of single filaments versus wavelength follows. Because of their long pulse duration, a quasi-steady-state theory of their propagation is possible. An eigenvalue approach leads to a steady-state field envelope that is compared to the Townes soliton. However, that solution is close enough to a Gaussian shape to justify a parametric evolution approach.

After this theoretical introduction, an experimental verification at 266 nm follows. Femtosecond UV filaments were generated with frequency-tripled Ti: sapphire sources and KrF amplifiers (Section 2.6.1). The source for long-pulse filaments is an oscillator–amplifier Nd–YAG Q-switched system, frequency doubled, compressed, and frequency doubled again to reach 170 ps pulses of 300 mJ energy (Section 2.6.2). The sub-nanosecond duration of the UV pulse may revive the debate as whether the filament is a moving focus or self-induced waveguide.

[1]Department of Physics, The University of New Mexico, Albuquerque, USA
[2]Center for High Technology Materials, Albuquerque, USA
[3]Southern Methodist University, Dallas, USA

This is addressed by focusing the high-energy pulse in vacuum and launching it into the atmosphere through an aerodynamic window (Section 2.7.2.1). Such beam preparation also transforms the super-Gaussian profile of the source into a parabolic profile more conducive to filamentation.

The development of intense sources of higher repetition rate opens the possibility of generating permanent optical waveguides in the atmosphere. The heat produced as the ions and electrons recombine creates a shock-wave that can be engineered into an acoustic waveguide. This shock-wave is discussed in Section 2.8.

Two applications of UV filaments are presented in the last two sections. It is shown that isotopically selective laser-induced breakdown spectroscopy (LIBS) is possible by exploiting the narrow dips observed in the emission spectrum. These dips are due to reabsorption by the material in the plume created by the impact of the filament on a solid. These absorption lines are only a few-pm wide and are exactly centered at the wavelength of a transition from ground state of the material, without any Stark shift or broadening.

A final application is laser-induced discharge, which is a guided discharge that follows exactly the path of the inducing UV filament. Laser-induced discharge may lead to the control of lightning, which is a topic of intense research in Europe [2].

2.2 Trends of filamentation versus wavelength

2.2.1 Multiphoton versus tunnel ionization

In the description of matter by an "index of refraction" or a "polarization," one tends to forget that the nature of light–matter interaction is simply reradiation of electrons driven by a combination of the applied electromagnetic field of the light and the field of other particles. The moving electrons radiate a field that adds to that of the light, resulting in phase and amplitude changes of the optical field.

The nature of the ionization process and the electron trajectories following ionization are quite dependent on the wavelength. At the UV wavelength the photon energy is a small fraction of the ionization potential of the gas molecules, and the electron plasma is created through multiphoton ionization (Figure 2.1(a)). If N is the minimum number of photons of energy $\hbar\omega$ required to exceed the ionization potential I_p, the photoelectron production rate is proportional to the Nth power of the light pulse intensity I. The electrons are created with a kinetic energy equal to the excess energy $N\hbar\omega I_p$. The electron energy spectrum after ionization shows peaks corresponding to "above threshold ionization" (ATI) at energies of $(N + m)\hbar\omega - I_p$ corresponding to the absorption of $N + m$ photons.

At longer wavelengths, the intensity required for N photons to stack themselves out of the potential well of the gas molecule gets comparable to the intensity required to tunnel out of the potential wall created by the distorted potential surface (Figure 2.1(b)). In contrast to the case of multiphoton ionization, the tunneled electron is released at the other side of the wall with zero velocity.

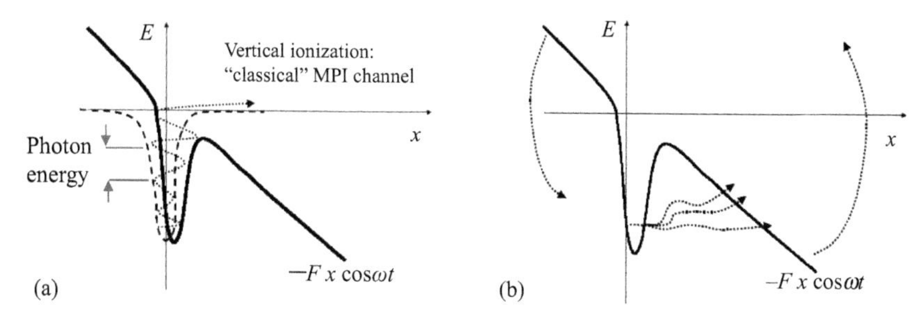

Figure 2.1 *(a) Multiphoton ionization: at short wavelengths, the relatively high-energy photons stack themselves out of the potential well of the molecule; (b) tunnel ionization dominates at lower photon energies. While the electron remains near the bottom of the well, the potential surface gets tilted by the field, creating a narrowing potential wall through which the electron can tunnel*

The distinction between multiphoton and tunnel ionization is quantified by the "Keldysh parameter" γ [3]:

$$\gamma = \sqrt{\frac{I_p}{2U_p}}, \tag{2.1}$$

where U_p is the ponderomotive energy or the average kinetic energy of a free electron oscillating in the laser field. If e and m_e are the charge and mass of the electron; ω is the (angular) frequency of the light field of amplitude E:

$$U_p = \frac{e^2 E^2}{4 m_e \omega^2}, \tag{2.2}$$

where U_p expressed in eV as a function of the light intensity I_ℓ in W/cm^2 and the wavelength λ in microns is

$$U_p = 9.33 \times 10^{-14} I_\ell \lambda^2. \tag{2.3}$$

In the "quasistatic limit" of $\gamma < 1$ the dressed Coulomb barrier is essentially static as seen by the electrons and the method of releasing the electrons is dominated by *tunneling*. For $\gamma > 1$, the electron release is most likely described by photon absorption, and *multiphoton* features are more dominant [4]. The difference between tunneling and multiphoton is experimentally recognized in measurements of velocity mapping imaging where the electron momentum distribution following ionization is measured [5]. These measurements show that, at 800 nm where $\gamma \approx 1$, the electron is released with approximately zero velocity and there is no evidence if ATI. One concludes therefore that tunnel ionization corresponds to $\gamma \leq 1$. A tunneled electron leaves its parent atom/molecule instantaneously along the direction of light polarization, at the moment of ionization, with zero velocity [6]. The electrons leave the atom from a Rydberg state that typically has an orbit radius one order of magnitude larger than the atomic radius. Formulae can be found in the literature for the tunneling rate and the ratio of electron production for various polarizations [7,8]. The trajectory of the electron released near a peak alternate of the laser field can be calculated classically. In the case of linear polarization, it is easily seen that the released electron remains in the vicinity of the molecule, which implies that recollision is possible, leading to attosecond pulse generation [9]. In the case of circular polarization, the electron drifts away from the parent molecule.

2.2.2 From photoionization to avalanche ionization

As a high-intensity beam above critical power starts to focus in the atmosphere, it creates a low-density plasma by multiphoton ionization. In the intense light field, the photoelectrons will gain energy by inverse Bremsstrahlung [10]. The rate of energy gain dW/dt depends on the optical frequency, being proportional to the square of the wavelength. The accelerated electrons can collide with neutral molecules, thereby increasing the plasma density by collisional ionization. This is the avalanche process leading to full ionization of the medium. The additional influx of electrons will perturb the delicate balance between self-focusing and defocusing that leads to a filament. A filament equilibrium between self-focusing and defocusing by photoionized electron is possible before the electron density gets dominated by collision ionization. A characteristic parameter is the time required for the electron energy to reach the ionization energy of oxygen.

From the electron–ion collision rate being $\nu_{ei} = 1.07 \times 10^{11}\ \mathrm{s}^{-1}$ [11], an electron will gain an energy U_p in the time interval $1/\nu_{ei}$ between collisions. The energy gain dW in a time interval dt due to inverse Bremsstrahlung can be estimated as

$$\frac{dW}{dt} \approx U_p \nu_{ei} \approx 10^{-14} I_\ell \lambda^2, \tag{2.4}$$

where we have used the units of (2.3) and expressed the collision time in ps. A rough estimate for the time Δt needed to reach the ionization potential of 12.2 eV (oxygen) is

$$\Delta t = \frac{1.22 \times 10^{15}}{I_\ell \lambda^2}. \tag{2.5}$$

For the typical 50 TW/cm^2 at 800 nm, (2.5) indicates that the pulse duration should be small compared to 40 ps to prevent collisional ionization from affecting the balance between self-focusing and plasma self-defocusing. This limit increases to 40 ns for 0.5 TW/cm^2 at 250 nm.

In the case of fs near-IR filaments, a number of phenomena can contribute to balancing the self-focusing effect: creation of an electron plasma, energy loss due to pulse-splitting or conical emission, higher order nonlinearities. In the case of long-pulse UV filaments, this balance is entirely provided by the creation of electrons due to multiphoton ionization that is a purely intensity-dependent mechanism.

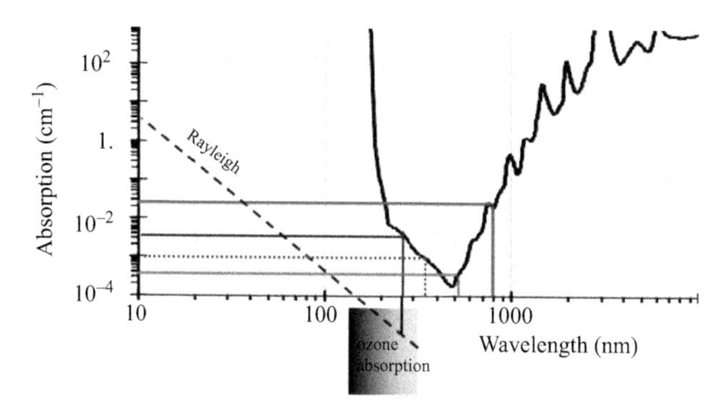

Figure 2.2 Attenuation lengths for various atmospheric factors. Solid line: water absorption. Dashed line: Rayleigh scattering. The red, green, dotted blue, and solid dark blue lines indicate wavelengths of 800, 532, 355, and 266 nm, respectively

Knowing that UV filaments consume 40 μJ/m of propagation [12], one should be able to increase the length of UV filaments by increasing the pulsewidth of the UV laser beam.

2.2.3 Filament stability versus order of ionization

To have a stabilizing effect on the beam (or filament) diameter, the plasma self-defocusing should have an intensity dependence of higher order than the Kerr effect. It is clear from the theory of Section 2.3 that three-photon ionization is the lower order possible. The wavelength dependence of filamentation has been studied in ZnSe [13]. The wavelength was varied from a condition of three-photon excitation of the conduction band to five photons. The spectral bandwidth of the filamented pulses as well as the length of the fluorescence track was seen to increase with the order of ionization. Multiphoton absorption was cited as a major contributing factor for increased losses at low order. The situation is different in dry air where an energy loss of only 40 μJ/m was measured [12] for 248 nm filaments with pulses of 1 ps duration. These measurements were made in the 5% humidity atmosphere of New Mexico. The situation is different in humid air, where a two-photon resonance-enhanced ionization was identified [14]. The initial excitation proceeds via the two-photon resonant transition in water vapor $\widetilde{C}^1B_1 \leftarrow \widetilde{X}^1A_1$. This excitation is followed by single-photon ionization from the $\widetilde{C}^1B_1$, the rate of which is enhanced by a resonance with the molecular-ionic transition $^2A_1(3a_1^{-1}) \leftarrow \widetilde{C}^1$ [14].

2.2.4 Single-filament propagation versus wavelength

Linear optical properties of the atmosphere should be a factor in choosing the wavelength that would propagate with the least attenuation. The attenuation length corresponding to various phenomena is plotted in Figure 2.2. The solid line shows water absorption, with a minimum near 400 nm. Closest to this minimum is the second harmonic of Nd:YAG at 532 nm. This wavelength is indicated by the green lines in Figure 2.2. Attempts of filamentation were made with Nd:YAG laser pulses that are frequency doubled and compressed down to 200 ps, at an energy of 1.2 J. The peak power was clearly above the self-trapping critical power. Plasma beads created by avalanche ionization were observed, indicating that 200 ps duration at that wavelength is too long for stable filament formation. At the wavelength of 266 nm indicated in the figure by the dark blue line, ozone absorption, water absorption and Rayleigh scattering are present. These factors are negligible at the third harmonic of the Nd:YAG laser of 355 nm, indicated by the dotted line.

Because of the plasma defocusing, the intensity of a single filament is clamped. Consequently, a 800 nm filament of about 100 fs will have its energy typically clamped around 1 mJ. Because of the high intensity in the filament core, it loses its energy through nonlinear effects (in particular by generation of broadband conical emission). These energy losses limit the propagation distance of the single 800 nm filament to a distance of the order of a meter. The "brute force" method to extend the range of filaments is to use a high-energy pulse as an initial condition. For instance, a pulse of 1 J energy will generate approximately 10^3 filaments distributed over the initial beam diameter and covering a distance that can exceed 100 m. An approach to extend the *range* of filaments, proposed initially for UV pulses [11], is to downchirp the initial pulse so that it can be compressed by propagating through the atmosphere that has positive dispersion. This method of postponing the start point of a filament is well adapted to broadband fs pulses and has been demonstrated over a distance of 1 km [15] with 800 nm pulses. While the dispersion of the atmosphere is larger in the UV, the narrow bandwidth of a 100 ps pulse precludes the application of this technique of temporal focusing on long UV pulses.

There has not been a systematic study of filament properties versus wavelength, because of limited access to high-power sources in multiple wavelength ranges. Figure 2.3 attempts to give a rough idea of the trend for filament parameters versus wavelength.

The longer the wavelength, the shorter the pulse duration required to prevent avalanche, and the intensity required for creating the defocusing plasma is higher. Because of intensity clamping, the energy of the filamented pulse decreases by three orders of magnitude between the UV and infrared. Losses are mostly linear in the UV (ozone absorption, Rayleigh scattering) and nonlinear in the infrared due to the high intensity.

2.3 Physical parameters relevant to UV filaments formation

2.3.1 Conditions particular to long UV pulse

The time to reach avalanche being of the order of tens of nanoseconds, as shown by (2.5), one might search for a condition where the interaction of light with air reaches a steady state. Under constant electric field of the light, the dynamics of the

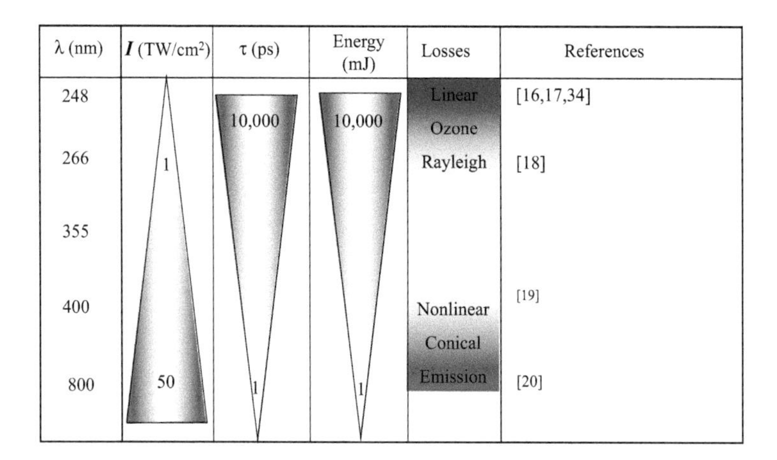

Figure 2.3 Trend in the wavelength range from 248 to 800 nm for single filaments. Second column: intensity range. Third column: range of energy/filamenting pulse. Fourth column: dominant loss mechanism. Fifth column: Zhao [16]; Smetanin [17]; Chalus [18]; Daigle [19]; Braun [20]; Zvorykin [34]

evolution of the electron density depend on the ionization cross section $\sigma^{(3)}$, the recombination rate β_{ep}, and the rate of attachment of electrons to oxygen γ. In the approximation that the number of electrons remains small compared to the number of neutrals, the equation for the generation of electrons is

$$\frac{dN_e}{dt} = N_0\sigma^{(3)}I^3 - \beta_{ep}N_e^2 - \gamma N_e, \tag{2.6}$$

where I is the laser field intensity, and N_0 is the density of oxygen molecules in air. Measurement and calculation of the parameters $\sigma^{(3)}$, β_{ep}, and γ are presented in Section 2.3.2 that follows.

Figure 2.4 shows the temporal evolution of the density of electrons as a function of time, for a step-function intensity of 0.5 TW/cm^2 (intensity measured in UV filaments [12,18,21]). The asymptotic value is

$$\begin{aligned} N_{eq} &= \frac{-\gamma + \sqrt{\gamma^2 + 4\beta_{ep}N_O\sigma^{(3)}I^3}}{2\beta_{ep}} \\ &= \frac{\sqrt{4\beta_{ep}N_O\sigma^{(3)}I^3}}{2\beta_{ep}}\left[1 + \frac{1}{2}\frac{\gamma^2}{\left(4\beta_{ep}N_O\sigma^{(3)}I^3\right)}\right] - \frac{\gamma}{2\beta_{ep}}. \end{aligned} \tag{2.7}$$

The last term of this expansion contains no field dependence and will be ignored since it contributes only to the linear polarization. Figure 2.4 indicates that

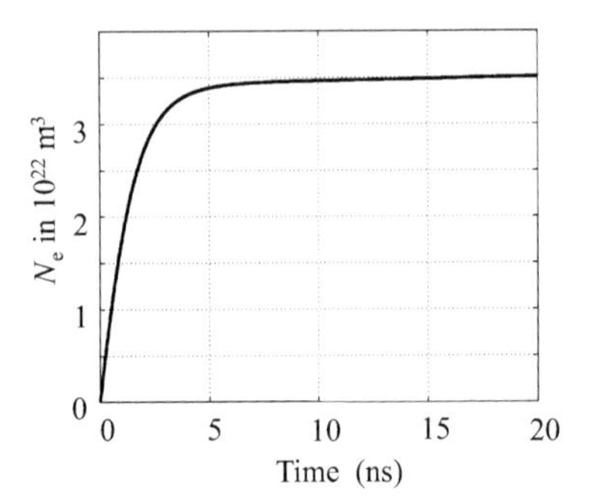

Figure 2.4 Evolution of the electron density N_e as a function of time for a step-function irradiation of intensity $I = 0.54\ \mathrm{TW/cm^2}$

75% of this value is reached after 2 ns. This justifies using a steady-state approach in Section 2.5 for calculating the energy density when pulses have a duration between 2 and 40 ns.

The nonlinear polarization amplitude contribution from the electron plasma is

$$\mathscr{P}_{NL} = \varepsilon_0 \frac{\omega_p^2}{\omega^2} \mathscr{E} = \frac{\varepsilon_0}{\omega^2} \frac{N_e e^2}{m_e \varepsilon_0} \mathscr{E}, \tag{2.8}$$

where ω_p is the plasma frequency and m_e the electron mass. One notes that, with three-photon absorption, the nonlinear index due to the plasma is only proportional to the power 1.5 of the intensity. Three-photon ionization is the lowest order multiphoton process to create a plasma stabilizing the $n_2 I$ of Kerr self-focusing. If the attachment coefficient to oxygen γ is neglected, the nonlinear polarization of (2.8) corresponds to a nonlinear index:

$$\Delta n = \frac{\omega_p^2}{\omega^2} = \frac{N_e e^2}{\omega^2 m_e \varepsilon_0} \approx \frac{\sqrt{\beta_{ep} N_O \sigma^{(3)} I^3}}{2\beta_{ep}} \frac{e^2}{m_e \varepsilon_0 \omega^2}. \tag{2.9}$$

Neglecting the attachment coefficient to oxygen will lead in Section 2.4 to a simple eigenvalue equation for the UV steady-state pulse, an extension of the Townes soliton. The parameter γ will be reintroduced in Section 2.5.2 where a complete simulation of the UV filament is presented in which losses are also included.

2.3.2 Determination of $\sigma^{(3)}$, β_{ep}, and γ

A self-defocusing associated with the photoelectron plasma is crucial in preventing the collapse of the beam through Kerr self-focusing. From the ionization energy of oxygen being 12.2 eV, at the wavelength of 266 nm (4.66 eV), three photons are needed to extract the electron from the oxygen:

$$O_2 + 3h\nu \rightarrow O_2^+ + e^-_{(\varepsilon=1.92\ \mathrm{eV})}. \tag{2.10}$$

At wavelengths shorter than 300 nm, the density of electrons is critically dependent on three parameters: the three-photon ionization cross-section of oxygen $\sigma^{(3)}$, the recombination rate of electrons β_{ep}, and the attachment coefficient γ of electrons to neutral oxygen to form a negative ion.

2.3.2.1 Negative oxygen ions

The negative oxygen ion (O^-) plays an important role in reducing the number of photoelectrons created by the ionizing pulse. It can be created by simple inelastic collision of the oxygen molecule with an electron of energy >6.5 eV:

$$O_2 + e^- \rightarrow O^- + O. \tag{2.11}$$

This two-body attachment requires that the plasma has been heated up to $\approx$ 6.5 eV by inverse Bremsstrahlung. A second process that results in the creation of O^- at low electron temperatures consists in a cascade of two reactions: a two-photon dissociation of O_2 in two oxygen atoms, followed by the attachment of the previously formed electron with one oxygen atom. This latter reaction, unlike the attachment of the electron with O_2, requires only an electron with kinetic energy of 1.5 eV. The reaction sequence is

$$O_2 + 2h\nu \rightarrow O + O \tag{2.12}$$

$$O + e^-_{(\varepsilon=1.92\mathrm{eV})} \rightarrow O^- + h\nu. \tag{2.13}$$

2.3.2.2 Pump–probe experiments on O^-

Attachment to oxygen

The conventional approach to photoionization measurements is to irradiate the molecules in a low pressure cell and to measure the charges collected on electrodes. Since the parameters are needed to model filaments *in air*, the approach chosen here is to study oxygen at a pressure as close as possible to atmospheric pressure [21].

In the experimental setup sketched in Figure 2.5, the concentration of O^- is monitored through the absorption of an He–Ne laser beam. The single-photon photodetachment spectrum of O^- indicates an absorption cross section of 6×10^{-18} cm^2 [22] in the wavelength range from 350 to 620 nm. The photodetachment reaction is

$$O^- + h\nu \rightarrow O + e^-. \tag{2.14}$$

In the experimental setup sketched in Figure 2.5(a), a 4.5 m cell filled with up to 2 atm of pure oxygen is used.[1] An ionizing pulse is provided by the fourth

[1]More experimental details can be found in [21].

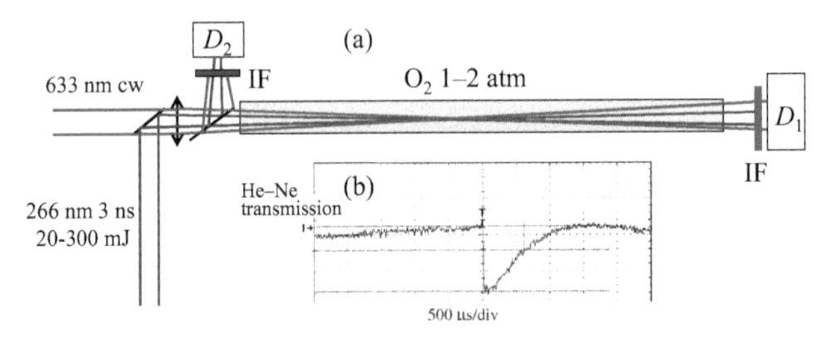

Figure 2.5 (a) Setup for measurement of absorption by O^- ($D_{1,2}$, detectors; IF, interference filter); (b) induced absorption trace on the He–Ne beam

harmonic (266 nm) of an Nd:YAG laser, with a pulse duration of 3 ns and an energy from 20 to 300 mJ at 2 Hz. The 633 nm intensity is monitored at the same time as the energy of the UV pulse with detectors D_1 and D_2, respectively, after passing through corresponding interference filters, and for two different pressures (1 and 2 atm). In the presence of ionizing radiation, the induced absorption coefficient $\alpha(t)$ can be recorded as a function of time:

$$\alpha(t) = -\frac{1}{L} ln \frac{D_1}{D_2} \approx \frac{1}{L}\left(1 - \frac{D_1}{D_2}\right), \tag{2.15}$$

where L is the length of interaction region between the focused He–Ne laser beam and the plasma. The measurement of the absorption coefficient α at the He–Ne wavelength leads to the concentration in O^- ions since $\alpha = \sigma N_{O^-}$, where $\sigma = 5.8 \times 10^{-18}$ cm^2 [22]. At 1 atm, with an ionizing laser power of 10 MW (intensity of 2.5×10^{11} W/cm^2), an attenuation of 34% of the probe He–Ne laser is measured. This leads to an estimate of $N_{O^-} = 5.6 \times 10^{16}$ cm^{-3} for the density of O^-.

A typical oscilloscope trace of the probe transient absorption is shown in Figure 2.5(b), indicating a $1/e$ decay of 620 μs. After deconvolution for the response time of the detection, the lifetime of the O^- is found to be $\tau_o = 590$ μs. At electron energies below 6.5 eV, the two-body attachment process of (2.12) and (2.13) dominates. Three-body attachment reactions involving nitrogen and oxygen can also be considered but are negligible above 1-eV electron temperature [21]. The contribution of the two-body attachment to the decay of the electron population is $\gamma = \eta_{att}[O_2]$ where $[O_2]$ is the concentration of oxygen and η_{att} is given by [23]

$$\eta_{att} = 2.75 \times 10^{-10}\, T_e^{-0.5} e^{-5/T_e} \text{ cm}^3/\text{s}, \tag{2.16}$$

where T_e is the temperature of the electrons expressed in eV. Choosing $T_e = 2$ eV leads to $\gamma = 1.5 \times 10^8$.

Photoionization cross section

The induced absorption, plotted versus the intensity of the UV beam, shows the expected cubic dependence (Figure 2.6). The induced absorption is proportional to the concentration of oxygen and photoelectrons. Since $N_e \geq N_{O^-}$ one can use the measurement of Figure 2.6 to get a rough upper estimate of the three-photon

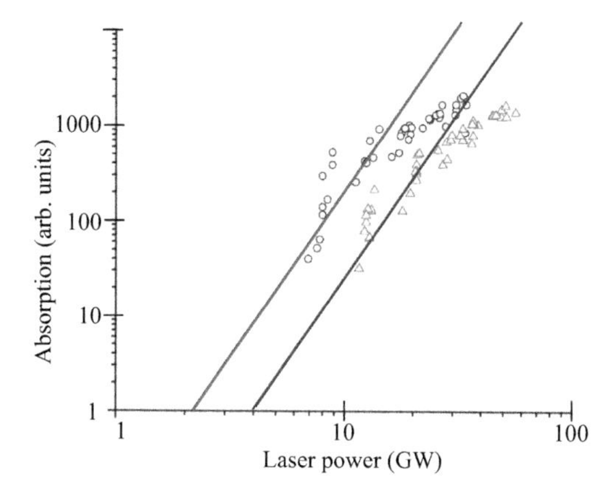

Figure 2.6 Absorption at 633 nm induced by the recombination of electrons with oxygen. The absorption coefficient is proportional to the concentration of O^- *ions, which is proportional to the number of photoelectrons. Circles: 2 atm pressure, triangles: 1 atm pressure. The straight lines indicate the slope corresponding to the UV laser power to the third power*

ionization cross section of oxygen. That estimate gives $\sigma^{(3)} = 4.1 \times 10^{-41}$ m^6 s^2/J^3, close to the theoretical value of 3.0×10^{-29} cm^6 s^2/J^3 [3] and published experimental data [12,24].

Summary of the parameters used for the plasma defocusing

Parameter	$\sigma^{(3)}$	β_{ep}	γ
Value	3×10^{-41}	1.3×10^{-14}	1.5×10^8
Unit	m^6 s^2/J^3	m^3/s	s^{-1}

2.4 Stationary UV filament solution versus Townes soliton

2.4.1 Eigenvalue equation

In this section, we derive a modified nonlinear Schrödinger equation, taking into account the change in index of refraction due to three-photon ionization. By neglecting losses and attachment to oxygen, one obtains a nonlinear equation that has an eigenvalue solution that can be compared to the Townes soliton [25]. We start from Maxwell's propagation equation:

$$\left[\Delta_{tr} + \partial^2_{zz} - \frac{n^2}{c^2}\partial^2_{tt}\right] \mathscr{E} e^{(i\omega t - kz)} = 0. \tag{2.17}$$

Consistent with a stationary regime defined in Section 2.3.1, the field E is only a slowly varying function of the propagation distance z and of the transverse coordinate r (cylindrical symmetry is assumed). The index of refraction has a linear part n_0 (≈ 1 for air), a Kerr effect component $n_2 I$, and a self-defocusing component $\Delta n(r,z)$ due to the electron plasma:

$$n = n_0 + n_2 I - \Delta n. \tag{2.18}$$

Making a slowly varying envelope approximation in (2.17) leads to

$$2ik\partial_z \mathscr{E} = \partial^2_{rr} E + \frac{1}{r}\partial_r E + \frac{\omega^2}{c^2} 2n_0(n_2 I - \Delta n)\mathscr{E}. \tag{2.19}$$

This equation can easily be normalized by choosing $1/k$ as unit distance ($z \rightarrow z$ and $\chi = kr$), and normalizing the field to $\mathscr{E}_0 = \sqrt{\eta_0/n_2}$:

$$2i\partial_z \mathscr{E}_r = \partial^2_{\chi\chi}\mathscr{E}_r + \frac{1}{\chi}\partial_\chi \mathscr{E}_r + |\mathscr{E}_r|^2 \mathscr{E}_r - \Delta n \mathscr{E}_r. \tag{2.20}$$

Substituting the expression for Δn from (2.9),

$$2i\partial_z \mathscr{E}_r = \partial^2_{\chi\chi}\mathscr{E}_r + \frac{1}{\chi}\partial_\chi \mathscr{E}_r + |\mathscr{E}_r|^2 \mathscr{E}_r - -\left(\frac{\mathscr{E}_0}{\mathscr{E}_c}\right)^3 \mathscr{E}_r^3 \mathscr{E}_r, \tag{2.21}$$

where a characteristic field amplitude E_c relating to the plasma production has been defined:

$$\frac{1}{E_c^3} = \frac{e^2 c}{2m_e\omega^2}\sqrt{\frac{\sigma^{(3)} N_0}{2\eta_0 \beta_{ep} n_0}}. \tag{2.22}$$

The eigenvalue β is found by making the ansatz $E/E_0 = E_r exp(-i\beta z)$:

$$2\beta E_r == \frac{d^2 E_r}{d\chi^2} + \frac{1}{\chi}\frac{dE_r}{d\chi} + E_r^3 - aE_r^4, \tag{2.23}$$

where we defined the coefficient $a = (E_0/E_c)^3$.

The solution of (2.23) with $a = 0$ is the well-known Townes soliton, characterized by a radius χ_0 and a peak field E_{0p}. There are an infinite number of solutions with the same product $E_{0p}^2 \chi_0^2$ which correspond to the critical power P_{cr}. In physical units,

$$P_{cr} = \frac{\lambda^2}{8\pi n_0 n_2} = 36\ \text{MW}, \tag{2.24}$$

for air at atmospheric pressure [18,26]. The Townes soliton is unstable: if its power exceeds the critical value, it will reduce in size, collapsing to a point. In this process however, the overall shape of the eigenfunction E_r is conserved [27]. With plasma defocusing ($\alpha > 0$), the infinite family of solitons is reduced to a single, unique

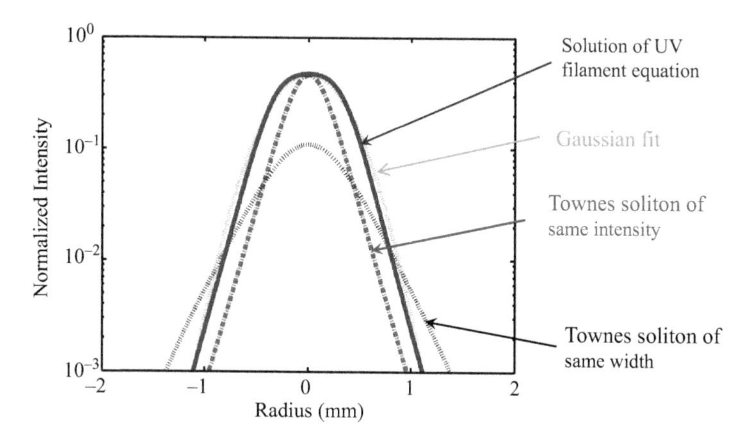

Figure 2.7 Solution of (2.23) (blue solid line) for a beam size of w = 600 μm, a power of 600 MW that corresponds to 13.8 times the critical power. For comparison, a Gaussian fit (green line), the Townes soliton of the same intensity (red dashed line), and the Townes soliton of same width (dotted gray line)

solution, shown in Figure 2.7. That solution corresponds to a filament with a waist of 600 μm and a power of 500 MW or 13.8 times the critical power defined in (2.24).

The shape of the eigenfunction of (2.23) (blue solid line in Figure 2.7) is very close to that of a Gaussian (blue solid line in Figure 2.7). For comparison, the Townes soliton of the same intensity (dotted red line in Figure 2.7) and the Townes soliton of the same width (dotted gray line in Figure 2.7) are plotted in the same figure. The Gaussian shape is clearly a better fit near the center of the beam. In the wings however, both the eigenfunction of (2.23) and the Townes soliton have an exponential decay, as can be seen from the linear dependence on the radius in the logarithmic plot of Figure 2.7. This is to be expected, since the two sets of propagation equations differ only by the highest order term in E^4 that vanishes in the wings.

2.5 Gaussian beam approximation

2.5.1 Lossless propagation equations

The excellent Gaussian fit to the UV filament stationary solution inspires to search for an analytical evolution equation, assuming a Gaussian shape for the dimensionless field in (2.21), i.e.,:

$$E_r(z,\chi) = \sqrt{\frac{\rho_0}{\rho(z)}}\mathscr{E}_{\chi,0}e^{i[-\chi^2/2q\phi(z)]}, \tag{2.25}$$

where the complex parameter q is defined as [26]

$$\frac{1}{q} = \frac{1}{R} - \frac{i}{\rho},$$

where R is the radius of curvature of the wavefront and the beam size w is defined through $\rho = w^2/2$, all dimensions being normalized to the inverse of the wave vector k. $\phi(z)$ is the phase factor on axis. All these *normalized* quantities $\phi(z)$, ρ, beam size $w(kz)$, and wavefront curvature R are linked to each other by Maxwell's equation. The relation between these functions will result from the substitution of (2.25) into the propagation equation (2.21), separating real and imaginary parts, and identifying terms of the same order in χ. The three spatial derivatives on the Gaussian beam (2.25) to be substituted in (2.21) are

$$\begin{aligned} 2i\frac{\partial \mathscr{E}_r}{\partial z} &= \left[\frac{i}{\rho}\sqrt{\frac{\rho_0}{\rho}}\frac{d\rho}{dz} + \chi^2\frac{d}{dz}\frac{1}{R} - i\chi^2\frac{d}{dz}\frac{1}{\rho} + 2\frac{d}{dz}\phi(z)\right] \\ E_r\frac{1}{\chi}\frac{\partial E_r}{\partial \chi} &= \left(-i\frac{1}{R} - \frac{1}{\rho}\right)E_r \\ \frac{\partial^2 E_r}{\partial \chi^2} &= \left[\left(-i\frac{1}{R} - \frac{1}{\rho}\right) - \left(\frac{\chi}{R}\right)^2 + \frac{\chi^2}{\rho^2} - 2i\frac{\chi^2}{R\rho}\right]E_r. \end{aligned} \tag{2.26}$$

Substituting into (2.21) yields the following real and imaginary parts:

$$2\left[\frac{d}{dz}\phi(z) + \frac{\chi^2}{2}\frac{d}{dz}\frac{1}{R}\right] = \frac{\chi^2}{\rho^2} - \frac{\chi^2}{R^2} - \frac{2}{\rho} + E_r^2 - aE_r^3 \tag{2.27}$$

$$\frac{1}{\rho}\sqrt{\frac{\rho_0}{\rho}}\frac{d\rho}{dz} + \frac{\chi^2}{\rho^2}\frac{d\rho}{dz} = \frac{2\chi^2}{\rho R} - \frac{2}{R}. \tag{2.28}$$

The zeroth-order term in the radial coordinate χ yields

$$\frac{1}{R} = \frac{1}{2\rho}\sqrt{\frac{\rho_0}{\rho}}\frac{d\rho}{dz} = \frac{1}{w}\frac{dw}{dz}, \tag{2.29}$$

which simply states the physical relation between the broadening of the beam and the wavefront curvature.

To include the nonlinear terms in (2.27), a first-order Taylor expansion of the field is made:

$$\mathscr{E}_r(z,\chi) = \sqrt{\frac{\rho_0}{\rho(z)}}\mathscr{E}_{\chi,0}e^{-i\chi^2/2q}e^{-i\phi(z)} \approx \sqrt{\frac{\rho_0}{\rho(z)}}E_{\chi,0}\left[1 - i\chi^2/2q\right]. \tag{2.30}$$

A differential equation for the evolution of the beam size is found by inserting (2.29), and (2.30) into (2.27), and replacing ρ by $w^2/2$ in these equations:

$$\frac{d^2w}{dz^2} = -\mathscr{E}_{r,0}^2\frac{w_0^2}{w^3}\left(2 - 3aE_{r,0}\frac{w_0}{w}\right) + \frac{4}{w^3}. \tag{2.31}$$

The successive terms on the right-hand side can easily be interpreted as a self-focusing term, a self-defocusing term in $\chi^{(4)}$ proportional to the cube of the field, the plasma parameter a, and a diffraction term $4/w^3$. The last equation is of the form $d^2y/dz^2 = f(y)$, which can be integrated as $(dy/dz)^2 = 2\int_{y_0}^{y} f(y')dy'$. For the particular case of (2.31), we have the following solution:

$$\left(\frac{dw}{dz}\right)^2 = \left(2E_{r,0}^2 w_0^2 - 4\right)\left[\frac{1}{w^2} - \frac{1}{w_0^2}\right] - 2aE_{r,0}^3 w_0^3\left[\frac{1}{w^3} - \frac{1}{w_0^3}\right] + \frac{w_0^2}{R_0^2}, \qquad (2.32)$$

where the last term is the initial value of $(dw/dz)^2$, if $R_0 \neq \infty$.

It is shown in Section 2.5.2 that this evolution equation is consistent with the steady-state eigenvalue solution presented in Section 2.4.1. A more complete set of equations is introduced in the following section, including attachment to oxygen and nonlinear losses.

2.5.2 Propagation equations including losses

Using the complete expression (2.7) for the electron density N_e redefines the index change Δn (Equation 2.9) relating to the plasma production:

$$\begin{aligned}\Delta n &= -\frac{N_e e^2}{m_e \varepsilon_0 n_0^2 \omega^2} \\ &= -\frac{e^2}{m_e \varepsilon_0 n_0^2 \omega^2}\left\{\sqrt{\left(\frac{\gamma}{2\beta_{ep}}\right)^2 + \frac{\sigma^{(3)} N_0 I^3}{\beta_{ep}}} - \left(\frac{\gamma}{2\beta_{ep}}\right)\right\} \\ &= -\sqrt{A^2 + B^2 E_r^6} + A,\end{aligned} \qquad (2.33)$$

where A and B are the dimensionless quantities:

$$A = \frac{e^2}{2m_e\omega^2}\frac{\gamma}{\beta_{ep} n_0^2 \varepsilon_0}, \qquad (2.34)$$

and

$$B^2 = \left(\frac{e^2}{m_e\omega^2}\right)^2 \frac{\sigma^{(3)} c^2 N_0 E_0^6}{8\eta_0 \beta_{ep} n_0}. \qquad (2.35)$$

Maxwell's equation, in dimensionless form, now becomes

$$2i\frac{\partial}{\partial(z)}\mathscr{E}_r = \partial_{\chi\chi}^2 \mathscr{E}_r + \frac{1}{\chi}\partial_\chi \mathscr{E}_r + \mathscr{E}_r^2 \mathscr{E}_r - \left(\sqrt{A^2 + B^2 \mathscr{E}_r^6} - A\right)\mathscr{E}_r. \qquad (2.36)$$

The presence of attachment to oxygen reduces the fourth-order dependence in the laser field seen in (2.21), by decreasing the density of the plasma, for a given plasma generation rate. Equation (2.36) leads to an eigenfunction equation with a solution close to that shown in Figure 2.7.

With the low-density plasma that is generated, we can neglect the losses due to plasma absorption. The dominant nonlinear losses are due to three-photon ionization. The beam depletion due to three-photon ionization is

$$\frac{dI}{dz} = -3\hbar\omega\left(\frac{dN_e}{dt}\right)_{3\text{photon absorption}} = -3\hbar\omega\sigma^{(3)}N_0 I^3. \tag{2.37}$$

For a Gaussian beam, the power $P = \pi w^2/(2I) = \pi w^2 E^2/(2\eta_0)$ evolves with distance according to

$$dP = 2\frac{dw}{w}P + 2P\frac{dE}{E}, \tag{2.38}$$

where the change in field amplitude due to a change in beam width is

$$dE = -\frac{E}{w}dw. \tag{2.39}$$

In dimensionless form, the field attenuation can be written as [18,21]

$$\frac{\partial E_r}{dz} = -\left(\frac{E0}{E_{3ph}}\right)^4 |E_r|^4 E_r - E_r\frac{1}{w}\frac{dw}{dz}, \tag{2.40}$$

where a characteristic field for three-photon absorption has been defined as

$$\frac{1}{E_{3ph}^4} = \frac{3\hbar\omega\sigma^{(3)}N_0 n_0^2}{8k\eta_0^2}. \tag{2.41}$$

The first term of (2.40) relates to energy losses due to three-photon absorption, while the second one expresses energy conservation as the beam diameter varies.

2.5.3 Simulations versus eigenfunction solution

The pulse evolution is defined by associating (2.40) with the equations of evolution of the Gaussian width, which, taking into account attachment to oxygen, is

$$\frac{d^2 w}{d^2 z^2} = -2E_{r,0}^2\frac{w_0^2}{w^3} + \frac{4}{w^3} + \frac{3B^2 E_{r,0}^6 w_0^6}{w^7\sqrt{A^2 + B^2 E_{r,0}^6 \frac{w_0^6}{w^6}}}. \tag{2.42}$$

As in the simplified (2.32), the first term of the right-hand side is the self-focusing term, the second the diffraction term, and the last one a plasma defocusing term.

Some samples of UV filament simulation are shown in Figure 2.8. In red, Figure 2.8(a), the conditions are the same as for the eigenvalue solution of Figure 2.7: a Gaussian beam size of 600 μm, and a power of 500 MW or 13.8 times the critical power. The simulation indicates indeed a stationary condition. For a much lower power (36.5 MW) only slightly above critical power, the beam reduces in size to half its initial value, before expanding again when a higher intensity

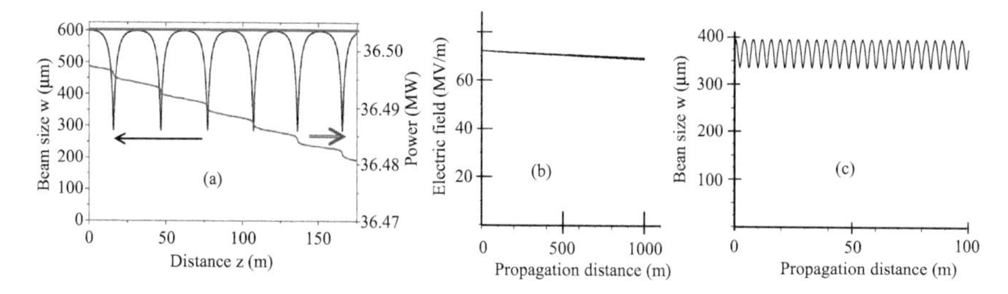

Figure 2.8 UV filament simulations. (a) Beam size and power versus distance for a Gaussian beam of $w_0 = 600\ \mu m$. *In red,* $P(z = 0) = 13.8P_{cr} = 500$ *MW. Both the beam size w and the power remain unchanged with distance. Solid black line: w versus distance for the same initial* $w(z = 0) = 600\ \mu\text{m}$, *but the initial power is 36.5 MW. Blue line corresponding beam power versus distance. (b) and (c) Propagation of a* $P = 72$ *MW,* $w_0 = 400\ \mu m$, *including three-photon ionization. (b) Electric field versus distance. (c) Beam size w versus distance*

corresponding to dominant defocusing is reached. The beam size appears to ricochet periodically on the 300 µm value. The power (right scale) decreases stepwise, each time the minimum diameter (maximum intensity) is reached, as shown by the blue curve. A simulation for an initial beam power of 72 MW, initial $w_0 = 400$ µm is illustrated in Figure 2.8(b) and (c). The plot of the electric field versus distance (b) indicates that the losses due to three-photon absorption are small, in agreement with experimental measurement of 300 µJ/m [12]. As in (a), the plot of Figure 2.8 (c) shows a periodicity in beam size due to successive self-focusing dominating at larger diameters, plasma defocusing taking over at the higher intensities.

2.6 Laser sources for UV filaments

2.6.1 Femtosecond UV sources for filaments

The first source of UV filaments was a Ti:sapphire oscillator–amplifier system at 744 nm, frequency tripled to 248 nm, and seeded in a three-path KrF amplifier [11]. This system produced a random array of 248 nm filaments of 1 ps duration, 200 µJ energy, and a diameter around 100 µm. With a total beam energy of 20 mJ, filaments were observed over a distance of 12 m. However, each single filament of 200 µJ energy had a length of ≈0.5 m, losing energy at a rate of 30 µJ/m through three-photon ionization of oxygen [12]. A clamping intensity of 0.5 TW/cm^2 was measured. A similar source was developed with comparable results by Tzortzakis *et al.* [28].

The UV filaments proved efficient at triggering and guiding electrical discharges [29]. Comparative measurements with 744 and 248 nm filaments show a 20× larger induced conductivity in air with the weaker UV filaments [16]. The weak point of the 248 nm source is the nonuniform KrF discharge resulting in a distorted wavefront. Considerable improvement was achieved by a two-cascade e-beam-pumped wide-aperture excimer KrF system [30,31] that boosted the energy to 0.2–0.5 J. The seed for the amplifiers was frequency tripled from a 744 nm Ti:sapphire system (90 fs, 4 mJ). The 300 μJ, 90 fs seed pulses were broadened by gain spectral filtering to 870 fs in the KrF amplifiers. In [30], the critical power for self-focusing was measured to be 75 MW for the ps pulses at 248 nm. Measurements at 266 nm for 300 ps pulses yielded a smaller critical power of 36 MW [18], difference that can be attributed to the pulsewidth dependence of the critical power [32].

In order to achieve a reproducible transverse pattern of filaments, the 10-cm wide collimated beam was transmitted through a mesh with three sizes of square cells of 4.5 mm [30]. With this size of mesh, one filament per cell was generated. As with prior UV filament work [29], one main motivation appears to laser-induced discharges [33,34].

2.6.2 Sources for long-pulse UV filaments

2.6.2.1 The oscillator

One property of UV filaments that is not exploited in the 248 nm implementations is the possibility to concentrate much higher energy in each filament by using longer pulses. The first possibility that comes to mind is to use the third or fourth harmonic of a Q-switched Nd:YAG laser at 1064 nm. It has been shown, however, that a power of 1 GW is required, which for typical pulse duration of 10 ns implies 10 J/pulse at 266 nm (in the case of fourth harmonic). Budget realities suggest to compromise to less energy/filament and a shorter duration pulse. One solution still being considered is to use as primary source a mode-locked Nd:YAG stabilized by passive negative feedback [35,36] generating 10 ps pulses. Amplification of these pulses to the GW power poses some challenges.

The solution that has been chosen is to start with a higher energy, longer wavelength and duration, compress, and make harmonics. The UV laser source is based on a 10 ns Nd:YAG oscillator Q-switched and seeded, followed by a chain of six amplifiers, frequency doubled, compressed by backward stimulated Brillouin compression, and frequency doubled in KDP to 266 nm. The final characteristics of the laser system are

Wavelength	266 nm
Pulse duration	170 ps
Pulse energy	300 mJ
Beam profile	Super-Gaussian
Repetition rate	1.25 Hz

In order to be compressed by stimulated Brillouin scattering (SBS), pulses should have a near-Gaussian temporal profile and be free of any modulation. Because of the low repetition rate (1.25 Hz), standard techniques such as the buildup-time (BUT) minimization method [37,38] cannot be used. BUT is based on the fact that the BUT reaches its minimum when the cavity is tuned at resonance

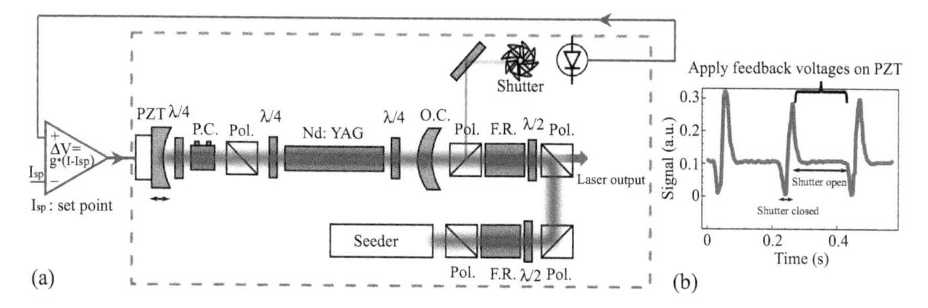

Figure 2.9 (a) Experimental setup—Pol, polarizer; PZT, piezo; P.C., Pockels cell; $\lambda/4$, quarter wave-plate; $\lambda/2$, half wave-plate; O.C., output coupler; F.R., Faraday rotator. (b) Real-time cavity reflected signal measured from leakage of a polarizer, after a synchronized shutter and with feedback on

with the injected frequency. For a high gain Q-switched cavity, it takes only a couple of round-trips to build up a giant pulse, with a pulsewidth of 6–10 ns. In this case, the BUT difference between resonance and off resonance case is only 1–2 round-trips, which is 3.5–7 ns for a cavity of 50 cm long. The correction is based on a measurement on previous pulses, information obsolete when operating at low repetition rates (10 Hz or less). In the real-time feedback control developed for this source, the cavity mode continuously tracks the seed wavelength in the time interval between pulses [39]. A feedback loop continuously locks a mode of the laser cavity to a seed laser, as shown in Figure 2.9.

A 60 mW continuous wave (cw) seed beam (linewidth <10 MHz) is injected into the Q-switched laser cavity through two isolators and the Gaussian output coupler. The purpose of the isolators is (i) to inject the seed laser via the output coupler and (ii) to decouple the seed cavity from the high-power Q-switched cavity. Two cross-axis quarter-wave-plates are placed on both sides of the Nd:YAG medium to create a "twisted mode" inside the gain medium for spatial-hole-burning elimination [40].

A clean single mode can only be achieved when the Q-switch speed is slowed down to a certain value. The switching speed is controlled by resistors in series with an electrode of the Pockels' cell. The cavity Q-switching time is measured by detecting the 60 mW seed laser after double passage through the quarter-wave-plate and the Pockels cell, and an analyzing polarizer. The intensity contrast of the resonant seeded mode over its adjacent mode increases almost by two orders of magnitude as the switching time is slowed down from 2.6 to 24.6 ns, as shown in Figure 2.10.

A measure of the purity of a single longitudinal mode is the normalized Fourier amplitudes of the temporal pulse shapes at 285 MHz (the longitudinal mode spacing of the cavity). A histogram taken with 104 pulses reveals that, with the real-time feedback control [39], $>95\%$ of the pulses have a modulation depth $\leq 6.3\%$, or an intensity contrast $>1{,}000$. All of the pulses show an intensity contrast >500.

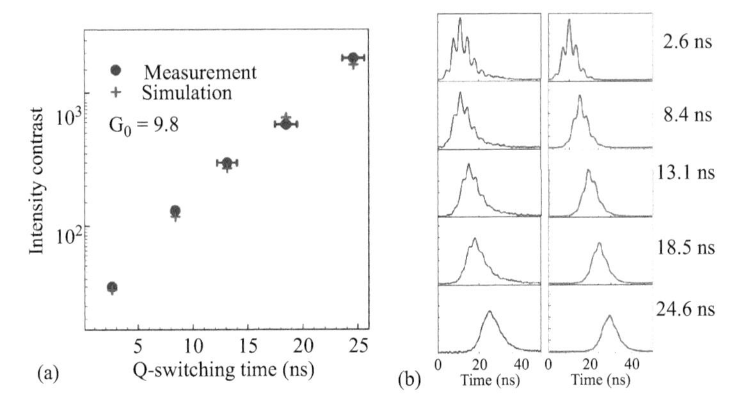

Figure 2.10 (a) Intensity contrast of the resonant seeded mode over its adjacent mode at different Q-switch opening times. (b) Temporal pulse shapes at different Q-switch opening time. The temporal positions of theses pulses show their relative arriving time. Left (blue): experimental measurement with zero detuning; right (red): simulation

One shortcoming of the unstable cavity design is that the beam profile is a super-Gaussian. The latter is advantageous for efficient amplification harmonic conversion across the beam and pulse compression by stimulated backward Brillouin scattering (the next section), but undesirable for filamentation, as detailed in Section 2.7.2.2. Furthermore, upon propagation through saturated amplifier, the center of the beam propagates at a superluminal velocity [41], leaving the edges of the beam behind [42,43].

2.6.2.2 Amplification, compression, and harmonic generation

Six stages of amplification boost the 100 mJ/pulse energy of the oscillator to 5–6 J, while maintaining a uniform beam profile, as shown in Figure 2.11. The pulse duration remains at 10 ns. An unavoidable feature of a large saturated amplifier chain is that the amplified pulse will ride on the exponential pulse rise, resulting in superluminal velocity [41]. This effect is much larger in the center of the beam than on the edges, resulting in a spatiotemporal profile shaper as a swallow, the center of the beam advancing at superluminal velocity leaving the edges of the beam behind [42,43].

The 1064 nm pulse is frequency doubled to 532 nm in an LBO crystal. The second harmonic reaches 3.5 J, with fluctuation less than 1%, for an input IR energy of 5 J. The corresponding conversion efficiency is as high as 70%. The pulsewidth is reduced with a small amount (less than 20%), while the frequency-doubled pulses inherit the spatiotemporal characteristics of fundamental pulses. A detailed analysis of the amplifier chain can be found in [43].

The next step of the source is to compress the 10 ns pulses down to 300 ps by SBS. Most of the studies on SBS pulse compression focus on the energy range of a few mJ to 100s of mJ. The two-stage SBS compressor sketches in Figure 2.12 can

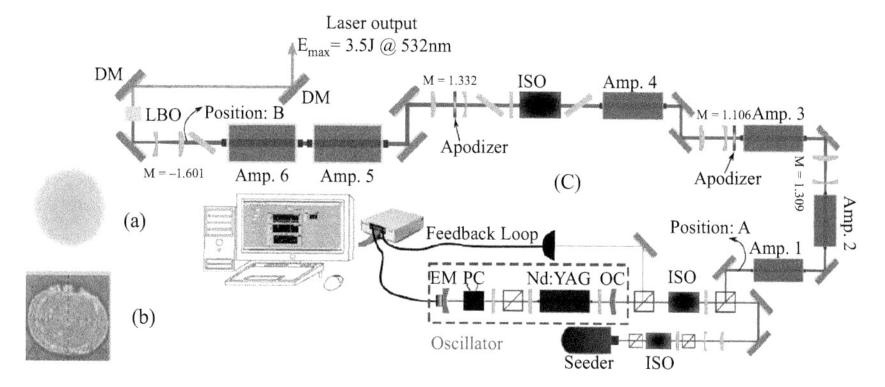

Figure 2.11 *(a) Profile of the Q-switched pulse amplifier to 6 J. For comparison, the profile of an 800 mJ Q-switched pulse from a Quantel advertisement is shown in (b). (c) Sketch of the Nd:YAG laser system. The seeded oscillator includes a concave end mirror (EM) and a convex output coupler (OC) with super-Gaussian reflectivity profile. (PC): Pockels cell for Q-switching. After the sixth amplifier the IR pulse is frequency doubled through in an LBO crystal with a conversion efficiency of 70%. Position A and B: locations for characterization of the oscillator/amplifier. The uncompressed green pulse after the second dichroic mirror is magnified by a telescope with* M = *3.31 (not shown). ISO: Faraday isolator; DM: dichroic (long pass) mirror;* M = *1.3 denotes that the telescope has a magnification of 1.3*

be scaled to the highest energy levels. A key factor is the choice of liquid. Fluorocarbons that are mostly used in SBS compression have to be abandoned because of their high-temperature dependence of the index dn/dT. Water was selected, partly because of its short phonon lifetime and chemical stability, but mainly its small dn/dT, becoming zero at 4 °C. Heavy water would be an even better choice, since it has advantages of zero dn/dT at 11 °C and lower absorption coefficient at 532 nm. Even given the small temperature dependence of the index of refraction, the water cell had to be double-walled, with temperature-stabilized liquid in the outer mantel to prevent beam distortion by convection.

The input beam diameter is chosen to be 30 mm (full width at $1/e^2$ of maximum intensity), limited by the 50 mm diameter of the cells. With this aperture limitation, the maximum pump energy that can be sent into the amplifier is 2.2 J, above which point the back-scattered signal from the pump increases exponentially. The transmittance through the cells is 85%, corresponding to a linear loss coefficient of 6.5×10^{-4} cm^{-1}. This is slightly larger than 4.5×10^{-4} cm^{-1} found in the literature [44] because of the extra scattering losses from different surfaces.

A systematic study of the energy-scalable generator–amplifier setup and its optimization can be found in [45].

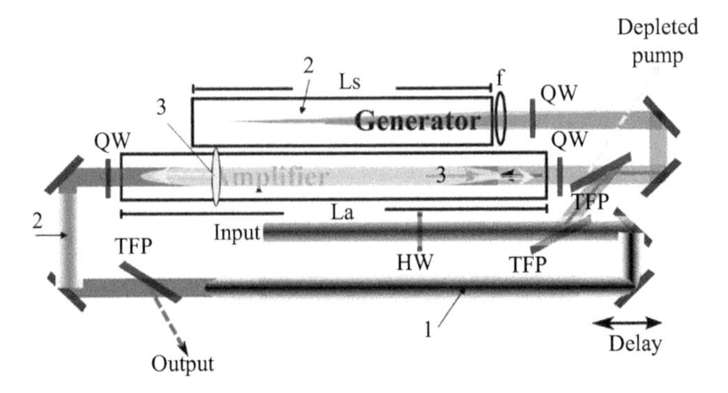

Figure 2.12 *Two-stage SBS compressor. The long input pulse at 532 nm [gray shaded area (1)] is split by a thin film polarized (TFP) between a weak pulse sent to the generator cell and a strong pump pulse directed to the amplifier cell. In green, at positions (2), the seed SBS pulse created in the generator, and the pump pulse propagating toward the amplifier cell. In yellow, position (3), the seed is amplified and compressed as it counterpropagates with the pump. HW, half wave-plate; QW, quarter wave-plates; both cells are 2.5 m long*

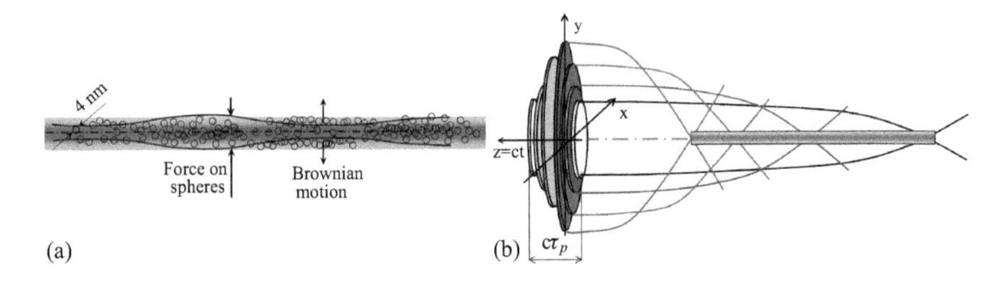

Figure 2.13 *(a) Sketch of the cw filamentation observed in suspensions of dielectric spheres [48]; (b) moving focus model*

2.7 Filament observations

2.7.1 *Self-induced waveguide versus moving focus*

In the case of femtosecond mid-IR filaments, multiple observations of a well-defined clamping intensity have clearly established [1] that these filaments involve a balance between Kerr focusing, diffraction, and plasma defocusing. There have been persistent claims that self-induced waveguide does not exist in the case of longer pulse, invoking the moving focus model [46,47]. Perhaps the most convincing example of self-induced waveguide has been achieved with low-power cw light. In that demonstration sketched in Figure 2.13(a), a laser beam is focused in a cell containing a suspension of dielectric spheres. The dipolar force on the spheres being proportional to the gradient of the field, spheres of higher index of refraction than the liquid migrate toward the center of the beam, creating a focusing effect similar to Kerr lensing. The Brownian motion counterbalances the light-induced concentration, akin to plasma defocusing in the case of fs filaments. The dielectric spheres are either latex spheres of the order of 100 nm diameter, or oil spheres of 40 nm in microemulsions. This experiment can be seen as a scaling-up in time and space of the optical filaments in air. The electrons responsible for the nonlinearity in air are replaced by microscopic spheres. Interestingly, the giant nonlinearity of the microemulsion has a time constant of the order of minutes, and one can observe in real time the collapse and channeling of the beam as it enters the liquid cell.

It remains that the moving focus model has explained satisfactorily the filament-like rosary of glass beads produced by self-focusing of a Q-switched pulse in a transparent material [46,49]. The "moving focus" model is explained by the sketch of Figure 2.13(b) by decomposing the laser pulse in successive temporal slices of different intensities. As the intensity increases along the leading edge of the pulse, it reaches the critical intensity (Equation 2.24), corresponding to focusing at infinity. The subsequent slices will reach focal points closer to the source as the intensity increases. The fourth slice corresponding to the peak intensity in the figure will focus at the shortest distance.

One may wonder whether the moving focus model should apply to the UV filament, which is closer to nanosecond than femtosecond. The experimental answer to this question is to enter the nonlinear medium with a beam focused down to the diameter of the filament. There is no possibility of moving focus if at the edge of the nonlinear medium the beam has already been focused. This experiment is described in Section 2.7.2.1, where the beam is focused in vacuum, and launched into the atmosphere through an aerodynamic window.

2.7.2 Beam preparation

The source presented in Section 2.6.2 presents two major challenges for the purpose of filamentation. First, the unstable cavity of the oscillator creates a super-Gaussian beam profile. Second, the spatiotemporal profile of the beam resulting from the oscillator and amplifier has a curved energy profile. The latter results in an undesirable energy spread along the propagation direction in a focused beam. The super-Gaussian profile is ideal for obtaining high energy from the oscillator and amplifiers, as well as for harmonic generation. Such a beam results in the formation of multiple filaments distributed along a circle [50]. This filament configuration remains the same for all focusing geometry, because the nonlinear lensing dominates before the far field pattern from diffraction is achieved.

If focused in vacuum, in the absence of nonlinearity, the beam profile will be the Fraunhofer diffraction of the super-Gaussian. This far-field diffraction having a parabolic profile on axis favors the formation of a single filament on axis. The challenge is to design a window between vacuum and air that can withstand the high intensity of the focused beam. It is shown in the next subsection that an "air window" can be created in the expansion chamber of a supersonic nozzle. Another property of this window is that it does not introduce beam-pointing instabilities and turbulence that would perturb the nonlinear focusing, defocusing, and guiding.

The purpose of using a focused beam in vacuum as initial condition is twofold. First, it serves as eliminating the high spatial frequencies prior to entering the nonlinear medium. Second, it serves to establish the existence of filaments as self-guided non-diffracting entities, as opposed to a moving focus.

2.7.2.1 Aerodynamic windows

Aerodynamic windows were first developed in the 1970s as windows for very high-power CO_2 lasers [51]. In an aerodynamic window, a supersonic flow of air is forced through a narrow nozzle. As the air flows through a curved expansion chamber, it creates a pressure gradient from vacuum to atmospheric pressure that results in the effective separation of the low vacuum chamber (pressure of about 10 Torr) from atmosphere [51,52]. The profile of the air channel cut in an aluminum block is shown in Figure 2.14(a). The red arrow indicates the direction of propagation of light, entering from the left (vacuum side) through a hole or slit, and exiting at atmospheric pressure to the right. A setup for the investigation of the optical quality of the window is shown in Figure 2.14(b). The pressure distribution in the expansion chamber of the aerodynamic window is shown by the color-coded profile. The supersonic flow is generated by an air compressor providing a flow of up to 10 m^3/min at a pressure of 8 kg/cm^2. The aerodynamic window provides a high damage threshold barrier between the atmosphere and the vacuum chamber, enabling linear propagation of the intense pulse up to the boundary of the nonlinear medium (air in this case).

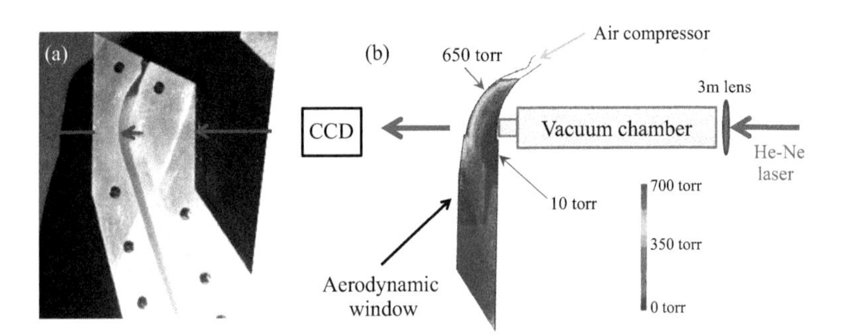

Figure 2.14 (a) Profile of the aerodynamic window nozzle and expansion chamber, cut in an aluminum block. The cut is 20 mm deep. The arrow indicates the direction of propagation of the light from vacuum to atmosphere. The light passes through a 3 mm diameter hole, or in another realization a 3 × 20 mm slot. (b) Setup for investigation of the stability of the beam profile of an He–Ne laser going through the aerodynamic window

2.7.2.2 Beam-pointing stability

The fact that a supersonic air flow is used to generate the pressure gradient in an aerodynamic window may raise suspicion about the stability of a laser beam going through such a window. In order to address this issue, an expanded He–Ne laser focused with a 3-m lens is used to investigate the beam-pointing stability through the aerodynamic window (Figure 2.14(b)). Successive images of the beam profile of the He–Ne laser are captured at 2.4 m from the exit point of the aerodynamic window by a high-resolution camera and saved for further processing. We compare the centroid or the center of gravity ($\bar{x}$) and the beam size (w) mean-square deviation [53,54] of the He–Ne laser in two cases of operational and nonoperational aerodynamic window. All other conditions are kept the same. For the center of gravity along x:

$$\bar{x} = \frac{\sum_{i,j=0}^{N} x_i I(x_i, y_j)}{\sum_{i,j=0}^{N} I(x_i, y_j)},$$

while the beam waist along the x direction is defines as

$$w_x = \sqrt{\frac{\sum_{i,j=0}^{N} (x_i - \bar{x})^2 I(x_i, y_j)}{\sum_{i,j=0}^{N} I(x_i, y_j)}}.$$

These quantities are calculated and plotted in Figure 2.15 for 48 successive images. Surprisingly, the fluctuations are reduced when the aerodynamic window is operating, indicating that the beam is more disturbed by the turbulence in the stagnant air of the 3 m tube, than by the window. The calculated standard deviations of the data sets for the measurement of the centroid of the beam in the cases of operational versus nonoperational aerodynamic window are 1.84 and 7.52 pixels,

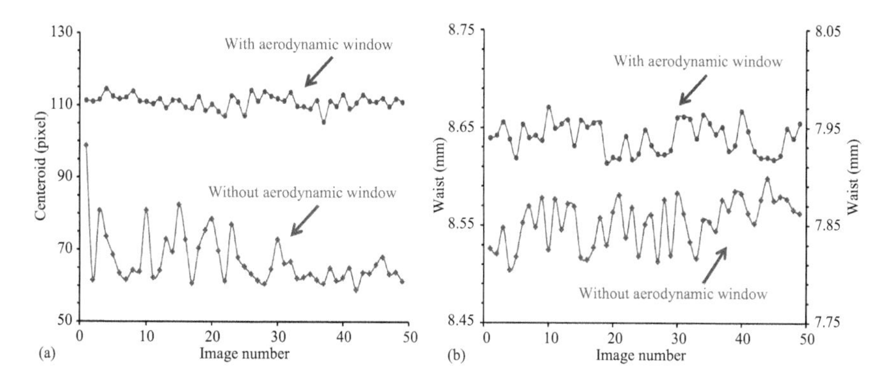

Figure 2.15 Comparison between the centroid (a) and the beam waist (b) of the beam profile of an He–Ne laser (for 49 successive images), with and without aerodynamic window. The expanded He–Ne laser beam is focused with a 3 m lens

respectively (one pixel is 15 μm). In addition, the calculated standard deviations of the data sets for the measurement of the beam waist in the cases of operational versus nonoperational aerodynamic window are 0.015 and 0.024 mm, respectively. Since all other conditions are the same, we can conclude that the stability of the laser beam going through the vacuum tube terminated by the operational aerodynamic window is improved and the beam profile has less fluctuations both in the location of its centroid and its beam waist.

2.7.3 UV filament spatial profile measurements

It has been established that properties of filaments such as length and plasma density are dependent on the focusing conditions used for the initiation of the filament [55,56]. Thus, various lenses with focal lengths from 0.5 up to 9 m have been used for filament experimental generation and detection. Results of the investigation of spatial profiles of the filaments generated by a 3 m focal distance lens are presented in the following.

Measuring the beam profile of a 300 mJ UV pulse of less than 1 ns in a submillimeter cross section is a challenge. The solution found to attenuate *linearly* the beam is to use a 15 cm diameter grazing incidence coated fused silica plate [52]. Because one linear dimension of the filament is stretched over 15 cm, the intensity can be kept below the damage threshold of the coating. A standard 355 nm maximum reflectivity coating for normal incidence is found to have less than 10^{-6} transmission in s-polarization when used at 89° incidence. The transmitted beam is expanded by a telescope and recorded with a CCD (Charge Coupled Device) camera (Figure 2.16). For the measurements of the spatial profiles of the UV filaments, the distance between the 3 m lens and the entrance window of the vacuum chamber is chosen in a way that the geometrical focus is located a few millimeters

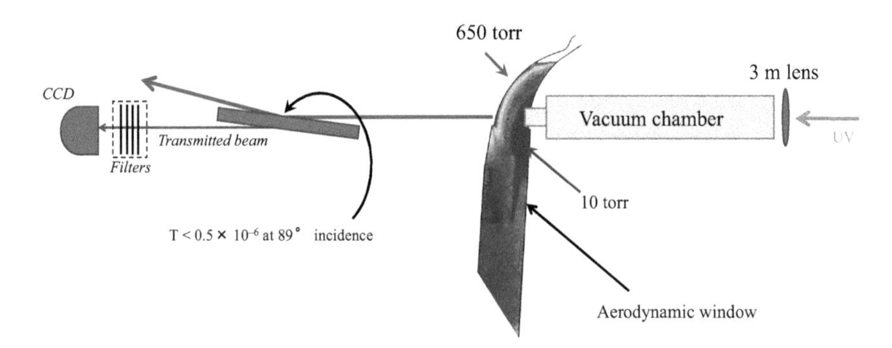

Figure 2.16 Experimental setup that is used for observing the spatial profile of the UV filaments. The beam is linearly attenuated by a grazing incidence plate (BK7, 15 mm thickness, 150 mm diameter coated for maximum reflectivity at 355 nm (normal incidence). The beam is thereafter further attenuated by neutral density filters before being recorded by a CCD

before the exit of the aerodynamic window. The aerodynamic window allows us to focus the UV pulse in vacuum before the onset of the UV filament, resulting in a different initial preparation phase for the filament.

In Figure 2.17(a) and (b), the beam profile of the UV filament for the case of focusing in vacuum using a 3-m lens (operational aerodynamic window) is presented. The UV filament has a Full Width at Half Maximum (FWHM) of about 375 μm. In Figure 2.17(b) the dashed line shows the profile calculated in Section 2.4 (see also [18]), which is in good agreement with the experimental results (solid line). A transverse image of a weak beam (sufficiently attenuated not to induce any nonlinear effect) is shown in Figure 2.17(c). An FWHM of 20 μm indicated on the figure corresponds to the waist of a $w = 13$ mm Gaussian beam after focusing by a 3 m focal distance lens; hence the initial condition for the filamenting beam exiting the aerodynamic window at the end of the 3 m vacuum tube. Transverse images of the UV filament that are taken using an image intensifier for same experimental conditions are shown in Figure 2.17(d). It can be seen that the FWHM of the

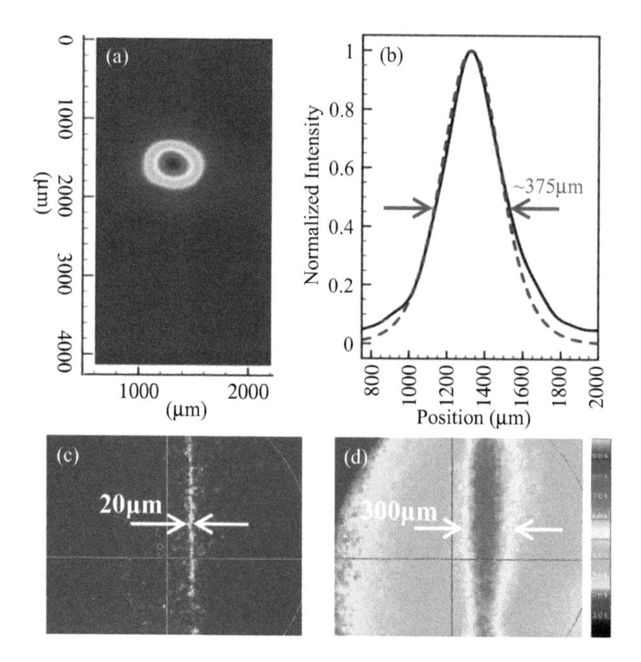

Figure 2.17 *(a) and (b) UV filament beam profile for the case of focusing with a 3-m lens in vacuum (operational aerodynamic window) [21]. The UV filament has FWHM of about 375 μm. The dashed line in (b) represents the simulated beam profile and is in good agreement with the experimental data. (c) Image of the focused UV laser beam at low powers taken with an image intensifier used for calibration; (d) image of the filament taken with the image intensifier, showing FWHM of about 300 μm*

filament in this image is 300 μm, which is in a good agreement with beam profile measurements. Figure 2.17 is another demonstration that the moving focus model [47] does not apply, even for these sub-nanosecond pulses. It is a rare example of a filament starting from a 20-μm waist to evolve into the quasi-steady-state filament shown in Figure 2.17(d).

2.8 Hydrodynamic waveguiding with filaments

Sudden heating of a gas with a high-intensity UV filament will create a low-density plasma channel. The impulsive heating of a gas creates shock-waves of cylindrical symmetry that propagate outward from the center line, leaving the low-density channel behind [57]. Initially the shock-waves propagate with velocities much larger than the speed of sound, i.e., supersonic, but their velocity rapidly decreases to the speed of sound as they propagate. The diameter of the low-density tube is determined by the deposited energy and the gas pressure [58,59].

2.8.1 Transient imaging of UV filaments via shadowgraphy

Shadowgraphy is an optical technique that can be used for the investigation of small amplitude changes in refractive index in transparent materials, which in the case of UV filaments is a consequence of the change in air density [60]. Shadowgraphy technique is used to investigate the characteristics of the mentioned low-density tube and shock-waves created by the UV filaments. The probe beam is a collimated and expanded (about 6 cm) green (532 nm) pulse with 10 ns temporal pulsewidth. This probe beam is sent perpendicular to the UV filament and images the area of interest on a screen located in a 1 m distance from the filament (Figure 2.18).

A CCD camera is used to capture the shadowgrams on the screen. An optical fiber is placed just below the UV filament, providing a reference for the size of shadowgrams. A master-clock is used to synchronize the UV laser, the green laser, and the camera. A delay generator controls the relative delay times between the

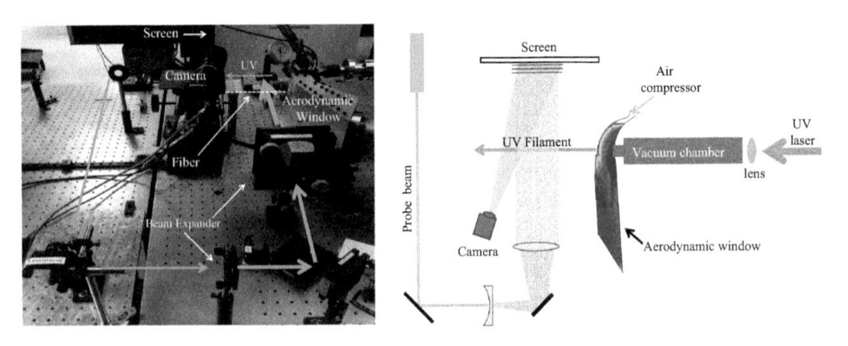

Figure 2.18 (Left) Picture of the shadowgraphy setup, (right) schematic diagram of the shadowgraphy setup

probe beam and the UV laser, making it possible to investigate the temporal evolution of the shock-waves and the low-density channel.

The shadowgrams of UV filaments in different focusing schemes were investigated. The results of focusing with 1 and 3 m lenses are presented here. In Figure 2.19, selected shadowgrams of UV filaments are displayed for different time delays between the UV laser and the probe beam for both focusing schemes.

From the shadowgrams, we can conclude that, as expected, the filament length is longer when focusing with a 3 m lens. In this case, the low-density tube is present all over the probe beam. This fact is also evident in Figure 2.20 (left), which is a direct image of the plasma channel of the filament taken with a digital camera.

The shock-waves created from tighter focusing are more intense. This is also expected, since in tighter focusing higher plasma electron densities are created. The higher intensity deposited by the filament [61,62] affects the width of the low-density tube. The widths of the low-density channels in both cases are measured using the image of the fiber as reference. These widths are 485 and 420 μm for the 3 and 1 m lenses, respectively. The widths of these channels are in agreement with the UV filament FWHM reported in Section 2.7.3. The shock-waves propagate initially with supersonic velocities, but their velocity rapidly decreases to the speed of sound, in good agreement with our measurements of the speed of shock-waves shown in in Figure 2.20 (right).

2.8.2 Hydrodynamic waveguiding

Another observation from the shadowgraphy data is that the generated waveguide is formed within 300 ns and remains up to 1 ms. As a result, a series of ionizing laser pulses will lead to the production of heat bursts, and each heat burst will

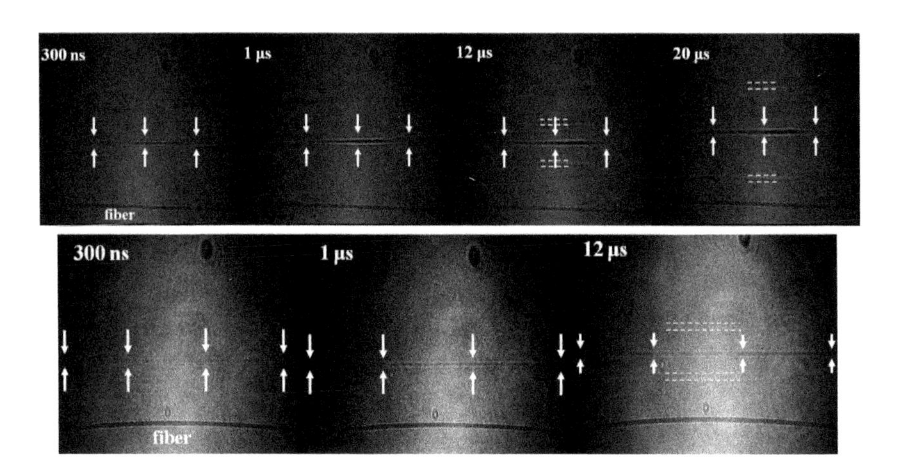

Figure 2.19 Shadowgraphs of UV filament versus time. (Top) Focused with a 1 m lens. (Bottom) Focused with a 3 m lens. The low-density tube and shock-waves are demonstrated by white arrows and dashed lines, respectively

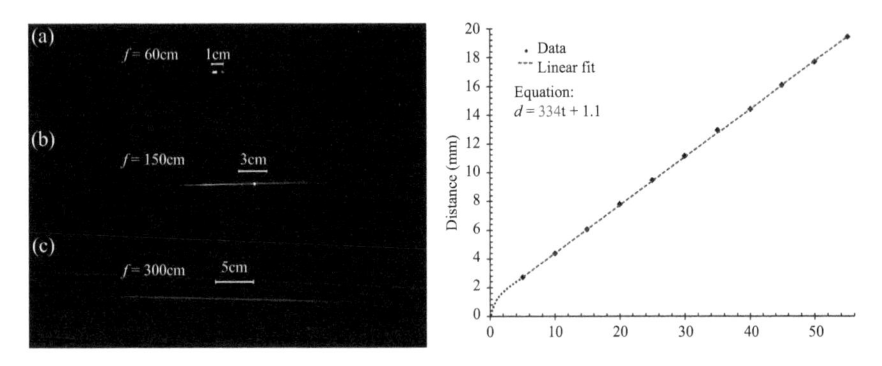

Figure 2.20 (Left) Image of the plasma channels taken with a digital camera from the side for the filaments generated with three focusing lenses of (a) 0.6, (b) 1, and (c) 3 m [43]. (Right) Shock-wave propagation versus time for shock-waves generated by UV filaments created by a 1 m lens

create a shock-wave that does not decay before the arrival of the next pulse. The generation of this waveguide, based on the shock-wave being long lived, can possibly result in a stationary waveguide at sufficiently high repetition rates. The stationary waveguides mentioned in Section 2.8 can be used for optical guiding of an another filamenting laser pulse [63–66]. This can have possible applications for remote sensing [67–69], directed energy applications [70,71], and triggering and guiding of electrical discharges [2,72–75].

In [63] the authors have investigated waveguides generated by fs IR filaments. Their experiments in the single filament scenario have shown that an initial density hole will grow for tens of nanoseconds as a shock-wave propagates outwards. Later after approximately 1–2 μs, the pressure inside the waveguide reaches equilibrium. This generated leaky waveguide will decay in ms timescales. These measurements reported in [57] are also in agreement with our shadowgraphy study using UV filaments reported in Section 2.8. Positive index profiles—hence dielectric waveguides—were achieved with multiple IR filaments forming a symmetric four-lobed pattern that created a long-lived, of order of ms, waveguide in air. It was also reported that the length of these waveguides depended only on the propagation distance of the inducing fs array of filaments [63]. A probe laser beam used to investigate the guiding property of the positive pressure waveguide indicated a peak guiding efficiency of 70% at $\simeq$600 μs that decreased to 15% at $\simeq$2 ms. In their latest publication [64], the authors reported successful guiding over lengths of up to 30 cm.

In applications to remote sensing, the generated waveguide can not only be used for guiding a second high peak power filament to irradiate a sample, but it has also been shown that these air waveguides can act as a broadband collection optics that enhances the returning signal from the sample that will improve the sensitivity in remote sensing [65]. Application of UV filaments for remote sensing is discussed in more detail in Section 2.9.

2.9 Applications in remote sensing

2.9.1 Laser-induced breakdown spectroscopy

Filaments allow for standoff spectroscopy of the radiation emitted from the plume produced by the impact of the high-intensity beam on a solid target. This emission spectrum is pressure and Stark broadened and has to be corrected for Stark shift and Stark broadening in order to identify the emission lines. There are structures within the emission line—essentially dips—that have been dismissed as an annoyance [76,77]. These dips are mostly observed in the spectrum delayed by typically 5 μs or more. In the case of irradiation by UV filaments, they are much more pronounced and appear almost immediately after irradiation [69]. These structures are interpreted as "self-absorption" of light emitted in the excited medium, by a portion of the plume in ground state [77]. Considerable efforts have been made to eliminate them either through experimental techniques or through simulation and data analysis [78–80]. These attempts to correct the emission spectrum are futile: the dips are much more accurate and precise signatures of the material interrogated by the laser beam than the emission spectrum. Indeed, in contrast to the emission, the dips are centered from the early times exactly at the tabulated frequency of the corresponding atomic transition.

2.9.2 Experiment

Experiments were performed with the laser source described in Section 2.6.2. The laser pulse is focused onto a solid target, creating a plume that is analyzed by a gated Double Echelle Monochromator (DEMON by Lasertechnik Berlin GmbH). Time resolution is realized with a gated ICCD Andor camera to record the spectrum. The optimum gate width of the Andor camera is chosen to be 500 ns as a compromise between best signal-to-noise ratio versus time resolution. Each measurement is an average of 100 images with spectral resolution of 6 pm at 300 nm to 8 pm at 400 nm. All the data are taken in air at atmospheric pressures.

Lines observed in emission are typically Doppler, pressure and Stark broadened, and also Stark shifted [77,81]. At atmospheric pressures, the broadening of the emission lines makes it impossible to distinguish isotopes as can be seen in the case of the two isotopes of lithium in Figure 2.21 (left). The broadening of emission lines also prevents the identification of emitting transitions in the presence of a high density of lines, as is the case in the spectrum of steel reproduced in Figure 2.21 (right). In contrast, the reabsorption dips seen in transitions from the ground state are neither affected by Stark shift nor by spectral broadening.

In the case of lithium, due to the fact that it is a very light atom, the emission lines are very broad in atmospheric pressures ($\simeq$3 nm FWHM). The isotopic shift between the two natural stable isotopes of Li is about 19 pm and these broad

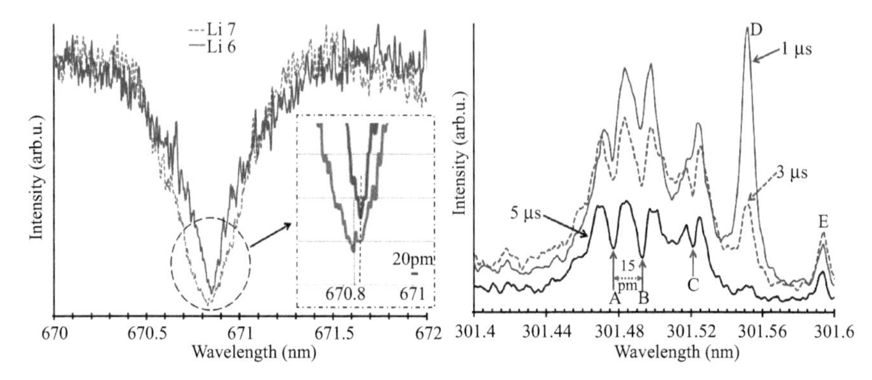

Figure 2.21 LIBS spectra of samples irradiated by a UV filament at atmospheric pressure. UV laser pulse is a 200 ps pulse of 200 mJ energy, focused by a lens of 40 cm focal distance. (Left) LIBS spectra of the two isotopes of lithium. The transition lines for these two isotopes are at 670.830 (Li7) and 670.849 nm (Li6), which indicates an isotopic shift of 19 pm. The emission line extends far beyond the boundaries of the figure. (Right) LIBS spectrum of steel taken at gate delays of 1, 3, and 5 µs. A broad emission structure can be seen between 301.44 and 301.54 nm, which includes three chromium transitions from the $[3d^4 4s^2\ (J = 1, 2, 3)]$ *ground state to the* $[3d^4(^5D)4s4p(^1P^\circ)\ (J = 2, 3, 0)]$ *levels at 301.476 (a), 301.491 (b), and 301.520 nm (c). Emission lines labeled (d) chromium transition from* $[3d^4(a^3F)4s\ (J = 7/2)]$ *to* $[3d^4(a^3F)4p\ \ (J = 7/2)]$ *at 301.550 nm and (e) iron transition from* $[3d^6 4s^2\ \ (J = 5)]$ *to* $[3d^7(^2H)4p\ (J = 4)]$ *at 301.592 nm do not show a reabsorption dip*

emission lines cannot be utilized for the detection of this isotopic shift. However, as demonstrated in Figure 2.21 (left), by using the self-absorption dips that are much narrower ($\simeq$0.3 nm FWHM), this isotopic shift is resolved [82]. The fact that self-absorption dips have higher resolution compared to the emission lines can also be observed in the case of the LIBS spectra from a steel sample in Figure 2.21 (right). Three transition lines of chromium (labeled A, B, and C) are forming a broad emission structure. By utilizing the self-absorption dips within the emission structure, two lines separated by as little as 15 pm can be resolved, thus a resolution sufficient for isotopic selectivity [83]. It should be mentioned that in both cases of lithium and chromium spectra, there does not appear to be any environmental shift affecting the self-absorption dips, as their frequencies match the values tabulated by NIST [84].

2.9.3 Nature of the self-absorption

The dips are interpreted as reabsorption by the material ejected upon impact from the filament. Contrary to the Lorenz-broadened emission line (indicating pressure and Stark broadening), the reabsorption dip fits a Gaussian shape and is considerably narrower. This is consistent with reabsorption from a lower density region, likely behind the shock-wave created by the impact of the filament on the metal. Relying on the absorption dip results in higher resolution, making even isotopic selective LIBS possible [69,82,83] at atmospheric pressures. An example of an LIBS spectrum of a uranium sample irradiated by UV filament at atmospheric pressures is shown in Figure 2.22(a). Resolution of about 9 pm for the self-absorption dip is observed, while the emission line has an FWHM of about 124 pm.

The delay-dependent emission from two samples of uranium and aluminum are presented in Figure 2.22(b) and (c). The fits presented in Figure 2.22(b) illustrate clearly the fact that the reabsorption dip central wavelength is fixed within a couple of pm, and that its width decreases with delay. The dashed vertical line shows that the absorption dips are perfectly lined up. The center of gravity of the Lorentzian emission spectra, however, does not match the transition wavelength and changes with time for both samples. The vertical arrows pointing to the peaks of the Lorentzians in Figure 2.22(b) show that both linewidth and Stark shift decrease with delay. Since the widths of the lines (absorption and emission) decrease with atomic mass of the elements, LIBS spectra for the 396.152 nm transition are shown in Figure 2.22(c) to better illustrate the wavelength dependence of the emission and absorption peaks. The center emission line at 225 ns delay has 120 pm Stark shift that is reduced to 50 pm at a delay of 1 μs.

2.9.4 Comparison of the emission and absorption lines

The temporal dependence of the main parameters of the emission line and self-absorption dip are presented in Figure 2.23. The peak of the Gaussian reabsorption remains at the same value for all delays within 1.5 pm (well within the 6-pm reported resolution of the DEMON at that wavelength), while the peak of the Lorentzian approaches asymptotically the transition wavelength, from an initial Stark shift of 50 pm (Figure 2.23(a)). The time evolution of the linewidth of the emission and absorbing features can be correlated to the expansion and collisional cooling of the plasma. A nearly linear decrease in Gaussian linewidth from 12 to 5 pm over 1.5 μs indicates a slower evolution of the absorbing region (Figure 2.23 (b)). The excited region exhibits a much faster expansion and cooling in Figure 2.23(c). It should be noted that the emission rides over a continuum background (plasma emission) decay with time. The continuum emission in our case is due to a plasma and decays to $1/e^2$ in 1.0 μs, while the much shorter decay reported for a lead sample in reference [85] is that of the filament emission.

2.10 Applications in laser-induced discharges

Triggering and guiding high voltage electrical discharges can have many applications. Among them, control of lightning has caught attention during the past years [2,72–74,86,87]. As discussed in Section 2.8.2, UV filaments have the potential of

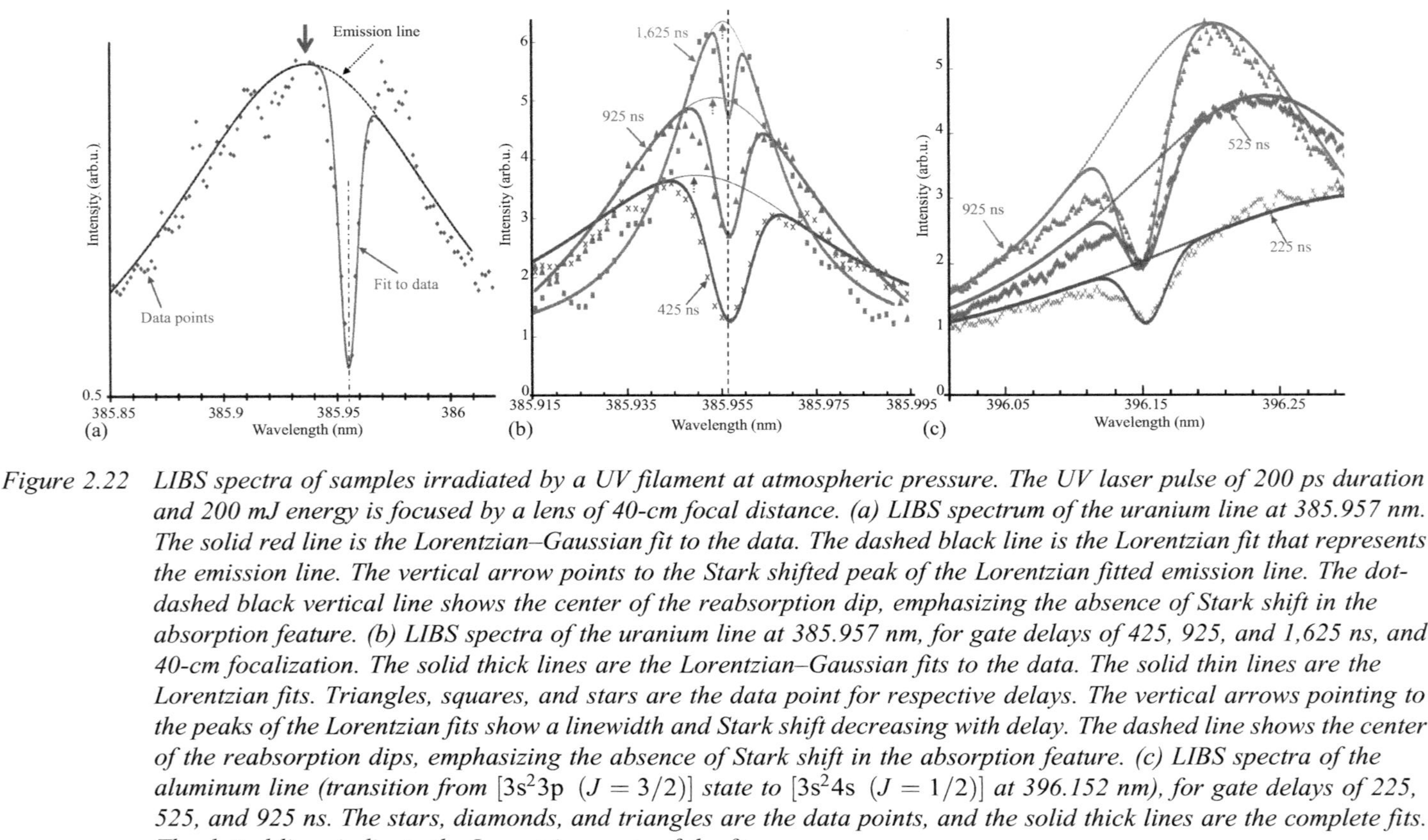

Figure 2.22 *LIBS spectra of samples irradiated by a UV filament at atmospheric pressure. The UV laser pulse of 200 ps duration and 200 mJ energy is focused by a lens of 40-cm focal distance. (a) LIBS spectrum of the uranium line at 385.957 nm. The solid red line is the Lorentzian–Gaussian fit to the data. The dashed black line is the Lorentzian fit that represents the emission line. The vertical arrow points to the Stark shifted peak of the Lorentzian fitted emission line. The dot-dashed black vertical line shows the center of the reabsorption dip, emphasizing the absence of Stark shift in the absorption feature. (b) LIBS spectra of the uranium line at 385.957 nm, for gate delays of 425, 925, and 1,625 ns, and 40-cm focalization. The solid thick lines are the Lorentzian–Gaussian fits to the data. The solid thin lines are the Lorentzian fits. Triangles, squares, and stars are the data point for respective delays. The vertical arrows pointing to the peaks of the Lorentzian fits show a linewidth and Stark shift decreasing with delay. The dashed line shows the center of the reabsorption dips, emphasizing the absence of Stark shift in the absorption feature. (c) LIBS spectra of the aluminum line (transition from* $[3s^2 3p \; (J = 3/2)]$ *state to* $[3s^2 4s \; (J = 1/2)]$ *at 396.152 nm), for gate delays of 225, 525, and 925 ns. The stars, diamonds, and triangles are the data points, and the solid thick lines are the complete fits. The dotted lines indicate the Lorentzian parts of the fit*

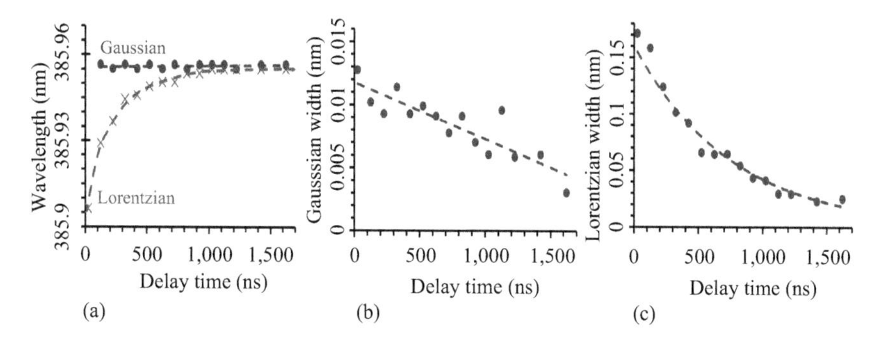

Figure 2.23 Comparison of the emission and reabsorption, as analyzed through the fits of Figure 2.22 versus gate delay. (a) Wavelength of the emission (at the peak of the Lorentzian) and of the reabsorption (at the peak of the Gaussian fit). The 50 pm Stark shift of the emission (Lorentzian) rejoins the absorption dip wavelength (Gaussian) for delays in excess of 1 μs. (b) Decrease of the width of the absorption dip versus delay. (c) The broader width of the emission line exhibits a faster decay rate, indicating a faster cooling/expansion of the excited region

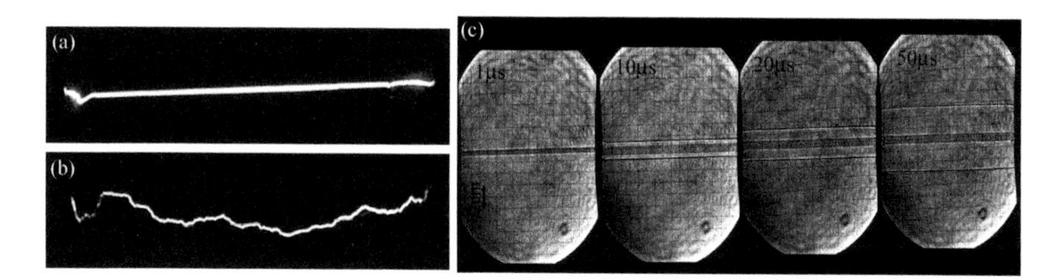

Figure 2.24 (a) Discharge guided by the laser filament, (b) unguided discharge, (c) shadowgrams of the laser induced discharge at different times after the initiation of discharge for a 16 cm gap

creating waveguides over extended distances; hence, they are ideal candidates for the purpose of lightning control via laser-induced discharge.

In Figure 2.24(a), an image, taken by a digital camera, of a discharge over a gap of 37.5 cm guided by the UV filament is presented. In contrast to the case of unguided discharge (Figure 2.24(b)), the guided discharge follows exactly the path of the inducing UV filament. Shadowgrams of the UV-filament-induced discharges over a 16 cm gap have been recorded using a shadowgraphy technique similar to the one mentioned in Section 2.8. The impulsive heating of air by the electrical discharge creates shock-waves traveling outward, leaving a low-density channel behind. In the shadowgrams in Figure 2.24(c) taken at different delays with respect to the initiation of the discharge, a low-density tube and a traveling shock-wave can be seen.

The lifetime of the electrons created in the UV filament induced plasma is of the order of tens of nanoseconds. Their number decays mainly due to electron attachment to O_2 and O and electron–ion recombination processes [88] as detailed in Section 2.3.2. Compared to this lifetime, the guided discharge formation is a relatively slower process, dependent on the separation distance between electrodes and also the electrode geometry [74]. For instance, in Figure 2.25(b), the delay time between the electrical discharge and the UV filament is 650 ns for a 37.5 cm gap that is approximately ten times larger than the plasma lifetime. Consequently, although guiding over short gaps via the plasma conductivity is possible, the laser-induced discharges over longer gaps are guided by air rarefaction, rather than by plasma conduction. As mentioned in Section 2.8, sudden heating of air by the UV filament will create a cylindrical shock-wave propagating outward leaving a long-lived lower density channel of air behind which makes guiding through air rarefaction possible.

UV filament-induced discharge between two electrodes was investigated for different gaps. The UV filament was created using the source presented in Section 2.6.2. A schematic diagram of the arrangement of the discharge circuit is shown in Figure 2.25(a). Two steel spheres (8 in. diameter) are used as electrodes and the laser beam passes through holes in both positive and negative electrodes. Two capacitor banks, each consisting of ten capacitors with equal capacitance of 2 nf, were discharged through the electrodes. The capacitor banks were designed for minimum inductance, resulting in a discharge rise-time of less than 20 ns. These capacitor banks were charged by two high voltage power supplies up to 300 kV with opposite polarities. The discharge current was monitored with a Pearson current probe. In Figure 2.25(b), a current waveform corresponding to the discharge over a 37.5 cm gap at potential difference of 372 kV is presented.

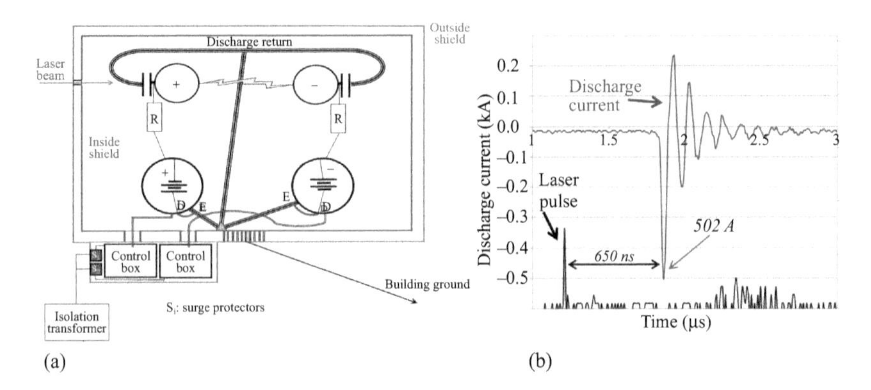

Figure 2.25 (a) A schematic diagram shows the circuit and its elements in the Faraday cage. (b) The discharge current and guiding filament pulse for a 372-kV potential difference between two electrodes with a 37.5 cm gap

In Figure 2.26(a), the discharge probability and the average delay between the UV filament and discharge for different applied voltages and for a 37.5 cm gap are presented. As expected, while the discharge probability increases with the applied voltage, the delay decreases. It should be mentioned that for all applied voltages, no breakdown was observed in the absence of a triggering UV filament. These experiments have been repeated in different environmental conditions, and from these experiments, it was concluded that absolute humidity is a key factor affecting both the delay and probability [73,74]. In Figure 2.26(b), it can be seen that as the humidity increases the discharge probability also increases while the delay decreases. Similar results have been reported by other research groups [14,73].

By using the circuit represented in Figure 2.27 (right), a set of measurements at different electrode separation distances has been performed in order to measure the resistivity of the UV filament induced plasma channel versus time. In Figure 2.27 (left), one of the measured voltage waveforms for the separation distance of 11.5 cm between the two electrodes at 37 kV is shown. The resistivity of the UV-filament-induced plasma channel is about 5 kΩ/cm, which is much less than the 900 kΩ/cm measured during a similar experiment with IR filaments (800 nm) [89]. The delay between the peak of the voltage waveform and the UV filament is about 25 ns and the plasma conductivity lasts for less than 50 ns. It should be mentioned that these measurements are performed for the cases in which separation distances between the electrodes are short enough so that the plasma filament covers the whole gap.

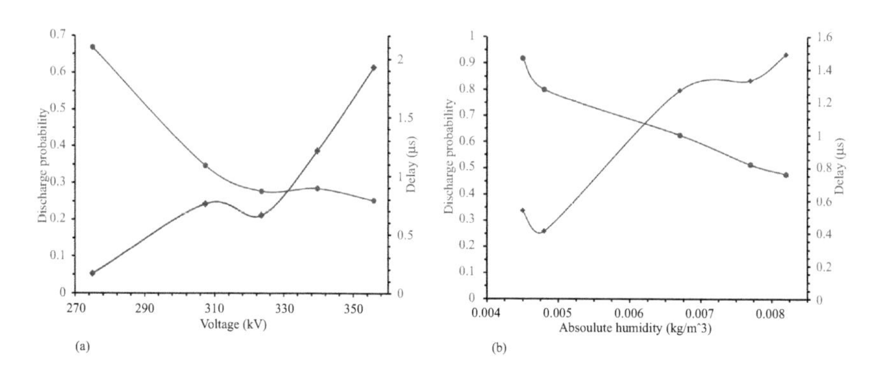

Figure 2.26 (a) Measurements of discharge probability and delay time between the UV filament and the discharge, versus the applied potential difference, for a 37.5 cm gap. Each point on the graph is average of 100 data points. (b) Delay and discharge probability versus absolute humidity for UV-induced discharges. Data are taken for a 37.5 cm gap between the electrodes and each point on the graph is average of 100 data points

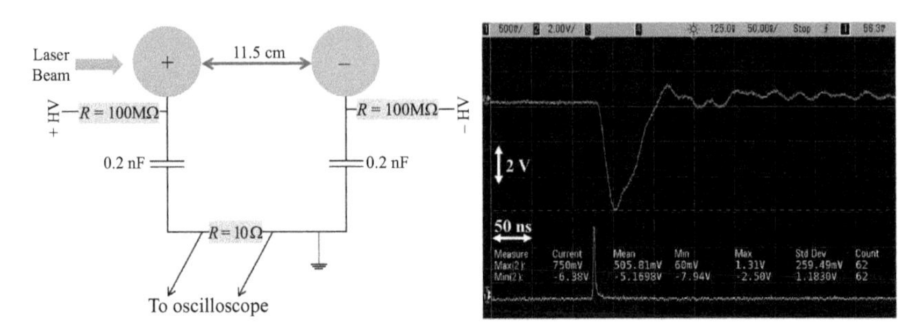

Figure 2.27 (Left) Schematic diagram of the circuit used for measuring the conductivity of the plasma channel. (Right) The voltage waveform from conductivity measurements. The upper waveform is the measured voltage and the lower waveform is the photodiode signal showing the laser pulse

Acknowledgments

This work is supported by the Army Research office (ARO: W911NF-19-1-0272) and US Department of Energy (DOE: DESC0011446).

References

[1] H. R. Lange, G. Grillon, J.-F. Ripoche, *et al.*, "Anomalous long-range propagation of femtosecond laser pulses through air: Moving focus or pulse self-guiding?," *Opt. Lett.*, vol. 23, pp. 120–122, 1998.
[2] T. Produit, P. Walch, C. Herkommer, *et al.*, "The laser lightning rod project," *Eur. Phys. J. Appl. Phys.*, vol. 93, no. 1, p. 10504, 2021.
[3] L. V. Keldysh, "Ionization in the field of a strong electromagnetic wave," *Sov. Phys. JET*, vol. 20, no. 5, pp. 1307–1314, 1965.
[4] V. S. Popov, "Tunnel and multiphoton ionization of atoms and ions in a strong laser field (Keldysh theory)," *Phys. Usp.*, vol. 47, p. 855, 2004.
[5] L. Arissian, C. Smeenk, F. Turner, *et al.*, "Direct test of laser tunneling with electron momentum imaging," *Phys. Rev. Lett.*, vol. 105, p. 133002, 2010.
[6] M. Y. Ivanov, M. Spanner, and O. Smirnova, "Anatomy of strong field ionization," *J. Mod. Phys.*, vol. 52, pp. 165–184, 2005.
[7] V. P. Krainov, "Ionization rates and energy and angular distributions at the barrier-suppression ionization of complex atoms and atomic ions," *J. Opt. Soc. Am. B*, vol. 14, p. 425, 1997.
[8] L. Arissian and J.-C. Diels, "Ultrafast electron plasma index; an ionization perspective," *J. Lasers Opt. Photonics*, vol. 1, pp. 107–111, 2014.
[9] P. B. Corkum, "Plasma perspective on strong-field multiphoton ionization," *Phys. Rev. Lett.*, vol. 71, pp. 1994–1997, 1993.

[10] Y. P. Raizer, "Breakdown and heating of gases under the influence of a laser beam," *Sov. Phys. Usp.*, vol. 8, no. 5, pp. 650–673, 1966.

[11] X. M. Zhao, J.-C. Diels, C. Y. Wang, and J. Elizondo, "Femtosecond ultra-violet laser pulse induced electrical discharges in gases," *IEEE J. Quantum Electron.*, vol. 31, pp. 599–612, 1995.

[12] J. Schwarz, P. K. Rambo, J.-C. Diels, M. Kolesik, E. Wright, and J. V. Moloney, "UV filamentation in air," *Opt. Commun.*, vol. 180, pp. 383–390, 2000.

[13] M. Durand, A. Houard, K. Lim, A. Durécu, O. Vasseur, and M. Richardson, "Study of filamentation threshold in zinc selenide," *Opt. Express*, vol. 22, pp. 5852–5858, 2014.

[14] A. V. Shutov, N. N. Ustinovskii, I. V. Smetanin, *et al.*, "Major pathway for multiphoton air ionization at 248 nm laser wavelength," *J. Chem. Phys.*, vol. 64, pp. 1733–1740, 1975.

[15] M. Durand, A. Houard, B. Prade, *et al.*, "Kilometer range filamentation," *Opt. Express*, vol. 21, pp. 26836–26845, 2013.

[16] X. M. Zhao, P. Rambo, and J.-C. Diels, "Filamentation of femtosecond UV pulses in air," in *QELS, 1995*, vol. 16, (Baltimore, MA), p. 178 (QThD2), Optical Society of America, 1995.

[17] I. Smetanin, A. Levchenko, A. V. Shutov, N. Ustinovskii, and V. D. Zvorykin, "Role of coherent resonant nonlinear processes in the ultrashort KrF laser pulse propagation and filamentation in air," *Nucl. Instrum. Methods Phys. Res., Sect. A*, vol. 369, pp. 87–91, 2016.

[18] O. J. Chalus, A. Sukhinin, A. Aceves, and J.-C. Diels, "Propagation of non-diffracting intense ultraviolet beams," *Opt. Commun.*, vol. 281, pp. 3356–3360, 2008.

[19] J.-F. Daigle, A. Jacon-Becker, S. H. T.-J. Wang, *et al.*, "Intensity clamping measurement of laser filaments in air at 400 and 800 nm," *Phys. Rev. A*, vol. 82, p. 023405, 2010.

[20] A. Braun, G. Korn, X. Liu, D. Du, J. Squier, and G. Mourou, "Self-channeling of high-peak-power femtosecond laser pulses in air," *Opt. Lett.*, vol. 20, pp. 73–75, 1995.

[21] O. Chalus, *Study of Nonlinear Effects of Intense UV Beams in the Atmosphere*. PhD thesis, University of New Mexico and University Louis Pasteur, Strasbourg, 2007.

[22] L. M. Branscomb, D. S. Burch, S. J. Smith, and S. Geltman, "Photodetachment cross section and the electron affinity of atomic oxygen," *Phys. Rev.*, vol. 111, pp. 504–513, 1958.

[23] A. W. Ali, "The electron avalanche ionization of air and a simple air chemistry model," NRL memorandum report 4794, Naval Research Laboratory, April 1982.

[24] J. Schwarz and J.-C. Diels, "Analytical solution for UV filaments," *Phys. Rev. A*, vol. 65, pp. 013806-1–013806-10, 2001.

[25] V. E. Zakharov and A. B. Shabat, "Exact theory of two-dimensional self-focusing and one-dimensional self-modulation of waves in nonlinear media," *Sov. Phys. JETP*, vol. 34, pp. 62–69, 1972.

[26] J.-C. Diels and W. Rudolph, *Ultrashort Laser Pulse Phenomena*. Boston, MA: Elsevier, ISBN 0-12-215492-4; second edition, 2006.

[27] K. D. Moll, G. Fibich, and A. L. Gaeta, "Self-similar optical wave collapse: Observation of the Townes profile," *Phys. Rev. Lett.*, vol. 90, p. 203902, 2003.

[28] S. Tzortzakis, B. S. Prade, M. A. Franco, and A. Mysyrowicz, "Time-evolution of the plasma channel at the trail of a self-guided IR femtosecond laser pulse in air," *Opt. Commun.*, vol. 25, pp. 1397–1399, 2000.

[29] X. M. Zhao, J.-C. Diels, A. Braun, *et al.*, "Use of self-trapped filaments in air to trigger lightning," in *Ultrafast Phenomena IX* (P. F. Barbara, W. H. Knox, G. A. Mourou, and A. H. Zewail, eds.), (Dana Point, CA), pp. 233–235, Springer Verlag, Berlin, 1994.

[30] D. E. Shipilo, N. A. Panov, E. S. Sunchugasheva, *et al.*, "Fifteen meter long uninterrupted filaments from sub-terawatt ultraviolet pulse in air," *Opt. Express*, vol. 25, pp. 2286–25391, 2017.

[31] V. D. Zvorykin, A. A. Ionin, A. O. Levchenko, *et al.*, "Multiple filamentation of supercritical UV laser beam in atmospheric air," *Nucl. Instrum. Methods*, vol. B 355, pp. 227–231, 2017.

[32] G. G. Luther, J. V. Moloney, A. C. Newell, and E. M. Wright, "Self-focusing threshold in normally dispersive media," *Opt. Lett.*, vol. 19, pp. 862–864, 1994.

[33] V. D. Zvorykin, A. O. Levchenko, and N. N. Ustinovskii, "Control of extended high-voltage electric discharges in atmospheric air by UV KrF-laser radiation," *Quantum Electron.*, vol. 41, pp. 227–233, 2011.

[34] V. D. Zvorykin, A. A. Ionin, A. O. Levchenko, *et al.*, "Extended plasma channels created by UV laser in air and their application to control electric discharges," *Plasma Phys. Rep.*, vol. 41, pp. 112–146, 2015.

[35] A. Agnesi, A. D. Corno, J.-C. Diels, *et al.*, "Generation of extended pulse trains of minimum duration by passive negative feedback applied to solid state Q-switched lasers," *IEEE J. Quantum Electron.*, vol. 28, pp. 710–719, 1992.

[36] A. Umbrasas, J. C. Diels, G. Valiulis, J. Jacob, and A. Piskarskas, "Generation of femtosecond pulses through second harmonic compression of the output of a Nd:YAG laser," *Opt. Lett.*, vol. 20, pp. 2228–2230, 1995.

[37] L. A. Rahn, "Feedback stabilization of an injection-seeded Nd:YAG laser," *Appl. Opt.*, vol. 24, pp. 940–942, 1985.

[38] X. Xu, *High Power UV Source Development and Its Applications*. PhD thesis, The University of New Mexico, Albuquerque, NM, 2015.

[39] X. Xu and J.-C. Diels, "Stable single-axial-mode operation of injection-seeded Q-switched Nd:YAG laser by real-time resonance tracking method," *Appl. Phys. B*, vol. 114, p. 579, 2014.

[40] V. Evtuhov and A. E. Siegman, "A "twisted-mode" technique for obtaining axially uniform energy density in a laser cavity," *Appl. Opt.*, vol. 4, no. 1, pp. 142–143, 1965.
[41] N. G. Basov, R. V. Ambartsumyan, V. S. Zuev, P. G. Kryukov, and V. S. Letokhov, "Nonlinear amplifications of light pulses," *Sov. Phys. JETP*, vol. 23, pp. 16–22, 1966.
[42] C. Feng, X. Xu, and J.-C. Diels, "Spatially-resolved pulse-front-tilt and pulse-width distribution of Q-switched pulses from an unstable Nd:YAG resonator," in *CLEO, 2015*, (San Jose, CA), p. JTh2A.89, 2015.
[43] C. Feng, *Sub-Nanosecond UV Filaments and Their Applications for Remote Spectroscopy and High-Voltage Discharges*. PhD thesis, University of New Mexico, Albuquerque, NM, 2016.
[44] R. M. Pope and E. S. Fry, "Absorption spectrum (380–700 nm) of pure water," *Appl. Opt.*, vol. 36, pp. 8710–8723, 1997.
[45] X. Xu, C. Feng, and J.-C. Diels, "Optimizing sub-ns pulse compression for high energy application," *Opt. Express*, vol. 22, pp. 13904–13915, 2014.
[46] Y. R. Shen, *The Principles of Nonlinear Optics*. New York, NY: John Wiley & Sons, 1984.
[47] A. Brodeur, C. Y. Chien, F. A. Ilkov, S. L. Chin, O. G. Kosareva, and V. P. Kandidov, "Moving focus in the propagation of ultrashort laser pulses in air," *Opt. Lett.*, vol. 22, pp. 304–306, 1997.
[48] E. Freysz, M. Afifi, A. Ducasse, B. Pouligny, and J. R. Lalanne, "Giang optical non-linearities of critical micro-emulsions," *J. Phys. Lett.*, vol. 46, pp. 181–187, 1985.
[49] Y. R. Shen and M. M. Loy, "Theoretical interpretation of small-scale filaments of light originating from moving focal spots," *Phys. Rev. A*, vol. 3, pp. 2099–2105, 1971.
[50] T. Grow, A. A. Ishaaya, L. T. Vuong, A. L. Gaeta, R. Gavish, and G. Fibich, "Collapse dynamics of super-Gaussian beams," *Opt. Express*, vol. 14, pp. 5468–5475, 2006.
[51] E. M. Parmentier and R. A. Greenberg, "Supersonic flow aerodynamic windows for high-power lasers," *AIAA J.*, vol. 11, no. 7, pp. 943–949, 1973.
[52] J.-C. Diels, J. Yeak, D. Mirell, *et al.*, "Air filaments and vacuum," *Laser Phys.*, vol. 20, pp. 1101–1106, 2010.
[53] T. S. Ross, *Laser Beam Quality Metrics*. Bellingham,WA, USA: SPIE, 2013.
[54] Y. Du, Y. Fu, and C. Zheng, "Beam quality M^2 factor matrix for non-circular symmetric laser beams," *Laser Phys.*, vol. 27, p. 025001, 2016.
[55] K. Lim, M. Durand, M. Baudelet, and M. Richardson, "Transition from linear- to nonlinear-focusing regime in filamentation," *Sci. Rep.*, vol. 4, p. 7217, 2014.
[56] F. Théberge, W. Liu, P. T. Simard, A. Becker, and S. L. Chin, "Plasma density inside a femtosecond laser filament in air: Strong dependence on external focusing," *Phys. Rev. E*, vol. 74, p. 036406, 2006.
[57] Y.–H. Cheng, J. K. Wahlstrand, N. Jhajj, and H. M. Milchberg, "The effect of long timescale gas dynamics on femtosecond filamentation," *Opt. Express*, vol. 21, p. 4740, 2013.

[58] M. N. Plooster, "Shock waves from line sources. Numerical solutions and experimental measurements," *Phys. Fluids*, vol. 13, pp. 2665–2675, 1970.

[59] K. Kremeyer, K. Sebastian, and C.-W. Shu, "Computational study of shock mitigation and drag reduction by pulsed energy lines," *AIAA J.*, vol. 44, p. 1720–1731, 2006.

[60] P. K. Panigrahi and K. Muralidharl, Chap. 2: Laser schlieren and shadowgraph. *Schlieren and Shadowgraph Methods in Heat and Mass Transfer*. New York, USA: Springer, 2012.

[61] X.-L. Liu, X. Lu, X. Liu, *et al.*, "Tightly focused femtosecond laser pulse in air: From filamentation to breakdown," *Opt. Express*, vol. 18, pp. 26007–26017, 2010.

[62] F. V. Potemkin, E. I. Mareev, A. A. Podshivalov, and V. M. Gordienko, "Laser control of filament-induced shock wave in water," *Laser Phys. Lett.*, vol. 11, p. 106001, 2014.

[63] N. Jhajj, E. Rosenthal, R. Birnbaum, J. K. Wahlstrand, and H. M. Milchberg, "Demonstration of long-lived high-power optical waveguides in air," *Phys. Rev. X*, vol. 4, p. 011027, 2014.

[64] B. Miao, L. Feder, J. E. Shrock, A. Goffin, and H. M. Milchberg, "Optical guiding in meter-scale plasma waveguides," *Phys. Rev. Lett.*, vol. 125, p. 074801, 2020.

[65] E. W. Rosenthal, N. Jhajj, J. K. Wahlstrand, and H. M. Milchberg, "Collection of remote optical signals by air waveguides," *Optica*, vol. 1, pp. 5–9, 2014.

[66] O. Lahav, L. Levi, I. Orr, *et al.*, "Long-lived waveguides and sound-wave generation by laser filamentation," *Phys. Rev. A*, vol. 90, p. 021801, 2014.

[67] M. Baudelet, M. Boueri, J. Yu, *et al.*, "Time-resolved ultraviolet laser-induced breakdown spectroscopy for organic material analysis," *Spectrochim. Acta, Part B*, vol. 62, no. 12, pp. 1329–1334, 2007. A Collection of Papers Presented at the 4th International Conference on Laser Induced Plasma Spectroscopy and Applications (LIBS 2006).

[68] K. Stelmaszczyk, P. Rohwetter, G. Méjean, *et al.*, "Long-distance remote laser-induced breakdown spectroscopy using filamentation in air," *Appl. Phys. Lett.*, vol. 85, pp. 3977–3979, 2004.

[69] A. Rastegari, M. Lenzner, J.-C. Diels, K. Peterson, and L. Arissian, "High resolution remote spectroscopy and plasma dynamics induced with UV filaments," *Opt. Lett.*, vol. 44, pp. 147–150, 2019.

[70] P. Sprangle, J. Penano, and B. Hafizi, "Optimum wavelength and power for efficient laser propagation in various atmospheric environments," *J. Dir. Energy*, vol. 2, no. 1, pp. 71–95, 2006.

[71] P. Sprangle, B. Hafizi, A. Ting, and R. Fischer, "High-power lasers for directed-energy applications," *Appl. Opt.*, vol. 54, pp. F201–F209, 2015.

[72] F. Théberge, J.-F. Daigle, J.-C. Kieffer, F. Vidal, and M. Châteauneuf, "Laser-guided energetic discharges over large air gaps by electric-field enhanced plasma filaments," *Sci. Rep.*, vol. 7, p. 40063, 2017.

[73] A. Rastegari, E. Schubert, C. Feng, *et al.*, "Beam control through nonlinear propagation in air and laser induced discharges," in *Laser Resonators and Beam Control XVIII, Photonics West, Conference 9727-51*, (San Francisco, CA), pp. 9727–51, SPIE, 2016.

[74] E. Schubert, A. Rastegari, C. Feng, *et al.*, "HV discharge acceleration by sequences of UV laser filaments with visible and near-infrared pulses," *New J. Phys.*, vol. 19, no. 12, p. 123040, 2017.

[75] A. Houard, C. D'Amico, Y. Liu, *et al.*, "High current permanent discharges in air induced by femtosecond laser filamentation," *Appl. Phys. Lett.*, vol. 90, no. 17, p. 171501, 2007.

[76] F. Bredice, F. Borges, H. Sobral, *et al.*, "Evaluation of self-absorption of manganese emission lines in laser induced breakdown spectroscopy measurements," *Spectrochim. Acta*, vol. B61, pp. 1294–1303, 2006.

[77] D. A. Cremers and L. J. Radziemski, *Handbook of Laser-Induced Breakdown Spectroscopy*. Chichester, UK: Wiley, 2013.

[78] J. M. Li, L. B. Guo, C. M. Li, *et al.*, "Self-absorption reduction in laser-induced breakdown spectroscopy using laser-stimulated absorption," *Opt. Lett.*, vol. 40, no. 22, pp. 5224–5226, 2015.

[79] J. Hou, L. Zhang, W. Yin, *et al.*, "Development and performance evaluation of self-absorption-free laser-induced breakdown spectroscopy for directly capturing optically thin spectral line and realizing accurate chemical composition measurements," *Opt. Express*, vol. 25, no. 19, pp. 23024–23034, 2017.

[80] H. Amamou, A. Bois, B. Ferhat, R. Redon, B. Rossetto, and M. Ripert, "Correction of the self-absorption for reversed spectral lines: Application to two resonance lines of neutral aluminium," *J. Quant. Spectrosc. Radiat. Transfer*, vol. 77, no. 4, pp. 365–372, 2003.

[81] S. Musazzi and U. Perini, *Laser-Induced Breakdown Spectroscopy; Theory and Applications*. Berlin, Germany: Springer, 2014.

[82] A. Rastegari, M. Lenzner, C. Feng, L. Arissian, J.-C. Diels, and K. Peterson, "Exploiting shock wave and self-absorption for high resolution laser induced breakdown spectroscopy," in *CLEO: 2017*, (San Jose, CA), p. JW2A.77, Optical Society of America, 2017.

[83] A. Rastegari, M. Lenzner, L. Arissian, J.-C. Diels, and K. Peterson, "Utilization of self-absorption for high resolution laser induced breakdown spectroscopy," in *CLEO: 2018*, (San Jose, CA), p. SM1O.7, Optical Society of America, 2018.

[84] W. Meggers, C. H. Corliss, and B. Scribner. *Tables of spectral-line intensities*. Part I – arranged by elements, Institute for Basic Standards National Bureau of Standards, Washington, DC, USA, 1975.

[85] H. L. Xu, J. Bernhardt, P. Mathieu, G. Roy, and S. L. Chin, "Understanding the advantage of remote femtosecond laser-induced breakdown spectroscopy of metallic targets," *J. Appl. Phys.*, vol. 101, p. 033124, 2007.

[86] T. Produit, G. Schimmel, E. Schubert, *et al.*, "Multi-wavelength laser control of high-voltage discharges: From the laboratory to Säntis mountain," in

Conference on Lasers and Electro-Optics, p. JM2E.5, Optical Society of America, 2019.

[87] A. Houard, V. Jukna, G. Point, *et al.*, “Study of filamentation with a high power high repetition rate ps laser at 1.03 μm,” *Opt. Express*, vol. 24, pp. 7437–7448, 2016.

[88] P. K. Rambo, J. Schwarz, and J. C. Diels, “High voltage electrical discharges induced by an ultrashort pulse UV laser system,” *J. Opt. A*, vol. 3, pp. 146–158, 2001.

[89] G. Point, *Interferometric Study of Low Density Channels and Guided Electric Discharges Induces in Air by Laser Filament*. PhD thesis, ENSTA/ Université Paris-Saclay, Saclay, France, 2015.

Chapter 3

Long-wave infrared filamentation in air: lasers and high-field phenomena

Sergei Ya. Tochitsky[1] *and Chandrashekhar Joshi*[1]

3.1 Background

The low-loss propagation of a laser beam through a turbulent atmosphere over many kilometers is highly desirable for a variety of applications spanning from LIDAR and free-space communications [1] to laser propulsion considered for deep-space travel [2,3]. Earth's atmospheric transmittance favors several spectral windows that have low molecular absorption among which the long-wave infrared (LWIR) window from 8 to 14 μm offers additionally the smallest molecular and aerosol scattering as well as a factor of 10 lower sensitivity to turbulence [4,5]. However, diffraction, which is proportional to the wavelength λ, results in strong beam spreading in space at longer wavelengths. If the diffraction compensated propagation of a large radius LWIR beam in the atmosphere were possible over many Rayleigh lengths ($Z_r = \pi w_0^2/\lambda$, where w_0 is the characteristic beam radius), it would allow for a long-range propagation of a high-power laser through the atmosphere perturbed by fog and clouds.

As was shown experimentally over the last 20 years, a short intense laser pulse with a peak power exceeding the critical power for Kerr self-focusing $P_{cr}3.77\lambda^2/8\pi n_0 n_2$ (where n_0 and n_2 are the linear and nonlinear refractive optical indices, respectively) can modify the optical properties of air and compensate for diffraction [6–8]. When the laser power P is greater than P_{cr}, a dynamic equilibrium between Kerr-driven self-focusing, diffraction, and defocusing caused by laser-ionized plasma results in the production of a high intensity "laser channel" in air which is often called a filament. Previous experiments have generated single filaments with diameters on the order of ~100 μm that extend over a distance of up to several meters but these filaments are limited by imperfect guiding (leaky channels) [8]. The vast majority of atmospheric filamentation research has been carried out using ~100 fs, 0.8–1-μm laser pulses where the guided energy in a single filament is only a few mJ, limiting the propagation distance in this regime. Recent results on filamentation in air using a femtosecond 3.9-μm laser has continued this

[1]Department of Electrical Engineering, University of California, Los Angeles (UCLA), Los Angeles, USA

trend [9,10]. If the laser power is increased to TW levels (PP_{cr}), the laser beam can self-channel over several hundred meters but unavoidably breaks up into a random pattern of multiple filaments, each having a power on the order of P_{cr} [11,12]. This deteriorates the coherence of the overall laser beam dramatically.

Some of the atmospheric applications such as lightning control [7,13] lasing in the atmosphere [7,14] and remote spectroscopy [7,15] may benefit from a high electron density and long life time plasma channels generated in multifilamentation [11,12] or superfilamentation [16] regimes. In addition to the previous applications, for microwave guiding in air [7,17], long-distance projection of high-energy pulses [2,3], and filament-induced water-cloud condensation in the subsaturated atmosphere [18], a single large-diameter filament is desirable. The main challenge in filamentation in air is how to scale up both the guided energy/power and transverse/longitudinal dimensions of the light channel without transverse breakup of the laser beam so as to preserve its coherence. Switching to LWIR pulses, which can have orders of magnitude higher P_{cr}, according to recent numerical studies, has the potential to revolutionize long-distance atmospheric propagation and its applications [19–21]. However, this requires TW power lasers at these long wavelengths that did not exist until recently.

Recent years have seen a renaissance of high-power mid-IR laser development, including sources both of intense mid-wave infrared with wavelength, $\lambda = 3-8$ μm and LWIR $\lambda = 8-14$ μm pulses [22–24]. The main motivation behind the development of these lasers is that longer wavelengths enable the exploitation of effects related to the ponderomotive scaling of electron motion in laser–matter and laser–plasma interactions. The former is of basic interest for nonlinear optics [25], generation of light from X-rays to THz waves [26,27], and time-resolved probing of molecular structures [28]. The latter is successfully utilized for the acceleration of electrons and ions from laser-ionized plasmas [29,30]. Note that historically many of the high-field physics effects in gases and solids have been initially studied with relatively low-power LWIR lasers at an intensity level of a few MW/cm^2 [31–33]. Laser–matter interactions at much higher intensities of 1–100 TW/cm^2 required 0.1–1-TW peak power and therefore, the amplification of much shorter pulses than nanosecond pulses previously used in the 1970s–80s was needed. At 0.8–1 μm, the advent of broadband solid-state lasers in combination with chirped pulse amplification (CPA) has made TW power available for high-field experiments. However until recently, the generation of high-peak-power LWIR pulses remained one of the main challenges to extending the short pulse laser technology to longer wavelengths.

At present, the majority of mid-IR high-peak-power sources rely on optical parametric amplifiers (OPAs) or optical chirped pulse parametric amplifiers (OPCPAs) pumped by 1–2-μm lasers [22–24]. However, these parametric amplifiers become prohibitively inefficient with an increase of the output wavelength to the LWIR range and as a result, 0.1–1-TW solid-state sources with a wavelength longer than 4 μm do not exist.

CO_2 lasers operating at 10 μm are currently the only viable candidates for reaching very high relativistic intensity ($>10^{16}$ W/cm^2) and high-energy pulses in the long wavelength portion of the IR spectrum because of their ability to store a great deal of energy within the active medium and relatively high wall-plug efficiency of $\geq$10%. Discharge-pumped atmospheric pressure CO_2 lasers are rather well developed, and an output energy of up to 100 J at a repetition rate of 10–100 Hz is achievable for nanosecond pulses. Reaching very high-peak power in a laser requires the amplification of the shortest possible pulses as we shall see this implies picosecond pulses in the case of the CO_2 active medium. The main difficulty in the development of picosecond CO_2 laser systems is the rather narrow bandwidth attributed to an individual rotational line at one atmosphere pressure. One method to circumvent this limitation is to operate at multiatmosphere gas pressure that increases the bandwidth available for amplification through collisional broadening. During the last decade, significant progress has been made in the amplification of few picosecond long pulses to the terawatt [34,35] and more recently to >10-TW power level [36] using a combination of pressure and power broadening in the CO_2 medium. Such powerful 10-μm lasers have been successfully applied for particle acceleration: generation of monoenergetic proton beams from a gas jet [30], acceleration of externally injected electrons to >50 MeV using laser beat-wave acceleration process [29,37], and acceleration of electron beams in a 10-μm inverse-free electron laser [38]. Availability of terawatt-power picosecond CO_2 lasers has also opened a door for using short 10-μm pulses for nonlinear propagation in the atmosphere.

This chapter is structured as follows. We first highlight the current status of picosecond 10-μm pulse generation and its amplification to terawatt power level using the CO_2 laser technology. We then present data on the first demonstration of self-guiding of a 3-ps CO_2 laser pulse in air and describe spatial and temporal characteristics of laser beam propagating in a self-induced single centimeter-diameter atmospheric channel. We also discuss the nonlinear optics and ionization physics phenomena behind such LWIR self-channeling in air and examine spectral broadening and odd-harmonic generation in the filament.

3.2 High-power picosecond CO_2 gas lasers

Terawatt and higher power laser systems typically operate around 1 μm and use the CPA technique to prevent damage to the solid-state active medium. As opposed to solid-state medium, a CO_2 gas laser has a very high damage threshold (limited only by gas ionization at an intensity of $>10^{12}$ W/cm^2) and can therefore in principle directly amplify short pulses to high powers. At present, to our knowledge, only two high-power picosecond CO_2 master oscillator-power amplifier (MOPA) systems exist in the world and they are actively being used for advanced particle acceleration studies. They are the 3-ps multiterawatt CO_2 laser at the UCLA Neptune Laboratory [36] and the 5-ps terawatt CO_2 laser at the Advanced Test Facility at Brookhaven National Laboratory (BNL) [35]. In the following, we describe the key elements of a picosecond CO_2 laser MOPA that are common to both these systems.

3.2.1 Current status of picosecond CO_2 lasers

Amplification of a picosecond pulse in a CO_2 laser is complicated because of the relatively narrow bandwidth of the CO_2 molecule for which the gain spectrum consists of discrete rotational lines. As shown in Figure 3.1, when the bandwidth of these lines is sufficiently broadened, they overlap, filling the gaps in the spectrum that result in a quasi-continuous bandwidth across the entire vibrational–rotational branch (~1.2 THz for the 10*P* branch) suitable for the amplification of $\geq$1-ps pulses. Thus 1–5-ps pulses can be amplified using high-pressure ($\geq$10 atm) CO_2 lasers, when a collisionally broadened linewidth becomes approximately 37 GHz ($\Delta\upsilon_{col} = 3.7$ GHz/atm) [39] and lines overlap. Ideally, one could increase the pressure until the individual rotational lines completely fill the gaps between them, which happens at around 20 atm. Unfortunately, it is technically extremely difficult to obtain a uniform glow discharge in a large aperture (>10 cm) module at a high pressure (~10 atm). As a result, the energy output of a multiatmosphere CO_2 amplifier is limited because of a rather small volume.

Field broadening is an alternative mechanism to collisional broadening for increasing the bandwidth necessary for picosecond pulse amplification. The presence of the large electrical field associated with laser radiation resonantly interacting with molecular levels results in line broadening due to the ac Stark effect. The effect of field broadening on the rotational line bandwidth is described by the following equation: $\Delta\upsilon_{field} = 2\Omega = 2\left(6.91 \cdot 10^6 \mu\sqrt{I}\right)$, where $\Delta\upsilon_{field}$ is the field-broadened linewidth, Ω is the Rabi frequency, μ is the dipole moment, and I is the laser pulse intensity in W/cm^2. For the 10.6-μm CO_2 transition, the dipole moment is equal to 0.0371 debye [40]. Because of this self-effect, the electric field of a 10-μm pulse with an intensity of 5 GW/cm^2 will broaden the 1-atm CO_2 medium to give a bandwidth of 37 GHz—i.e., comparable to collisional broadening induced bandwidth at 10 atm. Then the total bandwidth of each individual rotational line, $\Delta\nu = \Delta\upsilon_{col} + \Delta\upsilon_{field}$ becomes a sum of collisional and field-broadening components. Therefore, with a sufficiently intense 10-μm pulse, the field broadening can compensate for the lack of bandwidth in relatively low-pressure amplifiers. As demonstrated experimentally, the field broadening in a large aperture 2.5-atm

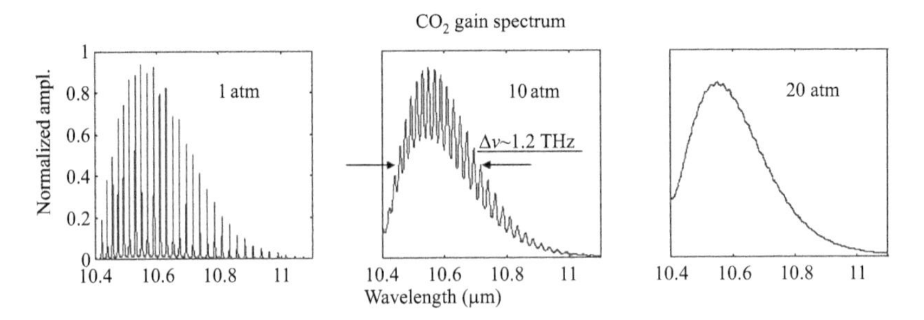

Figure 3.1 The gain spectrum of the P *branch of 001–100 band of a CO_2 molecule with a gas mixture of 1:1:14 CO_2:N_2:He at different total pressures*

amplifier allows for the amplification of pulses with a bandwidth much broader than that defined by collisional broadening alone, first to 1 TW [34] and in later to >10-TW power level [36]. Recently, we have demonstrated amplification of 3-ps pulses in an atmospheric pressure CO_2 module, in which the bandwidth was provided by the field broadening only [41].

Even though the broadening mechanisms creating the bandwidth in an active medium can be different, a short picosecond seed pulse needed for amplification is always generated using a short pulse solid-state laser. This is mainly due to the limited progress in mode-locking technology for high-pressure CO_2 lasers. It was P. Corkum who first demonstrated that a picosecond short pulse sliced from a long 10-μm pulse by a pair of semiconductor switches can be amplified in a multiatmosphere CO_2 laser to a GW power level [42]. Later this technique was scaled to a higher level of peak power both at UCLA and BNL by staging multiple high-pressure CO_2 amplifiers. The main strategy behind the high-power picosecond CO_2 MOPA is a two-step process of amplification. First, the amplification of a weak nJ seed pulse to mJ level in a multiatmosphere CO_2 preamplifier, in which the bandwidth is broadened collisionally, and second, the amplification of the mJ pulses to the Joule level and beyond that can be done both with high pressure or with a relatively low-pressure, large-aperture discharge modules. In the latter case, the field-broadening mechanism plays a major role in producing the bandwidth. This two-step approach to an amplification of a short 10-μm pulse is illustrated by a simplified scheme of the UCLA MOPA system shown in Figure 3.2.

In this scheme, the first stage master oscillator is based on a CO_2 laser. A 200-ns long pulse from a grating tunable single-longitudinal mode TEA CO_2 laser is used for this purpose. In order to generate a picosecond seed pulse, the polarization of a portion of a 10-μm, 100-ns beam is rotated by an ultrafast 1-μm pulse from an Nd:glass CPA system in a Kerr nonlinear medium [43]. This Kerr modulator is used in combination with a Ge semiconductor switch to increase the contrast of a sliced pulse. As a result of such pulse switching, a 1–3-ps long 10-μm seed with an energy of several nJ and a contrast of 10^3–10^4 is produced. This pulse is then amplified by a factor of 10^6 in an 8-atm TE CO_2 preamplifier. Because of the relatively low gain in gas lasers, such a strong increase in energy from nJ to mJ requires the use of a regenerative amplification scheme with 20–30 round trip

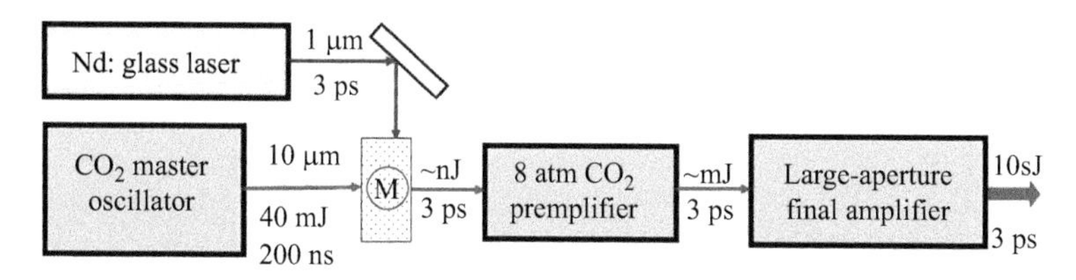

Figure 3.2 Simplified scheme of a three-stage CO_2 laser MOPA system. M *is an ultrafast modulator used for the production of a picosecond seed pulse*

passes through the medium. In the UCLA system, an injection mode-locking scheme with an external selection of a single 3–5 mJ pulse by a CdTe Pockels cell was used [34]. Note that a classical regenerative amplifier with two intracavity Pockels cells—one for trapping the seed pulse and the other for dumping the cavity—was successfully demonstrated by the BNL's group [35].

Regardless of the line-broadening mechanism in a CO_2 molecule, the insufficient overlap of these lines even at $\Delta\nu$~30–40 GHz (~10 atm) results in the residual modulation of the gain spectrum at 55 GHz (see Figure 3.1). When a short 3-ps pulse propagates in a medium with a periodically modulated gain spectrum, some frequencies within the initial pulse bandwidth will not be amplified efficiently and the inverse Fourier transform for such a frequency-modulated spectrum gives a pulse train with a pulse separation equal to 1/55 GHz, or 18.5 ps. A large number of passes through the regenerative amplifier result in significant gain narrowing and the output is a train of 3-ps pulses in which the duration of the pulse train envelope can be estimated from a well-known pulse broadening equation [44]. An elegant solution to such pulse splitting was experimentally demonstrated by using a mixture of CO_2 isotopes in the regenerative amplifier at BNL [35]. Here an asymmetric $^{16}O^{12}C^{18}O$ molecule, for which selection rules allow transitions between both odd and even rotational quantum states, doubles the line number making the gain spectrum at 8–10 atm analogous to 16–20 atm in the regular CO_2 laser mix. However, CO_2 isotopes are expensive and some of them have lower Einstein coefficients. This has limited the use of isotopic mixes in a final amplifier where discharge volume must be large in order to extract tens of Joules of energy.

Final amplification in Figure 3.2 occurs in a large aperture ($20 \times 35 \times 250$ cm^3) CO_2 amplifier [36]. It is an electron-beam-sustained module that has a small-signal gain of 2.5%/cm on the 10*P*(20) line at 2.5 atm. The gain bandwidth for a CO_2:N_2 mix (80:20) is around 14 GHz. During the three passes of amplification, the laser intensity for the last two passes increases from 4 to 140 GW/cm^2. At these intensities, field broadening augments the collisional-broadened manifold bandwidth and not only sustains 3-ps pulses but also decreases the duration of the overall pulse train envelope. As shown in Figure 3.3, the duration of the train of pulses generated in the CO_2 regenerative amplifier (left panel) was significantly shortened and only a few pulses were recorded after three passes of amplification in the large-aperture amplifier (right panel) [36]. A record >10-TW peak power was measured for the most powerful pulse in a train of 3.6-ps pulses containing a total energy of ~100 J in a 5″ beam.

A different X-ray preionized CO_2 module with a self-sustained discharge is used at BNL. Here to obtain a stable discharge at 8 atm, the amplifier aperture was decreased to 10×10 cm^2 and 5-ps pulses were amplified to 5-J energy with a relatively small pulse splitting due to use of CO_2 isotopes in the regenerative amplifier [35]. This is the current status of high-power CO_2 lasers. Now we discuss possible scaling of the pulse length and the repetition rate of a high-power laser system based on CO_2 gas.

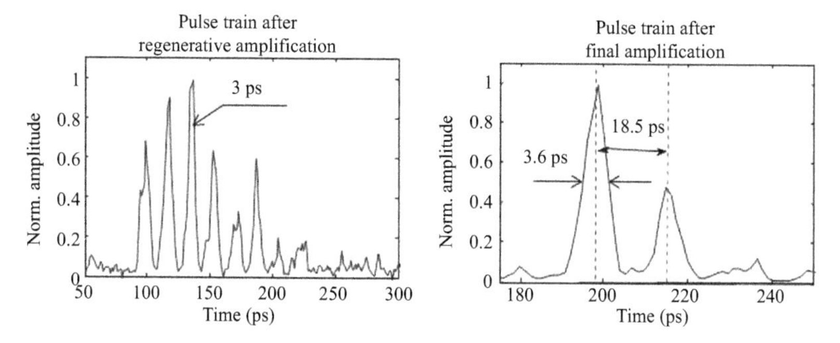

Figure 3.3 Temporal profiles of the CO_2 laser pulses after amplification in a MOPA system. A typical pulse train after the 8-atm regenerative amplifier (left panel) and a pulse train after three passes in the final amplifier as measured by a streak camera diagnostic

3.2.2 Picosecond CO_2 lasers: modern trends and future developments

It is interesting that new trends in short pulse CO_2 laser development are related to the progress in mid-IR solid-state lasers. In 2015, the BNL group pioneered a new approach for seed pulse production: generation of a seed pulse using an LWIR OPA pumped by an ultrafast 0.8-μm Ti:sapphire laser system [45]. Such systems can produce pulses as short as 0.35 ps at any central wavelength in the 9–11-μm range and with an output energy of ~10 μJ. The factor of 10^3 increase in the seed level in comparison with the traditional slicing scheme in Figure 3.2 is particularly important for minimizing pulse-broadening effect while subsequent amplification in the CO_2 active medium [44]. The main result here is the generation of a quasi-single picosecond pulse in a CO_2 regenerative amplifier. For subpicosecond seed pulses, the efficient energy extraction and minimization of nonlinear effects in optical elements required stretching of the pulse prior to and its compression after the amplification (CPA configuration). Note that at 10 μm the dispersion of diffraction gratings is much smaller than that in the near-IR range imposing additional challenges in the CPA design. In connection with our experiments on LWIR filamentation in air, later we will discuss this system in detail.

Even higher energy seed pulses can be generated in the mid-IR OPCPA when switching from a traditional ~1- to ~2-μm pump lasers [22,23]. This allows for more efficient nonlinear materials to be used for OPA. To date, 0.5-mJ ultrafast pulses have been generated at a wavelength around 7 μm [23]. In the near future, one can envision the perfect marriage of an OPA-based source with an output pulse energy of 0.1–1 mJ in the 9–11 μm range with a terawatt class ~1–3-J CO_2 gas amplifier. It should be emphasized that the high-pressure CO_2 discharge technology is still the only option for generating multi-Joule pulses at LWIR, especially when a kW level of average power is desired. Also note that the majority of mid-IR OPCPA systems so far were built using femtosecond parametric amplifiers, so the

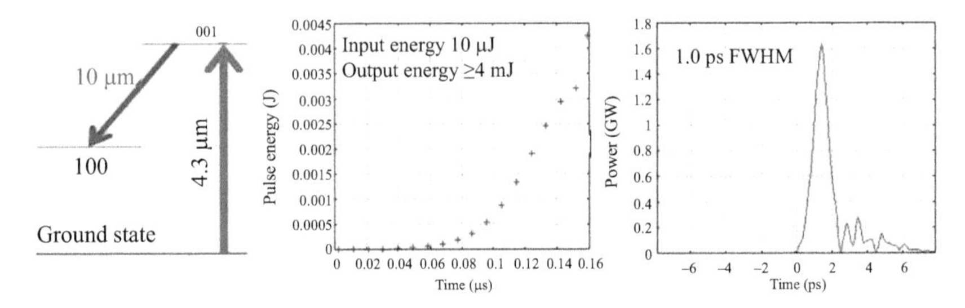

Figure 3.4 CO_2 laser level diagram showing an optical pump from the ground state to the 001 level at 4.3 μm and lasing in the 001–100 channel (10 μm). Simulations data on amplification of a 0.5-ps 10.2-μm pulse in a 20-atm multipass CO_2 cell as a function of propagation time (center) and the output pulse profile (right) pumped by a ~200 ns long pulse of an Fe:ZnSe laser at 4.32 μm. Modeling is a courtesy of Dr. M. Polyanskiy, BNL

generation of ~1-mJ picosecond pulses as an appropriate seed for a multiatmosphere CO_2 amplifier is yet to be demonstrated.

Another important development in short pulse CO_2 lasers is also related to the progress in the mid-IR solid-state lasers. Recent demonstration of energetic Fe: ZnSe lasers tunable around 4.3 μm [46] renewed the interest in an old idea of optical pumping of a CO_2 molecule. From spectroscopic studies of the CO_2 molecule, as depicted in Figure 3.4, it is known that 4.3-μm pumping of an upper laser level 001 is the most efficient way to create a high inversion of population in the 001–100 (10 μm) channel. In 1970s–80s, such a lasing scheme has been investigated by using a chemical HBr laser [47] and a 4.3-μm CO_2 laser [48,49] as a pump. In this scheme, up to 40% of the absorbed pump energy was converted to 10 μm at atmospheric pressure. As shown in Figure 3.1, at 20 atm, a complete overlap of rotational lines provides for the *P* branch of CO_2 a smooth 1.2-THz gain profile sufficient for picosecond pulse amplification. Moreover, at such high pressures, even vibrational branches are partially overlapped creating an opportunity to generate a much broader ~3-THz bandwidth by CO_2 gain tailoring. In Figure 3.4, we present an example of modeling when a 0.5-ps pulse is amplified in a few centimeter long 20-atm cell filled with a $^{12}CO_2$–$^{13}CO_2$-He mixture and pumped by an Fe:ZnSe laser pulse at 4.32 μm. According to the simulations, ~1-ps (full width at half maximum (FWHM)) pulse with up to ~4 mJ of energy can be extracted when a ~200 mJ of pump radiation is absorbed in a cell. Potential of generation of even shorter ~0.5-ps pulses in a compact optically pumped CO_2 laser clearly qualifies such a scheme as a new frontier in the LWIR source development. The main advantage of using an active medium instead of a parametric amplifier is its flexibility in the pump pulse duration. We would like to note that this technology is still in the early stages of development and an optical-to-optical efficiency of ~10% or higher, which would be far superior to parametric amplifiers, seems to be possible but yet to be demonstrated in a multiatmosphere CO_2 laser.

3.3 Nonlinear propagation of a 10-μm laser in air

The atmospheric attenuation of laser radiation is caused by aerosols and molecular constituents in the propagation path. For a variety of applications, interest in propagating LWIR laser pulses over long distances is driven by the relatively low molecular absorption, on the order of 0.1–0.2 km^{-1}, and the small aerosol absorption/scattering in this spectral range. In the past, several groups used high-energy $\geq$100 J pulses of a transversely excited atmospheric pressure CO_2 laser to study a propagation of 10-μm radiation in the atmosphere [50–52]. Here, a typical gain switched pulse consists of a 100–200-ns high-power spike followed by an energetic 1.5–2 microseconds long tail. The laser peak power 0.5–2 GW was insufficient for Kerr self-focusing in air and the laser beam was typically either expanded to compensate for diffraction in long-range experiments or focused by long focal length optics in order to create a high-intensity laser channel in the atmosphere. In the latter case, the optical breakdown of air occurred above the threshold intensity of around 100 MW/cm^2, causing the almost complete ionization of air due to the electron avalanche formation initiated on aerosol or dust particles [51–53]. Note that fully ionized air would correspond to the electron plasma density of ~3×10^{19} cm^{-3} slightly exceeding the critical plasma density for 10-μm radiation. It therefore introduces significant losses for the laser beam and practically terminates the beam propagation. It was absorption losses in such inhomogeneous collisional ionized plasma that prevented the use of high-energy nanosecond CO_2 lasers for the production of a conductive channel in the atmosphere considered for guiding electrical discharges and lightning control [54,55].

The problem of avalanche ionization of air can be alleviated by using pulses with durations shorter than the time required for the avalanche to build up the electron density at the atmospheric pressure [56]. Also the optical breakdown threshold increases with the decrease of the pulse length. For example, in our experiments using ~150-ps CO_2 laser pulses, the threshold of optical breakdown in air increased to ~10 GW/cm^2 [57]. Here, by measuring the spectral shift due to rapid expansion of the plasma, its electron density was estimated to be ~10^{16} cm^{-3}. The breakdown threshold of air for a single 3-ps CO_2 laser pulse has not been measured yet but previous considerations indicate that when such a short but intense (TW/cm^2) pulse is used, the resultant plasma may have an electron density much below the critical density. Another important factor related to propagation of short terawatt LWIR pulses in air is that the peak power can be larger than the critical power for Kerr self-focusing, thus opening for the first time opportunity to explore experimentally nonlinear self-guiding at $P>P_{cr}$ for diffraction compensation. However, any estimate of the P_{cr} value requires first of all knowledge of the nonlinear refractive index, n_2 of air, which had not been measured at 10 μm until recently.

3.3.1 Nonlinear optical characteristics of air at LWIR

Until recently for the CO_2 laser wavelengths, Kerr coefficients of gases that constitute our atmosphere were extrapolated from their known values near ~1 μm using the generalized Miller's rule [19]. This rule does not take into account any contribution of resonant absorption of water or other gases. For a long wavelength such as $\lambda = 10.6$ μm, the total extinction coefficient of air is dominated by the absorption of the "water continuum" [52].

In order to generate high precision set of data on n_2 of gases at 10 μm, a high repetition rate intense laser is needed. We have recently built a 1-Hz laser system that is capable of generating high-power 3–200-ps long laser pulses [41] and made measurements of Kerr coefficients of air and air constituents around 10 μm [58,59]. We have used collinear four-wave mixing (FWM) in gases of 200-ps long beat-wave obtained by our CO_2 laser operating on two lines. For this experiment, a beat-wave was produced by operating the laser oscillator simultaneously on the 10*P*(20) (10.59 μm) and 10*R*(16) (10.27 μm) CO_2 lines. The laser beam was focused in a ~2-m long gas-filled cell. After the cell, the generated sidebands were frequency resolved by a monochromator and analyzed by an HgCdTe detector. The experimental arrangement shown schematically in Figure 3.5(a) was used for the detection of FWM signals that were 10^{-7} relative to the pump.

To avoid the ionization contribution, manifested by drop in the FWM signal at pressures above 630 Torr, all n_2 measurements were performed in high purity gases at pressures less than or equal to 380 Torr. The measured values for effective nonlinear coefficients scaled to atmospheric pressure were 4.5 ± 0.9 for N_2, 8.4 ± 1.3 for O_2, and 5.0 ± 0.9 for dry air (in units of 10^{-19} cm^2/W). We would like to note that measurements in the laboratory air produced similar results pointing to the fact that within the experimental uncertainty, H_2O contribution was too small to detect. The results of the effective nonlinear index at 10 μm are comparable with values reported at 0.8 and 2.4 μm [60,61]. Since the magnitude of the electronic

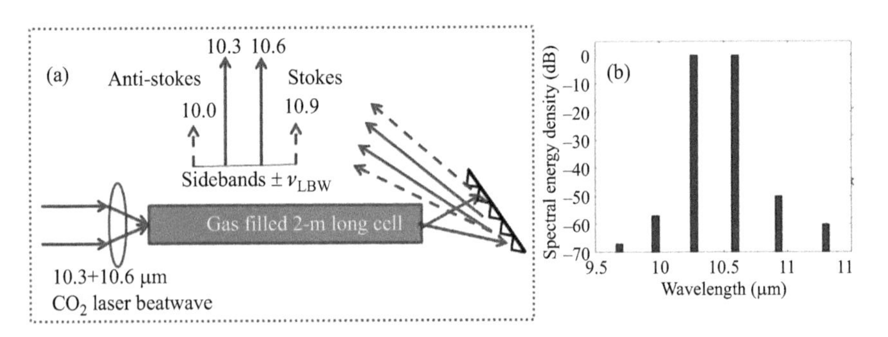

Figure 3.5 A simplified schematic of the experimental arrangement for n_2 *measurements in gases by using 10.3- and 10.6-μm lines of* CO_2 *laser beat-wave generating sidebands due to the FWM process (a) and a typical FWM spectrum recorded in the 380 Torr of laboratory air at an intensity of ~12 GW/cm*^{2} *(b)*

nonlinear response is expected to decrease for longer wavelengths, these measurements indicate the importance of the delayed molecular nonlinearity at longer wavelengths. The asymmetry observed in the FWM spectrum in Figure 3.5(b) indeed points to the fact that the molecular response plays an important role. To study this further, we have performed FWM measurements in Ar, Kr, and Xe, gases that have a comparable electronic nonlinearity as air but that lack a molecular response. Measurements of FWM in Kr and Xe produced a comparable FWM spectrum as produced in air; however, the FWM sidebands were symmetric and were generated with lower yield [59].

Based on the measured nonlinear refractive index of air for 10-μm radiation [58], the estimated value for P_{cr} lies in the range of 300–400 GW placing a lower bound on the CO_2 laser power necessary for self-channeling in air to be around ~1 TW. Note that the previous measurements were conducted using 200-ps pulses and for shorter duration pulses (1–3 ps), the n_2 value may deviate and be slightly smaller.

3.3.2 LWIR beam filamentation in laboratory air

The proof-of-principle experiments on picosecond CO_2 laser filamentation in air were carried out using a multistage CO_2 MOPA system at BNL. In Figure 3.6, we depict the details of the experimental setup. A single ~350-fs seed pulse at a central wavelength of 10.2 μm was generated by a commercial OPA pumped by a Ti: sapphire laser. After stretching and filtering the bandwidth using a grating stretcher, a 1–2-μJ seed pulse was amplified in the CO_2 regenerative amplifier by a factor of ~1,000 to a ~5-mJ level and recompressed to ~3-ps pulse length by another grating compressor [45]. As seen in Figure 3.7 (left panel), a mixture of CO_2 isotopes in the discharge module minimized the gain modulation and resulted in an almost single pulse profile after the regenerative amplifier. In the final large-aperture CO_2

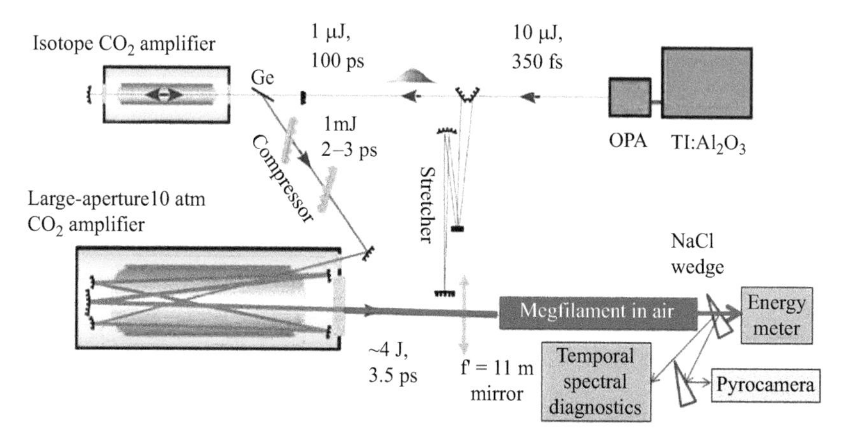

Figure 3.6 A schematic of the experimental arrangement for 10.3-μm picosecond CO_2 laser filamentation in air

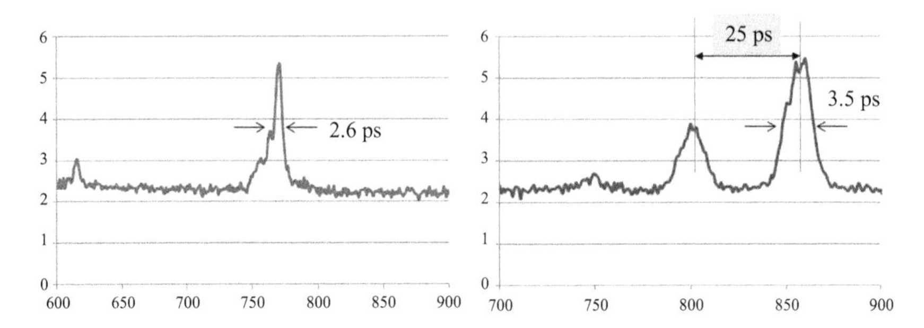

Figure 3.7 Temporal profiles of the laser pulses as measured by a streak camera diagnostic. A typical pulse after the 8-atm regenerative isotopic CO_2 amplifier (left panel) and a pulse train after five passes in the final amplifier filled with a regular CO_2 mix

amplifier, the pulse energy was increased by another factor of ~1,000, bringing the total energy in the pulse to 5.2 J. This large-aperture gain module was filled with a regular $^{16}C^{12}O_2$ mix, consequently the pulse experienced some modulation and a train of 3.5-ps pulses was measured by a streak camera (right hand panel in Figure 3.7). For the 10*R*-branch, this residual periodic modulation of the gain spectrum at 40 GHz resulted in a periodic modulation of the frequency spectrum of a broadband picosecond seed pulse that was amplified. This in turn generated a pulse train with a pulse separation of 1/40 GHz = 25 ps. As shown in Figure 3.7, in the experiment a typical 50–100-ps macropulse consisted of a leading terawatt power 3.5-ps long pulse and a few postpulses containing around 30% of the total energy. For atmospheric propagation, it is important that the first pulse in the pulse train, corresponding to the seed pulse, contains a few times the critical power for Kerr self-focusing in air (≥1-TW power or 4 J in 3.5 ps).

In order to accelerate the evolution of self-focusing of a 5″ diameter laser beam in the laboratory air, a spherical Cu mirror with a long focal length of 11 m was installed right after the final CO_2 amplifier. As shown in Figure 3.6, after the focus, the laser beam was intercepted by a pair of NaCl wedges that allowed for simultaneous measurements of the spatial and temporal or spectral properties of the beam during its propagation in air along with monitoring energy in the transported beam. The beam profile was analyzed by using a two-dimensional IR array (Pyrocam). Time and spectral parameters were measured by a streak camera (or an IR autocorrelator) or a grating spectrometer (or a set of narrow-band IR filters corresponding to harmonics of CO_2 laser with an energy meter), respectively.

By moving the NaCl wedges along the beam path, the CO_2 laser parameters were studied at a few fixed locations as a function of the peak power. We found that for low power ($P<P_{cr}$) pulses, the laser beam diverged from the focal plane of the mirror where the FWHM spot size and corresponding Rayleigh length were measured to be ~2.3 mm and 1.6 m, respectively. In Figure 3.8(a) (top panels) the beam profiles, recorded at distances z = 11.2 and 22.4 m, clearly indicate that even up to

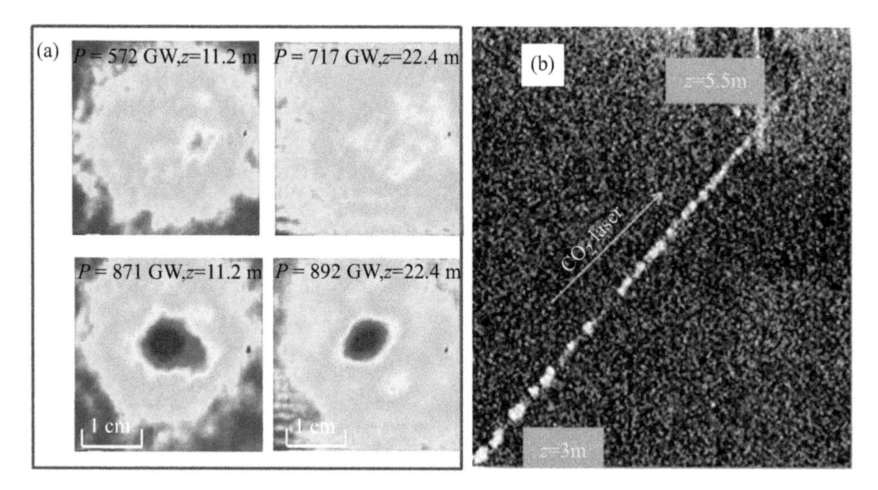

Figure 3.8 (a) Spatial beam profiles of the CO_2 laser as measured by a two-dimensional pyroelectric array (Pyrocam) at powers P *below (top panels) and above (bottom panels) the self-guiding threshold* ($P \geq 871$ *GW); (b) a CCD camera image showing an approximately 2.5-m long section of a 5.5-m long plasma channel taken at* $z = 3$ *m after the focus of the CO_2 beam*

a power of 717 GW, the CO_2 laser beam is still diverging, although the divergence angle could be affected by Kerr self-focusing. For CO_2 laser shots with a peak power in the first pulse above ~870 GW, this energetic laser pulse was suddenly seen to be self-guided producing an ~5-m long visible plasma channel followed by a many-meter long light channel with no visible plasma or fluorescence emission from air. In the bottom panels in Figure 3.8(a) for the same distances $z = 11.2$ and 22.4 m, we have observed a single centimeter-diameter filament. We have called it a *megafilament* because its cross-section is 10^4 times larger than a typical near-IR single filament and it confines a ~1-TW laser pulse with several Joules of energy [62]. For higher powers ($\geq$1 TW), this ~1-cm diameter channel persisted from the curved mirror focus up to $z = 32$ m ($20Z_r$) being limited only by the available laboratory space. At the maximum distance, the initial energy of 5.2 J in the entire laser pulse was reduced by approximately half. As seen from the CO_2 laser beam profile in Figure 3.8(a), the *megafilament* was surrounded by a larger diameter low-power halo indicating the dynamics of its formation and leaky nature of such self-induced waveguide in air.

In Figure 3.8(b), we show a portion of an ~5-m long plasma channel formed around the nonlinear focus of the self-guided CO_2 laser in air. The plasma density in this channel was estimated to be around 10^{15}–10^{16} cm^{-3}. The inhomogeneous structure of this plasma channel was similar to plasmas formed by optical breakdown of air due to the electron avalanche mechanism [56]. The plasma fluorescence gradually faded away after a 5–6-m distance even though the CO_2 laser beam was still self-guided over many Z_r. Note that a plasma similar in brightness and spatially beaded structure has been observed in our previous studies of an optical spark in air produced by a $\geq$150-ps CO_2 laser pulses in which the plasma density was limited to ~10^{16} cm^{-3} [57].

By using the streak camera diagnostic, we have studied how the temporal profile of the ~100-ps CO_2 laser macropulse in Figure 3.7 evolves while propagating in air. Note that in order to detect the long wavelength pulses, the CO_2 laser pulse has to be first upconverted to the visible range by mixing it with a red diode laser beam in an $AgGaS_2$ nonlinear crystal [45]. For the self-guided pulse with a power exceeding 1 TW, after the first 5 m of propagation we detected a significant drop in energy contained in the postpulses. Recall that the plasma fluorescence could only be seen over this distance. Clearly, the peak intensities of these postpulses are strongly attenuated by the plasma produced via the avalanche ionization and diffraction over a short distance (2–3Z_r) because their power is below the threshold for self-guiding. Both plasma related and diffraction losses explain why only the single leading pulse was detected further downstream of the focus, at z = 11.2 m and after. While disappearance of the postpulses can be explained by the experimentally observed plasma channel extended over the first 5 m, these observations do not shed light on how the free carrier generation necessary to prevent the Kerr-related collapse of the leading TW power 3.5-ps pulse self-guided in air occurs.

The leading self-guided pulse is expected to undergo pulse shortening, because both the head and the tail of the pulse have $P<P_{cr}$ and, therefore, experience significant diffraction losses. A single-shot autocorrelator was used for the high temporal resolution measurements of the pulse length [45]. It had a limited time window of ~17 ps but a high temporal resolution of <0.2 ps. The autocorrelator based on noncollinear second harmonic generation principle was sufficient to time resolve a single picosecond pulse but not the pulse train with 25-ps separation between pulses. As shown in Figure 3.9(a), when a 3.5-ps terawatt power CO_2 laser was self-guided in air, at z = 11.2 m we measured 10-μm pulses as short as 1.8-ps FWHM assuming the Gaussian pulse profile in the autocorrelation trace. Measurements of the pulse length in combination with the recorded spatial energy distribution in the CO_2 laser beam allowed us to estimate the "clamped" intensity on an axis of such LWIR *megafilament* in air. It was about 1 TW/cm^2 or about 20 times smaller than that in near-IR filaments [7,8]. The detected pulse shortening is the reason why, despite energy losses, the clamped intensity of ~1 TW/cm^2 (see Figure 3.9(b)) and the condition $P/P_{cr}>1$ was sustained in the *megafilament* over almost the entire measurable propagation distance.

Let us summarize the results of the spatial and temporal measurements of a 10.2-μm *megafilament* in air. For near-IR terawatt femtosecond pulses, the filamentation regime is usually characterized by the formation of a weakly ionized plasma channel followed by the second propagation stage (post-filamentation stage) in which the laser field is below the ionization threshold (also $P<P_{cr}$) and diffraction is only balanced by the Kerr self-focusing effect [8,11,12]. In our experiment, the clamped intensity of 10^{12} W/cm^2 in the filament was significantly

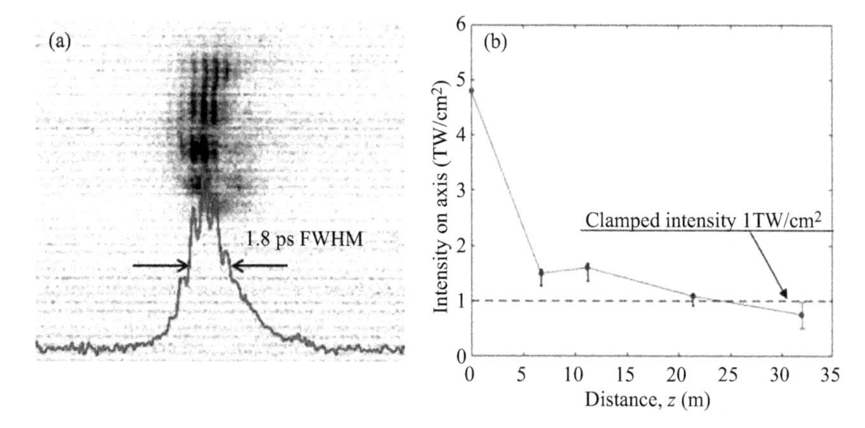

Figure 3.9 Pulse measurements in the megafilament. (a) Pulse length shortening of the initial 3.5-ps CO_2 laser pulse with an incident power P = *1.14 TW as measured by a single-shot autocorrelator at* z = *11.2 m. (b) CO_2 laser intensity on the axis of the megafilament as a function of propagation distance recorded at the initial peak power of 1.1–1.2 TW* (P/P_{cr}*~2–3). The larger error bar at* z = *32 m is due to an uncertainty in the pulse length that was seen to be between 0.5 and 1 ps*

below the tunnel ionization threshold of O_2/N_2 (~10^{13} W/cm^2). Note that for long-wavelength radiation, multiphoton ionization should not play any significant role, since the Keldysh parameter [63] is ~0.1. In the experiment, we clearly measure that the length of the plasma channel (electron density $\geq 10^{15}$ cm^{-3}), in which air fluorescence is visible, correlates well with the distance at which the $\geq$ 100-ps long CO_2 laser macropulse consisting of several postpulses evolves into a single picosecond pulse. Once, at $z > 5$ m, a single pulse with a peak power above the critical power for Kerr focusing is self-guided in air. At this point, the laser–air interaction time becomes too short for the avalanche to grow and apparently the plasma density drops dramatically (no fluorescence observed), being only just sufficient to compensate for the nonlinear focusing. At the beginning of such a self-guiding process, the laser beam has a power P/P_{cr}~2–3 (1–1.2 TW) that decreases to P/P_{cr}~1.5 at $z = 32$ m. Thus, we cannot consider a light channel at $z > 5$ m as simply a post-filament in which self-focusing is compensating partially for diffraction. As shown in Figure 3.9(b), the clamped intensity of ~1 TW/cm^2 was recorded for the entire propagation length and the *megafilament* persisted over the ~$20Z_r$ distance. It should be noted that in the experiment, failure to sustain the $P/P_{cr} > 1$ condition over the entire propagation immediately resulted in the strong divergence of the beam as seen at the maximum distance of $z = 32$ m.

Thus, there appears to be an ionization mechanism at work that has an onset at laser intensities below the tunnel ionization threshold for a single molecule when the laser pulse is propagating through a high density gas as atmospheric air. In our case this mechanism generates a very small level of fractional ionization of air but the defocusing effect of these free carriers apparently is sufficient to compensate for Kerr self-focusing. Furthermore, the saturation of the optical Kerr effect considered at elevated laser intensities tens of terawatts [64] is unlikely to play any role in our experiment where the observed clamped intensity is only ~1 TW/cm^2. Since a traditional combined multiphoton/tunnel ionization and impact ionization model could not describe a propagation of a 3.5-ps long CO_2 laser pulse in air, a new theoretical approach was necessary as described in the following.

3.3.3 Gas ionization model for an intense LWIR picosecond pulse

At ~1-TW/cm^2 intensity of 10.2-μm light measured in the *megafilament*, semi-classical electron trajectories extend out some ~15 nm from the parent molecule, allowing for various Coulombic interactions between the semi-free electron and other molecules separated in air by ~3 nm. Also a collisional mean-free path of 60 nm in air becomes comparable with the electron traveling distance. Therefore, an electron from one molecule is able to interact with many other molecules in its vicinity. Electrons in such a weakly ionized gas can be subsequently further multiplied due to avalanche ionization. Compared to near-IR wavelengths, the theoretical treatment of this process requires a consideration of many-body effects in dilute gases. For a theoretical description at this longer wavelength and relatively long pulse duration, we therefore resort to the many-body formalism proven in solid-state physics [65]. This allows us to account for the many-body Coulomb effects and describe the optically driven, strongly nonresonant polarization of the individual molecules [66].

The theoretical model, including many-body Coulomb effects in an atmospheric pressure gas, was developed by J. V. Moloney and S. W. Koch at the University of Arizona [21,66]. In Figure 3.10(a), we show how the many-body effects described by a generalized version of the optical Bloch equations [65] result in a low degree of ionization (red curve) well below the standard single-atom ionization due to the usually considered multiphoton absorption and tunneling mechanisms (blue curve). Here, the theoretical value of the ionization degree of around 10^{-7} obtained for a terawatt, 1-cm diameter CO_2 laser beam indicated by the black arrow in Figure 3.10(a).

At 1-atm pressure this degree of ionization would correspond to a free-electron density of 3×10^{12} cm^{-3}. However, since change in the polarizability due to free electrons scales as the square of the wavelength of the exciting field, even such a low ionization degree can have a significant impact on the net nonlinear refractive index of air partially ionized by the LWIR pulse. Indeed, as shown in Figure 3.10 (b) for 3-ps pulses at a peak intensity of 1.5×10^{16} W/m^2, the contribution of the positive Kerr lensing (purple curve) and the negative plasma lensing (green curve) partially compensate each other. The black curve in Figure 3.10(b) shows the net lensing experienced by the 10-µm pulse, including diffraction. It is apparent that the many-body-mediated negative lensing can dynamically balance the nonlinear Kerr self-focusing for LWIR pulses prohibiting collapse of the beam and thereby

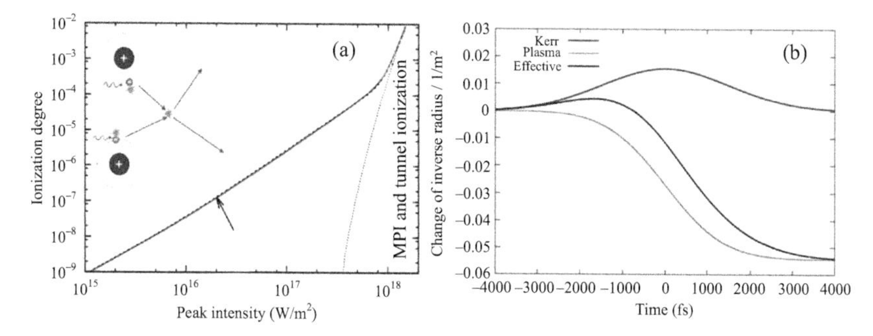

Figure 3.10 *Many-body ionization. (a) Calculated ionization degree in an atmospheric pressure gas versus intensity of a 10-μm Gaussian pulse. The inset illustrates that many-body interaction involves electrons from neighboring atoms. The red curve shows the results from the full microscopic theory taking into account many-body interactions, the blue dotted curve shows the ionization degree based on a single-atom ionization model. The arrow indicates the peak intensity reached in our experiment, and this remains below the threshold value 2×10^{17} W/m^2 for single-atom MPI/tunnel ionization. (b) Kerr lensing (purple curve) and plasma-related defocusing (green curve) as functions of time delay with respect to the center of a 3-ps CO_2 laser pulse at an intensity of $1.5 \times 1{,}016$ W/m^2. The black curve show the net lensing experienced by the pulse. Courtesy of J. V. Moloney*

can provide conditions for self-channeling of some portion of the laser pulse. It should be noted that the contribution of many-body-related ionization physics for near-IR wavelengths and for mid-IR pulses but with femtosecond durations was negligibly small [21]. Since our experiments are performed using laser intensities of around 1 TW/cm^2, i.e., well below the threshold for tunnel ionization, this full microscopic model, in which many-body Coulombic ionization is the dominant mechanism for the free carrier generation, was adopted for numerical analysis and interpretation of the picosecond CO_2 laser filamentation in air.

3.3.4 Numerical modeling of LWIR picosecond pulse propagation in air

To analyze the physics behind the *megafilament* formation using the previous microscopically based theory to include very weak ionization and to take into account odd-harmonic generation by the picosecond CO_2 laser pulse in air, the computational model of choice is the carrier-resolved unidirectional pulse propagator [67,68] in radial symmetric geometry. The modeled CO_2 laser beam propagation scenario mimics the experiment in a sense that the first high-power 3.5-ps pulse (Gaussian temporal profile) after 25 ps is followed by a low-power 3.5-ps pulse. As depicted in Figure 3.11(a), we simulated propagation starting just before

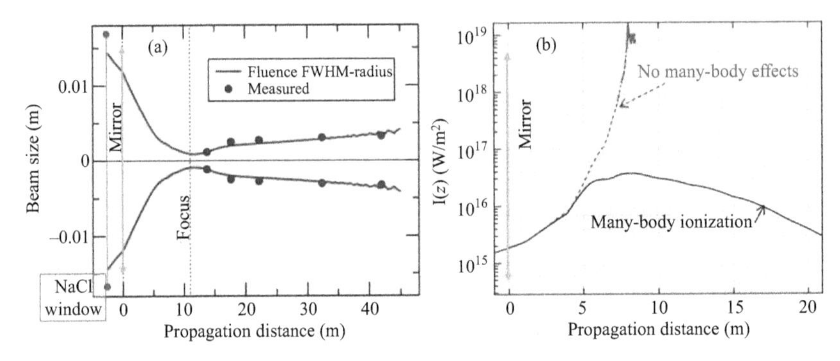

Figure 3.11 Dynamics of propagation in air. (a) Beam waist (FWHM) as a function of propagation distance for two 3.5-ps, 10.2-μm laser pulses separated by 25 ps with a power 823 GW (leading pulse) and 206 GW (postpulse) and a total energy of 3.6 J. (b) Peak laser intensity on axis of the filament as a function of propagation distance recorded with (black curve) and without (red dashed curve), including the many-body ionization mechanism. Courtesy of Paris Panagiotopoulos from the University of Arizona

the 10-cm-thick NaCl window of the final CO_2 amplifier, followed by a 2.3-m section of air, and finally the reflection of the focusing mirror, and the propagation up to the focus and beyond. The diameter of the beam incident at the NaCl window was varied in the range of 25–30 mm (FWHM) with a super-Gaussian spatial profile of order 2. Additionally, the third-order spherical aberration was applied to the incident field to model aberrations that can arise throughout the optical system. Both the beam diameter and the spherical aberration were used as control parameters to match the experimental results. The linear properties of NaCl and air were imported along with the measured Kerr coefficients [58,69]. The model for air takes into account the full HITRAN database for describing both dispersion and absorption, including all main gases: O_2, N_2, CO_2, CO, CH_4, argon, and water vapor at 30% relative humidity [70]. Temperature is 25°C and pressure is 1 atm. The nonlinear refractive index was taken to be 4×10^{-19} cm^2/W [58]. The initial electron generation is described by the many-body Coulomb effects [21], while the avalanche ionization is modeled by a two-temperature model (see details in [62]). This model provides a current source term for the field propagation and fully accounts for transient avalanche ionization on the picosecond timescales considered here.

The main result of these simulations is shown in Figure 3.11(a), which depicts the FWHM of the laser beam size as a function of propagation distance (red lines), the blue circles being the experimental data. The data supports the notion of formation of a *megafilament* in air in that the beam size stays largely constant over around $20Z_r$. Simulations were able to produce the *megafilament* using a broad range of initial focusing conditions, and it appears that for its launching a smooth approach to the focus is needed to allow the Kerr and plasma defocusing effects to accumulate in tandem. The relatively low plasma density of $\geq 10^{12}$ cm^{-3} created by many-body Coulombic interactions effectively decreases the molecular polarizability during the 3.5-ps laser pulse sufficiently to arrest Kerr self-focusing and provide for self-balancing mechanism to channel light in the LWIR *megafilament*. Moreover, avalanche formation seeded by this Coulomb ionization plays a minor role and only during the falling edge of the pulse. Further analysis shows that the *megafilament* arises from the main pulse when, as in the experiment, an incident pulse train is considered. In particular, including a secondary pulse of lesser peak intensity 25 ps after the main pulse shows that the secondary pulse is strongly spatially expanded by plasma defocusing due to electron avalanche buildup. Modeling also shows that the peak electron density of ~5×10^{15} cm^{-3}, comparable with the value inferred from the experiment, is reached later in time only when the free electrons seeded by the first pulse are cascaded by the second CO_2 pulse. This is consistent with the experimental observation of the pulse train evolution during propagation described earlier. Furthermore, the high density plasma in simulations extends for a distance of around 5 m correlating well with the visible plasma channel observed in the experiment.

We have studied the role of plasma defocusing on arresting Kerr self-focusing during the first 3.5-ps pulse. The results of modeling in Figure 3.11(b) clearly indicate that without 10^{12}–10^{13} cm^{-3} concentration of electrons in air, the CO_2 laser beam would collapse to a small spot size resulting in very high field intensities. As seen here, the laser beam without plasma defocusing (red curve) collapses due to Kerr self-focusing resulting in a laser intensity of 1.6×10^{19} W/m^2 that is 100 times above the tunnel ionization threshold of O_2/N_2. Such a field, if generated, could fully ionize the air molecules producing an electron density number close to or even greater than the critical plasma density for 10-μm radiation. It would have a deleterious effect on the CO_2 laser beam propagation due to (a) ionization-induced refraction; (b) strong back reflection from the high-density plasma, and (c) very efficient plasma absorption of the laser postpulses in the focal region. None of these effects were observed in the experiment. Note that the intensity deduced from Figure 3.11(b) (black curve) agrees well with the measured intensity in a *megafilament* shown in Figure 3.9(b).

3.3.5 *Spectral broadening of a self-guided in air LWIR pulse*

Several third-order nonlinear optic effects such as self-phase modulation, stimulated Raman scattering, and FWM are typically involved in the high-intensity laser–air interactions and are considered to be responsible for supercontinuum (SC) generation in the near-IR atmospheric filaments [7]. The results of the spectral broadening measurements in the 8–14-μm LWIR window are presented in Figure 3.12(a). Here, the conversion efficiency dropped dramatically from 2×10^{-3} at 11 μm to 2×10^{-5} at 13.5 μm. We also observed a significant asymmetry in the broadened spectrum favoring the red-shifted light. This could be attributed to the importance of the delayed molecular contribution to n_2 for air favoring the red side as

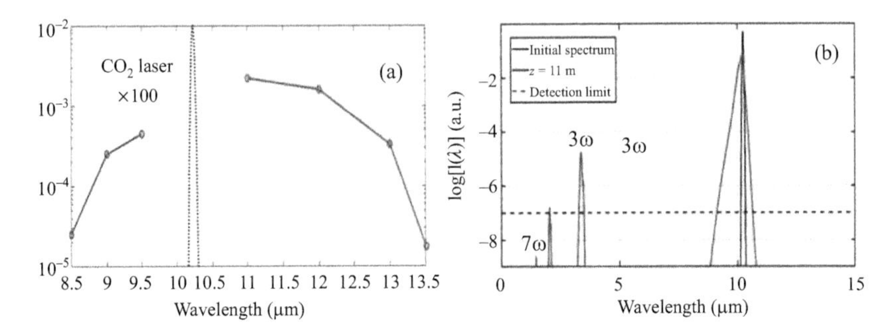

Figure 3.12 Spectral measurements. (a) Spectral broadening of an initial 3.5-ps CO_2 laser pulse self-guided in air recorded at z = 22.4 m measured by the use of a grating spectrometer in combination with a room temperature mercury cadmium telluride detector and (b) calculated spectrum of the 10.2-μm laser pulse at z = 11.2 m

Table 3.1 Results of measurements of efficiency of odd-harmonic generation in a CO_2 laser megafilament *in the atmosphere as a function of distance calculated as an average value for 5–7 shots with powers in the range of 1.1–1.3 TW*

Distance (m)	Efficiency (3ω)	Efficiency (5ω)	Efficiency (7ω)
$z = 11.2$	$1.88 \pm 0.04 \times 10^{-4}$	BDS	BDS
$z = 22.4$	$3.55 \pm 0.14 \times 10^{-4}$	~10^{-5}	BDS

BDS refers to signals to be below the detection sensitivity.

was observed in 10 μm FWM in the atmosphere shown in Figure 3.5(b). Thus efficient SC generation with a short pulse CO_2 laser discussed in literature [71] is suppressed in the experiment by a competing third-order nonlinear effect. To estimate the possible role of χ^3-induced odd-harmonic generation on CO_2 laser self-guiding in air, at different distances we measured the energy inside the *megafilament* contained in the harmonics: third—3.41 μm, fifth—2.05 μm, and seventh—1.46 μm. The spectral content of the self-guided CO_2 laser beam in air was analyzed using a set of broadband-pass filters centered around odd-harmonics of the central wavelength of 10.2 μm along with different types of energy meters. Note that only the central part of the CO_2 laser beam close to the axis was collected for these measurements. The measured efficiencies of harmonic generation at $z = 11.2$, 22.4 m are listed in Table 3.1. We found that the harmonic energy increased with the distance and the harmonic conversion efficiency at $z = 22.4$ m reached $3.55 \pm 0.14 \times 10^{-4}$ and 10^{-5} for the third and fifth harmonic, respectively. The signal for the seventh harmonic was below the detector sensitivity. The practically dispersionless infrared atmospheric window makes the harmonic generation process rather efficient.

The results of calculations of spectral broadening are presented in Figure 3.12 (b). In good agreement with the experiment, spectral broadening of the 10.2-μm laser pulse is small and affected by odd-harmonic generation in air in the range of 10^{-3}–10^{-4}, on the same order-of-magnitude as the values in Table 3.1. Dispersion and absorption are described by a 1-atm HITRAN air model with a 30% relative humidity [70].

Thus, when a 3.5-ps long CO_2 laser pulse is self-guided in air, the generation of a broadband LWIR radiation due to nonlinear optics of the atmosphere seems to be not very efficient process in comparison with femtosecond filaments in near-IR. However, recent numerical studies indicate that nonlinear propagation of a TW power laser beam over hundreds of meters can result in self-compression of a 10-μm pulse and consequently production of the SC spectrum that covers almost the entire spectral window from 8 to 14 μm [72]. For laser remote sensing, the potential of such an LWIR "backlighter" generated at a certain distance from the ground-based transmitter is great, since practically all minor air molecular constituents as well as many pollutants can be optically excited and, therefore, detected. Experimental verification of this aspect of LWIR filamentation in the atmosphere is a matter of near future.

3.4 Conclusion

It was shown that a self-guided terawatt picosecond CO_2 laser beam forms a single centimeter-scale-diameter *megafilament* in air, which has a cross-section four orders of magnitude larger as compared with a short-wavelength laser filament and guides many Joules of pulse energy. This is due to the terawatt level of critical power of Kerr self-focusing in the atmosphere at ~10 μm and a relatively long ~3-ps CO_2 laser pulses successfully used for propagation. The nonlinear guiding of a terawatt picosecond 10.2-μm pulse in a self-induced waveguide in air over ~32 m ($20Z_r$) described in this chapter indicates the possibility of transporting a coherent multi-Joule laser beam in a single channel over multiple Rayleigh ranges (from many meters to many kilometers). One of the possible scenarios for long-range propagation is self-channeling of the entire collimated 3–5-cm diameter TW CO_2 laser beam with a power slightly above P_{cr} over km-scale distances [62]. The LWIR atmospheric window with its low extinction coefficient makes low-loss propagation of such a high-power, short-pulse laser in the troposphere possible and potentially useful for multiple atmospheric applications.

The measured clamped intensity in the LWIR *megafilament* of around 10^{12} W/cm^2 is significantly below the tunnel ionization threshold of O_2/N_2. In combination with the negligibly small contribution of the multiphoton ionization at long wavelengths, the regular field ionization mechanism does not play any significant role here. Furthermore, a self-guided pulse with a duration of a few picoseconds seems to be too short for an electron avalanche to grow especially assuming a very low concentration of free carriers in the ambient air. Thus a new ionization model was developed to describe the observed phenomena theoretically.

In this chapter, we discuss that this *megafilament* arises from the balance between self-focusing, diffraction and defocusing caused by free carriers generated via many-body Coulomb-induced ionization that effectively decrease the molecular polarizability during the long-wavelength laser pulse. This is a new paradigm in filamentation physics because the weak plasma with an electron density of $<5 \times 10^{13}$ cm^{-3} is generated due to a very different mechanism of Coulomb electron–electron scattering during the laser pulse [21,66]. This mechanism is well adopted in solid-state physics but was never applied before to the atmospheric propagation experiment. Such a low-density plasma as theorized in the paper can be produced at $\leq 10^{12}$ W/cm^2 intensities, much below the tunnel ionization threshold of air constituents, and on a few picosecond timescale, much shorter than the time required for electron avalanche in air to grow. In effect, one may introduce a new hybrid field+many-body ionization regime in a high-pressure gas, which is unexplored but potentially could play a strong role in LWIR pulse propagation. Understanding where this new ionization regime plays a role, below the threshold of the well adopted single-atom tunnel ionization in gases, will require a set of dedicated high precision measurements and could become a new frontier in strong-field interactions.

In this chapter, we also describe the remarkable progress in development of terawatt-power CO_2 lasers predominately in the USA. Here the optimal approach in designing a high-peak-power LWIR laser system will be combining of an mJ-class mid-IR OPCPA solid-state system as a seed source and a multiatmosphere CO_2 gas laser as an efficient power amplifier of such a seed pulse to a terawatt level (~1 J in ~1 ps). The efficient generation of a mJ 0.5–1-ps energetic seed at 9–11 μm is a true challenge but seems to be feasible in the future. Note that emerging new technologies based on high-pressure CO_2 lasers optically pumped by a Fe:ZnSe laser pulses can become an alternative to the OPCPA systems. A compact, palm-size CO_2 amplifier that can ultimately have a remarkable multi-THz bandwidth suitable for subpicosecond pulse amplification will have a fascinating future. In general, combining of an ultrafast solid-state 1–2 μm laser with a high-power LWIR gas laser in one system opens many interesting opportunities in studying nonlinear optical properties of atoms and molecules.

Clearly, there is a wealth of different phenomena that can be studied using a high-power picosecond LWIR pump pulse and an ultrafast probe pulse deterministically synchronized to this pump with a femtosecond accuracy. We have touched upon of only one of them related to LWIR filamentation in air.

Acknowledgments

We are grateful to Dan Haberberger, Ken Marsh from the UCLA Neptune laboratory for numerous contributions to picosecond TW CO_2 laser development, Jeremy Pigeon and Eric Welch for nonlinear optics measurements in gases, Drs. Igor

Pogorelsky and Misha Polyanskiy from BNL, and Dr. Paris Panagiotopoulos, Profs. Miroslav Kolesik, Jerome V. Moloney, and Stephan W. Koch from the University of Arizona for contributions to experimental and theoretical studies of LWIR filamentation in the atmosphere.

References

[1] Strohbehn, J.W. (editors) *Laser Beam Propagation in the Atmosphere*. Springer Series: Topics in Applied Physics v. 25 (Springer, Berlin, 1978).
[2] Lubin, P., Hughes, G.B., Bible, J., *et al.* "Toward directed energy planetary defense". *Opt. Eng.*, 2014, 53, 025103.
[3] Daukantas, P. "Breakthrough starshot". *Opt. Photonics News*, 2017, 28, 26–33.
[4] Nicholls, R.W. "Wavelength-dependent spectral extinction of atmospheric aerosols". *Appl. Opt.*, 1984, 23, 1142–1143.
[5] Andrews, L.C. and Phillips, R.L. *Laser Beam Propagation through Random Media* (SPIE Optical Engineering Press, Spokane, Washington, 1998).
[6] Braun, A., Korn, G., Liu, X., Squier, J., Du, D., and Mourou, G. "Self-channeling of high-peak-power femtosecond laser pulses in air". *Opt. Lett.*, 1995, 20, 73–75.
[7] Couairon, A. and Mysyrowicz, A. "A femtosecond filamentation in transparent media". *Phys. Rep.*, 2007, 44, 47–189.
[8] Chin, S.L. *Femtosecond Laser Filamentation*. Springer Series on Atomic, Optical and Plasma Physics v. 55 (Springer, Berlin, 2010).
[9] Mitrofanov, A.V., Voronin, A.V., Sidorov, D.A., *et al.* "Mid-Infrared laser filaments in the atmosphere". *Sci. Rep.*, 2015, 5, 8368–8673.
[10] Shumakova, V., Aliskauskas, S., Malevich, P., *et al.* "Filamentation of mid-IR pulses in ambient air in the vicinity of molecular resonances". *Opt. Lett.*, 2018, 43, 2185–2188.
[11] Mechain, G., Couairon, A., Andre, Y.B., *et al.* "Long range self-channeling of infrared laser pulses in air: a new propagation regime without ionization". *Appl. Phys. B*, 2004, 79, 379–382.
[12] Durand, M., Houard, A., Prade, B., *et al.* "Kilometer range filamentation". *Opt. Express*, 2013, 21, 26836–26845.
[13] Zhao, X.M., Diels, J.-C., Wang, C.Y., and Elizondo, J.M. "Femtosecond ultraviolet laser pulse induced lightning discharges in gases". *IEEE J. Quantum Electron.*, 1995, 31, 599–612.
[14] Kartashov, D., Alisauskas, S., Andriukatis, G., *et al.* "Free-space nitrogen gas laser driven by a femtosecond filament". *Phys. Rev. A*, 2012, 86, 033831.
[15] Stelmaszczyk, K, Rohwetter, P., Mejeean, G., *et al.* "Long-distance remote laser-induced breakdown spectroscopy using filamentation in air". *Appl. Phys. Lett.*, 2004, 85, 3977–3979.
[16] Point, G., Brelet, Y., Houard, A., *et al.* "Superfilamentation in air". *Phys. Rev. Lett.*, 2014, 112, 223902.

[17] Chateauneuf, M, Payeur, S., Dubois, J., and Kieffer, J.-C. "Microwave guiding in air by a cylindrical filament array waveguide". *Appl. Phys. Lett.*, 2008, 92, 091104.
[18] Rohwetter, P., Kasparian, J., Stelmaszczyk, K., *et al.* "Laser-induced water condensation in air". *Nat. Photonics*, 2010, 4, 451–456.
[19] Geints, Y.E. and Zemlyanov, A.A. "Single and multiple filamentation of multi-terawatt CO_2 laser pulses in air: numerical simulations". *J. Opt. Soc. Am. B*, 2014, 31, 788–797.
[20] Panagiotopoulos, P., Schuh, K., Kolesik, M., and Moloney, J.M. "Simulations of 10 μm filaments in a realistically modeled atmosphere". *J. Opt. Soc. Am. B*, 2016, 33, 2154–2161.
[21] Schuh, K., Kolesik, M., Wright, E.M., and Moloney, J.M. "Self-channeling of high-power long-wave infrared pulses in atomic gases". *Phys. Rev. Lett.*, 2017, 118, 063901.
[22] Andriukaitis, G., Balciunas, T., Alisasuskas, S., *et al.* "90 GW peak power few-cycle mid-infrared pulses from an optical parametric amplifier". *Opt. Lett.*, 2011, 36, 2755–2758.
[23] Sanchez, D., Hemmer, M., Baudisch, M., *et al.* "7 μm, ultrafast, sub-millijoule-level mid-infrared optical parametric chirped pulse amplifier pumped at 2 μm". *Optica*, 2016, 3, 147–150.
[24] Grafenstein, L., Bock, M., Ueberschaer, D., *et al.* "5 μm few-cycle pulses with multi-gigawatt peak power at a 1 kHz repetition rate". *Opt. Lett.*, 2017, 42, 3796–3799.
[25] Mitrofanov, A.V., Voronin, A.A., Rozhko, M.V., *et al.* "Self-compression of high-peak-power mid-infrared pulses in anomalously dispersive air". *Optica*, 2017, 4, 1405–1408.
[26] Weisshaupt, J., Juve, V., Holtz, M., *et al.* "High-brightness table-top hard X-ray source driven by sub-100-femtosecond mid-infrared pulses". *Nat. Photonics*, 2014, 8, 927–930.
[27] Koulouklidis, A.D., Golner, C., Shumakova, V., *et al.* "Observation of Strong THz Fields from Mid-Infrared Two-Color Laser Filaments" Conference of Lasers and Electro-Optics, 13–18 May, 2018, San Jose, CA, USA, 2018, paper FF1E.2.
[28] Pullen, M.G., Wolter, B., Le, A.T., *et al.* "Imaging an aligned polyatomic molecule with laser-induced diffraction". *Nat. Commun.*, 2015, 6, 7262.
[29] Tochitsky, S.Ya., Narang, R., Filip, C.V., *et al.* "Enhanced acceleration of injected electrons in a laser-beat-wave-induced plasma channel". *Phys. Rev. Lett.*, 2004, 92, 095004.
[30] Haberberger, D., Tochitsky, S., Fiuza, F., *et al.* "Collisionless shocks in laser-produced plasma generate monoenergetic high-energy proton beams". *Nat. Phys.*, 2012, 8, 95–99.
[31] Wynne, J.J. "Optical third-order mixing in GaAs, Ge, Si, and InAs". *Phys. Rev.*, 1969, 178, 1295.
[32] Kildal, H. and Deutsch, T.F. "Infrared Third-harmonic generation in molecular gases". *IEEE J. Quantum Electron.*, 1976, 12, 429–435.

[33] Osipov, A.I., Panchenko, V.Ya., and Filippov, A.A. "Self-focusing of laser radiation in molecular gases". *Sov. J. Quantum Electron.*, 1985, 15, 465–468.
[34] Tochitsky, S.Ya., Narang, R., Filip, C., Clayton, C.E., Marsh, K.A., and Joshi, C. "Generation of 160-ps terawatt-power CO_2 laser pulses". *Opt. Lett.*, 1999, 24, 1717–19.
[35] Polyanskiy, M.N., Pogorelsky, I.V., and Yakimenko, V. "Picosecond pulse amplification in isotope CO_2 active medium". *Opt. Express*, 2011, 19, 7717–7725.
[36] Haberberger, D.J., Tochitsky, S.Ya., and Joshi, C. "Fifteen terawatt picosecond CO_2 laser system". *Opt. Express*, 2010, 18, 17865–17877.
[37] Filip, C.V., Narang, R., Tochitsky, S.Ya., *et al.* "Nonresonant beat-wave excitation of relativistic plasma waves with constant phase velocity for charged particle acceleration". *Phys. Rev. E*, 2004, 69, 026404.
[38] Musumeci, P., Tochitsky, S.Ya., Boucher, S., *et al.* "High energy gain of trapped electrons in a tapered, diffraction-dominated inverse-free-electron laser". *Phys. Rev. Lett.*, 2005, 94, 154801.
[39] Brimacombe, R.K. and Reid, J. "Accurate measurements of pressure-broadened linewidth in a transversely excited CO_2 discharge". *IEEE J. Quantum Electron.*, 1985, 19, 1668–1673.
[40] Brimacombe, R.K. and Reid, J. "Influence of the dynamic Stark effect on the small-signal gain of optically pumped 4.3-μm CO_2 laser". *J. Appl. Phys.*, 1985, 58, 1141–1145.
[41] Tochitsky, S.Ya., Pigeon, J.J, Haberberger, D.J., and Joshi, C. "Amplification of multi-gigawatt 3 ps pulses in an atmospheric CO_2 laser using ac Stark effect". *Opt. Express*, 2012, 20, 13762–13765.
[42] Corkum, P.B. "Amplification of picosecond 10 μm pulses in multiatmosphere CO_2 lasers". *IEEE J. Quantum Electron.*, 1985, 21, 216–232.
[43] Filip, C.V., Narang, R., Tochitsky, S.Ya., Clayton, C.E., Marsh, K.A., and Joshi, C. "Optical Kerr switching technique for the production of a picosecond, multiwavelength CO_2 laser pulse". *Appl. Opt.*, 2002, 41, 3743–3747.
[44] Siegman, A.E. *Lasers* (University of Science Books, Mill Valley, CA, 1986).
[45] Polyanskiy, M.N., Babzien, M, and Pogorelsky, I.V. "Chirped-pulse amplification in a CO_2 laser". *Optica*, 2015, 2, 675–679.
[46] Mirov, S., Fedorov, V., Martyshkin, D., Moskalev, M., Mirov, M., and Vasilyev, S. "Progress in mid-IR lasers based on Cr and Fe doped II-VI chalcogenides". *IEEE J. Sel. Top. Quantum Electron.*, 2015, 21, 1601719.
[47] Chang, T.Y. and Wood, O.R. "Optically pumped continuously tunable high-pressure molecular lasers". *IEEE J. Quantum Electron.*, 1977, 13, 907–915.
[48] Auyeung, R.C.Y. and Reid, J. "High vibrational temperatures in optically-pumped CO_2". *IEEE J. Quantum Electron.*, 1988, 24, 573–579.
[49] Petukhov, V.O., Tochitsky, S.Ya., and Churakov, V.V. "Investigation of the output parameters of a transversely excited CO_2 laser in the wavelength range 4.2-4.5 μm (10^01-10^00 band)". *Sov. J. Quantum Electron.*, 1990, 20, 602–608.

[50] Gebhardt, F.G. "High power laser propagation". *Appl. Opt.*, 1976, 15, 1479–1493.

[51] Autric, M., Caressa, J.P., Bournot, P., Dufresne, D., and Sarazin, M. "Propagation of pulsed laser energy through the atmosphere". *AIAA J.*, 1981, 19, 1415–1421.

[52] Zuev, V.E., Zemlyanov, A.A., Kopytin, Y.D., and Kuzikovskii, A.V. *High-Power Laser Radiation in Atmospheric Aerosols* (Riedel, New York, 1984).

[53] Smith, D.C. and Brown, R.T. "Aerosol-induced air breakdown with CO_2 laser radiation". *J. Appl. Phys.*, 1976, 46, 1146–1154.

[54] Greig, J.R., Koopman, D.W., Fernsler, R.F., Pechacek, R.E., Vitkovsky, I. M., and Ali, A.W. "Electrical discharges guided by pulsed CO_2 laser radiation". *Phys. Rev. Lett.*, 1978, 41, 174–177.

[55] Shindo, T., Aihara, Y., Miki, M., and Suzuki, T. "Model experiments of laser-triggered lightning". *IEEE Trans. Power Delivery*, 1993, 8, 311–317.

[56] Raiser, Y.P. *Gas Discharge Physics* (Springer, New York, NY, 1987).

[57] Tochitsky, S.Ya., Filip, C., Narang, R., Clayton, C.E., Marsh, K.A., and Joshi, C. "Efficient shortening of self-chirped picosecond pulses in a high-power CO_2 amplifier". *Opt. Lett.*, 1995, 26, 813–815.

[58] Pigeon, J.J., Tochitsky, S.Ya., Welch, E.C., and Joshi, C. "Measurements of the nonlinear refractive index of air, N_2 and O_2 at 10 μm using four-wave mixing". *Opt. Lett.*, 2016, 41, 3924–3927.

[59] Pigeon, J.J., Tochitsky, S.Ya., Welch, E.C., and Joshi, C. "Experimental study of the third-order nonlinearity of atomic and molecular gases using 10-μm laser pulses". *Phys. Rev. A*, 2018, 97, 973801.

[60] Nibbering, E.T.J, Grillon, G., Franco, M.A., Prade, B.S., and Mysyrowicz, B. "Determination of the inertial contribution to the nonlinear refractive index of air, N_2, and O_2 by use of unfocused high-intensity femtosecond laser pulses". *J. Opt. Soc. Am. B*, 1997, 14, 650–660.

[61] Zahedpour, S., Wahlstrand, J.K., and Milchberg, H.M. "Measurements of the nonlinear refractive index of air, constituents at mid-infrared wavelengths". *Opt. Lett.*, 2015, 40, 5794–5797.

[62] Tochitsky, S., Welch, E., Polyanskiy, M., *et al.* "Megafilament in air formed by self-guided terawatt long-wavelength infrared laser". *Nat. Photonics*, 2019, 13, 41–46.

[63] Keldysh, L.V. "Ionization in the field of a strong electromagnetic wave". *Sov. Phys. JETP*, 1965, 20, 1307–1314.

[64] Bree, C., Demircan, A., and Steimeyer, G. "Saturation of the all-optical effect". *Phys. Rev. Lett.*, 2011, 106, 183902.

[65] Kira, M. and Koch, S.W. *Semiconductor Quantum Optics* (Cambridge University Press, Cambridge, UK, 2012).

[66] Schuh, K., Moloney, J.M., and Koch, S.W. "Influence of many-body interactions during the ionization of gases by short intense optical pulses". *Phys. Rev. E*, 2014, 89, 033103.

[67] Kolesik, M. and Moloney, J.V. "Nonlinear optical propagation simulation from Maxwell's to unidirectional equations". *Phys. Rev. E*, 2004, 70, 0366604.

[68] Couairon, A., Brambilla, E., Corti, T., Majus, D., Ramirez, O., and Kolesik, M. "Practitioner's guide to laser pulse propagation models and simulation". *Eur. Phys. J. Spec. Top.*, 2011, 199, 5–76.

[69] Adair, R., Chase, L.L., and Payne, S.A. "Nonlinear refractive index of optical crystals". *Phys. Rev. B*, 1989, 39, 3337–3349.

[70] Rothman, L.S., Gordon, I.E., Babikov, Y. *et al.* "The HITRAN 2012 molecular spectroscopic database". *J. Quant. Spectrosc. Radiat. Transfer*, 2013, 130, 4.

[71] Geints, Y.E. and Zemlyanov, A.A. "Dynamics of CO_2 laser filamentation in air influenced by spectrally selective molecular absorption". *Appl. Opt.*, 2014, 53, 5641–5648.

[72] Zheltikov, A.M. "Laser induced filaments in the mid-infrared". *J. Phys. B: At. Mol. Opt. Phys.*, 2017, 50, 092001.

Chapter 4

Extreme temporal compression of ultra-broadband mid-infrared pulses

Ugaitz Elu,[1] Luke Maidment,[1] Lenard Vamos,[1] Francesco Tani,[2] David Novoa,[2] Michael H. Frosz,[2] Valeriy Badikov,[3] Dmitrii Badikov,[3] Valentin Petrov,[4] Philip St. J. Russell,[2,5] and Jens Biegert[1,6]

Ultrafast mid-infrared sources [1] are useful for a wide variety of applications [2] ranging from environmental monitoring and remote sensing [3], medical diagnostics [4] and quality inspection in the food and pharmaceutical industries [5], to chemical reaction dynamics [6] and strong field physics [7]. For instance, the oscillator strength of absorption is much larger in the mid-infrared compared with the visible or ultraviolet (UV) that makes the mid-infrared ideal for extremely sensitive and element-specific spectroscopy measurements. As tabletop and even portable coherent mid-infrared sources have matured, compact spectroscopy systems have demonstrated applications in atmospheric monitoring and security, sensing pollutant gases and hazardous substances over long paths [8–11]. Coherent sources have far higher brightness than incoherent sources and can be focused to the diffraction limit [12], overcoming the main barrier to routine clinical use of infrared spectroscopy, the long data acquisition times. For instance, a compact imaging microscope has recently demonstrated highly sensitive and specific classification of cancerous tissue in human patients, with an acquisition time of just a few minutes [13].

Sources producing a coherent broad mid-infrared spectrum can make high sensitivity concentration measurements of many molecular species simultaneously [14], and short output pulse durations can be used for time-resolved measurements of molecular vibrations on a femtosecond timescale [15]. The information available

[1]ICFO – Institut de Ciencies Fotoniques, The Barcelona Institute of Science and Technology, Barcelona, Spain
[2]Max-Planck Institute for Science of Light, Erlangen, Germany
[3]High Technologies Laboratory, Kuban State University, Krasnodar, Russia
[4]Max-Born-Institute for Nonlinear Optics and Ultrafast Spectroscopy, Berlin, Germany
[5]Department of Physics, Friedrich-Alexander-Universität, Erlangen, Germany
[6]ICREA, Pg. Lluis Companys 23, Barcelona, Spain

from absorption spectroscopy can be extended from simply molecular composition to vibrational mode couplings. Thus, ultra-broadband few-cycle infrared pulses are key to investigate molecular dynamics through multidimensional spectroscopy techniques such as two-dimensional infrared spectroscopy (2DIR) [16]. Combining 2DIR with spectral coverage reaching down to the UV allows the investigation of dynamics of complex molecules such as DNA and proteins [17]. In solid-state physics the combination of ultra-broadband radiation with attosecond to femtosecond temporal resolution allows the study of electron dynamics such as electron–electron, electron–phonon and phonon–phonon interactions [18,19].

The infrared beamlines at synchrotrons [20] are typically used sources of broad mid-infrared radiation, but recently developed optical parametric sources have demonstrated similar spectral coverage with higher brightness and shorter pulse durations. High repetition rate, high average power sources are favorable to enable fast data acquisition and increase the signal-to-noise ratio in measurements, reducing both detector noise and shot noise contributions [21]. That is why high repetition rate sources of few cycle mid-infrared pulses [22] with sufficient intensity to access the strong field physics regime are in demand for experiments on laser-induced electron diffraction [7] and high harmonic generation (HHG) of attosecond soft X-ray pulses [23]. The cutoff photon energy that scales with the driving wavelength for HHG radiation has pushed the development of new lasers providing high peak powers at high repetition rates above 3 μm [24,25] with the aim of realizing tabletop coherent X-ray sources with above kilo-electron-volt photon energies. Difference frequency generation (DFG) providing carrier envelop phase stability combined with optical parametric amplification has been widely used to generate few-cycle high-average-power pulses at 3-μm wavelength [24–29].

Post-compression typically relies on spectra broadening via self-phase modulation (SPM), achieved in solid core photonic crystal fiber for nanojoule level pulse energies [30] and propagation in bulk materials or gases for microjoule and millijoule pulse energies, respectively. In the near- and mid-infrared, filamentation in dielectric [31] such as sapphire, YAG and fused silica has been exploited during the last decade to achieve sub-3-cycle microjoule level pulses [32]. Unfortunately, the ionization process required to achieve a quasi-stable filament induces high-order propagation loses that can reduce the compression efficiency down to 20%. In the near-infrared regime, hollow core fibers (HCFs) or capillaries are routinely used for post-compression processes [33] with low transmission loss. The loss of conventional gas-filled capillary scales as λ^2/a^3, where a is the fibre core radius. Moreover, to achieve significant broadening and post-compression, a fiber length close to the SPM characteristic length $L_{SPM} = \lambda/(2\pi n_2 I_0)$ is required. Therefore, the difficulty to post-compress mid-infrared pulses in an HCF faces a significant challenge as the characteristic length increases linearly with the wavelength, and the losses increase quadratically. One solution to reduce losses at mid-infrared wavelengths is to increase the core size. Then higher pulse energies are required to keep the characteristic length short. For instance at 3.2-μm wavelength, a 250-μm core diameter capillary exhibits over 80% loss through 1 m of propagation, while using a 1-mm core diameter capillary the loss is reduced down to under 10% [34].

Two experimental demonstrations have shown that this approach can be used to post-compress 3.2- [34] and 4.0-μm [35] pulses from around 100 to around 20 fs with few-mJ pulse energy. These factors mean that HCFs are suitable for mid-infrared post-compression, but long fiber lengths and high input pulse energies are required, which inherently limits this approach to low repetition rate sources.

The development of antiresonant-reflection photonic crystal fibers (ARR-PCF) [36–39] consisting of single-ring based antiresonant capillaries surrounded by a central hollow core with engineerable properties allowed ARR-PCFs with high transmission in certain mid-infrared regions to be designed. Antiresonant reflection [40] inhibits coupling to the cladding elements due to destructive interference in the thin satellite capillary walls, confining the guided light to the central hollow channel with very little overlap into the glass cladding. The resonant coupling between the core modes and modes guided in the satellite capillary walls is further reduced through a significant phase-velocity mismatch. This way ARR-PCFs provide high transmission efficiencies in the mid-infrared regime even at sub-100-μm core sizes. Figure 4.1 shows the design of an ARR-PCF designed to provide a loss-free window around 3.2 μm.

Unlike dielectric materials, noble gases have high ionization potentials that imply low ionization probabilities, and the nonlinear coefficient can be tuned easily by varying the pressure where the fiber sits. In free space the dispersion introduced to a visible/infrared ultrashort pulse propagating through a noble gas is positive, which adds with the positive nonlinear dispersion introduced by the SPM process. Therefore, after the SPM is produced, a post-compression scheme introducing negative dispersion is required.

ARR-PCF can be designed to achieve effective negative dispersion through the fiber, enabling self-compression and soliton generation [22,41,42]. In Figure 4.1, in gray the loss estimated during propagation through the ARR-PCF is shown. The

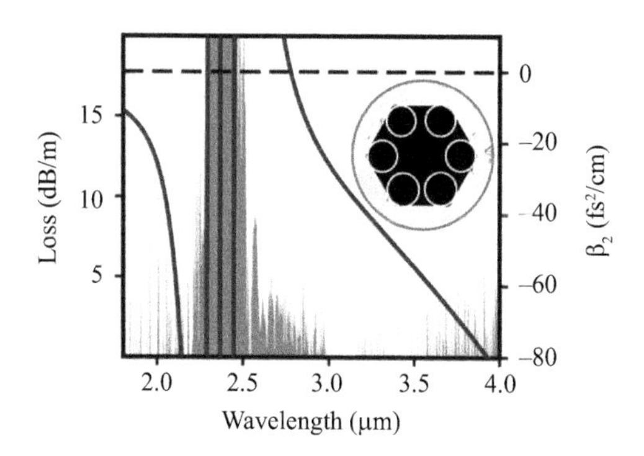

Figure 4.1 Measured transmission loss in the mid-infrared regime (gray) together with the dispersion estimated dispersion in vacuum (blue) for an 88-μm core with 1.2-μm ring wall thickness. The inset shows the scanning electron microscope image of the ARR-PCF-cleaved surface. Adapted from [22]

designed ARR-PCF with a core size of ~88 μm provides excellent transmission efficiencies in the mid-infrared regime enabled by the antiresonant ~1.2-μm thin rings. The calculated dispersion using the finite-element method [43,44] predicts negative dispersion in the designed mid-infrared transmission window regime, while close to the resonance around 2.4 μm the dispersion sign becomes positive rising exponentially to infinity.

Soliton self-compression is achieved when the soliton-order $N = L_D/L_{SPM} > 1$ where $L_d = T_0^2/\beta_2$ is the characteristic dispersion length (T_0 is the input pulse duration and β_2 is the GVD) [42,45,46]. Higher compression factors are expected at higher soliton-orders. Unfortunately, pulse compression quality is also affected at higher soliton-orders by the uncompensated higher order dispersion terms. Therefore, soliton-orders $N > 5$ are avoided to keep the pulse intensity envelope clean.

4.1 High-power single-cycle mid-infrared pulses

The soliton compression in an ARR-PCF gives unique opportunities to tailor the compression process by choosing the optimum noble gas, pressure, fiber length and input pulse energy. The noble gas selected determines the ionization probability during the ultrafast pulse propagation, which delimits the useable pressure range, fiber length and input pulse energies required for the SPM process. The tunability provided enables control over spectral broadening to allow enough broadening to achieve single-cycle pulses while minimizing the pulse sensitivity to higher order dispersion terms. This way, useable near-single-cycle self-compressed pulses can be generated [22], propagated in free space, temporally characterized and used for strong-field experiments.

The OPCPA source used produces high intensity 3.2-μm wavelength mid-infrared pulses at 160-kHz repetition rate with pulse energy up to 118 μJ. Mid-infrared seed pulses for amplification are generated using difference frequency mixing of the two outputs from a two-color Er:fiber laser, ensuring passive carrier envelope phase stability of the amplified pulses [24]. Booster amplifier stages amplify the mid-infrared pulses to 131 μJ, with high-efficiency compression in sapphire to 97-fs pulse duration. For coupling to the ARR-PCF, a diamond pinhole spatial filter was included to prevent damage to the fiber facet from potential spatial fluctuations of the beam, since the fiber and interior capillaries are made from fused silica with high absorption above 2.7 μm. This reduced the available pulse energy to 75 μJ (12-W average power), coupled into the ARR-PCF with coupling efficiency >90%.

The output window of the gas-cell is generally the main source of dispersion that limits the shortest pulses that can be measured and use in free space. Relatively thin (<500 μm) windows can be used for low gas pressures, but at tens of bars thicker windows of a few millimeters are required. While the aperture of the window can be reduced to decrease its thickness, high pulse-peak-powers and possible self-focusing-induced damage limits the minimum useable aperture. Fused silica in the near-infrared and CaF_2 or MgF_2 in the mid-infrared and in the UV respectively are commonly used, balancing between the dispersion introduced and the hardness of the material. Thus, in the UV the pulses will be chirped with positive dispersion and in the mid-infrared with negative dispersion when dielectric bulks such as CaF_2 or MgF_2 are used for the output window. In the visible and near-infrared regimes, broadband chirp mirrors [47] can be used to precisely compensate the dispersion introduced by the gas-cell output window and any additional optics in the beam path. In the UV and mid-infrared, more primitive dispersion compensation optics such as prisms in the UV or thin semiconductors such as silicon or germanium in the mid-infrared must be used and are typically not able to fully compensate dispersion.

Figure 4.2 shows the soliton generation and propagation at different argon pressures injected inside the gas-cell where the ARR-PCF sits [22]. Increasing the pressure in the gas-cell, the nonlinear coefficient required to produce SPM is amplified and so the mid-infrared pulses are broadened spectrally (see Figure 4.2 (a)). The linear negative dispersion introduced during the mid-infrared pulse propagation through the ARR-PCF is compensated with the positive nonlinear dispersion induced by the SPM process. Consequently, the spectrally broadened mid-infrared pulses are self-compressed in time in the absence of a significant amount of accumulated dispersion.

Figure 4.3 shows the temporal (in blue) and spectral (in red) characterization of the near single-cycle mid-infrared pulses and the corresponding input pulse spectral and temporal profiles (in black), measured after the 3-mm thick CaF_2 gas-cell window and a 0.5-mm thin Si wafer for dispersion compensation. The mid-infrared pulses from the OPCPA are efficiently compressed down to 13.5 fs in a single compact ~17-cm long ARR-PCF [22]. An output power of 9.6 W (60-μJ pulse energy) was measured after the fiber. The temporal intensity is presented in Figure 4.4(b). The front part of the pulse shows a pedestal induced by the uncompressed high-order dispersion terms and a post-pulse generated typically from a high-order soliton breakup. The output pulse duration is 14.5 fs, which corresponds to 1.35 optical cycles and peak power of 3.9 GW.

The high efficiency through the ARR-PCF and the scheme's exceptional tunability enables efficient self-compression of the mid-infrared high-power pulses.

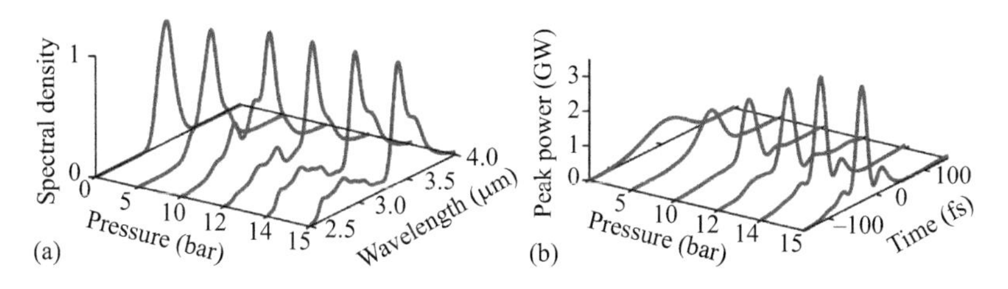

Figure 4.2 Spectra (a) and temporal (b) soliton behavior when the gas-cell where the ARR-PCF sits is filled with different argon gas pressures. Adapted from [22]

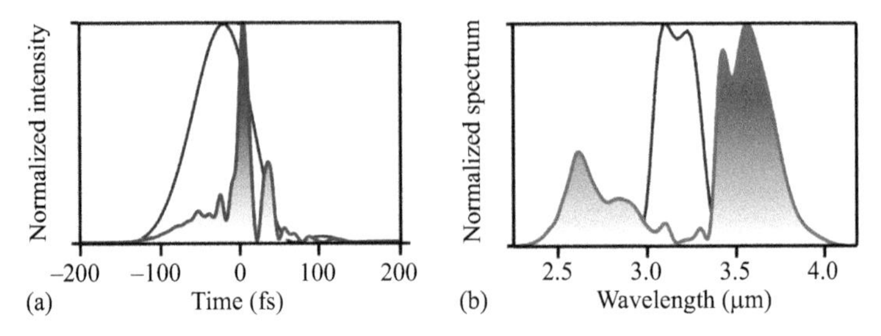

Figure 4.3 (a) Retrieved via SHG-FROG input (black) and self-compressed (shaded blue) pulse envelope; (b) measured input (black) and output (shaded red) 3.2-μm spectrum. Adapted from [22]

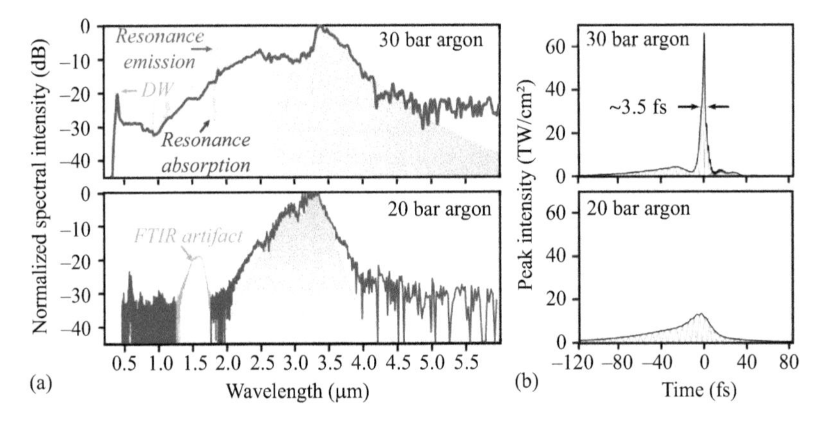

Figure 4.4 (a) Measured (blue) and simulated (shadowed blue–green) spectral broadening in ARR-PCF achieved at 20 and 30 bar of argon pressure. (b) Simulated corresponding temporal pulse profile of the mid-infrared optical shock waves. Adapted from [50]

Therefore, this technique provides unique feasibilities to boost the peak power of the OPCPA, essential for strong-field physics and HHG experiments. For instance, the high-power 3.2-μm pulses delivered from the high repetition rate OPCPA were enhanced from 1.25- to 3.9-GW peak powers.

4.2 UV dispersive wave and extreme transient soliton optical-shock-wave generation

As mentioned before, the propagation of $N > 5$ solitons in free space maintains the pulse shape, and the peak power becomes challenging and sometimes impractical. Still, inside ARR-PCF, a flat dispersion window can be designed to generate solitons above the $N > 5$ and increase the mid-infrared peak power further. Taking the soliton generation to the extreme can trigger highly nonlinear ultrafast dynamics with extreme pulse-peak intensities that can be key for differentially pumped HHG generation, strong-field research or ultra-broadband multidimensional spectroscopy. These dynamics enable the generation of dispersive waves (DWs) in the UV regime [48–50] or optical shock waves coupled with the soliton that steepens the tail of the pulse so strongly that soliton transients with optical cycles at FWHM down to 0.32 (3.5 fs at 3.2 μm) can be generated [50,51].

Using the same mid-infrared source with a different ARR-PCF design, spectral broadening and self-compression of the input pulse occur simultaneously with efficient transfer energy into a DW in the UV. Figure 4.4 shows the generation of the extreme ultrafast dynamics when the pressure inside the gas-cell where the ARR-PCF sits is tuned from 20 bar of argon to 30 bar. This ARR-PCF has the resonance shifted to ~1.8 μm (see resonance absorption and emission labels in Figure 4.4(a)) to maintain a negative dispersion area as width and flat as possible and achieve cleaner soliton propagation. The 3.2-μm spectrum broadens from an octave of bandwidth at 20 bar to several octaves at 30 bar. In the time domain, the optical shock wave couples with the soliton, steepening the pulse's tail and achieving soliton transients with record peak intensities enhanced from ~1 up to ~65 TW/cm^2 (Figure 4.4(b)). Unfortunately, such a broad and compressed soliton transient cannot be propagated in free space without destroying the temporal shape induced by any minor high-order dispersion. Still, a differentially pumped gas cell could be used to direct the generated soliton transients in vacuum and exploit them for new strong-field studies or characterize them via streaking [52].

Moreover, inside the ARR-PCF and at the highest soliton peak intensities, DW generation can be triggered in the UV [50]. Driving the process with near-infrared pulses, conversions up to 10% have been demonstrated [53]. At longer wavelengths, DW absorption in high-order resonances inside the ARR-PCF limits the achievable conversion efficiency. Consequently, modest conversion efficiency from mid-infrared to UV DW of around 0.2% has been demonstrated to date, with milliwatt average power [50]. This efficiency was reached using 35-bar Ar pressure, with the spectrum shown in Figure 4.5. Still, the milliwatt power and tens of nanojoule UV pulse energies are typically enough for key experiments where molecules like *trans*-azobenzene [54] are required to be ionized or to study excitons in, for instance, hexagonal boron nitrate [55].

4.3 Multi-octave bright femtosecond source

Intra-pulse DFG (IP-DFG) can be used to further expand the spectrum of the ultra-broadband radiation generated inside the ARR-PCF toward longer wavelengths in the mid-infrared. The broad output spectrum and the tunability of the spectral width provided by the ARR-PCF based self-compression make the mid-infrared pulses ideal for IP-DFG—tuning the broadening can tune the center wavelength of the IP-DFG spectrum.

Typically, noncollinear geometry in DFG or thin nonlinear crystals in IP-DFG are used to achieve broadband phase-matching to produce long wavelength infrared

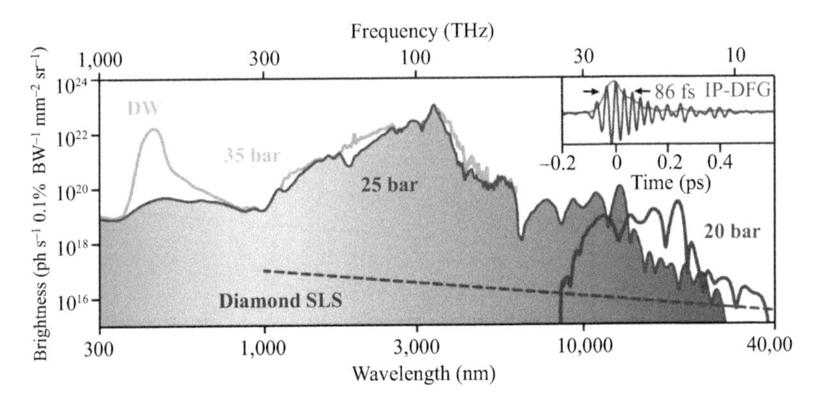

Figure 4.5 *Spectral brightness (estimated following the method from Pupeza* et al. *[21]) of our source at different Ar pressures, combining the broadened mid-infrared pulse and DW output from the ARR-PCF with the IP-DFG spectra. For comparison, values from the diamond synchrotron light source are overlaid [20], showing that our source exceeds it by several orders of magnitude. The inset shows the electric field waveform and intensity profile of the IP-DFG pulse, measured with EOS. Adapted from [50]*

and THz pulses. Birefringent nonlinear crystals with flat dispersion in the mid-infrared regime where signal and idler waves propagate with similar group velocity can provide unique phase-matching properties for efficient IP-DFG [56]. $BaGa_2GeSe_6$ (BGGSe) is a novel mid-infrared nonlinear crystal with a high non-linear coefficient and favorable collinear phase-matching over a longer crystal length, enabling broadband and efficient IP-DFG [57,58]. ZGP, CSP and GaSe are some of the most used mid-infrared nonlinear crystals for the generation [50,58–60] and amplification of [61] long-wavelength infrared radiation. Unfortunately, the limited transmission window in the case of ZGP and CSP, and the fragile nature of GaSe limits their application in generating bright and ultra-broadband coherent sources. Instead, the mid-infrared nonlinear crystals are typically swapped depending on the experimental requirements [59]. A novel mid-infrared nonlinear crystal, $BaGa_2GeSe_6$ (BGGSe), possesses a high nonlinear coefficient and favorable collinear phase-matching over a longer crystal length, enabling broadband and efficient IP-DFG [57,58].

Mid-infrared pulses extending from 8 to 40 μm are demonstrated with 2.6 optical-cycle duration from IP-DFG in a BGGSe crystal driven by the few-octave 3.2-μm pulses from the ARR-PCF when the gas-cell where the fiber sits is filled by 20 bar of argon (see Figure 4.5). Tuning the gas pressure to higher pressures increases the 3.2-μm broadening in the ARR-PCF, and consequently, the IP-DFG spectrum is shifted to the blue. Figure 4.5 shows the coherent seven-octave spanning spectrum generated by combining the self-compressed mid-infrared pulses, DW and the IP-DFG. The exceptional brightness of the coherent ultra-broadband source achieved is orders of magnitude larger than provided by a standard synchrotron infrared beamline.

4.4 Conclusion

We have shown that efficient and clean compression of mid-infrared pulses to the single optical cycle is straightforwardly achieved with soliton self-compression in gas-filled ARR-PCF. This is the most efficient post-compression technique suitable for high repetition rate and high average power mid-infrared pulses demonstrated today. The achieved combination of high peak and average power is ideal for nonlinear interactions such as HHG of coherent X-rays or multimodal spectroscopies. The excellent mode from the ARR-PCF and the high efficiency is also ideal for low pulse energies but high repetition rates, or applications in which near diffraction limited modes are desired. The sub-cycle soliton transient at the fiber exit can be output directly to vacuum to drive strong field processes. The UV spectrum from the DW is suitable for inducing structural changes and investigating dynamics in complex molecules.

On the more exotic side of applications, the availability of waveform-controlled intense long-wavelength radiation will allow to field-resolve the extreme nonlinear dynamics of optical shock front and DW generation during the self-compression process. Overall, the availability of sources of pulsed coherent radiation in the anomalous dispersion regime with waveform-control, and focusable to high intensity, unlocks an incredible wealth of applications ranging from fundamental science to applications in environment, industry, security and many more.

Funding

J.B. and group acknowledge financial support from the European Research Council for ERC Advanced Grant “TRANSFORMER” (788218) and ERC Proof of Concept Grant “miniX” (840010), the European Union’s Horizon 2020 for FET-OPEN “PETACom” (829153), FET-OPEN “OPTOlogic” (899794), Laserlab-Europe (EU-H2020 871124), Marie Sklodowska-Curie grant No 860553 (“Smart-X”), MINECO for Plan Nacional FIS2017-89536-P; AGAUR for 2017 SGR 1639, MINECO for “Severo Ochoa” (SEV-2015-0522), Fundació Cellex Barcelona, CERCA Programme/Generalitat de Catalunya, the Alexander von Humboldt Foundation for the Friedrich Wilhelm Bessel Prize.

References

[1] O. Chalus, P. K. Bates, M. Smolarski, and J. Biegert, “Mid-IR short-pulse OPCPA with micro-Joule energy at 100 kHz,” *Opt. Express* 17, 3587–3594 (2009).

[2] H. Pires, M. Baudisch, D. Sanchez, M. Hemmer, and J. Biegert, “Ultrashort pulse generation in the mid-IR,” *Prog. Quantum Electron.* 43, 1–30 (2015).

[3] T. Gardiner, J. Helmore, F. Innocenti, and R. Robinson, "Field validation of remote sensing methane emission measurements," *Remote Sens.* 9, 956 (2017).
[4] A. B. Seddon, T. M. Benson, S. Sujecki, *et al.*, "Towards the mid-infrared optical biopsy," *Proc. SPIE* 970302 9703 (2016).
[5] R. H. Wilson and H. S. Tapp, "Mid-infrared spectroscopy for food analysis: Recent new applications and relevant developments in sample presentation methods," *TrAC, Trends Anal. Chem.* 18, 85–93 (1999).
[6] M. A. Abbas, Q. Pan, J. Mandon, S. M. Cristescu, F. J. M. Harren, and A. Khodabakhsh, "Time-resolved mid-infrared dual-comb spectroscopy," *Sci. Rep.* 9, 17247 (2019).
[7] B. Wolter, M. G. Pullen, M. Baudisch, *et al.*, "Strong-field physics with Mid-IR fields," *Phys. Rev. X* 5, 021034 (2015).
[8] M. C. Phillips, B. E. Bernacki, S. S. Harilal, J. Yeak, and R. J. Jones, "Standoff chemical plume detection in turbulent atmospheric conditions with a swept-wavelength external cavity quantum cascade laser," *Opt. Express* 28, 7408–7423 (2020).
[9] G. Ycas, F. R. Giorgetta, J. T. Friedlein, *et al.*, "Compact mid-infrared dual-comb spectrometer for outdoor spectroscopy," *Opt. Express* 28, 14740–14752 (2020).
[10] S. Coburn, C. B. Alden, R. Wright, *et al.*, "Regional trace-gas source attribution using a field-deployed dual frequency comb spectrometer," *Optica* 5, 320–327 (2018).
[11] O. Kara, F. Sweeney, M. Rutkauskas, C. Farrell, C. G. Leburn, and D. T. Reid, "Open-path multi-species remote sensing with a broadband optical parametric oscillator," *Opt. Express* 27, 21358–21366 (2019).
[12] F. Mörz, R. Semenyshyn, T. Steinle, *et al.*, "Nearly diffraction limited FTIR mapping using an ultrastable broadband femtosecond laser tunable from 1.33 to 8 μm," *Opt. Express* 25, 32355–32363 (2017).
[13] C. Kuepper, A. Kallenbach-Thieltges, H. Juette, A. Tannapfel, F. Großerueschkamp, and K. Gerwert, "Quantum cascade laser-based infrared microscopy for label-free and automated cancer classification in tissue sections," *Sci. Rep.* 8, 7717 (2018).
[14] A. V. Muraviev, V. O. Smolski, Z. E. Loparo, and K. L. Vodopyanov, "Massively parallel sensing of trace molecules and their isotopologues with broadband subharmonic mid-infrared frequency combs," *Nat. Photonics* 12, 209–214 (2018).
[15] A. A. Lanin, A. A. Voronin, A. B. Fedotov, and A. M. Zheltikov, "Time-domain spectroscopy in the mid-infrared," *Sci. Rep.* 4, 1–8 (2014).
[16] P. M. Donaldson, G. M. Greetham, D. J. Shaw, A. W. Parker, and M. Towrie, " A 100 kHz pulse shaping 2D-IR spectrometer based on dual Yb: KGW amplifiers," *J. Phys. Chem. A* 122, 780–787 (2018).
[17] R. Borrego-Varillas, A. Nenov, L. Ganzer, *et al.*, "Two-dimensional UV spectroscopy: A new insight into the structure and dynamics of biomolecules," *Chem. Sci.* 10, 9907–9921 (2019).

[18] B. Buades, D. Moonshiram, T. P. H. Sidiropoulos, *et al.*, "Dispersive soft x-ray absorption fine-structure spectroscopy in graphite with an attosecond pulse," *Optica* 5, 502–506 (2018).
[19] B. Buades, A. Picón, E. Berger, *et al.*, "Attosecond state-resolved carrier motion in quantum materials probed by soft X-ray XANES," *Appl. Phys. Rev.* 8, 011408 (2021).
[20] G. Cinque, M. D. Frogley, and R. Bartolini, "Far-IR/THz spectral characterization of the coherent synchrotron radiation emission at diamond IR beamline B22," *Rend. Lincei* 22, S33–S47 (2011).
[21] I. Pupeza, D. Sánchez, J. Zhang, *et al.*, "High-power sub-two-cycle mid-infrared pulses at 100 MHz repetition rate," *Nat. Photonics* 9, 721–724 (2015).
[22] U. Elu, M. Baudisch, H. Pires, *et al.*, "High average power and single-cycle pulses from a mid-IR optical parametric chirped pulse amplifier," *Optica* 4, 1024–1029 (2017).
[23] F. Silva, S. M. Teichmann, S. L. Cousin, M. Hemmer, and J. Biegert, "Spatiotemporal isolation of attosecond soft X-ray pulses in the water window," *Nat. Commun.* 6, 6611 (2015).
[24] A. Thai, M. Hemmer, P. K. Bates, O. Chalus, and J. Biegert, "Sub-250-mrad, passively carrier–envelope-phase-stable mid-infrared OPCPA source at high repetition rate," *Opt. Lett.* 36, 3918–3920 (2011).
[25] M. Baudisch, M. Hemmer, H. Pires, and J. Biegert, "Performance of MgO: PPLN, KTA, and KNbO3 for mid-wave infrared broadband parametric amplification at high average power," *Opt. Lett.* 39, 5802–5805 (2014).
[26] C. Erny, K. Moutzouris, J. Biegert, D. Kühlke, F. Adler, A. Leitenstorfer, and U. Keller, "Mid-infrared difference-frequency generation of ultrashort pulses tunable between 3.2 and 4.8 μm from a compact fiber source," *Opt. Lett.* 32, 1138–1140 (2007).
[27] O. Chalus, P. K. Bates, and J. Biegert, "Design and simulation of few-cycle optical parametric chirped pulse amplification at mid-IR wavelengths," *Opt. Express* 16, 21297–21304 (2008).
[28] N. Thiré, R. Maksimenka, B. Kiss, *et al.*, "Highly stable, 15 W, few-cycle, 65 mrad CEP-noise mid-IR OPCPA for statistical physics," *Opt. Express* 26, 26907–26915 (2018).
[29] M. Mero, Z. S. Heiner, V. Petrov, *et al.*, "43 W, 1.55 μm and 12.5 W, 3.1 μm dual-beam, sub-10 cycle, 100 kHz optical parametric chirped pulse amplifier," *Opt. Lett.* 43, 5246–5249 (2018).
[30] J. M. Dudley, G. Genty, and S. Coen, "Supercontinuum generation in photonic crystal fiber," *Rev. Mod. Phys.* 78, 1135–1184 (2006).
[31] A. Couairon, J. Biegert, C. P. Hauri, *et al.*, "Self-compression of ultra-short laser pulses down to one optical cycle by filamentation," *J. Mod. Opt.* 53, 75–85 (2006).
[32] M. Hemmer, M. Baudisch, A. Thai, A. Couairon, and J. Biegert, "Self-compression to sub-3-cycle duration of mid-infrared optical pulses in dielectrics," *Opt. Express* 21, 28095–28102 (2013).

[33] T. Nagy, P. Simon, L. Veisz, and T. Nagy, “High-energy few-cycle pulses: Post-compression techniques techniques,” *Adv. Phys. X* 6, 1845795 (2021).
[34] G. Fan, T. Balčiūnas, T. Kanai, *et al.*, “Hollow-core-waveguide compression of multi-millijoule CEP-stable 3.2 μm pulses,” *Optica* 3, 1308–1311 (2016).
[35] P. Wang, Y. Li, W. Li, *et al.*, “26 mJ/100 Hz CEP-stable near-single-cycle 4 μm laser based on OPCPA and hollow-core fiber compression,” *Opt. Lett.* 43, 2197–2200 (2018).
[36] A. D. Pryamikov, A. S. Biriukov, A. F. Kosolapov, V. G. Plotnichenko, S. L. Semjonov, and E. M. Dianov, “Demonstration of a waveguide regime for a silica hollow – Core microstructured optical fiber with a negative curvature of the core boundary in the spectral region > 3.5 μm,” *Opt. Express* 19, 1441–1448 (2011).
[37] A. N. Kolyadin, A. F. Kosolapov, A. D. Pryamikov, A. S. Biriukov, V. G. Plotnichenko, and E. M. Dianov, “Light transmission in negative curvature hollow core fiber in extremely high material loss region,” *Opt. Express* 21, 9514–9519 (2013).
[38] F. Yu, W. J. Wadsworth, and J. C. Knight, ”Low loss (34 dB/km) silica hollow core fiber for the 3 μm spectral region,” in Advanced Photonics Congress OSA Technical Digest (online) (Optical Society of America, 2012), paper SM3E.2.
[39] P. Uebel, M. C. Günendi, M. H. Frosz, *et al.*, “Broadband robustly single-mode hollow-core PCF by resonant filtering of higher-order modes,” *Opt. Lett.* 41, 1961–1964 (2016).
[40] C. Wei, R. Joseph Weiblen, C. R. Menyuk, and J. Hu, “Negative curvature fibers,” *Adv. Opt. Photonics* 9, 504–561 (2017).
[41] M. Cassataro, D. Novoa, M. C. Günendi, *et al.*, “Generation of broadband mid-IR and UV light in gas-filled single-ring hollow-core PCF,” *Opt. Express* 25, 7637–7644 (2017).
[42] J. C. Travers, W. Chang, J. Nold, N. Y. Joly, and P. J. St. Russell, “Ultrafast nonlinear optics in gas-filled hollow-core photonic crystal fibers [Invited],” *J. Opt. Soc. Am. B* 28, A11–A26 (2011).
[43] F. Brechet, J. Marcou, D. Pagnoux, and P. Roy, “Complete analysis of the characteristics of propagation into photonic crystal fibers, by the finite element method,” *Opt. Fiber Technol.* 6, 181–191 (2000).
[44] A. Cucinotta, S. Selleri, L. Vincetti, and M. Zoboli, “Holey fiber analysis through the finite-element method,” *IEEE Photonics Technol. Lett.* 14, 1530–1532 (2002).
[45] G. P. Agrawal, Nonlinear Fiber Optics. In: P. L. Christiansen, M. P. Sørensen, and A. C. Scott (eds) *Nonlinear Science at the Dawn of the 21st Century*. Lecture Notes in Physics, vol 542. Springer, Berlin, Heidelberg (2000). Available from: https://doi.org/10.1007/3-540-46629-0_9.
[46] W. J. Tomlinson, R. H. Stolen, and C. V Shank, “Compression of optical pulses chirped by self-phase modulation in fibers,” *J. Opt. Soc. Am. B* 1, 139–149 (1984).
[47] F. X. Kärtner, N. Matuschek, T. Schibli, *et al.*, “Design and fabrication of double-chirped mirrors,” *Opt. Lett.* 22, 831–833 (1997).

[48] I. Cristiani, R. Tediosi, L. Tartara, and V. Degiorgio, "Dispersive wave generation by solitons in microstructured optical fibers," *Opt. Express* 12, 124–135 (2004).
[49] F. Belli, A. Abdolvand, W. Chang, J. C. Travers, and P. S. J. Russell, "Vacuum-ultraviolet to infrared supercontinuum in hydrogen-filled photonic crystal fiber," *Optica* 2, 292–300 (2015).
[50] U. Elu, L. Maidment, L. Vamos, *et al.*, "Seven-octave high-brightness and carrier-envelope-phase-stable light source," *Nat. Photonics* https://doi.org/10.1038/s41566-020-00735-1 (2020).
[51] A. A. Voronin and A. M. Zheltikov, "Sub-half-cycle field transients from shock-wave-assisted soliton self-compression," *Sci. Rep.* 10, 12253 (2020).
[52] S. L. Cousin, N. Di Palo, B. Buades, *et al.*, "Attosecond streaking in the water window: A new regime of attosecond pulse characterization," *Phys. Rev. X* 7, 041030 (2017).
[53] F. Köttig, F. Tani, C. M. Biersach, J. C. Travers, and P. S. J. Russell, "Generation of microjoule pulses in the deep ultraviolet at megahertz repetition rates," *Optica* 4, 1272–1276 (2017).
[54] M. Quick, A. L. Dobryakov, M. Gerecke, *et al.*, "Photoisomerization dynamics of trans–trans, cis–trans, and cis–cis diphenylbutadiene from broadband transient absorption spectroscopy and calculations," *J. Phys. Chem. B* 118, 8756–8771 (2014).
[55] F. Ferreira, A. J. Chaves, N. M. R. Peres, and R. M. Ribeiro, "Excitons in hexagonal boron nitride single-layer: A new platform for polaritonics in the ultraviolet," *J. Opt. Soc. Am. B* 36, 674–683 (2019).
[56] V. Petrov, "Frequency down-conversion of solid-state laser sources to the mid-infrared spectral range using non-oxide nonlinear crystals," *Prog. Quantum Electron.* 42, 1–106 (2015).
[57] V. V. Badikov, D. V. Badikov, V. B. Laptev, *et al.*, "Crystal growth and characterization of new quaternary chalcogenide nonlinear crystals for the mid-IR: $BaGa_2GeS_6$ and $BaGa_2GeSe_6$," *Opt. Mater. Express* 6, 2933–2938 (2016).
[58] U. Elu, L. Maidment, L. Vamos, *et al.*, "Few-cycle mid-infrared pulses from $BaGa_2GeSe_6$," *Opt. Lett.* 45, 3813–3815 (2020).
[59] D. M. B. Lesko, H. Timmers, S. Xing, A. Kowligy, A. J. Lind, and S. A. Diddams, "A six-octave optical frequency comb from a scalable few-cycle erbium fiber laser," *Nat. Photonics* https://doi.org/10.1038/s41566-021-00778-y (2021).
[60] P. G. Schunemann, K. T. Zawilski, L. A. Pomeranz, D. J. Creeden, and P. A. Budnui, "Advances in nonlinear optical crystals for mid-infrared coherent sources," *J. Opt. Soc. Am. B* 33, 36–43 (2016).
[61] U. Elu, T. Steinle, D. Sánchez, *et al.*, "Table-top high-energy 7 μm OPCPA and 260 mJ Ho:YLF pump laser," *Opt. Lett.* 44, 3194–3197 (2019).

Chapter 5
Interaction of optical beams in nonlinear media

Shermineh R. Fairchild[1,2], Danielle Reyes[1], Jessica Peña[1] and Martin Richardson[1]

5.1 Introduction

The theoretical notion of laser-induced waveguides was introduced as early as 1969 [1] based on the concept of creating a channel with an index of refraction gradient to decrease spreading of a laser beam propagating along the channel. The early studies were based on using a screen or plate to block the center of the beam, or a "BAGEL" beam [2] to create a lower, on-axis intensity and consequently a refractive index variation.

Using plasma waveguides for transferring electromagnetic waves was investigated by Shen [3] in 1991. It was shown theoretically that a vacuum core with plasma cladding induced by a hollow laser beam can behave similar to a dielectric waveguide. Gurevich *et al.* [4] explored the possibility of steering microwave signal by forming artificially ionized layers in the atmosphere.

Filaments can be modeled as conductive channels due to their induced plasma that can persist for as long as a few kilometers [5,6]. Using this ionized channel in air to trigger or guide electrical discharges and possibly lightning was introduced very early on after its discovery [7–9]. Free electron concentration in the filament-induced plasma lowers the local index of refraction, creating an index gradient with the non-ionized air surrounding the filament. Radar or microwave radiation can be guided through the air core in the center of a circular array of filaments serving as the waveguides cladding [10].

Even though the first experimental demonstration of a filament waveguide confining a 10-GHz signal was only over 16 cm and 10 ns [11], it made the previous theoretical models [12] a reality. It has been shown that the efficiency of these waveguides is dependent on the size of the structure, number and distance between the individual channels [10,11]. The thickness of the walls also should be more than the characteristic penetration depth of the guided wave. These are all parameters that can widely vary if the filaments are not generated and positioned intentionally.

[1]Laser Plasma Laboratory and Center for Directed Energy, College of Optics and Photonics, University of Central Florida, Orlando, FL, USA
[2]NanoSpective Inc., Orlando, FL, USA

Different configurations of filaments have been shown to guide long wavelength electromagnetic waves in radio, microwave [13–16], infrared and visible spectral range [17]. Many of the earlier studies were performed in multifilamentation regime, a random formation of laser filaments leading to disordered filaments if their interactions are left uncontrolled. This makes it nearly impossible to further optimize the waveguide parameters for high efficiency transferring the radar and microwave signals.

Successful generation of large, ordered filament arrays with high stability and efficiency requires accurate phase and intensity manipulation and controlling the interactions between adjacent filament plasma channels [18–20]. At first glance, using these "engineered arrays" seems like a dream come true for free-space optical communications. However, the propagation of these arrays and their capability of guiding the coupled signal present many challenges such as optimizing the size of the structure, number of filaments, their relative distance, limited conductivity and temporal existence of the waveguide that are determined by the clamped electron density and lifetime of the laser-induced plasma channel. Another major challenge is controlling the interaction between neighboring filaments and its effects on the characteristics and integrity of the waveguide along propagation which will be discussed in this chapter.

5.2 Power dependence of the interaction

Two parallel optical beams in a Kerr medium can repel or attract each other depending on their separation, initial power and relative phase [19–25]. Interaction between parallel beams and the importance of their relative phase and separation has been studied theoretically in extensive details.

Bergé *et al.* introduced a critical separation value, δ_c, below which in-phase beams each carrying $P_{cr}/4 < P < P_{cr}$ fuse together [20,21]. They showed that the interaction between two beams in a Kerr media strongly depends on their powers. Two beams each carrying less than quarter of the critical power ($P_{cr}/4$) scatter and diffuse along their propagation path. If the beams each have enough power to self-focus, they create a filament individually without any cross talk, unless the distance between them is less than $\delta_N = 2\rho$ (ρ being their "transverse width") where the two filaments can merge into a central peak intensity.

Beams with π phase difference have shown to be repulsive and can be used as a mediator for the amalgamation of two in-phase beams separated by larger than δ_c, by pushing them toward each other [20].

Similar behavior, attraction and fusion, repulsion, annihilation, power exchange and spiraling were observed in spatial solitons [26–32]. Ishaaya *et al.* [22] studied these interactions for two beams near P_{cr}. Their theoretical model, integration of the (2+1)D nonlinear Schrödinger equation, showed that for a fixed total power of $2.5P_{cr}$, when the beams are largely separated, their collapse dynamics are independent of each other for both in- and out-of-phase beams (Figure 5.1). At a critical distance of $\Delta_0 \approx 1.9$, (normalized initial separation), the in-phase beams fuse with a 40% increase in the collapsing distance, which they associated with "the phase transition between independent collapse and fusion."

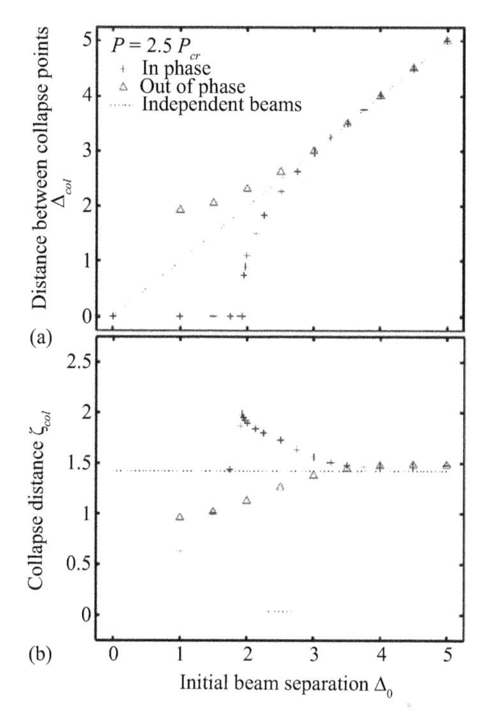

Figure 5.1 Numerical calculation of the (a) distance between collapse points and (b) collapse distance, for in-phase and out of phase two Gaussian beams in a Kerr medium versus their initial separation. Adapted with permission from [22] © The Optical Society

Out-of-phase beams responded opposite and started repelling each other at a smaller distance and the collapse distance was decreased. They also showed that for in-phase beams, as the total power in the beams increases, they will merge at smaller separations. Their experimental results in BK7 confirmed that the collapse pattern of the two beams is heavily dependent on their relative phase where they are at the optimum spatial separation (Figure 5.2). Attraction between two beams started at around 1.75 ω_0 resulting in a total power of $4P_{cr}$. Repulsion between the two beams kept the separation between the two out-of-phase beams above 1.8 ω_0.

The dependence of the interaction between two beams on power and separation was also the basis of another study, more focused on combining beams below P_{cr}. The dynamics between 2-fs beams propagating collinearly as a function of their initial separation was investigated (Figure 5.3) [19]. Their experimental and theoretical results were in agreement with previous studies, showing attraction and fusion of two in-phase beams, each carrying less than P_{cr}, when they got closer than ~330 μm (Figure 5.3). Even though each beam did not have enough power to self-focus individually, filamentation and its by-products such as supercontinuum generation were observed. The significance of this study was to be able to start with

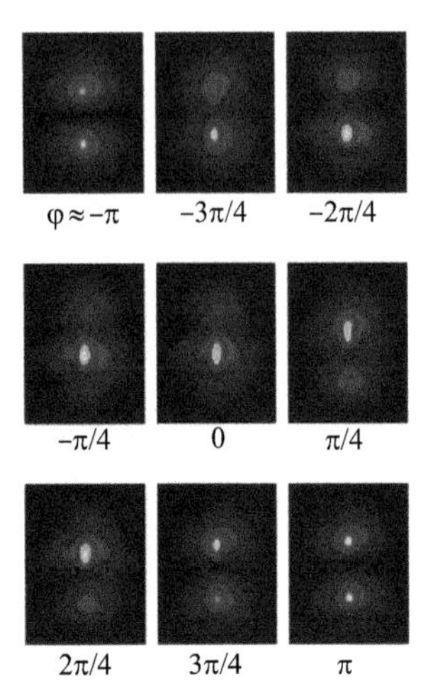

Figure 5.2 Collapse patterns of two beams measured in BK7 with a fixed initial separation, 1.4 ω_0, as their relative phase was varied. Adapted with permission from [22] © The Optical Society

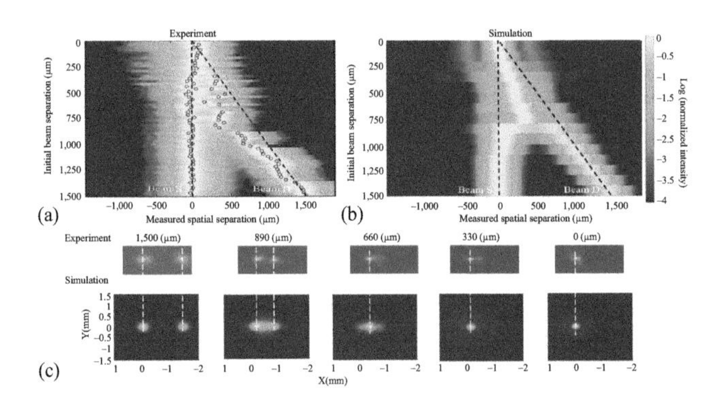

Figure 5.3 (a) Experimental and (b) simulated normalized intensity profile of two beams propagating collinearly. Separation between the two beams was changed before their propagation (Black dashed lines) and measured at the geometrical focus (z = 5 m). Red circles show the local intensity maxima for the stationary beam (Beam S) and dynamic beam (Beam D). (c) Experimental (top row) and simulated (bottom row) beam profiles for 1,500, 890, 660, 330 and 0 μm initial separation. White dashed lines indicate the position of the local intensity maxima in the simulation results [19]

"subcritical" beams for advantages such as avoiding damages by managing the energy deposition on optics and create a filament at a predetermined location by controlling the phase, in this case spatial separation, between the beams. Rostami Fairchild *et al.* showed that the starting point and length of the filament can be

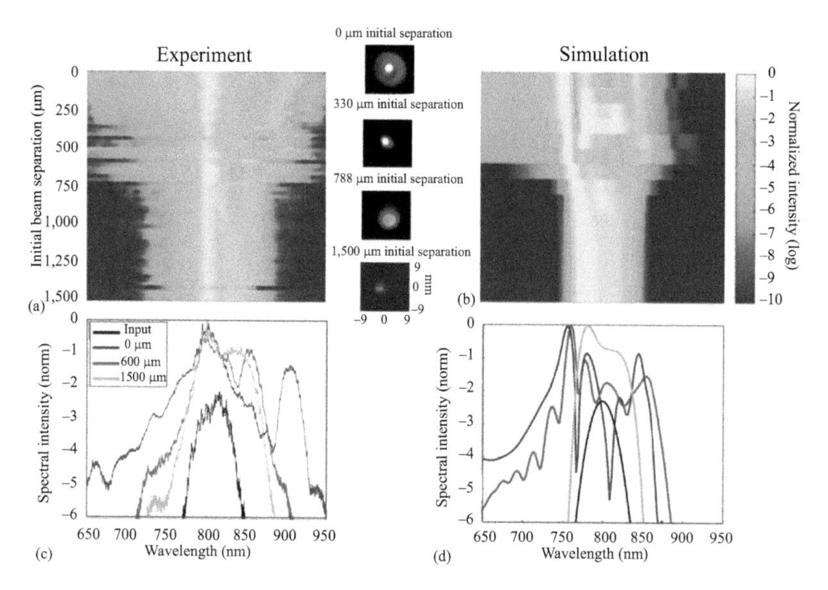

Figure 5.4 (a) Measured and (b) separation spectral intensity of combined beams as a function of initial beam. Pictures of the measured conical emission at selected initial beam separations are shown in the middle. (c) Measured and (d) calculated spectrum for selected initial beam separations compared to the spectrum of the individual subcritical beams. Adapted with permission from [19] © The Optical Society

controlled with this method. A single filament created with nonlinear combining of two beams was observed to be longer and more stable than a single filament created from a single beam carrying the same power as the combination of the two sub-critical beams [19]. This was also shown by numerical simulations performed by Xi *et al.*, on the stability of a channel formed by a single-light bullet with more energy compared to one formed by the fusion of two light bullets [25]. Spectral broadening was also shown to be controllable by varying the distance between the beams (Figure 5.4).

5.3 Interacting filaments

Filamentation is a complicated nonlinear process that puts additional stress on the importance of having control over these interactions. The interaction between in-phase filaments can present its effects through a change in the local index of refraction, affecting Kerr, plasma and many other nonlinear aspects of the filamentation process. The cross talk between parallel and crossed filaments has been used to control different phenomena induced by the filamentation process. The relative position of the two filaments was realized as a mean to control the polarization of a new THz radiation, different from "transition-Cherenkov" based on "bifilamentation" of 2 fs pulses spatiotemporally interacting in air. Liu *et al.* showed that this linearly polarized THz signal radiated in a small angle and was one order of magnitude more intense than their THz signal from the plasma filament [33]. They associated this enhancement to the similarity of the geometry of 2 fs pulses with a relative temporal delay and spatial separation, ~100 μm between their centers, to transmission lines. This structure supports the longitudinal plasma wave and an electromagnetic TM mode.

Cai *et al.* [23] explored the origin of interactions between parallel filaments to distinguish between Kerr, plasma and molecular-alignment-induced effects which can modulate the local index of refraction. Alignment of the molecules can also affect the plasma density since the ionization cross section varies with their relative orientation with respect to the polarization of the laser pulse. They investigated these effects in a pump–probe scheme by comparing the displacements of the two, when they originated at various delays and different initial separations. Figure 5.5 shows that for −200 fs (delay A), when the probe was ahead of the pump pulse, two filaments were separated by the same initial amount (70 μm for Figure 5.5(a) and 140 μm for (b)) indicating no filament cross talk. This is changed around delay D (pump filament ahead of the probe filament) where the filaments were repulsed to 90 and 145 μm. As the filaments got closer to each other, the probe filament was affected more intensely due to the higher intensity created by the pump, resulting in a larger "plasma-induced repulsion" compared to the same separation in negative delays. Cross-phase modulation was observed for delays within the pulse duration (−35 to 35 fs) and seen clearly around zero delay, resulted in the fusion of the filaments when separated by ~70 μm (Figure 5.5(a)). This Kerr-induced fusion was

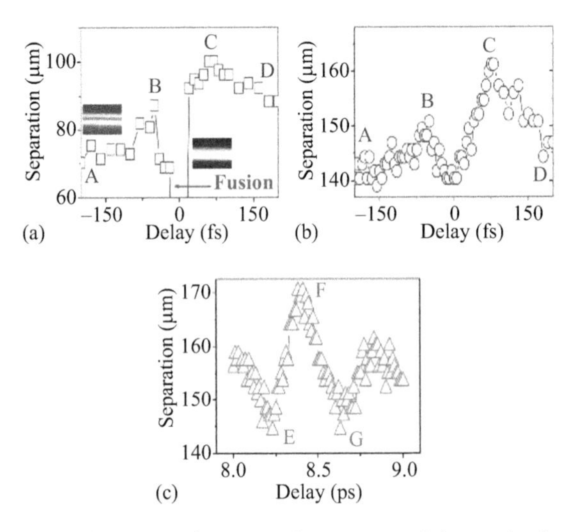

Figure 5.5 Measured distance between the pump and the probe for various delays for initial separations of (a) 70 μm and (b) 140 μm, and (c) larger delays. Adapted with permission from [23] © The Optical Society

not observed when the filaments were separated by 140 μm as shown in Figure 5.5 (b). The counterbalance between attraction and repulsion caused by Kerr and plasma effects when the filaments are temporally overlapped propagating at larger spatial separations keeps them in place. At delays B and C, the two filaments with negligible temporal overlap were still repulsed which Cai *et al.* claimed to be due to molecular alignment, hence the delayed response. Comparing these interactions induced by molecular alignment at two different separations (delays B and C in Figure 5.5(a) and (b)) confirmed that they still are effective at larger separations compared to plasma- and Kerr-induced interactions and were also seen periodically around revivals (Figure 5.5(c)) [23].

This attraction, repulsion and fusion were also observed in interaction between a fundamental wave filament and its second harmonic with relative parallel and perpendicular polarizations (Figure 5.6). Wu *et al.* observed fusion between the two parallel filaments with different wavelengths within a narrower time window around zero delay for orthogonal polarizations due to the difference in XPM-induced Kerr $n_{2,\parallel}^{XPM} = 3n_{2,\perp}^{XPM}$ [34].

In multifilamentation regime [35], it has also been shown that filaments can break up and merge along propagation depending on their separation. Measuring the fluorescence backscattered signal has shown an enhancement for closer "initial perturbations" [36] and smaller original beam size [37] leading to the generation of more filaments and a longer and more stable plasma channel when it comes to interaction and competition of multifilaments.

Interaction of crossed filaments has certainly received a lot of interests as well. Xi *et al.* showed that two filaments crossing at small angles (~0.01°) interact like parallel filaments, while those with larger crossing angles (~0.1°) go through each

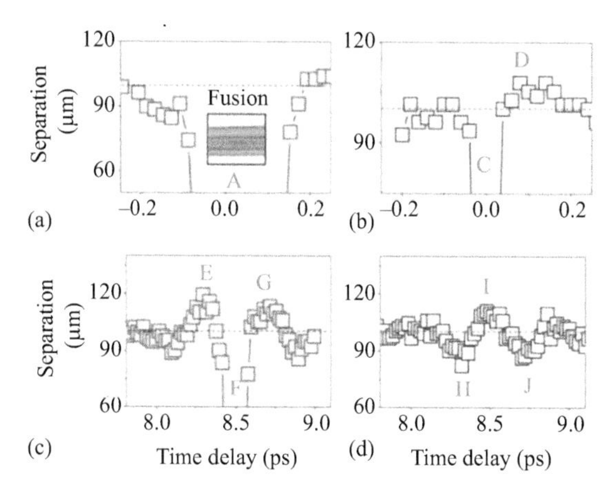

Figure 5.6 Measured separation of the second harmonic and the fundamental filaments for different time delays. Two beams have parallel polarizations in (a) and (c) and perpendicular in (b) and (d). Adapted with permission from [34] © The Optical Society

other and disperse [25]. They also observed a spiral propagation, similar to solitons [26], for two in-phase filaments in different planes [25]. This was also experimentally observed by Shim *et al.* [24]; and in helical plasma channels investigated by Barbieri *et al.* [38]. Bernstein *et al.* [39] concluded that the nonlinear impulsive Raman is responsible for a 7% energy exchange between two crossed (~0.6°) filaments. This amplification of one beam at the expense of the other can be controlled by the delay between the two in the tens of femtoseconds range. They performed this experiment for low intensity beams as well as filaments by measuring the percent of energy gained by beam 1 from beam 2, two-beam coupling (TBC signal), as a function of the delay between beams 1 and 2 when they start with nearly the same amount of energy (Figure 5.7). Comparing Figure 5.7(a) and (c) shows that plasma generation is an insignificant coupling mechanism at presented timescales since there was no plasma present in the case of two low-intensity crossed beams (Figure 5.7(a)).

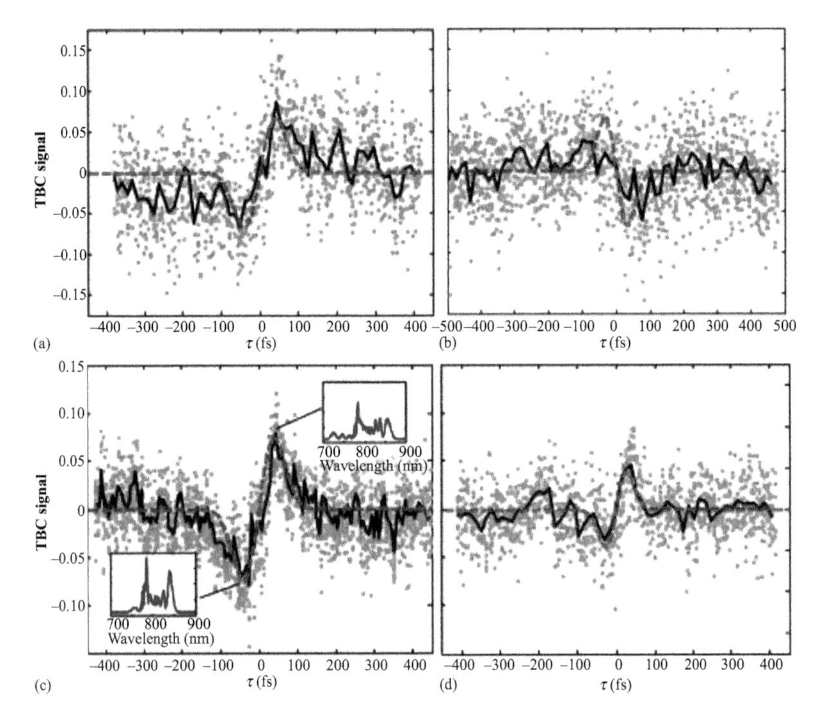

Figure 5.7 TBC signal for different time delays between two crossed beams for (a) low intensity, non-filamenting, beams positively chirped to 270 fs, (b) low intensity beams negatively chirped to 179 fs, (c) filaments positively chirped to 270 fs and (d) filaments positively chirped to 90 fs. Each black line is the average of 10–20 laser shots and the red dashed line is a calculated fit. Insets in (c) show the spatially averaged spectra. Adapted with permission from [39] © The Optical Society

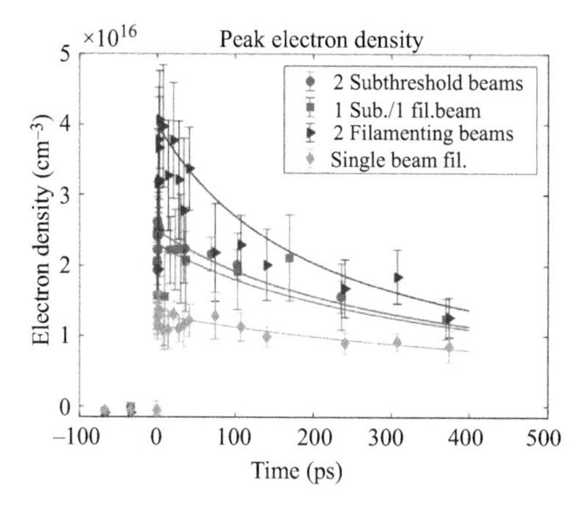

Figure 5.8 The on-axis electron density for two colinearly propagating beams with different power distributions. Two beams are positioned at a fixed separation of 180 μm to ensure interaction

Aside from the importance of keeping the integrity of the structure of the arrays when using them as a waveguide for long distances, conductivity and temporal life of plasma channels are also of concern.

Electron distribution along the plasma channel changes in both space and time, which will directly affect the spectral bandwidth supported by the plasma waveguide and the guiding efficiency. Plasma frequency $\omega_{pl} = \sqrt{N_e e^2/m_e \varepsilon_0}$ (N_e being the electron density, e and m_e the electric charge and mass, respectively, and ε_0 the free space permittivity) determines the upper limit of the frequency that can be transmitted through the waveguides, which is dependent on the electron density. Electron density in a filament plasma is "clamped" at around $N_e = 10^{15}$–10^{16} cm^{-3}, limiting the conductivity, spectral range and efficiency of the waveguides. The efficiency of confinement of the microwave radiation within these plasma structures primarily depends on the number and arrangement of the filaments, the filament diameters, in addition to the optimal separation between the conductive channels (electrical width) depending on the effective electron density.

There is a trade-off between having the plasma strings too close or too far apart from each other, which needs to be carefully considered when designing these structures. These interactions can be balanced out with optimized spatial separation and relative phase for stability of the waveguide [40], considering the fill fraction, for long-range guiding of the desired wavelength. The synergy created in this case can result in an enhanced conductivity of the structure, which will have a great impact on the guiding efficiency of the plasma arrays.

Reyes *et al.* have investigated the spatiotemporal dynamics of the plasma affected by the interaction between two colinearly propagated beams in different power distributions between the two [41]. As can be seen in Figure 5.8, introduction of a subcritical beam adjacent to a filament doubles the peak electron density without breaking into multiple filaments. The breakup would occur as a result of increasing the power of a single beam beyond its critical power. When the two beams carried enough powers to individually filament, they fused and increased the electron density by approximately 200% compared to a single-beam filament. These experimental data were also in agreement with the theoretical air chemistry model as well as NLSE. This study presented a significant increase in the plasma survival time, from 0.98 ns for a single-beam filament to 1.565 ns for two filaments positioned 180 μm from each other. Short plasma lifetime is another significant challenge when using filaments as a waveguide. Merging of the filaments did not show any change in the filament core diameter or intensity but an increase in the fluence. This can be an indication of the change in the pulse duration caused by the merging of the filaments [41].

5.4 Summary

Filament applications strongly rely on the spatiotemporal progression of filaments in the propagation medium. This dynamics in high-power regimes, often used in their most in-field applications, is affected by the interaction between neighboring plasma channels as multifilamentation is inevitable. Understanding these interactions, their onset and characteristics can present us with the ability to control and ultimately taking advantage of them.

Energy exchange, fusion and repulsion of two beams have been shown for different power regimes, even below critical power for filamentation, presenting a direct dependence to their relative phase. The origin of these interactions is linked to the modulation of the local index of refraction that can be from Kerr, plasma, molecular alignment or impulsive Raman depending on the timescale of interest.

Some applications of the cross talk between optical beams are the creation of a more “efficient” single filament from merging two subcritical beams [19], spiral beam propagation for two in-phase filaments in different planes [24,25], controlling the conical emission by the interaction between two “filament forming beams” [39], and extraordinary electron density created by two filaments intentionally positioned in space and time [41].

Acknowledgments

The authors acknowledge fruitful discussions and collaborations with Jean-Claude Diels, Matthieu Baudelet, Natalia Litchinitser, Wiktor Walasik, Daniel Kepler and support of Army Research Office W911NF12R001202, W911NF110297, W911NF2110274), the HEL-JTO (MRI FA05501110001) and the state of Florida.

References

[1] G. A. Askar’yan, “Waveguide properties of a tubular light beam,” *Soviet Physics JETP*, vol. 28, no. 4, pp. 732–733, 1969.

[2] L. A. Newman and D. C. Smith, "Laser-induced waveguide propagation concept," *Applied Physics Letters*, vol. 38, no. 8, p. 590, 1981.
[3] H.-M. Shen, "Plasma waveguide: a concept to transfer electromagnetic energy in space," *Journal of Applied Physics*, vol. 69, no. 10, p. 6827, 1991.
[4] A. V. Gurevich, N. D. Borisov and G. M. Milikh, "Physics of microwave discharges: artificially ionized regions in the atmosphere," Amsterdam: Gordon and Breach, 1997.
[5] M. Rodriguez, R. Bourayou, G. Méjean, *et al.*, "Kilometer-range nonlinear propagation of femtosecond laser pulses," *Physical Review E*, vol. 69, p. 036607, 2004.
[6] D. Thul, R. Bernath, N. Bodnar, *et al.*, "1 kilometer atmospheric propagation studies of a 5 TW ultrashort pulse laser," in *Imaging and Applied Optics*, 2019, (COSI, IS, MATH, pcAOP), OSA Technical Digest (Optical Society of America, 2019), paper JW2A.37.
[7] H. Pépin, D. Comtois, F. Vidal, *et al.*, "Triggering and guiding high-voltage large-scale leader discharges with sub-joule ultrashort laser pulses," *Physics of Plasmas*, vol. 8, no. 5, p. 2532, 2001.
[8] M. Rodriguez, R. Sauerbrey, H. Wille, *et al.* "Triggering and guiding megavolt discharges by use of laser-induced ionized filaments," *Optics Letters*, vol. 27, no. 9, p. 772, 2002.
[9] B. La Fontaine, D. Comtois, C.-Y. Chien, *et al.*, "Guiding large-scale spark discharges with ultrashort pulse laser filaments," *Journal of Applied Physics*, vol. 88, p 610, 2000.
[10] R. R. Musin, M. N. Shneider, A. M. Zheltikov and R. B. Miles, "Guiding radar signals by arrays of laser-induced filaments: finite-difference analysis," *Applied Optics*, vol. 46, no. 23, p. 5593, 2007.
[11] M. Châteauneuf, S. Payeur, J. Dubois and J.-C. Kieffer, "Microwave guiding in air by a cylindrical filament array waveguide," *Applied Physics Letters*, vol. 92, p. 091104, 9, 2008.
[12] A. E. Dormidonov, V. V. Valuev, V. L. Dmitriev, S. A. Shlenov and V. P. Kandidov, "Laser filament induced microwave waveguide in air," in *SPIE: International Conference on Lasers, Applications, and Technologies*, Minsk, 2007.
[13] Y. Ren, M. Alshershby, J. Qin, Z. Hao and J. Lin, "Microwave guiding in air along single femtosecond laser filament," *Journal of Applied Physics*, vol. 113, p. 094904, 9, 2013.
[14] M. Alshershby, Z. Hao and J. Lin, "Guiding microwave radiation using laser-induced filaments: the hollow conducting waveguide concept," *Journal of Physics D: Applied Physics*, vol. 45, no. 26, p. 265401, 2012.
[15] Y. Ren, M. Alshershby, Z. Hao, Z. Zhao and J. Lin, "Microwave guiding along double femtosecond filaments in air," *Physical Review E*, vol. 88, p. 013104, 1, 2013.
[16] Z. A. Kudyshev, M. C. Richardson and N. M. Litchinitser, "Virtual hyperbolic metamaterials for manipulating radar signals in air," *Nature Communications*, vol. 4, p. 2557, 2013.

[17] M. Alshershby, Z. Hao, A. Camino and J. Lin, "Modeling a femtosecond filament array waveguide for guiding pulsed infrared laser radiation," *Optics Communications*, vol. 296, p. 87, 2013.

[18] T. D. Grow and A. L. Gaeta, "Dependence of multiple filamentation on beam ellipticity," *Optics Express*, vol. 13, 12, p. 4594, 2005.

[19] S. Rostami Fairchild, W. Walasik, D. Kepler, M. Baudelet, N. M. Litchinitser and M. Richardson, "Free-Space Nonlinear Beam Combining for High Intensity Projection," *Scientific Reports*, vol. 7, p. 10147, 2017.

[20] L. Bergé, R. M. Schmidt, J. Juul Rasmussen, P. L. Christiansen and K. Ø. Rasmussen, "Amalgamation of interacting light beamlets in Kerr-type media," *Journal of the Optical Society of America B*, vol. 14, no. 10, p. 2550, 1997.

[21] L. Bergé, "Coalescence and instability of copropagating nonlinear waves," *Physical Review E*, vol. 58, no. 5, p. 6606, 1998.

[22] A. Ishaaya, T. D. Grow, S. Ghosh, L. T. Vuong and A. L. Gaeta, "Self-focusing dynamics of coupled optical beams," *Physical Review A*, vol. 75, no. 2, p. 023813, 2007.

[23] H. Cai, J. Wu, P. Lu, X. Bai, L. Ding and H. Zeng, "Attraction and repulsion of parallel femtosecond filaments in air," *Physical Review A*, vol. 80, no. 5, p. 051802, 2009.

[24] B. Shim, S. E. Schrauth, C. J. Hensley, *et al.*, "Controlled interactions of femtosecond light filaments in air," *Physical Review A*, vol. 81, no. 6, p. 061803, 2010.

[25] T. T. Xi, X. Lu and J. Zhang, "Interaction of light filaments generated by femtosecond laser pulses in air," *Physical Review Letters*, vol. 96, no. 2, p. 025003, 2006.

[26] G. I. Stegeman and M. Segev, "Optical spatial solitons and their interactions: universality and diversity," *Science*, vol. 286, no. 5444, p. 1518, 1999.

[27] J. Meier, S. I. George, Y. Silberberg, R. Morandotti and J. S. Aitchison, "Nonlinear optical beam interactions in waveguide arrays," *Physical Review Letters*, vol. 93, no. 9, p. 093903, 2004.

[28] S. Gatz and J. Herrmann, "Soliton collision and soliton fusion in dispersive materials with a linear and quadratic intensity depending refraction index change," *IEEE Journal of Quantum Electronics*, vol. 28, no. 7, p. 1732, 1992.

[29] V. Tikhonenko, J. Christou and B. Luther-Davies, "Three dimensional bright spatial soliton collision and fusion in a saturable nonlinear medium," *Physical Review Letters*, vol. 76, no. 15, p. 2698, 1996.

[30] W. Snyder and A. P. Sheppard, "Collisions, steering, and guidance with spatial solitons," *Optics Letters*, vol. 18, no. 7, p. 482, 1993.

[31] W. Królikowski and S. A. Holmstrom, "Fusion and birth of spatial solitons upon collision," *Optics Letters*, vol. 22, no. 6, p. 369, 1997.

[32] A. Stepken, F. Kaiser, M. R. Belić and W. Królikowski, "Interaction of incoherent two-dimensional photorefractive solitons," *Physical Review E*, vol. 58, no. 4, p. R4112, 1998.

[33] Y. Liu, A. Houard, B. Prade, S. Akturk, A. Mysyrowicz and V. T. Tikhonchuk, "Terahertz radiation source in air based on bifilamentation of femtosecond laser pulses," *Physical Review Letters*, vol. 99, no. 13, p. 135002, 2007.

[34] J. Wu, Y. Tong, X. Yang, H. Cai, P. Lu, H. Pan and H. Zeng, "Interaction of two parallel femtosecond filaments at different wavelengths in air," *Optics Letters*, vol. 34, no. 20, p. 3211, 2009.

[35] S. Tzortzakis, L. Bergé, A. Couairon, M. Franco, B. Prade and A. Mysyrowicz, "Breakup and fusion of self-guided femtosecond light pulses in air," *Physical Review Letters*, vol. 86, no. 24, p. 5470, 2001.

[36] S. A. Hosseini, Q. Luo, B. Ferland, *et al.*, "Competition of multiple filaments during the propagation of intense femtosecond laser pulses," *Physical Review A*, vol. 70, no. 3, p. 033802, 2004.

[37] O. G. Kosareva, N. A. Panov, N. Akozbek, *et al.*, "Controlling a bunch of multiple filaments by means of a beam diameter," *Applied Physics B*, vol. 82, p. 111, 2006.

[38] N. Barbieri, M. Weidman, G. Katona, *et al.*, "Double helical laser beams based on interfering first-order Bessel beams," *Journal of the Optical Society of America A*, vol. 28, no. 7, p. 1462, 2011.

[39] C. Bernstein, M. McCormick, G. M. Dyer, J. C. Sanders and T. Ditmire, "Two-beam coupling between filament-forming beams in air," *Physical Review Letters*, vol. 102, no. 12, p. 123902, 2009.

[40] W. Walasik and N. M. Litchinitser, "Dynamics of large femtosecond filament arrays: possibilities, limitations, and trade-offs," *ACS Photonics*, vol. 3, no. 4, p. 640, 2016.

[41] D. Reyes, J. Peña, W. Walasik, N. Litchinitser, S. Rostami Fairchild and M. Richardson, "Filament conductivity enhancement through nonlinear beam interaction," *Optics Express*, vol. 28, no. 18, p. 26764, 2020.

Chapter 6

Nonlinear processes in coupled light filaments: numerical studies

Alejandro Aceves,[1] Alexander Sukhinin,[2] and Edward Downes[1]

The study of light filamentation and the processes associated with it is currently a very active area of research. The formation of a light filament allows self-guided propagation of high-intensity laser light over long distances. Within a filament, several nonlinear processes can occur, including stimulated Raman scattering (SRS), nonlinear wave mixing and the generation of plasma. To help understand this phenomenon, careful numerical studies capable of dealing with numerous processes at different spatial, temporal and frequency scales are necessary.

In this chapter, we will highlight theoretical and computational approaches to the study of coupled light filaments. The chapter primarily considers two-color filament interactions, and we also touch on the numerical implementation of backscattering of light by stimulated Raman processes. The choices of models and methods aim at closely reflecting recent and proposed experiments.

6.1 Introduction

As seen throughout these chapters, during the propagation of high-intensity laser pulses in nonlinear media, such as air, there are several nonlinear interactions that, when balanced, have a prolonged self-guiding effect for the beam [1,2]. This self-guided propagation over large distances is one of the main characteristics of a light filament.

An emerging process is that of filament interactions. Here, we first present results on the coexistence of two-color filaments. The last decade has seen active research centered around the potential applications of coupled multicolored filaments, mainly focused on the generation of terahertz (THz) [3–5]. Here, we look to numerically study the waveguiding effect of multiple filaments with the idea of increased stable propagation distances. Specifically, a long (nanoseconds)

[1]Department of Mathematics, Southern Methodist University, Dallas, USA
[2]Department of Mathematics and Statistics, University of North Carolina Greensboro, Greensboro, USA

ultraviolet (UV) pulse co-propagating with a much shorter (picoseconds) infrared (IR) pulse interacts with each other due to the four-wave mixing mechanism induced by the Kerr effect. At first, by assuming the long UV pulse to be a continuous wave (CW) that effectively acts as a waveguide for the IR filament, results shed light on whether the presence of the UV serves as a stabilizing mechanism to the well-known spatiotemporal instabilities seen on single intense IR filaments' propagation in air.

Within a light filament, there are many more interesting interactions besides what is mentioned previously, one of which is SRS. There are countless uses of SRS; for example, for decades, SRS has been utilized in tunable laser development and high-energy pulse compression [6,7]. Similarly, in recent years, there have been many biological applications of SRS focused on noninvasive imaging of human tissue relating to the detection and treatment of diseases and tumors [8,9]. The main advantage of SRS microscopy in the medical community is the instantaneous 3D imaging that can be produced to accurately detail internal areas of the body that are not freely accessible. In [10,] the authors describe the successful use of SRS to detect tumor-infiltrated brain tissue, as well as uses during surgeries to reveal tumors that were previously undetectable with existing techniques. At the opposite end of the human torso, research in [11] outlines how SRS is used effectively to monitor the diffusion and concentration of drug treatments for diseases in human nails.

In this chapter, for each process, we briefly present suitable models and turn to numerical studies of such models. For the UV–IR filament interaction, our starting point is the classical two (in space) plus one (in time) dimensional nonlinear Schrödinger equation (NLSE), coupled with the plasma rate equations. We briefly explain the approximations made leading to the model where we present a sample of numerical simulations.

The second part is on SRS studies triggered by counter-propagating filaments, a model based on Maxwell's equations and the density matrix approach. From this model, numerical simulations of SRS within an UV filament in the air (assumed to be nitrogen) are presented. While not comprehensive, assumptions and simplifications made will be explicitly stated, along with full details of the numerical methods and schemes used.

Finally, we should point out that this effort stemmed from collaborations with the experimental group led by Dr. J.-C. Diels at the University of New Mexico.

6.1.1 Model

To model the nonlinear interactions of the two-color filament propagation, the first approximation is to consider the UV pulse as a CW. Given the UV pulse is in the nanosecond range and the IR pulse is on the picosecond scale, this assumption is reasonable.

The idea behind the stabilization of the IR filament's propagation is that the UV pulse will generate an initial plasma that can build up into a plasma channel, causing a defocusing effect, theoretically leading to stable prolonged propagation.

Furthermore, due to the nonlinear Kerr effect, the UV filament creates an effective transverse waveguide to the IR filament, which should act as an additional stabilizing mechanism.

In the absence of plasma or other guiding effects, as well as neglecting temporal dispersion, the propagation of the IR filament is the classical (2+1)D NLSE:

$$i\partial_z E_{IR} + \Delta_\perp E_{IR} + |E_{IR}|^2 E_{IR} = 0 \tag{6.1}$$

where E_{IR} is the amplitude of the laser light and $\Delta_\perp$ is the transverse diffraction of the focused filament, $\Delta_\perp = (\partial^2/\partial x^2) + (\partial^2/\partial y^2)$.

The results shown later represent a somewhat incomplete (by neglecting higher order linear and nonlinear dispersion effects), but still relevant, first improvement of the model that accounts for the presence of the UV filament and the ionization of the medium resulting from the fact that these are intense fields.

This classical model accounts for the self-focusing of high-power laser light, transverse diffraction and temporal (quadratic) dispersion light waves. When considering the interaction between two (IR/UV) filaments, necessary additional terms to be included in the NLSE are as follows:

$$\begin{aligned} i\partial_z E_{IR} + \Delta_\perp E_{IR} - \varepsilon_2 \partial_{tt} E_{IR} + (|E_{IR}|^2 + c_1|E_{uv}|^2)E_{IR} - NE_{IR} &= 0 \\ \partial_t N - \varepsilon_1 |E_{IR}|^{16} &= 0 \end{aligned} \tag{6.2}$$

The first equation in (6.2) is the extension of the NLSE, where we have added normal temporal dispersion and the self-focusing effects of the CW UV pulse, taking the form of a guiding potential. N represents the plasma density that produces a defocusing effect on the IR beam. The second equation is a rate equation to model the changes in the plasma during propagation and is dependent only on the amplitude of the IR pulse. What follows are numerical simulations on (6.2) to study the effect of the UV waveguide on the stability and propagation distance of the IR pulse.

6.1.2 Numerical approach

Equations (6.2) represent an extension of the $(3+1)$D NLSE so the popular spectral technique of the split-step Fourier method (SSFM) is employed for our numerical simulations. A brief description of the method is provided here, as well as numerical considerations for efficient coding.

Like all splitting methods, the SSFM is built on the idea that a complex problem can be "split" into two or more simpler problems, and if a *small* step size h is taken, the simpler problems can be treated separately with only a small numerical error.

In our case, the first equation in (6.2) is split into linear and nonlinear parts:

$$\begin{aligned} \mathscr{L} &= \Delta_\perp - \varepsilon_2 \partial_{tt} \\ \mathscr{N} &= |E_{IR}|^2 + c_1|E_{uv}|^2 - N \end{aligned} \tag{6.3}$$

where $\mathscr{L}$ and $\mathscr{N}$ are the linear and nonlinear operators, respectively. For a constant UV pulse, the nonlinear part will have the analytical solution:

$$i\mathscr{N}(E_{IR}) = E_0 e^{i\left(|E_{IR}|^2 + c_1|E_{uv}|^2 - N\right)h} \tag{6.4}$$

where E_0 is the initial condition $E_{IR}(z=0) = E_0$. The linear operator $\mathscr{L}$ can be transformed to and from the Fourier domain using the Fourier transform (FT) and inverse Fourier transform (IFT). Essentially, this leads to partial derivatives changing to complex frequency multiplication, i.e.,

$$\partial_x \to ik_x, \quad \partial_y \to ik_y, \quad \partial_t \to i\omega.$$

Once transformed to the Fourier domain, the linear part has the exact solution as follows:

$$\widehat{E}_{IR} = \widehat{E}_0 e^{i\left(k_x^2 + k_y^2 - \omega^2\right)h}, \tag{6.5}$$

where $\widehat{E}_0$ and $\widehat{E}_{IR}$ are the FTs of the initial condition E_0 and E_{IR}, respectively. With the use of the forward and backward FTs, we can therefore solve both the linear and nonlinear parts of (6.2).

Several splitting options are available but we choose the second-order accurate method of Strang [12]. This involves solving the linear part, first at a half-step $h/2$, then using this as the initial condition to solve the nonlinear part at a full-step h, and finally solving the linear part again at another half-step $h/2$. This splitting method is also commonly referred to as "symmetric splitting" and will be second-order accurate method if the following criteria are satisfied [13]:

$$h \leq \frac{2[\min(\Delta x, \Delta y, \Delta t)]^2}{\pi^2}, \tag{6.6}$$

where $\Delta x, \Delta y, \Delta t$ is the grid spacing in the given direction.

The loop for the numerical method is summarized later.

($\mathcal{F}$ is the forward FT and $\mathcal{F}^{-1}$ is the inverse/backward FT).

Given initial condition E_0;

1. Take FT of E_0: $\widehat{E}_{old} = \mathcal{F}(E_0)$
2. Compute linear part: $\widehat{E}^* = \widehat{E}_{old} e^{i(k_x^2 + k_y^2 - \omega^2)h/2}$
3. Take IFT of $\widehat{E}^*$: $E^* = \mathcal{F}^{-1}(\widehat{E}^*)$
4. Compute nonlinear part: $E^{nl} = E^* e^{i(|E|^2 + c_1|E_{uv}|^2 - N)h}$
5. Take FT of E^{nl}: $\widehat{E}^{nl} = \mathcal{F}(E^{nl})$
6. Compute linear part: $\widehat{E}_{new} = \widehat{E}^{nl} e^{i(k_x^2 + k_y^2 - \omega^2)h/2}$

The solution has now been evolved from $z \to z + h$. Steps 2–6 can be repeated to evolve the solution to the desired distance.

Our code for this algorithm is written in the C++ language, and the FT and IFT are handled by the *Fastest Fourier Transform in the West* (FFTW) library. This library estimates the full FT using the discrete Fourier transform and its respective inverse. It is worth noting to maximize the efficiency of the FFTW code, we choose grids of size 2^k ($k \in \mathbb{N}$), i.e., $[32 \times 32 \times 32]$, $[64 \times 64 \times 64]$, $[128 \times 128 \times 128]$, etc.

The second equation in (6.2) describes the evolution of the plasma buildup in the temporal direction. This ordinary differential equation (ODE) is approximated by numerical quadrature, in particular by composite Simpson's rule. Rearranging the second equation in (6.2), we can obtain

$$N(x,y,t) = \varepsilon_1 \int_{t_0}^{t} |E_{IR}(x,y,\tau)|^{16} d\tau. \tag{6.7}$$

Given that the values of E_{IR} are known at all the discretization points, it is simple to use our chosen numerical quadrature regime to approximate the integral in (6.7). For each x, y, the integral in the t direction is evaluated by adding successive evaluations of composite Simpson's rule, essentially "marching" through the t direction evaluating this integral as we go. It is worth noting the composite Simpson's rule does require an even number of subintervals, so for all the calculations where the number of subintervals is odd, we use Simpson's rule for all but the last subinterval, on which a lower order trapezoid method is used. Given that this trapezoid method is always at the end of our interval, errors from the lower order do not propagate with the calculations as the calculation immediately following the trapezoid method is always a full Simpson's method evaluation.

At this point, the mathematical methods that are required to study the interactions of multicolored filaments in our model have been covered, and numerical simulations can be produced.

6.1.3 Results

As stated before, our objective is to run simulations that help us determine if indeed the UV filament acts as a stabilizing factor in the propagation of the IR component. The initial condition for the IR pulse is essentially a "light bullet," defined as a 3D Gaussian. The UV pulse is a transverse 2D Gaussian that is constant in the temporal and propagation directions:

$$E_{IR}(x,y,t,z=0) = a_{IR}\exp\left(-\frac{x^2+y^2}{2\omega_0^2}\right)\exp\left(-\frac{t^2}{2\omega_1^2}\right)$$

$$E_{uv}(x,y,t,z=0) = a_{UV}\exp\left(-\frac{x^2+y^2}{2\omega_2^2}\right)$$

The initial plasma has the same 2D Gaussian shape as the UV pulse but can be of vastly different amplitude:

$$N(x,y,t,z=0) = a_{plas}\exp\left(-\frac{x^2+y^2}{2\omega_2^2}\right)$$

In the following simulations, the initial amplitude of the IR pulse is taken as $a_{IR} = 1.75$, and the pulse is symmetric in the transverse and temporal dimensions ($\omega_0 = \omega_1$). The amplitude of the UV pulse and the initial plasma is to be varied to study the effects of the multi-filament interactions and the plasma ionization.

For the center of the light bullet (sliced at $y = 0, x = 0$), Figure 6.1 displays a significant increase in the intensity of the IR pulse at longer propagation distances when the UV filament is present (purple and green). The pulse profile loses a little of this temporal symmetry, possibly due to the plasma effects on the leading and trailing edges, but overall, the pulse still holds its topological shape.

In the transverse direction (sliced at $y = 0, t = 0$), there is a similar result as in the temporal direction with the addition of a slight "*splitting*," seen for the double amplitude UV case. The peak of the pulse separates slightly in the first two plots of Figure 6.2, but the pulse appears to refocus back to its regular shape by the final distance.

Observing the full color map of the twice amplitude UV case, we can see a significant alteration to the shape of the pulse. Figure 6.3 displays a narrowing in the transverse direction in contrast to an elongated shape in the t direction. Again the differences in the trailing and leading edges are evident, while a fairly symmetric transverse shape is maintained.

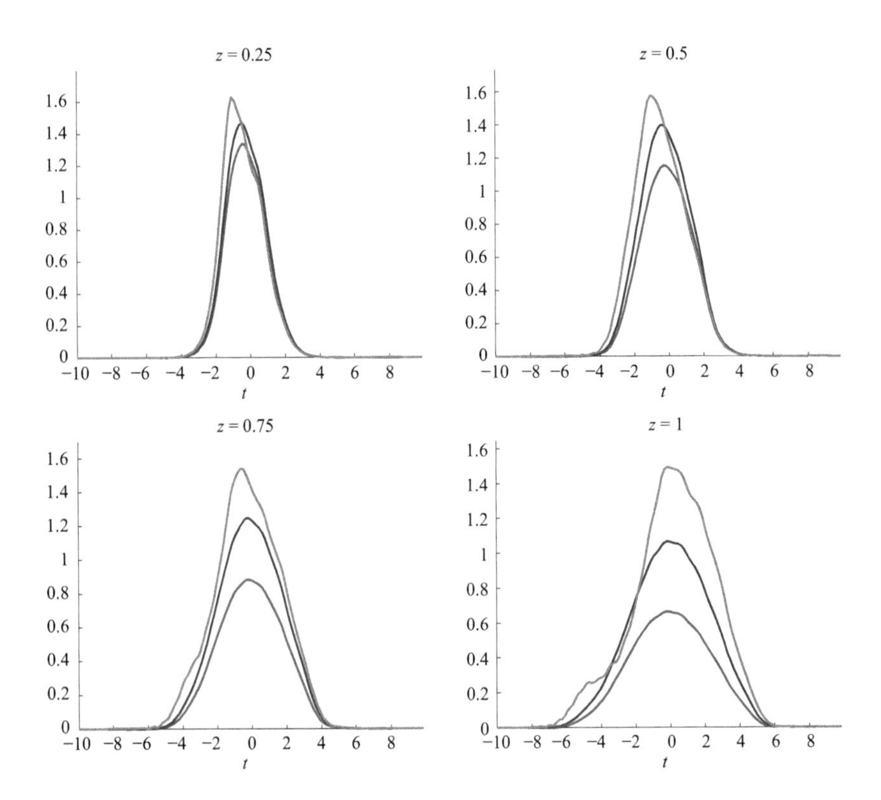

Figure 6.1 Temporal propagation profile (red—IR only, purple—with UV waveguide, green—with double amplitude UV)

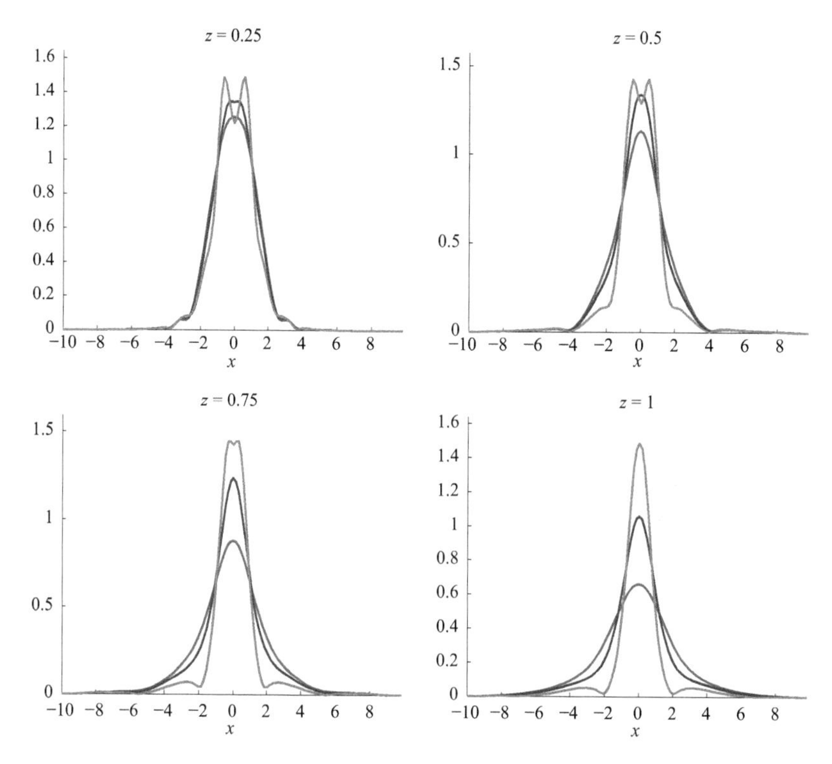

Figure 6.2 Transverse propagation profile (red—IR only, purple—with UV waveguide, green—with double amplitude UV)

In these simulations, there is clear evidence of the UV filament altering the amplitude and profile of the propagating IR pulse. While these are only very early observations, they provide an exciting insight into possibilities of co-propagating multicolored filaments. The next step is to build upon this foundation and apply realistic physical parameters to simulate experimentally achievable filaments.

6.2 SRS process

The second physical process in light filament science briefly described in this chapter considers the modeling of the stimulated Raman process and the numerical methods and schemes employed. Shown here are simulations that closely reflect experimental conditions.

6.2.1 Deriving Maxwell–Bloch model

The derivation of the mathematical model for the energy level diagram shown in Figure 6.4 begins with a trichromatic (three frequency) laser pulse described as

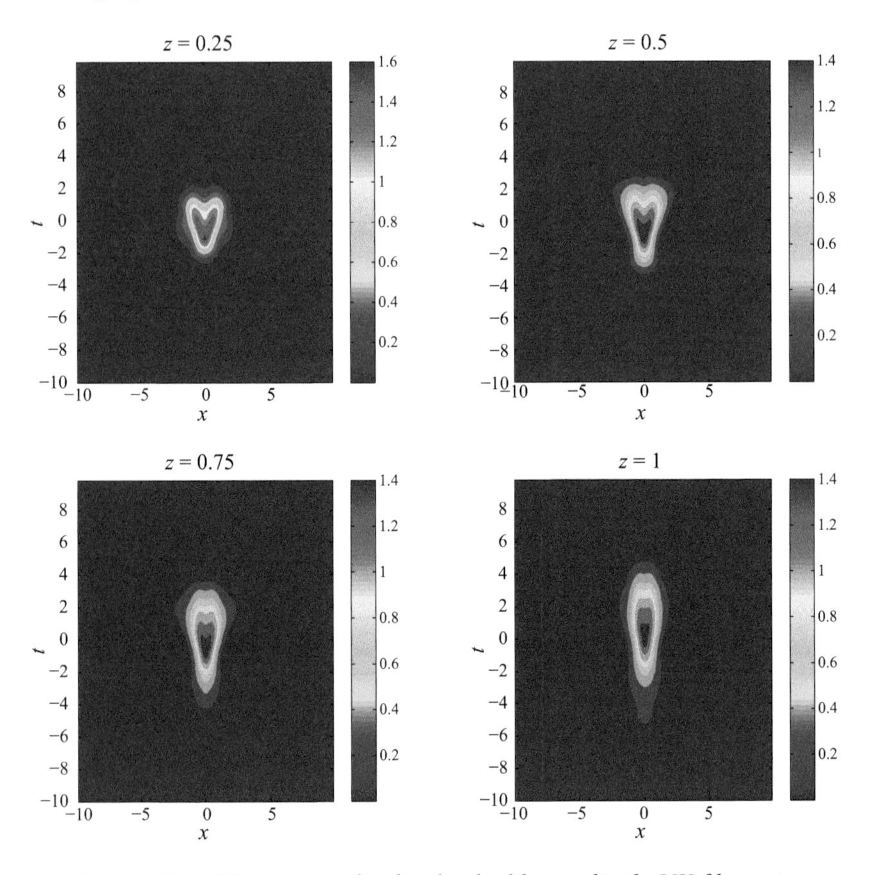

Figure 6.3 2D contour plot for the double amplitude UV filament

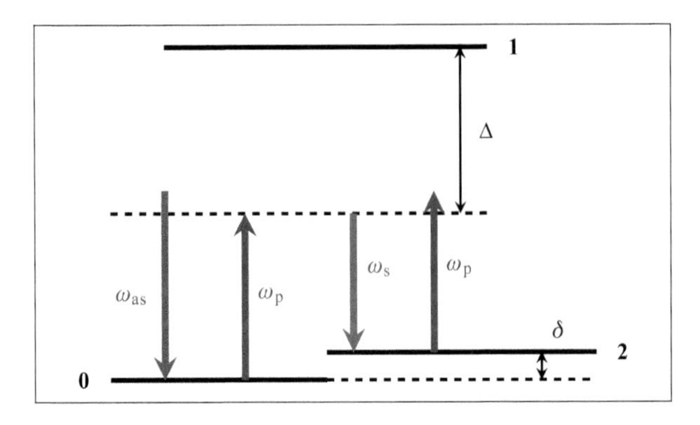

Figure 6.4 Sketch showing the three-level system and three resonant frequencies

follows:

$$E = E_p e^{i\omega_p t} + E_s e^{i\omega_s t} + E_{as} e^{i\omega_{as} t} + c.c., \tag{6.8}$$

referring to the pumping frequency ω_p, the frequency of Stokes emission ω_s, and the frequency of anti-Stokes ω_a.

Assuming the pump frequency is far-off resonance, i.e.,

$$\omega_p \ll \omega_1 - \omega_0,$$

and defining the detunings

$$\begin{aligned} \delta &= \omega_2 - \omega_0 = \omega_p - \omega_s = \omega_{as} - \omega_p, \\ \Delta &= \omega_p - (\omega_1 - \omega_0), \end{aligned} \tag{6.9}$$

and the auxiliary field

$$E_u = E_p + E_s e^{-i\delta t} + E_{as} e^{i\delta t}. \tag{6.10}$$

one can easily verify that $E = E_u e^{i\omega_p t}$, using the detunings in (6.9).

The equations to describe the changes in energy levels are derived from the Hamiltonian dipole interaction:

$$\frac{da_k}{dt} = -i\omega_{0k} a_k + \sum_{j=0}^{2} \frac{i}{2\hbar} p_{kj} a_j E(t), \tag{6.11}$$

where $p_{kj} = p_{jk}$ is the transition strength between two levels "k" and "j." Expanding (6.11) produces the ODEs:

$$\begin{aligned} \frac{da_0}{dt} &= \frac{i}{2\hbar} p_{01} a_1 E, \\ \frac{da_1}{dt} &= -i\omega_{01} a_1 + \frac{i}{2\hbar} p_{10} a_0 E + \frac{i}{2\hbar} p_{11} a_1 E + \frac{i}{2\hbar} p_{12} a_2 E, \\ \frac{da_2}{dt} &= -i\omega_{02} a_2 + \frac{i}{2\hbar} p_{21} a_1 E + \frac{i}{2\hbar} p_{22} a_2 E, \end{aligned}$$

under the assumption that transitions between levels "0" and "2" are forbidden, $p_{20} = p_{02} = 0$, and $\omega_{00} = 0$.

Next, a rotating wave approximation with rotation at pump frequency ω_p is used to replace the "a" eigenvalues with "c"s. The details of the rotating wave approximation are as follows:

$$a_0 = c_0, \; a_1 = c_1 e^{-i\omega_p t}, \; a_2 = c_2 e^{-i(\omega_{as} - \omega_p)t} = c_2 e^{-i\delta t}. \tag{6.12}$$

Substituting into (6.12), the ODEs become

$$\begin{aligned} \frac{dc_0}{dt} &= \kappa_{01} E_u c_1, \\ \frac{dc_1}{dt} &= -i\Delta c_1 + \kappa_{01} E_u^* c_0 + \kappa_{12} E_u^* c_2 e^{-i\delta t}, \\ \frac{dc_2}{dt} &= \kappa_{12} E_u c_1 e^{i\delta t}, \end{aligned} \tag{6.13}$$

Now given that level 1 is far-off resonance, it can be assumed to be in a steady state, known as an adiabatic approximation. Solving $dc_1/dt = 0$ for c_1,

$$c_1 = \frac{1}{i\Delta}\left(\kappa_{10}c_0E_u^* + \kappa_{12}c_2E_u^*e^{-i\delta t}\right)$$

then (6.13) becomes

$$\begin{aligned}\frac{dc_0}{dt} &= \frac{\kappa_{01}E_u}{i\Delta}\left(\kappa_{10}c_0E_u^* + \kappa_{12}c_2E_u^*e^{-i\delta t}\right),\\ \frac{dc_2}{dt} &= \frac{\kappa_{12}E_u}{i\Delta}\left(\kappa_{10}c_0E_u^*e^{i\delta t} + \kappa_{12}c_2E_u^*\right),\end{aligned} \tag{6.14}$$

where we have defined $\kappa_{jk} = p_{jk}/2\hbar$. The three-level system shown in Figure 6.4 has essentially been collapsed down into a two-level system.

Now using the ideas from a density matrix approach, one can define pseudo-vectors forming a Bloch sphere:

$$iQ = 2c_0c_2^*e^{i\delta t}, \qquad W = c_2c_2^* - c_0c_0^*. \tag{6.15}$$

Q represents the transition rate between levels 0 and 2, and W is the population difference in the two levels.

The time derivatives of Q and W can be computed from (6.14) and (6.15):

$$\begin{aligned}\dot{Q} &= -i\left[|E_b|^2 - |E_a|^2 - \delta\right]Q + 2E_aE_bW - \frac{Q}{T_2},\\ \dot{W} &= -\left[E_aE_b^*Q^* + E_a^*E_bQ\right]\end{aligned} \tag{6.16}$$

defining $a = \kappa_{01}E_u/\sqrt{\Delta}$ and $b = \kappa_{12}E_u/\sqrt{\Delta}$, to simplify the equations significantly. It is worth noting that W is a real-valued quantity, which is consistent with the differential equation $\dot{W}$ that is being real-valued too.

From the definitions of Q and W and in the absence of relaxations, the Bloch vectors lie on a sphere, i.e., they satisfy

$$\begin{aligned}|Q|^2 + W^2 &= \mathit{constant},\\ \Rightarrow Q\frac{dQ^*}{dt} + Q^*\frac{dQ}{dt} + 2W\frac{dW}{dt} &= 0.\end{aligned}$$

6.2.2 *Propagation equations*

The equations to describe the traveling wave electric fields for the pump and scattered fields are derived from Maxwell's equations [14,15]:

$$\frac{\partial^2 E}{\partial z^2} + \frac{1}{c^2}\frac{\partial^2 E}{\partial t^2} = \frac{\partial^2 P}{\partial t^2}. \tag{6.17}$$

The polarization is defined as $\partial^2 P/\partial t^2 = -\alpha QE$. Substituting each electric field from (6.8) into (6.17), and under the slowly varying amplitude approximation, second partial derivatives of amplitude are considered negligible.

For the right-hand side of (6.17), only the terms with the same phase as the field on the left-hand side are retained; i.e., for the pump, we only keep terms proportional to $e^{i\omega_p t}$. The pump field is given by

$$\frac{\partial E_p}{\partial z} e^{i\omega_p t} + \frac{1}{c}\frac{\partial E_p}{\partial t} e^{i\omega_p t} = -\alpha_p Q e^{i\delta t}(E_p e^{i\omega_p t} + E_s e^{i\omega_s t} + E_{as} e^{i\omega_{as} t} + c.c.)$$

$$\Rightarrow \frac{\partial E_p}{\partial z} + \frac{1}{c}\frac{\partial E_p}{\partial t} = -\alpha_p (QE_s + Q^* E_{as}) + \text{higher order terms}$$

$$\Rightarrow \frac{\partial E_p}{\partial z} + \frac{1}{c}\frac{\partial E_p}{\partial t} = -\alpha_p (QE_s + Q^* E_{as})$$

After substituting each electric field and following the same procedure shown before, the following propagation equations are obtained:

$$\begin{aligned} \frac{\partial E_p}{\partial z} + \frac{1}{c}\frac{\partial E_p}{\partial t} &= -\alpha_p (QE_s + Q^* E_{as}) \\ \frac{\partial E_s}{\partial z} + \frac{1}{c}\frac{\partial E_s}{\partial t} &= -\alpha_s Q^* E_p \\ \frac{\partial E_{as}}{\partial z} + \frac{1}{c}\frac{\partial E_{as}}{\partial t} &= -\alpha_a Q E_p \end{aligned} \tag{6.18}$$

This ends the derivation of the propagation equations. The full Maxwell–Bloch model consisting of (6.16) and (6.18) is stated here for completeness:

$$\begin{aligned} \dot{Q} &= -i\Big[|E_b|^2 - |E_a|^2 - \delta\Big] Q + 2E_a E_b W - \frac{Q}{T_2}, \\ \dot{W} &= -\big[E_a E_b^* Q^* + E_a^* E_b Q\big], \\ \frac{\partial E_p}{\partial z} + \frac{1}{c}\frac{\partial E_p}{\partial t} &= -\alpha_p (QE_s + Q^* E_{as}) \\ \frac{\partial E_s}{\partial z} + \frac{1}{c}\frac{\partial E_s}{\partial t} &= -\alpha_s Q^* E_p \\ \frac{\partial E_{as}}{\partial z} + \frac{1}{c}\frac{\partial E_{as}}{\partial t} &= -\alpha_a Q E_p. \end{aligned} \tag{6.19}$$

The previous two subsections detail a derivation of a set of ODEs for the Bloch equations and a set of partial differential equations (PDEs) for the propagation of the electric fields. As a whole, (6.16) and (6.18) form a coupled system for the full Maxwell–Bloch model. Some results and details on the numerical approach used now are as follows.

6.2.3 Stimulated Raman scattering processes

Realistic experimental conditions to generate Stokes and anti-Stokes scattering are as follows:

- A sharp IR laser pulse is sent through a relaxed medium. Pulse has a duration of ≈ 60 fs $= 0.06$ ps.

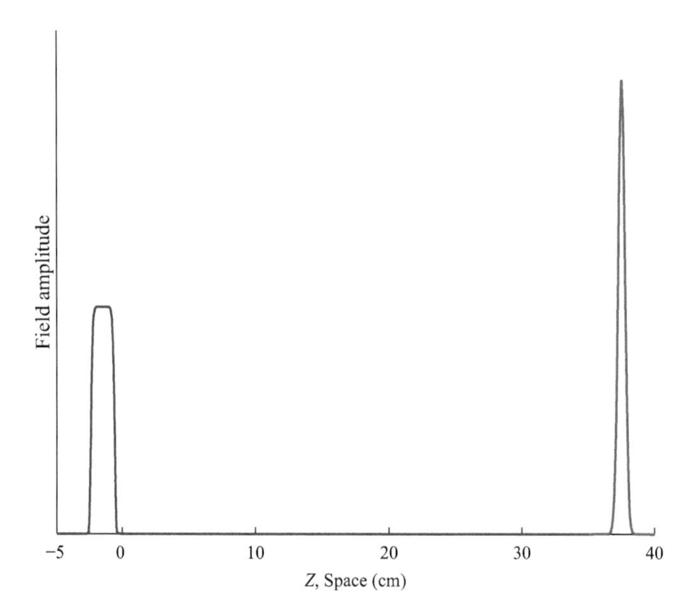

Figure 6.5 Diagram of UV and IR pulses with 1,000 ps *delay (propagation direction is from left to right)*

- This pulse essentially excites the medium and "prepares" it for the scattering to occur.
- Following a 1,000 ps delay, the UV pump pulse passes through the excited medium and the scattering begins.

Figure 6.5 shows a schematic of the two pulses and the prepared medium between the two, with the details for the individual IR stage and the UV stage to be covered next. The IR pulse is assumed to be Gaussian:

$$E_a = E_b = E_0 e^{-(t/\tau_0)^2},$$

where $\tau_0 = 0.0510$ (0.06 ps at full width half maximum). The amplitude of the IR pulse is $0 = \sqrt{15.66\theta}$, where θ ranges from $[0.1, 2\pi]$. We experiment with various θ values to see the effects on the amount of excitation (Q) generated.

Since we are only concerned with the changes in the medium during this stage (Q and W), we only need to evolve (6.16) as the IR pulse passes through. This is just a system of ODEs, so we can use our favorite solver, in this case, an explicit Runge–Kutta method. The changes in the medium (Q and W) during this stage provide the initial conditions for the stimulated scattering, which is the computationally expensive part to solve.

Using the values from the first stage, we numerically integrate the full Maxwell–Bloch system, (6.19). Assuming initially there is zero scattering,

$$E_s(0,z) = E_{as}(0,z) = 0.$$

The UV pump is a square field profile that is approximated by a "super-Gaussian" with a duration of 200 ps and initial amplitude $E_p^2(0, z) = 10^{-3}\ \text{ps}^{-1}$.

Now that the experimental aspects have been covered, the next subsection will discuss the numerical techniques used for the coupled system of Maxwell–Bloch equations.

6.2.4 *Numerical setup*

6.2.4.1 Numerical methods

The two equations in (6.16) represent a system of ODEs that are simple to integrate numerically using an explicit Runge–Kutta method, detailed later in this section. However, (6.18) bring about some numerical challenges that need to be addressed here.

To begin, a uniform spatial discretization is made and second-order finite difference schemes are used to approximate spatial derivatives. At this point consideration of suitable boundary conditions needs to be discussed. With propagating waves, we must truncate the infinite spatial domain with some form of absorbing boundary condition, to reduce reflections off the boundary back into the computational domain. As there is only one spatial dimension, a first-order sponge layer is sufficient to achieve this, if it is carefully applied. Essentially the electric fields are modified to include a damping term, $\eta(z)$, that is increased gradually throughout the sponge layer. It is worth noting the artificial sponge layer matches the computational domain at the boundary exactly. Bérenger [16] proved a quadratic or cubic damping is sufficient to reduce reflections from the boundary to an insignificant amplitude, if the boundary layer has a thickness of at least half the incoming wavelength.

$$
\begin{aligned}
\dot{Q} &= -i\left[|_b|^2 - |_a|^2 - \delta\right]Q + 2_a b\, W - \frac{Q}{T_2}, \\
\dot{W} &= -\left[a_b^* Q^* +_a^* bQ\right], \quad \frac{\partial E_p}{\partial z} + \frac{1}{c}\frac{\partial E_p}{\partial t} = -\alpha_p Q E_s - \eta(z) E_p \\
\frac{\partial E_s}{\partial z} + \frac{1}{c}\frac{\partial E_s}{\partial t} &= -\alpha_s Q^* E_p - \eta(z) E_s \quad \frac{\partial E_{as}}{\partial z} + \frac{1}{c}\frac{\partial E_{as}}{\partial t} = -\alpha_a Q E_p - \eta(z) E_{as}.
\end{aligned}
\tag{6.20}
$$

Equation (6.20) shows the damping parameter can be coupled to the electric fields. The damping is dependent on the spatial position z and is zero in the computational domain and therefore increases as the absorbing layer progresses.

To retain the undamped problem in the computational domain, we select one-sided finite difference stencils ("upwind" and "downwind" schemes), to ensure spatial derivatives in the direction of propagation are not approximated using values from inside of the sponge layer, as shown in Figure 6.6. Specifically, in this case, downwind schemes are implemented to ensure the correct stability for the respective propagation direction [17].

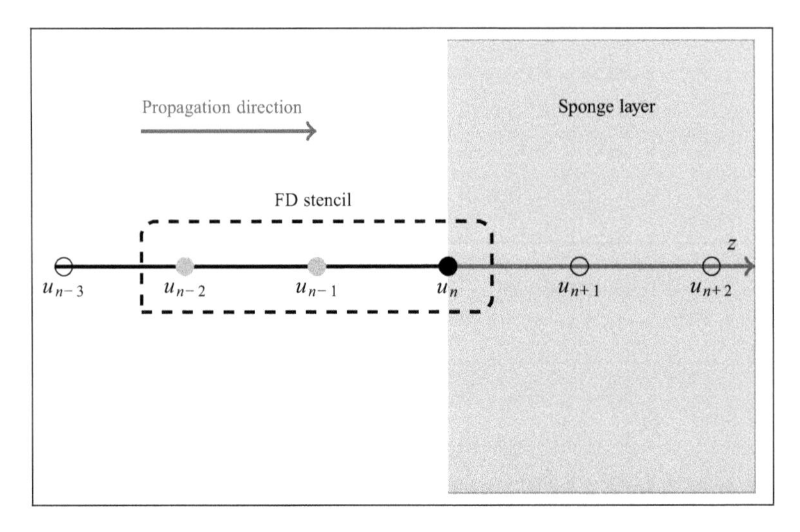

Figure 6.6 Finite difference stencil for " u_n "

For a discretization with N nodes $(0, 1, 2, \ldots, N-1)$, at any interior node n, the schemes are as follows:

$$\frac{\partial E_{p,n}}{\partial z} = \frac{3E_{p,n} - 4E_{p,n-1} + E_{p,n-2}}{2\Delta z} + (\Delta z^2), \tag{6.21}$$

At the boundaries of the domain, where the second-order schemes cannot be used, the accompanying first-order schemes are applied and then a zero-flux condition at the actual boundary:

$$\frac{\partial E_{p,N-1}}{\partial z} = \frac{E_{p,N-1} - E_{p,N-2}}{\Delta z} + (\Delta z), \tag{6.22}$$

These boundaries will be deep within the sponge layer, so any incoming waves should already be of negligible amplitude, meaning the change to a lower order scheme will have a little-to-no effect.

Using these finite difference stencils, we can approximate the PDEs in (6.18) by ODEs and use a Runge–Kutta method to integrate forward in time. Given that spatial discretization of N nodes will lead to a system of $7N$ where ODEs are to be evaluated, a fairly low-order Runge–Kutta method is chosen to decrease runtime and be comparative with the chosen finite difference methods. The Bogacki–Shampine method [18] is selected as it is explicit and third-order accurate method, with a second-order embedding for step size adaptivity (Table 6.1).

This method uses 25% less function evaluations than a typical method of this order. The final stage $K_4 = f(t_n + h, y_{n+1})$ is identical to the first step of the next stage $K_1 = f(t_n, y_n)$, commonly referred to as FSAL (first same as last). Thus, with appropriate storage of the stages, only three function evaluations are needed per step, rather than the typical four. This is another factor for choosing this method,

Table 6.1 Butcher table for Bogacki–Shampine and Runge–Kutta Methods [18]

0				
1/2	1/2			
3/4	0	3/4		
1	2/9	1/3	4/9	
	2/9	1/3	4/9	0
	7/24	1/4	1/3	1/8

given the large size of our right-hand-side function:

$$\begin{aligned} y_{n+1}^{(2)} &= y_n + \frac{2}{9}hk_1 + \frac{1}{3}hk_2 + \frac{4}{9}hk_3, \\ y_{n+1}^{(3)} &= y_n + \frac{7}{24}hk_1 + \frac{1}{4}hk_2 + \frac{1}{3}hk_3 + \frac{1}{8}hk_4. \end{aligned} \tag{6.23}$$

Using the second-order embedding (Equation 6.23), a lower order solution can be obtained at essentially no additional cost; no extra function evaluations are needed just for some vector sums. The error can be computed using these two solutions:

$$err = \left\| \frac{y^{(3)} - y^{(2)}}{\|y^{(3)}\|_\infty \times rtol + atol} \right\|_\infty . \tag{6.24}$$

If $err \leq 1$, the step is deemed successful, likewise an unsuccessful step is defined when $err > 1$. The step size h is adjusted by

$$h_{new} = safety \times h_{old} \times err^{-1/k} = 0.9 \times h_{old} \times err^{-1/3}. \tag{6.25}$$

Hence, if the step was successful ($err < 1$), the step size is increased as $1/\sqrt[3]{err} > 1$, and if the step was unsuccessful ($err > 1$) $\Rightarrow 1/\sqrt[3]{err} < 1$, the step size will be decreased. Additionally, step size adaptivity is always maintained within the interval of our specified $(h_{\min}, h_{\max})$.

6.2.4.2 Absorbing boundary testing

Before running full simulations on the Maxwell–Bloch equations, the absorbing sponge layer on the boundaries of the domain was thoroughly tested. The initial condition is a pulse at the center of the domain and a simple 1D wave equation is used for propagation, with no coupling or nonlinearities. Figures 6.7 and 6.8 show the results for propagation in both directions. It can be seen in both figures a Gaussian pulse is moving in the desired direction, without the loss of amplitude. Upon reaching the absorbing layer, the pulse is fully absorbed with zero reflections back into the computational boundary. This displays the desired behavior and puts us in a position to run full simulations on the two-stage stimulated Raman models.

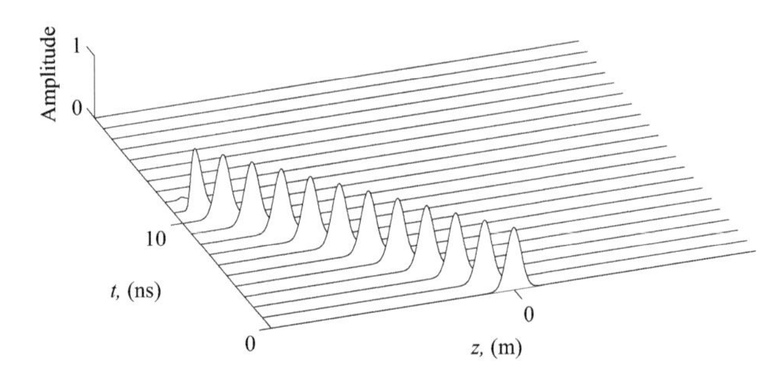

Figure 6.7 Absorbing boundary condition: left propagation

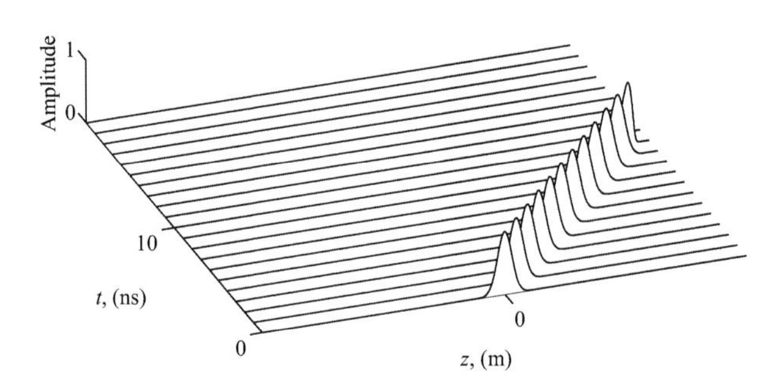

Figure 6.8 Absorbing boundary condition: right propagation

6.2.5 Results

This subsection begins by detailing experimental conditions provided to us to justify the initial conditions used. Results presented correspond to the impulsive and SRS process.

6.2.5.1 Impulsive excitation stage

Table 6.2 shows the initial conditions and parameter values to be used in (6.15). These values were supplied in collaboration with Jean Claude Diels and the "*Diels Research Group*" at the University of New Mexico.

Figure 6.9 shows the results of the initial impulsive Raman excitation process for four various θ values varying the IR pulse amplitude. The top plot shows the change in Q, the middle is the change in W and the bottom plot is a schematic of the IR pulse relative to the other two plots. From this impulsive stage, the final Q and W values from $\theta = 4.2221$ (pink) are selected, as this is the θ value that produces the most excitation Q.

Table 6.2 Parameters for impulsive Raman

Symbol	Value
$Q(t=0)$	0
$W(t=0)$	-1
t	$[-0.2, 0.2]$(ps)
T_2	20, 000

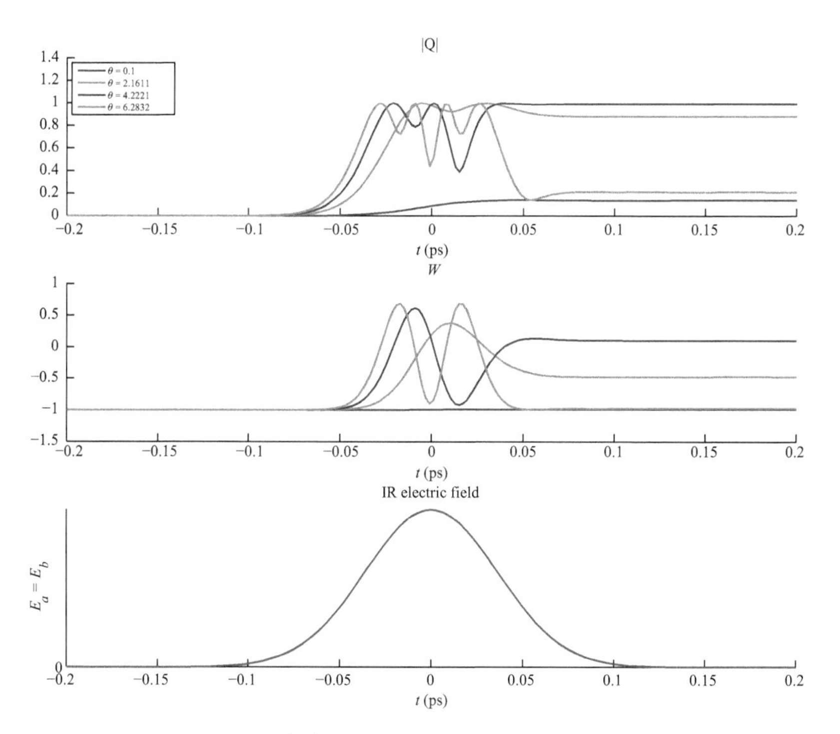

Figure 6.9 Top: $|Q|$, middle: W, bottom: electric field $E(t)$

6.2.5.2 Stimulated Raman stage

Using the values from the previous stage as the initial conditions for Q and W inside the excited medium, (6.20) are numerically integrated using the finite difference and Runge–Kutta methods detailed in the previous section (Table 6.3).

Using a "super-Gaussian" to approximate the longer duration of UV pulse, with a duration of $\tau = 200$ ps and amplitude $E_{p,0}^2 = 10^{-3}$:

$$E_p\left(0, z\right) = E_{p,0}e^{-(z/z_0)^{10}}$$

Initially, assume there is zero Stokes and anti-Stokes scattering, $E_s(0,z) = E_{as}(0,z) = 0$. The excited medium is initially set for all $z > 0$.

Table 6.3 Parameters for stimulated Raman

Symbol	Value
$Q(t=0)$	From impulsive stage
$W(t=0)$	From impulsive stage
t	$[0, 1000]$ (ps)
z	$[-5, 35]$ (cm)
W_0	-1
T_2	20,000
α_p	10
α_s	10
α_a	10

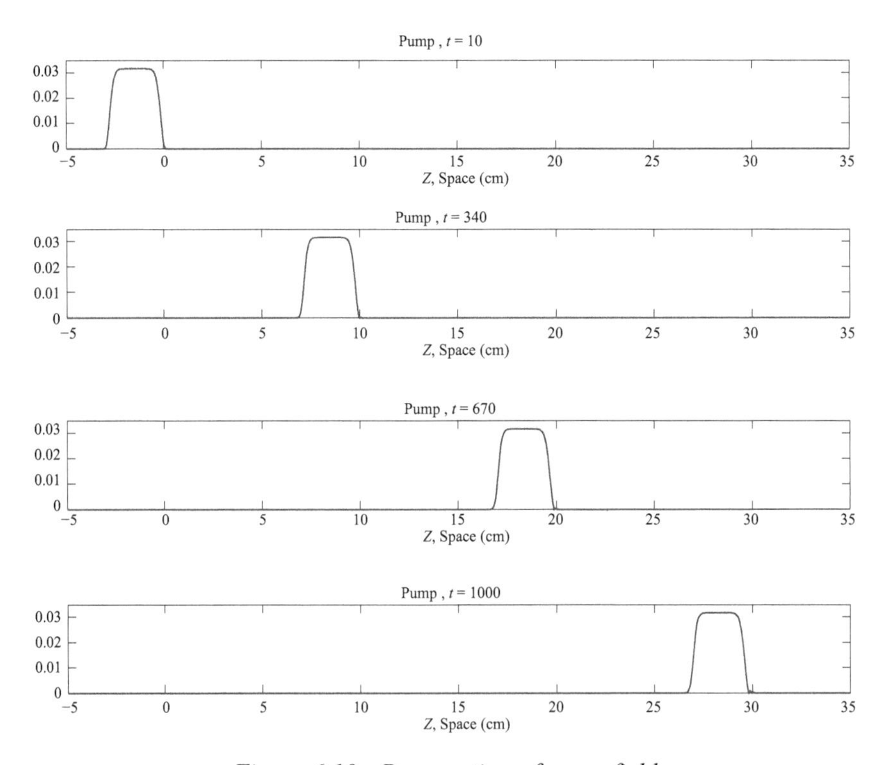

Figure 6.10 Propagation of pump field

Therefore, the pumping laser is located to enter the excited medium as soon as the simulation begins. As the pump (Figure 6.10) enters the medium, we see in Figures 6.11 and 6.12 that scattering begins to occur and continues to move through the medium where the Stokes and anti-Stokes fields propagate at the same rate. The scattered fields continue to increase in amplitude but only to about three orders of magnitude, smaller than the pump field.

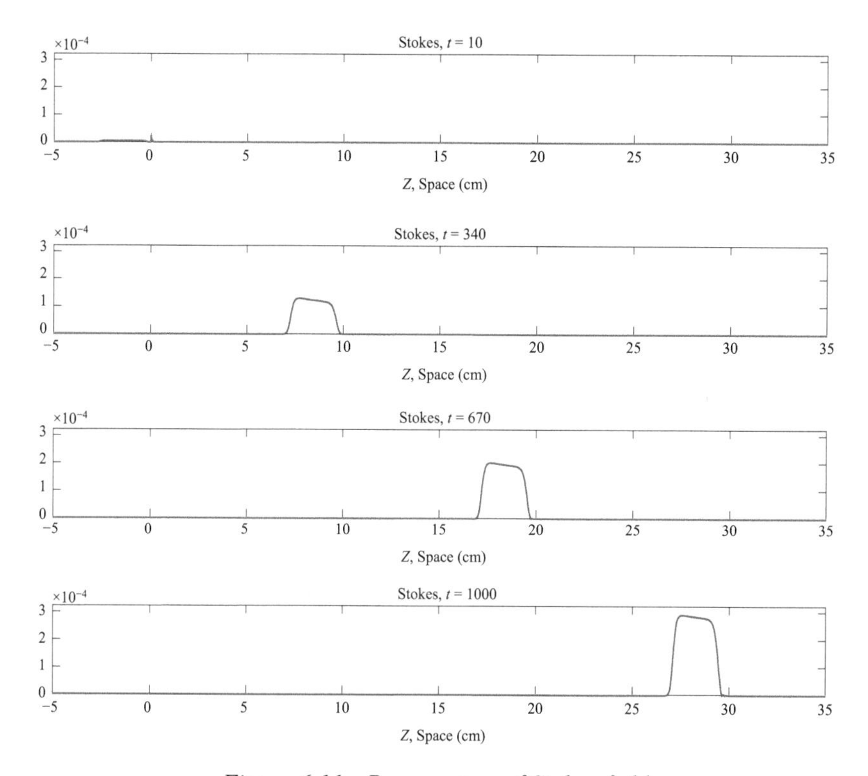

Figure 6.11 Propagation of Stokes field

It can be seen that while the Stokes and anti-Stokes fields have the same amplitudes throughout the propagation, we observed in our simulations that the phase of each field,

$$\varphi = \arctan\left(\frac{Im(E)}{Re(E)}\right),$$

always maintained opposite values at each spatial location. The two fields being out of phase by π suggest the following relation:

$$E_s = E_{as}^* \quad \Leftrightarrow \quad E_{as} = E_s^*. \tag{6.26}$$

One can prove analytically that (6.26) is a plausible outcome from (6.18).

Proof: First show that E_p remains real-valued.

$$\begin{aligned}
\frac{\partial E_p}{\partial z} + \frac{1}{c}\frac{\partial E_p}{\partial t} &= -\alpha_p(QmE_s + Q^* E_{as}) \\
\Rightarrow \qquad &= -\alpha_p\left(QE_s + Q^* E_s^*\right) \\
\Rightarrow \qquad &= -2\alpha_p \Re\{QE_s\}
\end{aligned}$$

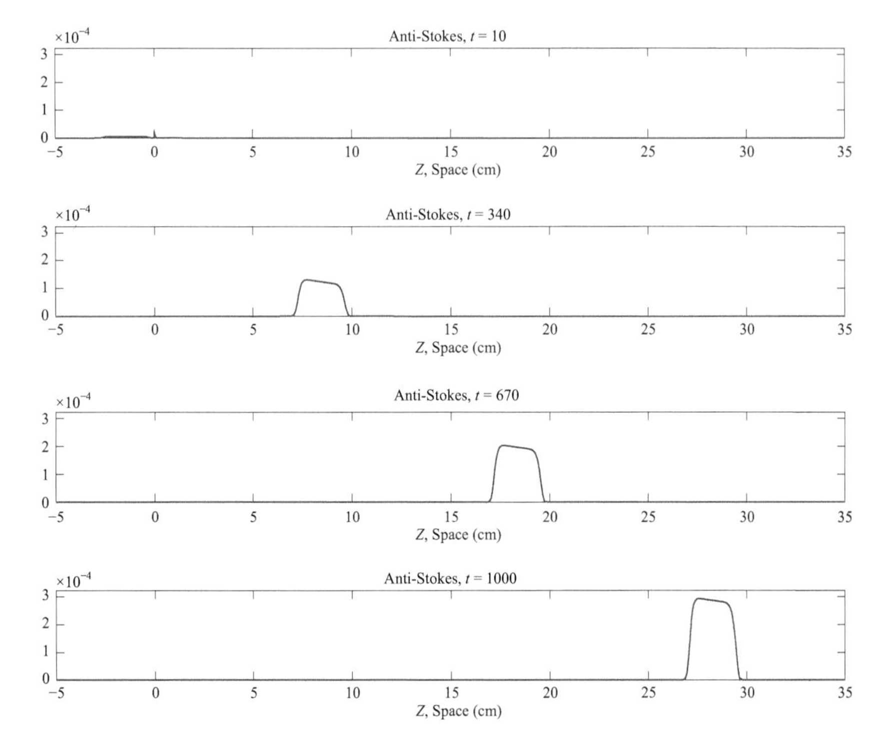

Figure 6.12 Propagation of anti-Stokes field

as the sum of a complex number and its conjugate is equal to two times the real part. Assuming E_p is initially real, it will remain real-valued as it propagates.

$$E_p = E_p^*$$

Next it can be shown that the equations for E_s and E_{as} are equivalent to the equations of E_{as}^* and E_s^*, respectively. Assuming $\alpha_p = \alpha_s = \alpha_a$, as has been the case for all our simulations:

$$\frac{\partial E_s}{\partial z} + \frac{1}{c}\frac{\partial E_s}{\partial t} = -\alpha_s Q^* E_p$$
$$\Rightarrow \quad \frac{\partial E_s}{\partial z} + \frac{1}{c}\frac{\partial E_s}{\partial t} = -\alpha_s Q^* E_p^*$$
$$\Rightarrow \quad \frac{\partial E_s}{\partial z} + \frac{1}{c}\frac{\partial E_s}{\partial t} = \frac{\partial E_{as}^*}{\partial z} + \frac{1}{c}\frac{\partial E_{as}^*}{\partial t}$$

Similarly, for E_{as},

$$\frac{\partial E_{as}}{\partial z} + \frac{1}{c}\frac{\partial E_{as}}{\partial t} = -\alpha_a Q E_p$$
$$\Rightarrow \quad \frac{\partial E_{as}}{\partial z} + \frac{1}{c}\frac{\partial E_{as}}{\partial t} = -\alpha_a Q E_p^*$$
$$\Rightarrow \quad \frac{\partial E_{as}}{\partial z} + \frac{1}{c}\frac{\partial E_{as}}{\partial t} = \frac{\partial E_s^*}{\partial z} + \frac{1}{c}\frac{\partial E_s^*}{\partial t}$$

It has been shown the equations are consistent with the claim $E_s = E^*_{as}$. This supports the numerical results already presented.

6.3 Parallel computing

For future studies, it may be necessary to run simulations for a longer time duration and hence will require a larger spatial domain. This presents the challenge of increasing the number of spatial nodes needed to maintain our desired accuracy, consequently increasing the size of the system of equations we must repeatedly evaluate dramatically.

6.3.1 Parallel Runge–Kutta

By far, the largest computational cost in the Runge–Kutta method, for a system of this size, is the right-hand-side function evaluations. Utilizing the well-known message passing interface (MPI), a domain decomposition may be used to divide the spatial domain into sub-domains of approximately equal size. Each processor is responsible for a single sub-domain, and only the values at the outer two nodes on each sub-domain must be passed to neighboring processors to compute finite difference approximations. With such a small amount of information needed to be sent and received (in comparison to the overall size of each sub-domain), there is excellent potential for parallel speedup.

By using non-blocking MPI commands, it is possible to fully optimize the time spent evaluating the right-hand-side function and reduce any time a processor that is idle or "waiting." The main structure of the message passing routine for each processor is as follows:

1. Pass information for two nodes on ends of subdomain to the processors that require it.
2. Compute right-hand-side function for all nodes on the interior of the domain (no message passing required).
3. Check the data that we received to compute spatial derivatives for the edge nodes.
4. Compute right-hand-side function for edge nodes.

To study how our problem scales as more processors are employed, we plot its strong scaling. Fixing the problem size and increasing the number of processors, a perfect scaling would show linear relationship, with a slope of -1, on a log–log scale. In other words, doubling the number of processors used should cut the run-time in half.

Testing the strong scaling is done on two problem sizes, one with 10,000 spatial nodes (system of 70,000 ODEs) and one with 20,000 nodes (140,000 ODEs). These tests run on the SMU high-performance computer "*SMUHPC*," with $2, 4, 8, 16, 32$ and 64 processors. The slopes for the line of the best fit are calculated using MATLAB® built-in "*polyfit*" function, with the "perfect" strong scaling having a slope of $r = -1$.

Figures 6.13 and 6.14 indicate that our problem scales are exceptionally well in this sense. Considering that only the evaluation of right-hand-side function was parallelized, and not any of the vector operations (add, subtract, multiply) within the Runge–Kutta method, this is significant speedup. This is certainly encouraging

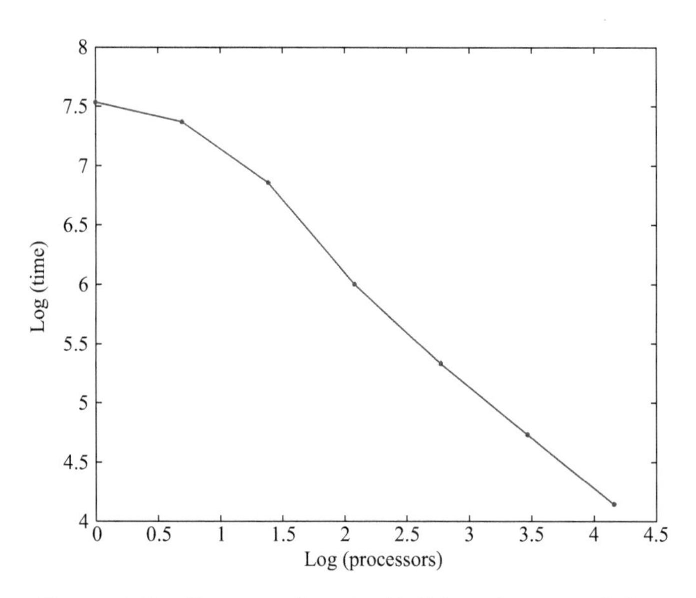

Figure 6.13 Strong scaling for 10,000 *nodes,* $r = -0.874$

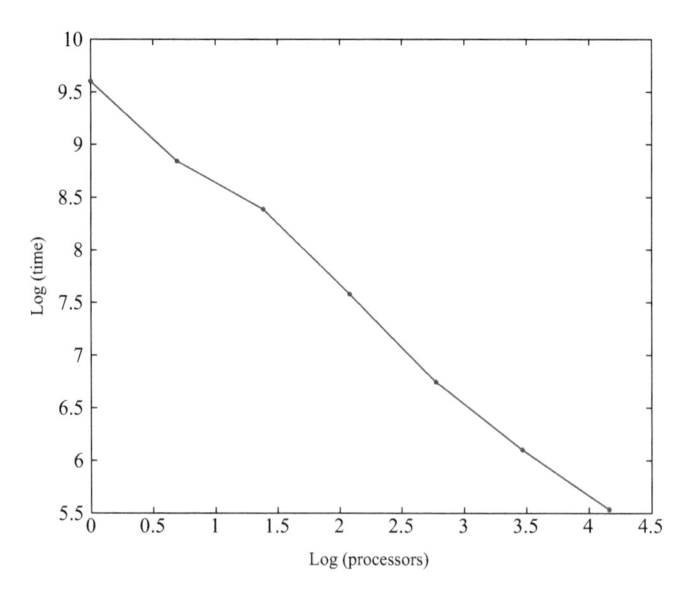

Figure 6.14 Strong scaling for 20,000 *nodes,* $r = -0.995$

for us going forward, and if larger simulations are required, parallel computing can significantly reduce longer times.

6.3.2 Parallel split-step Fourier method

As evidenced in literature [19], there are significant speedup benefits when suitable domain decomposition is utilized with the SSFM. In [19], the authors present a 2D+1 form of the NLSE, and with 4–20 parallel threads, they achieve excellent parallel efficiency of around 95%. Our work details strong scaling results and provides additional confirmation of the effective use of parallel computing in ultrashort pulse simulations.

To parallelize the SSFM algorithm the aim is to use multiple processors to decompose the 3D domain. Thankfully the FFTW library has multiple-processor MPI routines that once linked and initialized successfully can be called in the same manner as the serial transforms. Processor allocation is handled by the FFTW library as the 3D domain is decomposed in the primary direction (x-direction), with each processor then allocated a set of "*rows*" and all the "*columns*" and "*pages*" in those rows.

Calculations for the SSFM require zero/little contact between processors as each processor stores and owns the information they need for their chunk. A little care has to be taken in the storage of solutions (storing by rows rather than the traditional storage by columns) and the unpacking of solutions for plotting, but this is merely a formality.

Again the strong scaling is studied for a fixed problem size of $128 \times 128 \times 128$ (now on SMU's new supercomputer "*Maneframe*") and seen from Figure 6.15, the problem scales well as more processors are added. This is

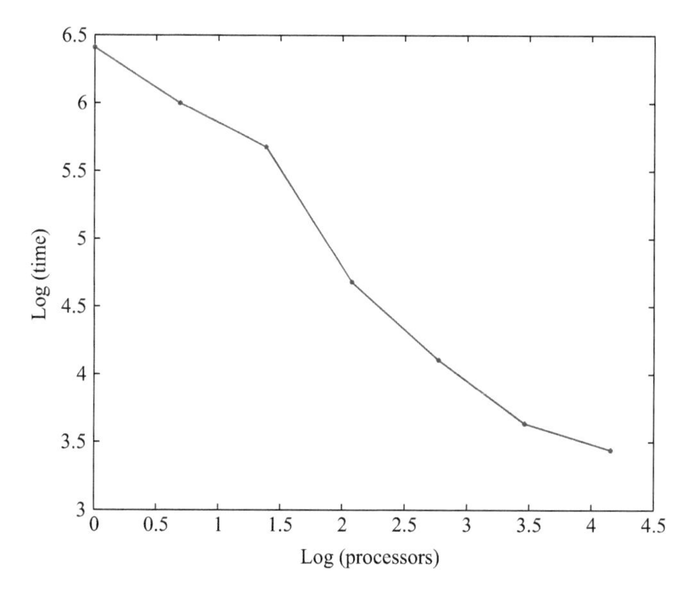

Figure 6.15 Strong scaling for problem size $[128 \times 128 \times 128]$, $r = -0.784$

encouraging as larger domains in three dimensions that will require significantly increased computing may be required to solve to our desired resolution.

In addition to the MPI parallelization, a "*hybrid*" MPI/OpenMP approach was also tested, utilizing both the MPI of MPI and the threading teams of OpenMP. Our goal was to squeeze any additional performance from our parallelization efforts. As before the MPI routines decompose the 3D problem, however, each node of the supercomputer is assigned one portion of the domain only. This allows each individual node to spawn up to eight OpenMP threads to further parallelize the calculations in the y and t directions of each sub-domain.

In smaller simulations, it was found that two threads per node produced optimum results, so we transferred this to a larger problem size of [256 × 256 × 256]. As Figure 6.16 shows combining the two parallel computing approaches can decrease the runtime of our simulations further. Using either 1 or 2 MPI tasks per node improves the speedup for all processor combinations from 2 to 64. However, it should be mentioned that to run with 1 MPI task per node requires at least eight times the total nodes used in the MPI only case. For the 2 MPI tasks per node, it will be four times the nodes in the MPI case. The trade-off between the speedup achieved and the significant increase in the nodes required for the hybrid code indicates a fairly small reward for much larger computing commitments. The

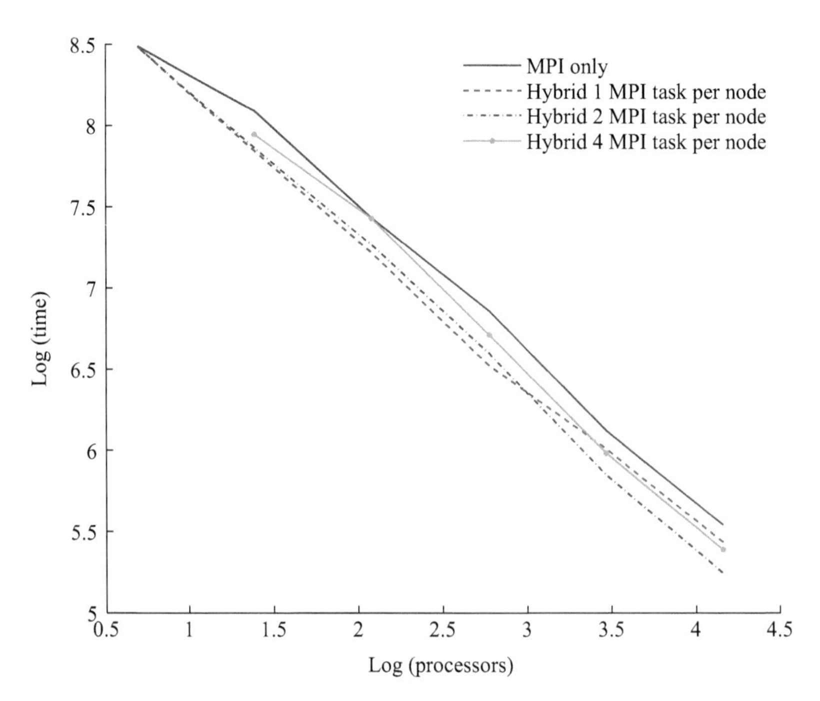

Figure 6.16 Strong scaling for problem size [256 × 256 × 256]*; all MPI tasks use 2 OpenMP threads each*

Table 6.4 Strong scaling results for parallel SSFM

	Hybrid	**Hybrid**	**Hybrid**
MPI only $r = -0.876$	1 MPI task per node $r = -0.885$	2 MPI tasks per node $r = -0.945$	4 MPI tasks per node $r = -0.950$

hybrid code does show improved strong scaling comparable to MPI only code as seen in Table 6.4.

Essentially, all options of MPI only and hybrid codes scale very well and should be effective when scaled up for much larger simulations of 100s–1,000s of MPI processes.

In summary, the results presented have shown excellent strong scaling of this problem with the MPI routines for two different size domains. For larger 3D domains, parallel computing techniques can be used efficiently and effectively to reduce runtime. Second, there appears to be evidence of additional performance benefits available using the hybrid code approach, providing there are sufficient computing resources to run with only a few MPI processors per node.

6.4 Conclusions

In this chapter, we highlighted numerical studies of two important interactions that can occur within light filaments. First, we covered forward SRS by a UV filament by a careful derivation of the Maxwell–Bloch equations that highlight dominant effects in both the impulsive and stimulated scattering stages. Appropriate numerical schemes were applied to integrate this normalized and reduced system of equations. In addition, suitable boundary conditions for extended timescale simulations have been developed. Our simulations clearly show the creation of Stokes and anti-Stokes fields in the presence of a high-power laser pulse as the pump for stimulated scattering. Both scattered fields propagate with the pump field and have the same amplitude as each other. However, their amplitude is still significantly smaller than the UV pulse, around three orders of magnitude. The scattered fields are always of opposite phase, as seen in our numerical results, and have been proved analytically.

Second, a fully classical model for the interaction of multicolored filaments was derived, specifically the potential effects of a UV filament acting as a waveguide for the unstable IR filament. This model was numerically integrated with the SSFM and results indicate clear differences in the amplitude and characteristics of the IR pulse during propagation. The pulse also has drastically different temporal and transverse behavior with and without the UV waveguide. The presence of the UV pulse and the plasma channel appears to maintain the transverse symmetry of the pulse while prompting a lengthening of the temporal width. Early evidence suggests a UV filament can provide a successful waveguide to an IR pulse, resulting in reduced losses in energy due to diffraction and temporal dispersion at longer propagation distances.

Finally, the parallel scaling of both problems was also discussed, emphasizing should there be a need to increase the spatial domain in either case for future simulations. The strong scaling is very good, so as one looks to increase the number of spatial nodes, employing additional CPUs should be an effective tactic and keep our runtimes reasonable.

References

[1] A. Braun, G. Korn, X. Liu, D. Du, J. Squier, and G. Mourou. Self-channeling of high-peak-power femtosecond laser pulses in air. *Opt. Lett.* 20(1), 73–75, 1995.

[2] A. Brodeur, C. Chen, C. Ilkov, S. Chin, O. Kosareva, and V. Kandidov. Moving focus in the propagation of ultrashort laser pulses in air. *Opt. Lett.* 22(5), 304–306, 1997.

[3] J. F. Daigle, F. Théberge, M. Henriksson *et al.* Remote THz generation from two-color filamentation: long distance dependence. *Opt. Express* 20(6), 6825–6834, 2012.

[4] L. Bergé, S. Skupin, C. Köhler, I. Babushkin, and J. Herrmann. 3D numerical simulations of THz generation by two-color laser filaments. *Phys. Rev. Lett.* 110(7–15), 2013.

[5] Y. Minami, T. Kurihara, K. Yamaguchi, M. Nakajima, and T. Suemoto. High-power THz wave generation in plasma induced by polarization adjusted two-color laser pulses. *Appl. Phys. Lett.* 102, 2013.

[6] R. Wyatt and D. Cotter. Tunable infrared generation using the 6s–5d Raman transition in caesium vapour. *Appl. Phys.* 21, 199, 1980.

[7] T. R. Loree, R. C. Sze, D. L. Barker, and P. B. Scott. *IEEE J. Quantum Electron.* 15, 337, 1979.

[8] Y. Ozeki, W. Umemura, Y. Otsuka *et al.* High-speed molecular spectral imaging of tissue with stimulated Raman scattering. *Nat. Photonics* 6, 845–851, 2012.

[9] B. G. Saar, L. Rodrigo Contreras-Rojas, X. Sunney Xie, and R. H. Guy. Imaging drug delivery to skin with stimulated Raman scattering microscopy. *Mol. Pharmaceutics* 8(3), 969–975, 2011.

[10] M. Ji, D. A. Orringer, C. W. Freudiger *et al.* Rapid, label-free detection of brain tumors with stimulated Raman scattering microscopy. *Sci. Transl. Med.* 5(201), 2013.

[11] W. S. Chiu, N. A. Belsey, N. L. Garrett, J. Moger, M. B. Delgado-Charro, and R. H. Guy. Molecular diffusion in the human nail measured by stimulated Raman scattering microscopy. *PNAS* 112(25), 7725–7730, 2015.

[12] G. Strang. On the construction and comparison of difference schemes. *SIAM J. Numer. Anal.* 5(3), 506517, 1968.

[13] J. A. C. Weideman and B. M. Herbst. Split-step methods for the solution of the nonlinear Schrodinger equation. *SIAM J. Numer. Anal.* 23(3), 485–507, 1986.
[14] Y. Shen. *The Principles of Nonlinear Optics*. Wiley-Interscience, New York, NY, 1984.
[15] R. W. Boyd. *Nonlinear Optics*, Second Edition. Academic Press, Cambridge, MA, 2003.
[16] J. P. Bérenger. A perfectly matched layer for the absorption of electromagnetic waves. *J. Comput. Phys.* 114(1), 185–200, 1994.
[17] J. W. Thomas. *Numerical Partial Differential Equations*. Springer, New York, 1998.
[18] P. Bogacki and L. F. Shampine. A 3(2) pair of Runge-Kutta formulas. *Appl. Math. Lett.* 2(4), 321–325, 1989.
[19] C. Ma and W. Lin. Parallel simulation for the ultra-short laser pulses' propagation in air. arXiv: 1507. 05988, 2015.

Chapter 7

Filamentation of femtosecond laser pulses under tight focusing

Andrey Ionin[1], Daria Mokrousova,[1] and Leonid Seleznev[1]

7.1 Introduction

The rapid development of femtosecond laser physics has opened up wide opportunities for the use of femtosecond lasers in the processing of various materials. Currently, such lasers are used in eye surgery, neurosurgery, and dentistry to create surface structures, including nanoscale ones; in drilling metals; cutting and marking diamonds, fabricating nanoparticles, etc. (see, e.g., reviews [1–6]). Converging (focused) laser beams are usually used to increase the laser energy density. In the case of medical applications, the interaction of laser pulses with biological tissues occurs under normal conditions, i.e., in air. Technological processes (drilling, cutting, etc.) sometimes require the use of a nonoxidizing medium, for example, inert gases (argon, nitrogen, etc.). One of the effective ways of producing nanoparticles at present is the action of femtosecond laser pulses on a metal or semiconductor surface in water or another solvent [7–9]. Thus, in most cases of such applications, even before exposure to the processed object, a high-intensity converging laser beam interacts with a gaseous or condensed medium, where it propagates. In this case, the beam self-focusing due to the Kerr effect can take place and, as a consequence of high intensity, ionization of the propagation medium can occur. As a result, the cross-sectional profile of the laser beam significantly changes even before reaching the processed object, i.e., processing conditions change. In addition, the plasma generated during the propagation of such a beam also affects the object. Thus, for correct femtosecond laser processing, it is necessary to take into account the features of nonlinear propagation of converging laser beams in transparent media, in particular, self-focusing and plasma formation, i.e., filamentation mode of propagation.

The study of filamentation started with consideration of a collimated or weakly focused laser beam in the near-IR spectral range. The parameters of filaments in air in the case of the collimated near-IR laser beam were well studied [10–12]. The

[1]Centre for Laser and Nonlinear Optics Technologies of N. G. Basov Quantum Radiophysics Division, P. N. Lebedev Physical Institute of Russian Academy of Sciences, Moscow, Russia

distance z_{fil} from a laser system (in experiments with femtosecond pulses usually from the output grating of a compressor) to the point of the beam collapses due to the beam self-focusing or, which is the same, before filamentation begins, depends on the laser pulse peak power P and is described by the so-called semiempirical Marburger formula [13]:

$$z_{\text{fil}} = \frac{0.367 L_{\text{DF}}}{\left\{ \left[(P/P_{\text{cr}})^{1/2} - 0.852 \right]^2 - 0.0219 \right\}^{1/2}},$$

Where L_{DF} is diffraction length for a given laser beam ($L_{\text{DF}} = ka_0/2$, a_0 is beam waist, k is wave number, P_{cr} is the critical power of self-focusing). With additional geometrical focusing of the beam, the distance z' to the beginning of the beam collapse shifts from the geometrical focus toward the propagating pulse in accordance with the formula [14]:

$$\frac{1}{z'} = \frac{1}{z_{\text{fil}}} + \frac{1}{f'},$$

where f is focal length, and z_{fil} is distance to the laser beam collapse point without geometric focusing. Thus, even in the case of very tight focusing ($f \ll z_{\text{fil}}$), the beginning of the filament is located before the geometric focus. The completion of filamentation does occur behind the geometric focus, when the geometric beam divergence prevails over the self-focusing process. On the other hand, in the case of $f \ll z_{\text{fil}}$, it can be shown that z' approaches the value of f, as a result of which it was asserted in some papers (see, e.g., [15]) that under tight focusing ionization of the medium occurs near the focus, and filamentation does not take place. Therefore, this chapter considers the propagation of femtosecond laser pulses under tight focusing, the formation of plasma channels under these conditions, the features of the energy reservoir behavior, post-filamentation mode, and other processes occurring under conditions of such propagation.

7.2 Filamentation under tight focusing: laser intensity, plasma density, geometric characteristics of a plasma channel

Even though the filamentation of collimated or weekly focused beams was thoroughly studied [10–12], there are far less papers devoted to the study of tightly focused pulse filamentation. For instance, in [16], the behavior of a tightly focused laser beam in a condensed medium (namely, in water) was experimentally and numerically studied. It was shown that such a beam causes avalanche ionization near its waist. In more rarefied media, for instance, in air, the time of an electron flight from its ion to a neighboring molecule is significantly longer than an ultrashort pulse duration, i.e., the avalanche mechanism of ionization is impossible and tunnel ionization takes place (at least, for UV, visible and near-IR laser pulses). Nevertheless, it was claimed in [15] that “Under the tightest focusing condition, filamentation is prevented and only a strong plasma volume appears at the geometrical focus," i.e., the very possibility of filamentation for tightly focused laser beam was denied (Figure 7.1 shows the plasma glow along the filament and in the waist region).

In the middle of the 2000s, several papers studying the filamentation of focused laser pulses appeared [17–19]. In these papers, the method of longitudinal diffractometry was used. In [17], the temporal behavior of a plasma channel formed under the filamentation of a laser beam focused by a lens with the focal length of 0.4 m was measured. It was found for the plasma electron density to be $\sim 4\times 10^{16}$ cm^{-3}. It was also experimentally demonstrated in [18] that the average diameter of the plasma channel depended on the peak power and increased from 20 to 80 μm with an increase of the relative power P/P_{cr} from 1.5 to 4 (focal length 0.5 m), the channel length being ~2 cm. Numerical simulations, the results of which were also presented in that paper, demonstrated slightly larger diameter: an increase from ~50 to 85 μm with P/P_{cr} growth from one to three. Paper [19] was devoted to experimental and numerical study of additional focusing influence on filamentation. Lenses with the focal lengths from 0.1 to 3.8 m (numerical apertures from 0.04 to 11×10^{-4}, respectively) were used. The nonlinear focus for the laser pulses used in these experiments was ~6 m, i.e., under these conditions, focusing with a lens with the focal length of 0.1 m can be considered as tight. It was shown in that paper (Figure 7.2) that an increase in the numerical aperture of the converging laser beam resulted in a significant (by several orders of magnitude in the experiments, Figure 7.2(a)) enhancement of the peak plasma density. It was also noted that in this case, the plasma channel diameter decreased in the experiments (insert in Figure 7.2(a)). However, the most significant increase in the plasma density and a significant decrease in the channel diameter were observed only with one of the

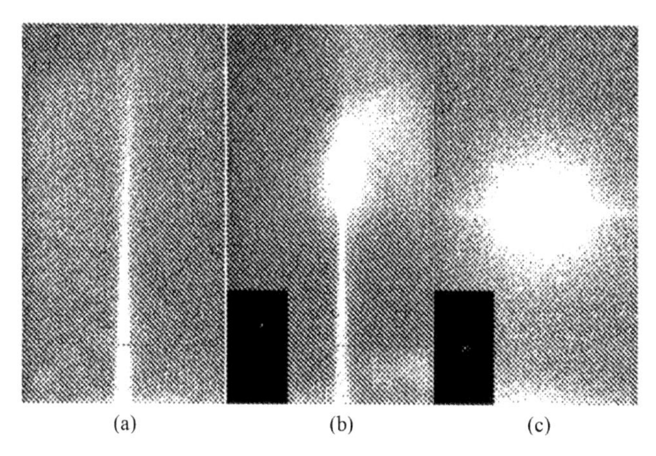

Figure 7.1 Plasma glow under laser beam focusing by lenses with the different focal lengths (a) 100, (b) 30, and (c) 5 cm. Pulse energy 5 mJ, pulse duration 42 fs, and central wavelength 810 nm. The inserts in (b) and (c) show a contrast plasma image near the geometric focus [15]

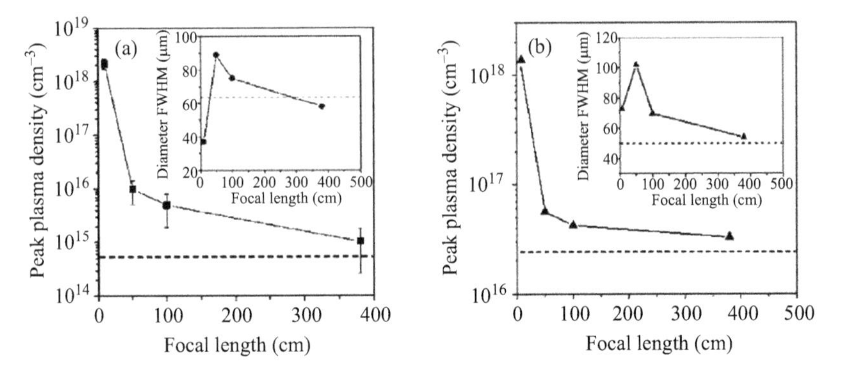

Figure 7.2 Peak plasma density and plasma channel diameter (in the inserts) versus the focal length: experiment (a) and calculations (b) [19]

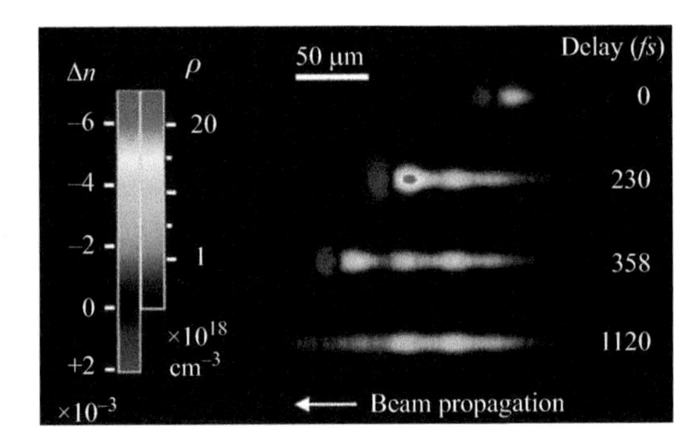

Figure 7.3 Spatial distribution and evolution of the nonlinear refractive index and plasma electron density under nonlinear propagation of a tightly focused laser beam [22]

most tight focusing (only one point). Thus, these articles demonstrated that the filamentation of tightly focused laser pulses did occur. In later paper [20], the filamentation and multifilamentation of tightly focused laser pulse were also observed (see Section 7.4 for details).

Numerical simulation of self-focusing and filamentation of ultrashort laser pulses in converging beams were published in [21]. With an enhancement of the numerical aperture from 10^{-3} to 10^{-2}, the filament radius fell down from ~45 to 5 μm, its length going down from 1.5 to 20 cm.

In the paper of [22], by applying the method of holographic microscopy, spatial distribution and evolution of the nonlinear refractive index and plasma electron density were measured for nonlinear propagation of a tightly focused (numerical aperture 0.11) laser beam (Figure 7.3). It should be noted that in these experiments, the pulse energy and duration were 35 μJ and 35 fs, respectively, i.e., the peak power did not exceed 1 GW that is almost an order of magnitude less than the critical power of self-focusing for the wavelength of 800 nm.

Paper in [23] reported on the study of filamentation in experiments where a laser beam was focused by an off-axis parabolic mirror with a numerical aperture of 0.12. In these experiments, multiple filamentation was observed by using shadow images. The diameter of a single filament was estimated as 10 μm. This paper also reported on the evaluations of the laser intensity during filamentation that was $2.5–5\times10^{14}$ W/cm^2. However, the authors informed about a possible overestimation of the intensity measured in these experiments. Using the interferometry, plasma density distributions in the filamentation region were obtained in this paper. The maximum density detected in these experiments was about 5×10^{19} cm^{-3}, which corresponded to almost double ionization of air and resulted in some doubts about the reliability of these results. In later paper of [24] by the same authors, the supercontinuum generation in a converging beam was studied. The plasma density estimations based on the supercontinuum spectral broadening gave the value of 10^{19} cm^{-3} for the maximum plasma density.

In [25], the filamentation of laser pulses at two wavelengths, 800 and 400 nm, was studied. When focusing a laser beam by a lens with the focal length of 15 cm, the laser pulse intensity in the filament was estimated as $(3–5.5)\times10^{13}$ W/cm^2 for 800 nm and $(1–2.5)\times10^{13}$ W/cm^2 for 400 nm. It is quite interesting to note that in these experiments, the laser pulse intensity during filamentation in the case of a fairly tightly focused laser beam (unfortunately, this paper does not indicate the beam diameter and/or the numerical aperture, which according to indirect data can be estimated as NA ~0.02) was practically the same as during the filamentation of a collimated laser beam.

In the paper of [26] published at about the same time, numerically estimated on the base of experimental data peak intensity of a laser filament under tight geometrical focusing (NA = 0.1) exceeded 10^{15} W/cm^2.

In most of the abovementioned papers, filamentation was studied at a fixed numerical beam aperture. In [27], filamentation of laser pulses was studied at various numerical apertures (NA = 0.2–0.008), and some results were compared with ones in [19]. In [27], an image of the plasma channel (Figure 7.4(a)) was obtained with an optical microscope and a CCD matrix. In this image of the plasma formation, a single plasma channel was selected. Analyzing the transverse profile of this image, the authors determined an average diameter (FWHM) of 8 μm for the plasma channel (Figure 7.4(c)).

The dependences of the plasma channel average radius R_{pl} on the numerical beam aperture NA presented in [27] are shown in Figure 7.5. An increase in the numerical aperture of the converging beam led to a decrease in the transverse size of the plasma channels formed during filamentation. At a numerical aperture of 0.05 or higher, the transverse size of the plasma channel R_{pl} remained practically unchanged at the level $R_{\mathrm{pl}} \approx 2–4$ μm. The authors attributed the stabilization of the radius R_{pl} to the strong refraction of laser pulses on the plasma formed in the laser beam waist, which was observed at laser peak powers even significantly lower than the critical self-focusing one. The dependence of the plasma channel length L_{pl} on

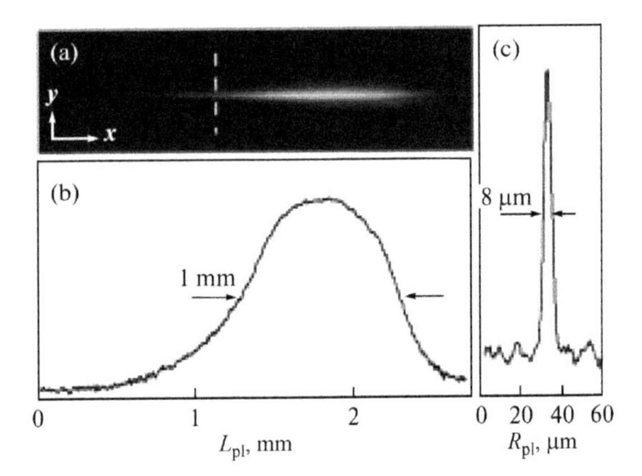

Figure 7.4 Plasma formation observed at NA = 0.2, $P_0 = 6.5$ GW (a). Plasma channel profiles: (b) longitudinal one (along the optical axis, i.e., along the x-coordinate) and (c) transverse one (y-coordinate) [27]

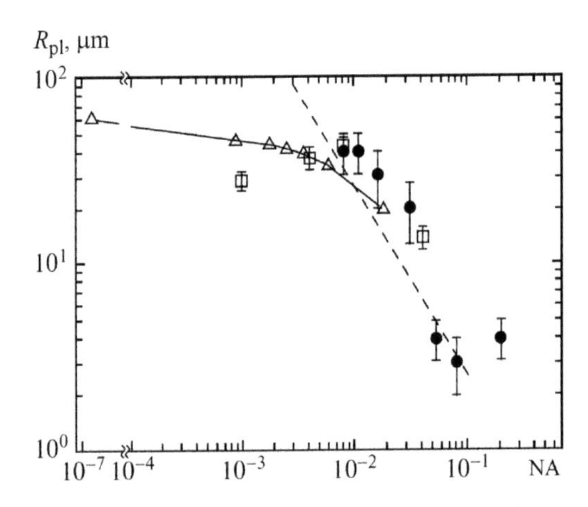

Figure 7.5 Dependences of the plasma channel radius on the numerical aperture for converging laser beam in experiment (dots) and simulations (triangles) and data from [19] (squares). The dotted line is the waist radius for a focused Gaussian beam without self-focusing [27]

the numerical aperture of the converging beam given in the paper of [27] is demonstrated in Figure 7.6.

Just as for the radius of the plasma channel, a decrease of its average length L_{pl} with rising the focusing system numerical aperture was observed. The authors noted that at high numerical apertures, the plasma channel length significantly, more than an order of magnitude, exceeded the Gaussian beam waist length (dashed line 5 in Figure 7.6). In the paper of [27], the authors also estimated the

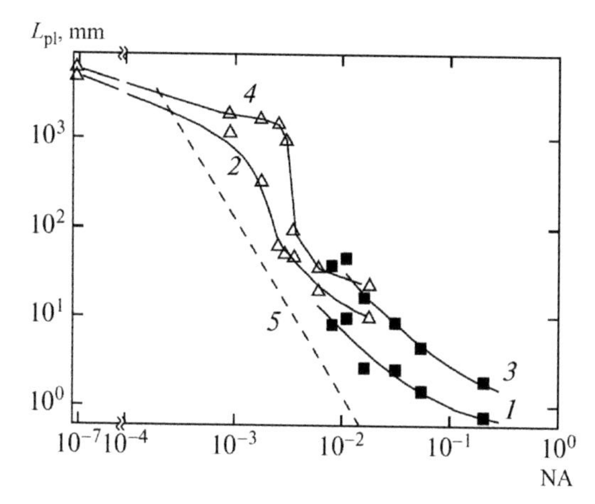

Figure 7.6 Dependences of the plasma channel average length L_{pl} on the converging beam numerical aperture NA obtained in the experiment (1, 3) and calculations (2, 4) for two laser peak powers: $P \approx P_{cr}$ (1, 2) and $P \approx 10P_{cr}$ (3, 4); dashed line (5)—the focal waist length of the Gaussian beam L_F [27]

plasma electron density from the laser pulse refraction angle. To do this, the authors applied the Drude model and obtained values of the radius R_{pl} and length L_{pl} of the plasma channel (Figures 7.5 and 7.6). The dependences of the plasma density averaged over plasma channel on the numerical aperture of the converging beam obtained by different methods [27], and the results of [19] are shown in Figure 7.7. At a laser beam numerical aperture of 0.04, the experimental estimate of the plasma density was $2\text{–}3 \times 10^{18}$ cm^{-3}. It should be noted that the filamentation of a weakly focusing or collimated laser beam is accompanied by the formation of plasma channels with a significantly lower electron density in comparison with tightly converging beams [19,27].

Filamentation was also demonstrated in a later paper [28] to exist under tight focusing, being significantly different from the collimated case. The regime of small numerical apertures was called as the "nonlinear" focusing mode. In this propagation regime, the Kerr self-focusing is crucial, and the laser pulse spectrum broadens in both directions. At the shorter focal lengths, the so-called "linear" focusing takes place, under which geometric convergence and laser plasma play a decisive role. This, in turn, results in the spectrum broadening predominantly into the short-wave region. A criterion for the transition from nonlinear to linear focusing was also proposed: if the nonlinear phase difference between the edges and the center of the laser beam (a "sag") due to the Kerr nonlinearity becomes compared with the geometrically determined phase difference earlier than one due to plasma, nonlinear focusing is realized. Otherwise, linear one takes place. The analysis of many published papers on filamentation at different numerical apertures carried out by these authors is in a good agreement with the proposed criterion. In [29], parameters of filaments and plasma channels were considered in detail under

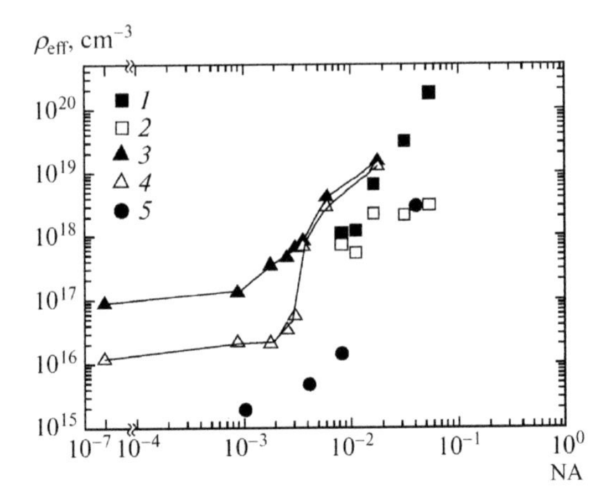

Figure 7.7 Dependences of the average (1, 2, 3, 5) and peak (4) electron density on the numerical aperture of the converging laser beam: 1—estimates by the angle of refraction, experiment ($P \approx 10P_{cr}$); 2—emission spectroscopy, experiment ($P \approx 1.5P_{cr}$); 3, 4—numerical calculations ($P = 10P_{cr}$); 5—data from [19] ($P \approx 3P_{cr}$) [27]

the "linear" and "nonlinear" focusing. It was noted that under weak geometric focusing, the filament and plasma channel parameters are practically independent of the laser pulse energy, while in the case of "linear" focusing, this dependence becomes very significant and noticeable. Higher intensity of a laser beam focused over a short focal length, in turn, can lead to a strong contribution of tunneling ionization, which makes the filamentation of collimated and tightly focused laser beam even more different from each other.

Thus, in these not very numerous papers, it was clearly demonstrated that with tight focusing, the filamentation of laser pulses does take place, an extended—as compared to the laser beam waist—plasma channel is formed, the geometric dimensions and the plasma density of which fall down and uprise, respectively, with an increase of focusing tightness.

7.3 Energy reservoir around the focal point

Under the filamentation of a collimated femtosecond laser beam in its cross section, a high-intensity paraxial region (light filament) is surrounded by electromagnetic field of relatively low intensity—an energy reservoir. A transverse profile of a laser beam in which the filament and energy reservoir are distinguishable is presented, for instance, in paper [30] (Figure 7.8). According to a modern concept of the moving foci dynamic model [30,31], in transversal distribution of self-focusing beam, the filament is formed by time slices of the laser pulse, the nonlinear foci of which lie at a given point or near it. The energy reservoir, in its turn, is formed by

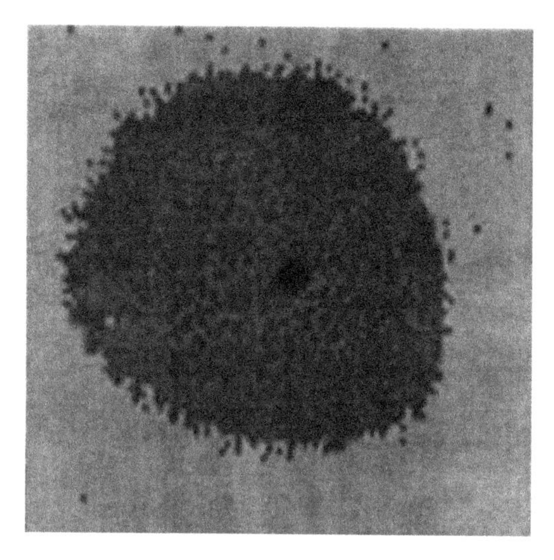

Figure 7.8 Typical burn pattern produced by a laser beam propagating in filamentation regime. The pattern shows the central darker spot (filament) surrounded by a larger weaker spot (energy reservoir) [30]

the rest of the pulse temporal layers with foci far from the given point or with peak powers that are not enough for self-focusing. Furthermore, the energy reservoir contains ~80–90 percent of all the laser pulse energy. It is important to note that there is no generally accepted boundary separating a laser filament from an energy reservoir. In various researches, the size of a diaphragm detaching the filament from the energy reservoir corresponded to the beam diameter at the level changing from ~1/5 to ~1/60 of the maximum energy density [30]. The most common estimate is ~1/10 of the maximum energy density in the filament [32,33].

The influence of the energy reservoir is so great that it was proposed in paper [34] to use an artificially added laser beam as an external energy reservoir to extend the filamentation region, which was also later studied in [35]. Paper [36] demonstrates that by changing the laser beam shape with a deformable mirror and, thus, controlling an energy flow from the energy reservoir to the filament, the shape of the conical emission can be influenced.

The energy reservoir plays an important role in the interaction of filaments during collimated beam propagation. The energy reservoir diffracted by a plasma channel inside the filament forms ring structures in the beam, the interference of which leads to the generation of secondary filaments [37]. An influence of the energy reservoir on filaments interaction within a focused laser beam will be discussed in more detail in Section 7.4 of this chapter. In addition, under diffraction of the energy reservoir on the filament plasma channel, specific light structures are formed in the beam—so-called post-filamentation channels [38]. More details about the post-filamentation channels in focused laser beams can be found in the corresponding section of this chapter.

The need to take into account the background emission surrounding the filament in numerical simulation was explicitly stated in paper [39]. It was demonstrated in paper [40] for such nonlinear processes as energy localization, dynamic stabilization of filament parameters, spectrum transformation, formation of ring structures in the cross section to take place over the entire cross section of the laser pulse, that is, in the region significantly exceeding the filament diameter.

Subsequent studies [32,33] did confirm the crucial role of the energy reservoir for the existence of an extended filament. It was shown that blocking laser beam by a diaphragm, which allowed the light filament to pass through it, but absorbed surrounding energy reservoir, resulted in an abrupt termination of the filamentation behind the diaphragm. This fact opens up an opportunity of interrupting the filamentation of collimated laser beam by introducing a diaphragm that transmits only the central part of the beam. In this case, it is possible to measure the dependence of some physical quantity (like third harmonic energy and spectral blue shift) on the distance without changing other parameters of the filament except for its length. However, the validity of extrapolating this procedure to the case of focused laser beam was needed to be verified.

The study of the energy reservoir propagation under tight—compared to the nonlinear self-focusing distance—geometric focusing was described in paper [41]. In this paper, experiments and numerical simulation of an influence of a diaphragm placed into a focused laser beam on its filamentation were carried out at a wavelength of 744 nm. The laser beam was focused by mirrors with the focal lengths of 26 and 51 cm, which corresponded to numerical apertures of the order of 10^{-2}. The distance to the nonlinear self-focusing point was longer than 10 m at the given energy and beam diameter, so these geometrical focusing conditions could be considered as quite tight.

Filament characterization in the experiment was carried out with a capacity probe. The filament was considered to be an area in which its plasma channel density was high enough to be detected by a sensor. A diaphragm with a diameter of ~300 μm was situated around the axis of the laser beam and absorbed its periphery. After the diaphragm, the beam energy and linear plasma channel density were measured. Numerical simulation was carried out for the same conditions, and additionally for a diaphragm with a diameter of 200 μm.

The results of numerical simulations of filamentation with a diaphragm of 300 μm in diameter are shown in Figure 7.9. Placing the diaphragm at the distance of 1 and 2 cm in front of the geometric focus—lines 2 (blue) and 3 (red), respectively—did not lead to a change in the length of the plasma channel at all in comparison with the free propagation (without diaphragm), line 1 (black). The difference between insertion of the diaphragm 1 cm before geometrical focus and free propagation is so negligible that the black line in Figure 7.9 is situated mostly behind the blue one. Moreover, placing the diaphragm of 200 μm in diameter 1 cm afore the geometrical focal point causes no noticeable differences in plasma formation behind the diaphragm either. Thus, we can conclude that the diameter of the energy reservoir near the geometric focus does not exceed 200 μm, which is comparable to the diameter of the filament for these conditions (about 50 μm). On the contrary, in

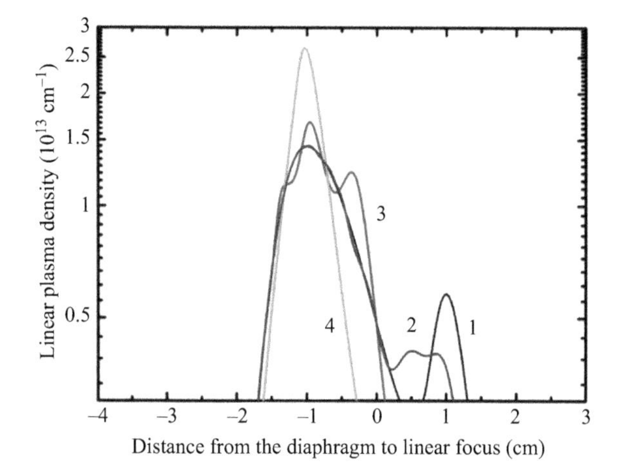

Figure 7.9 Calculated linear plasma density along the plasma channel for free propagation (line 1) and diaphragm of 300 μm in diameter placed 1, 2, and 3 cm in front of the geometrical focus (lines 2, 3, and 4, respectively) [41]

the case of a collimated laser beam (Figure 7.8), the energy reservoir diameter is about 12 mm and the filament diameter is far less than 1 mm.

While placing a diaphragm of 300 μm in diameter 3 cm in front of the geometric focus, a noticeable shortening of the plasma channel was observed. However, there was no abrupt termination of filamentation behind the diaphragm. In this case, only ~50 percent of the laser pulse energy went through the diaphragm.

The results of the experimental studies are in good agreement with the results of numerical simulations. When the diaphragm was placed 1 or 2 cm before the geometric focus, the length of the plasma channel remained unchanged in comparison with free propagation. With a diaphragm placed 3 cm off the focus, the filament length decreased, but its abrupt termination after the diaphragm was not observed.

From the results of this experimental study and numerical simulations, one can conclude that under geometric focusing the energy reservoir is localized near the optical axis of the laser beam. Moreover, only about 65 percent of the total laser pulse energy is involved in maintaining filamentation behind the nonlinear focus. This is confirmed by the following fact: when the diaphragm placed into the focused laser beam transmitted at least 65 percent of the beam energy, the length of the detected plasma channel practically did not change. It is important to note that under conditions of tight focusing, there is no noticeable dividing of the focused laser beam into an energy reservoir and filaments, since their intensities and diameters are comparable. This observation distinguishes the filamentation of focused radiation from the case of a collimated laser beam. It also means that terminating a filament of light under a laser beam focusing by placing a diaphragm is not always reasonable and appropriate. While trying to interrupt the filament under external focusing in such a way, it is crucial to make sure that the filament is really terminated and that the diaphragm is not burnt by relatively intense energy reservoir.

7.4 Multiple filamentation and filament fusion

During the propagation of a laser pulse, peak power of which is many times higher than the critical one for the laser beam self-focusing ($P \gg P_{cr}$), the latter as a whole does not occur. The laser beam becomes unstable because of local perturbations in nonlinear medium and electromagnetic field, which results in small-scale self-focusing and filamentation at various points of the beam cross section. Just in the initial paper, where self-focusing of nanosecond pulses in liquids was observed [42], it was noted that under certain conditions the formation of several light filaments occurred. Analytically, such an instability was considered in [43], where the propagation of a plane wave in a strongly nonlinear medium—nonlinear liquid—was studied. Numerical simulations of a laser beam propagation peculiarities for high-power femtosecond pulses in air were published in [39]. Its authors demonstrated that, as a result of multiple filamentation, a laser pulse was scattered on the resulting plasma with subsequent self-focusing of different time slices of the pulse. In this case, the scenario of pulse propagation with minimal energy losses was realized. In the same year, the experimental study of multiple filamentation in air was for the first time reported [44]. The authors observed the propagation of a weakly focused (focal length—8 m, beam diameter—70 mm) laser pulse with energy up to 160 mJ and pulse duration of 100 fs. Later on, many papers appeared, where authors studied both experimentally and numerically a high-power laser pulse propagation in the multiple filamentation mode [37,45–48]. In these papers, a mutual competition between two emerging filaments [20], effect of air turbulence [46], generation of conical emission [45] and supercontinuum [47], and appearance of extended plasma channels [48] were considered under conditions of multiple filamentation.

In all these papers, just collimated or weakly focused laser beam were analyzed. In a later paper [20], different stages of filament formation and transition to the regime of multiple filamentation for tight-focused laser pulses were presented. In these experiments, the converging beam numerical aperture (NA) was $\approx$0.08. With an increase in the laser pulse power above the critical value, filamentation started, and a glowing plasma channel appeared. In the case of $P \geq 2P_{crit}$, two plasma channels were observed (Figure 7.10(a), one plasma channel is out of the photosystem focus) with similar dimensions. A further increase in the laser pulse peak power (up to 68 GW in the given experiments) led to the growth of one filament out of the beam waist region toward the pulse propagation along the optical axis, which, as it developed, was surrounded by several, shorter ones over the periphery of the laser beam (Figure 7.10(b)–(e)). As a result, under high pulse powers, a spindle-shaped plasma formation consisting of many filaments was in the general case formed in front of the beam waist. Analyzing the transverse profile of this plasma formation, the authors determined the minimum diameter D of a single plasma channel corresponding to the light filament: $D \approx 6$ μm, which is in good agreement with the data of a later paper of the same authors

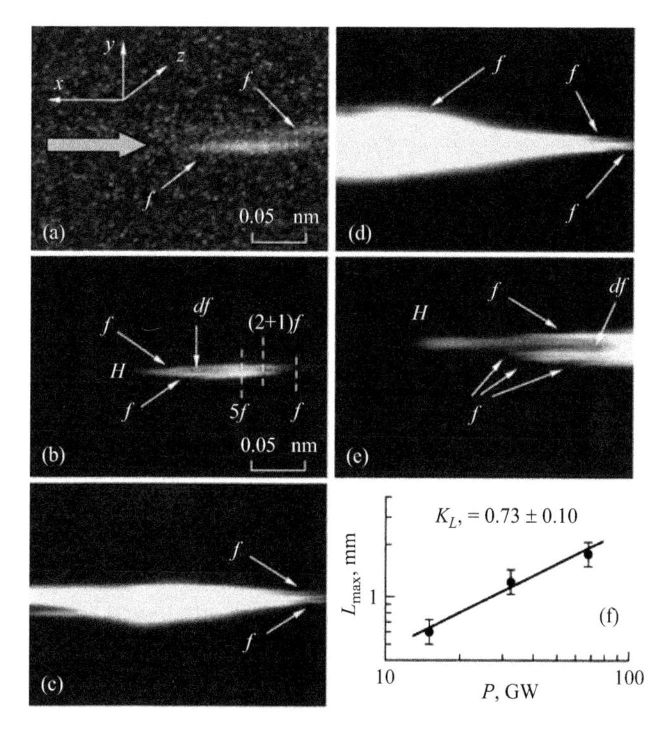

Figure 7.10 Filament photos in a microscope with a magnification of 10× at P ≈ (a) 6.4, (b) 15, (c) 33, and (d, e) 68 GW. The thin arrows indicate single or multiple filaments (f, nf) in the defocusing region (df), and H marks the head of the filamentation region. The thick arrow in panel (a) indicates the direction of the propagation of ultrashort laser pulses for panels (a)–(e). The $\log L_{max}(\log P)$ plot, where L_{max} is the filamentation region length and P is the power, and its linear approximation with the slope $K_L = 0.73 \pm 0.10$ (f) [20]

[26]. It should be noted that the diameter of a single plasma channel in these experiments did not depend on the pulse power.

A similar multiple filamentation was observed by the authors of [49]. A Ti: sapphire laser with a peak pulse power of about $230P_{cr}$ was applied in the experiments. Numerical aperture of 3×10^{-3} was more than an order of magnitude lower than in [20]. Figure 7.11 demonstrates the luminescence of experimentally observed plasma channels and the image profiles integrated along the optical axis. Also, as in [20], the existence of several plasma channels, i.e., a multiple filamentation, was clearly visible. However, the evolution of filaments near the geometric focus was quite different in these papers. So, in [49] both experimental and calculated results were given, where immediately in front of the focus many filaments merged, and a significant increase in laser intensity and, accordingly, plasma density was observed (Figure 7.12). The authors called the result of such a fusion as a superfilament.

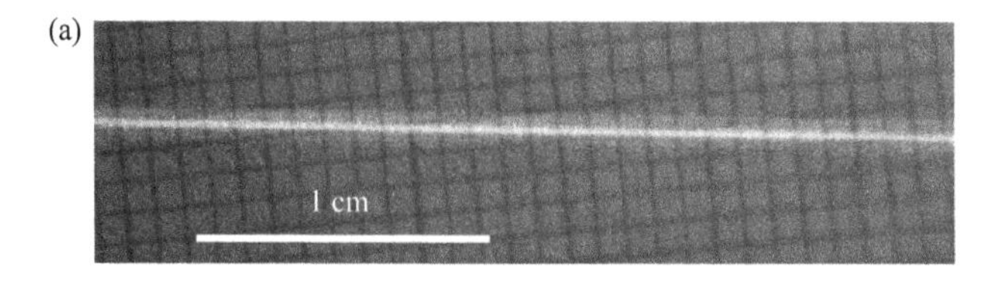

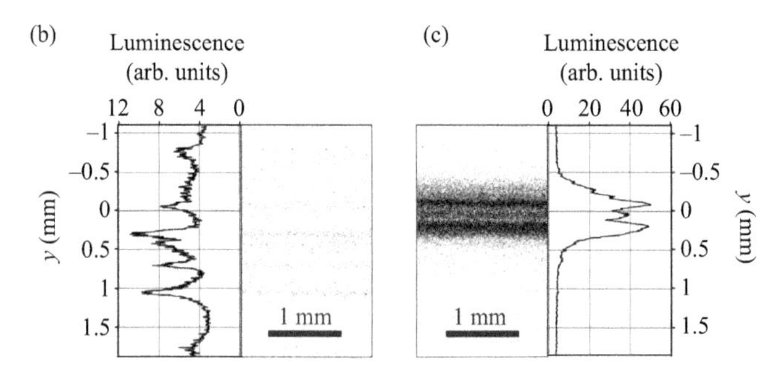

Figure 7.11 Luminescence of the plasma channels: (a) side-view picture of the filament bundle in the focal zone, (b) side-view ICCD images of the plasma luminescence centered at a distance from the lens z = 4.40 m and corresponding z-integrated over distance luminescence profile, (c) other ICCD picture centered at z = 4.90 m and corresponding z-integrated luminescence profile [49]

It should be noted that, as mentioned earlier, the numerical aperture in [20] was nearly an order of magnitude larger than in [49]. Accordingly, the area of filamentation and interaction of filaments were also smaller [26]. This is probably why such a phenomenon as superfilamentation was not observed in [20]. This is evidenced by Figure 7.13 from [20]. The figure demonstrates the energy density distribution behind the filamentation region. At low powers (up to 1 GW), after passing through a laser beam linear waist, a geometrically diverging Gaussian distribution was recorded (Figure 7.13(a)). At high powers, a ring-shaped emission was observed due to the laser beam diffraction on plasma channels (Figure 7.13(b) and (c)). A further increase in power (Figure 7.13(d)) resulted in multiple filamentation, which was reflected in the interference of coherent conical emission received from multiple sources (filaments). Thus, the formation of a superfilament did not occur under the conditions of paper [20] (see next).

An interaction of filaments in a collimated beam was considered in detail in paper [37]. It was numerically demonstrated for the formation of child filaments to take place due to the interference of annular spatial structures—diffracted energy reservoir—of individual filaments. In addition, it was noted that the filamentation pattern change from pulse to pulse could not be explained by laser fluctuations. For a more detailed study of filaments interaction, a numerical simulation of a pair of filaments propagation under an angle to each other was carried out in [50,51]. Attraction or repulsion of the filaments was claimed to occur depending on their

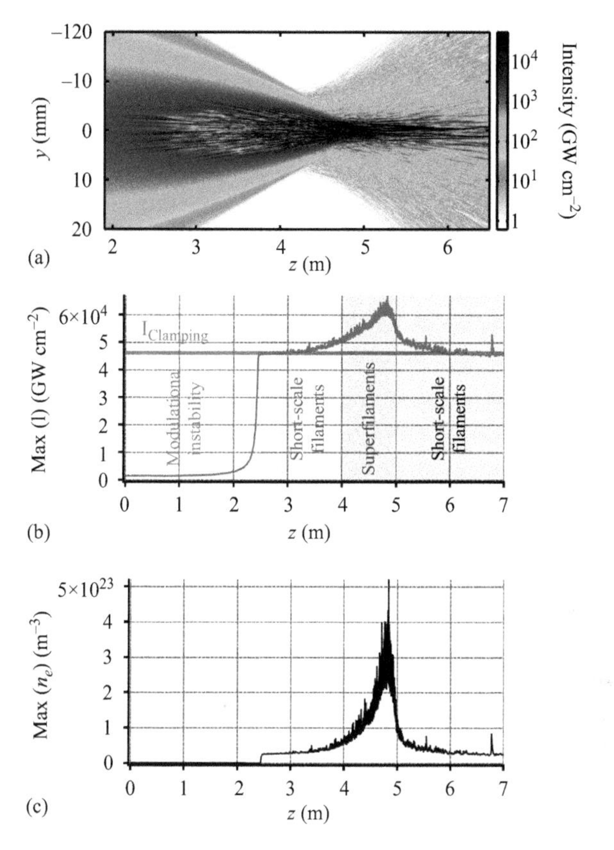

Figure 7.12 Simulated evolution of the multifilament bundle: Side view of the intensity map (a), evolution of the maximum intensity (b) and maximum electron density (c) in the transversal plane versus the distance [49]

relative phase. If the two filaments with close phases converged under a small angle, they were fused into a single one. This indicates the influence of a numerical aperture (i.e., an angle of convergence for individual filaments) on the interaction of filaments in a geometrically focused beam. Control of the interaction of a pair of filaments by changing the relative phase and convergence angle was experimentally demonstrated in [52]. It is important to note that when two beams were added together, the double increase in intensity was not observed, i.e., the linear optics approximation could not be used to calculate the resulting value. This effect is analogous to the clamping of intensity in the case of collimated laser beam and takes place due to a strong dependence of the plasma density on the electric field. As a result of filament addition, an increase in peak intensity of 30–40 percent can be expected compared to the case of a single filament [53,54]. A recent paper also demonstrated a

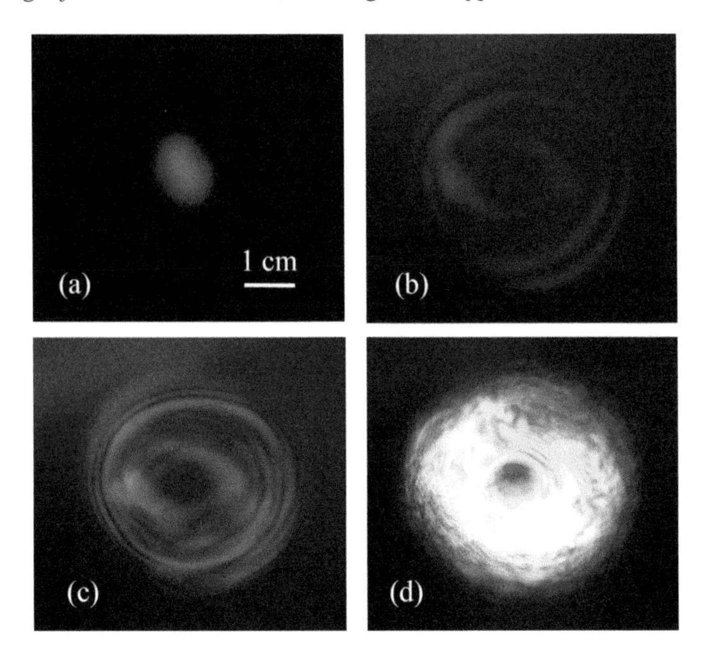

Figure 7.13 Transverse profiles of transmitted emission obtained by shooting the screen in the far-field region (45 cm from the filamentation region) for P = (a) 1, (b) 2, (c) 6.4, and (d) 68 GW. In paper [20], the pictures are in gray scale

noticeable increase in the conductivity of a plasma channel for a filament formed by combining two beams compared to the case of a single beam [55].

It was proposed in paper [56] to consider the interaction of several laser beams (4 or 6) as a more representative part of the multiple filamentation pattern. After passing through an amplitude mask with an appropriate number of diaphragms (4 or 6), the laser beam experienced geometric focusing. A sum on-axis filament appeared in front of the geometric focus (up to the point where the laser beams converged geometrically under the linear propagation mode).

In this case, the peak intensity and plasma density were higher than in the case of a single filament and depended on the initial number of beams. In addition, in the region behind the geometric focus, a bright post-filamentation channel appeared on the axis, which was absent in the case of linear propagation. Figure 7.14 demonstrates the fluence distribution at the distance twice longer than the geometrical focal length for the case of quasi-linear (i.e., for the under-critical power of a laser pulse) and filament fusion regime (Figure 7.14(b) and (c)). Transversal profile after "linear" propagation almost replicates initial amplitude mask (Figure 7.14(a)). On the contrary, after the filaments' interaction and sum filament formation, a brand new bright maximum on the optical axis emerges both in the experiments (Figure 7.14(b) and simulation (Figure 7.14(c)).

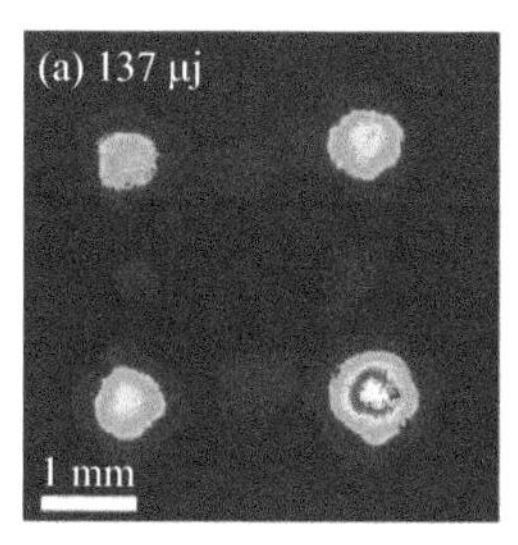

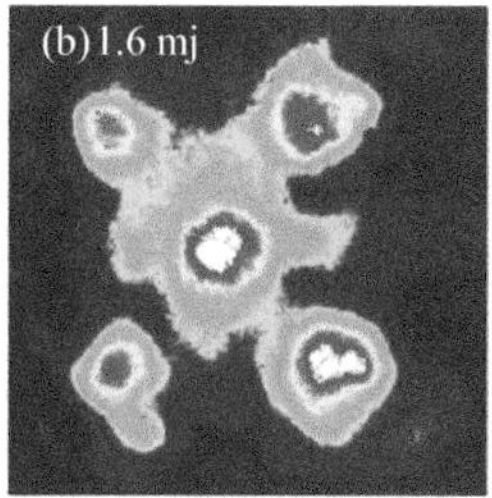

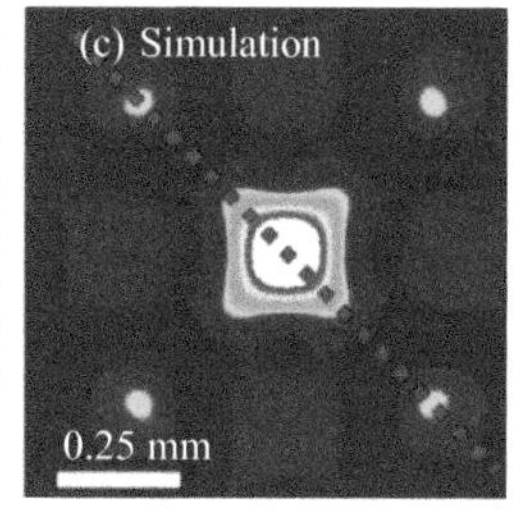

Figure 7.14 Fluence distributions at z≈2f: quasi-linear (a) and nonlinear (b) propagation regime (experiment) and (c) nonlinear propagation (numerical simulations) [56]

Similar results were obtained by regularizing a TW laser pulse in the case of weaker geometric focusing [57]. It was found that the linear energy density of light near the geometric focus in the case of superfilament formation was almost 20 times higher than in a single filament. This, in particular, is associated with a more efficient "collection" of energy from the reservoir into the paraxial region in the superfilamentation mode. The effect of superfilamentation can also be used to increase the energy input into the medium of propagation [58]. In this case, the linear energy density in the superfilament formed by filaments fusion in a regularized laser beam significantly exceeded the corresponding value for a single filament. A superfilament formed from a stochastic multiple filamentation pattern also had a plasma channel with higher plasma density and laser intensity as compared to single filaments, but lower ones in comparison with the case of a regularized beam.

It is important to note that by introducing not amplitude, but phase modulation, a superfilament can be prevented from the appearance. In [59], the laser beam was regularized by introducing a phase mask consisting of four sectors with a phase difference of π between adjacent sectors. After the mask, the laser beam structure corresponded to the TEM_{11} Hermite–Gaussian mode. In such a beam, all four filaments were formed independently and remained separated even in the beam waist.

A comparison of multiple filamentation under focusing of a laser beam with a numerical aperture of 10^{-2} and 4×10^{-3} was carried out in paper [60] that demonstrated for the superfilamentation to be sensitive to geometric focusing. The number of discernible plasma filaments in the superfilament was higher, and the increase in plasma density was more pronounced for tighter focusing.

The available set of data, both experimental and computer simulation, allow us to assert that, by adding filaments, it is possible to increase the intensity of a filament and achieve a higher plasma density, higher energy input into the medium, as compared to a single filament. In this case, both the numerical aperture and the number of initial filaments can affect the interaction process and the resulting intensity and density of the plasma. Moreover, a regularized beam has advantages over a laser beam with stochastically distributed filaments. By introducing phase modulation, it is possible, on the contrary, to avoid the interaction of filaments and to ensure the preservation of the initial number of individual beams even in the laser beam waist.

7.5 Post-filamentation channels

After filamentation of a laser pulse cease, areas of increased intensity often remain in the laser beam—so-called post-filamentation channels (PFC) of light looking like "hot spots" over the beam cross section. Such light structures are characterized by low divergence and relatively high intensity (in most cases insufficient, however, for effective plasma formation). In collimated laser beams, they were observed many times: for example, wide filaments in [48,61,62], which according to modern concepts are exactly post-filamentation channels of light. The mechanism of their formation was discussed in [38,63]. It was found that post-filamentation channels of light were formed due to diffraction of the energy reservoir by the plasma channel accompanying the filament. This plasma affects the energy reservoir practically like a small opaque circular screen and transforms the beam into a symmetric Bessel–Gaussian mode of light. The latter is subjected to the Kerr nonlinearity that just prevents linear diffraction. The resulting divergence of the light channels measured in [38] was 0.03 mrad. There was noted in a subsequent paper [64] that even in the case of a single filament, a set of post-filamentation channels of light could be formed. It was also demonstrated by putting an aperture around a post-filamentation channel that the low-intensity emission surrounding the "hot spot" helps to maintain the post-filamentation channel divergence at such a low level.

These characteristics of post-filamentation channels formed after the termination of filamentation for a collimated laser beam are important for delivering high-intensity laser beam to a distant target. For instance, in paper [62], the laser beam intensity in such a channel was up to 0.1 TW/cm^2 at a distance from the laser system of about 2 km. Moreover, the beam intensity in the post-filamentation channels can be so high that the transformation of the pulse spectrum still continues in them (see, e.g., [65]).

For geometrically focused laser beam, studies of post-filamentation channel parameters were carried out in papers [66,67]. A laser beam with a central wavelength of 744 nm underwent geometric focusing with a numerical aperture from 0.001 to ~0.02 (that corresponded to the focal lengths from 18 cm to about 3 m). The energy of a 100-fs pulse was 0.4 mJ (peak power ~1.5P_{cr}), 1 mJ (~3P_{cr}), or 3.9 mJ (~12P_{cr}). In the experimental part of this research, the laser beam transverse profile was measured after the termination of filamentation. The data obtained were used to determine both the radius of the laser beam as a whole and of the post-filamentation channels.

Since the dependence of both the radius of a laser beam as a whole and the radius of the post-filamentation channel on the distance is close to linear (Figure 7.15), the beam and the post-filamentation channel (a "hot spot" in the beam cross section) can be characterized by the coefficient of angular divergence.

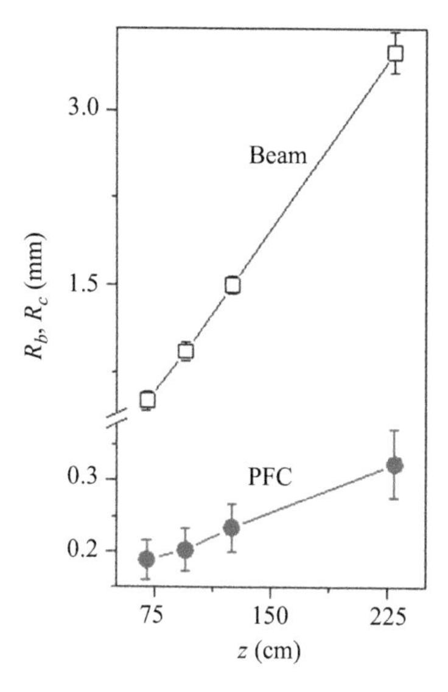

Figure 7.15 *Laser beam effective radius R_b (1) and post-filamentation channel (PFC) radius R_c (2) as functions of a distance from the linear focus z after the filamentation of a laser pulse with $E_0 = 3.9$ mJ, $f = 52$ [66]*

For the studied conditions (pulse energy—3.9 mJ, the focal length—52 cm), the divergence of the post-filamentation channel γ_c was about 0.07 mrad, the angular divergence of the laser beam γ_b—about 1.8 mrad. On the one hand, the angular divergence of the post-filamentation channel is higher than one in the case of collimated laser beam [38]. This fact is due to the geometric defocusing of a laser beam after its passing off the focal region. On the other hand, the experimentally obtained divergence γ_c was far less than both the angular divergence of the whole beam γ_b and the divergence γ_0 (~1 mrad) caused by geometric focusing of the subcritical laser beam propagating in the quasi-linear regime. The divergence of the whole beam γ_b, in its turn, exceeded the quasi-linear divergence γ_0.

For all studied laser beam parameters (namely, numerical apertures and energies), the same interrelation took place: the divergence γ_c of a post-filamentation channel was less than the quasi-linear divergence γ_0, and the divergence of a laser beam as a whole γ_b was higher than γ_0 (Figure 7.16). We would also like to note that the divergence of a post-filamentation channel depended on both the focusing tightness and the pulse energy. And the minimum divergence of 0.05 mrad obtained in the given experiments was achieved under the weakest focusing (f about 3 m) and still exceeded the divergence in the collimated case (~0.03 mrad).

One of the important reasons for the low divergence of a post-filamentation channel is the effect of the Kerr nonlinearity. Although self-focusing in this channel is not strong enough to result in refocusing and re-filamentation, it partially balances the diffraction. This leads to the preservation of the channel small radius and diminishing its angular divergence.

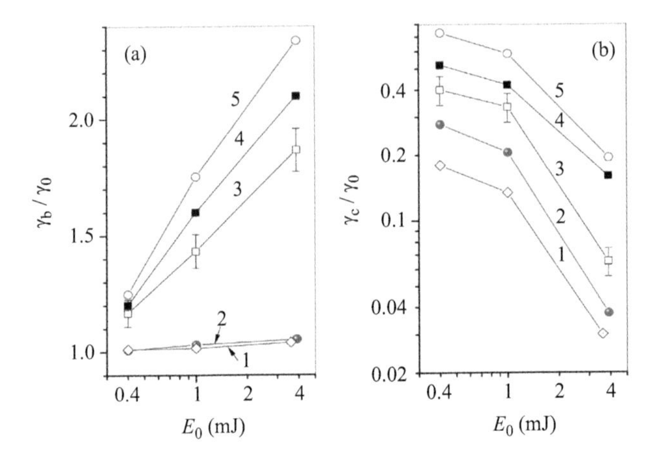

Figure 7.16 Experimental angular divergence of the whole laser beam γ_b (a) and of the post-filamentation channel γ_c (b) divided by the quasi-linear angular divergence γ_0 versus initial laser pulse energy for the focal length $f = 18$ (1); 25 (2); 52 (3); 110 (4); 295 cm (5) [66]

It is important to note that the post-filamentation channel radius growth rate goes down with the laser pulse energy rising, while the whole beam divergence, on the contrary, increases. This happens due to more active transformation of the laser beam transverse profile at higher energy, namely, into a ring structure around the filament. This ring structure is similar in shape to the Bessel–Gaussian mode.

In the numerical simulations [67], transverse profiles of the energy density (fluence) after the termination of the laser beam filamentation were compared for the cases of a twofold and tenfold excess of the pulse power over the critical one (Figure 7.17). At higher power, the observed number of rings in the transverse profile, as well as their relative intensity, increased. An examination of the diffraction of a Bessel–Gaussian transverse profile demonstrated that an increase in the number of rings in the distribution resulted in both an enlargement of the outer radius of the beam and a decrease in the radius of the central part. This conclusion is in excellent agreement with the experimentally observed tendency of the angular divergence of the whole beam γ_b and the number of rings rising with the energy enhancement.

For a post-filamentation channel, the same reasoning explains lower angular divergence at higher energies. It can be said that the circular light structure in the laser beam cross section plays the role of an energy reservoir and, by performing diffraction retention, contributes to maintaining a small radius of the "hot spot" in the transverse distribution of the beam. This hypothesis was experimentally tested (by analogy with [64]) in [67]. When this ring structure was removed by a diaphragm, the angular divergence of the post-filamentation channel was approximately doubled.

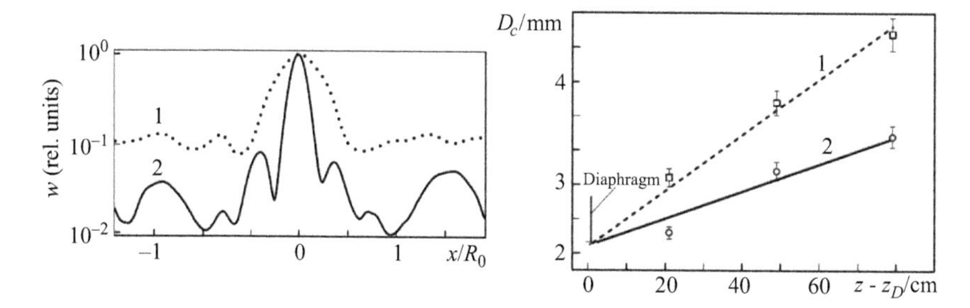

Figure 7.17 Transverse profile of the normalized pulse energy density (fluence) at $P_0/P_{cr} = 2$ (1) and 10 (2) (left). Dependences of an isolated post-filamentation channel diameter (1) and of the same channel as a part of the beam (2) on the distance at $E_0 = 1.4$ mJ and $f = 52$ cm (right) [67]

Subsequent studies of post-filamentation channels under a laser beam focusing demonstrated the dependence of such a light channel spectrum on the initial beam energy [68]. With an increase in the initial laser pulse energy, the spectrum of the post-filamentation channel became quite broader with a pronounced shift to the red region. In this case, the energy contained in the post-filamentation channel did not depend on the initial pulse energy. Taking into account a reduction of the angular divergence, this constancy of the energy in the post-filamentation channel leads to the conclusion that the intensity in it goes up with an increment of the initial pulse power. This conclusion, in turn, indicates a rising influence of the Kerr-effect self-focusing on the post-filamentation channel emission with an increase in the initial energy, which, along with the influence of the ring structure, results in a decline of the channel angular divergence.

Let us summarize what is known about the post-filamentation channels formed during the filamentation of a focused beam. First, the angular divergence of the post-filamentation channels formed after the termination of filamentation of a geometrically focused beam is not constant. It depends on both the focusing tightness and initial laser pulse energy. In this case, due to the presence of geometric defocusing, this divergence is expected to be greater than the divergence measured for collimated radiation. Second, this divergence is always less than the divergence due to geometric focusing and the divergence of the beam as a whole. Third, both the Kerr nonlinearity and the formation of an annular structure (a Bessel–Gaussian mode, i.e., an analogue of energy reservoir) around a "hot spot" in the beam cross section play an important role in maintaining a low angular divergence of a channel. An increase in the influence of both of these effects under initial energy growth leads to a reduction of the post-filamentation channel divergence. Fourth, at higher initial pulse energy, the intensity of the post-filamentation channel also rises and its spectral composition becomes quite richer (especially in the long-wavelength region).

7.6 Applications: third harmonic and terahertz emission

There are many various applications of filamentation under a laser beam geometric focusing. On the one hand, a geometrically focused laser beam is often more convenient than collimated one. Controlling the onset of filamentation, a short path through air to the start of a filament, less influence of the initial laser pulse energy on the latter, and the absence of necessity for a high excess over the critical power for self-focusing make geometric focusing quite convenient for carrying out model experiments under laboratory conditions. On the other hand, an increase in a laser pulse peak intensity in the filament of light and density of plasma in a corresponding plasma channel with enhancement of focusing tightness results in a higher efficiency of various nonlinear processes. Therefore, the overwhelming majority of researches on self-phase modulation, generation of harmonics, and terahertz emission were carried out in the presence of external geometric focusing.

For instance, the third harmonic generation under the propagation of tightly focused femtosecond laser beam (NA ≈ 0.07) in air was the subject of the paper in [69]. In the same paper, spectrum broadening and supercontinuum generation were considered. It should be noted that the authors claimed such a regime as air self-breakdown, but not as filamentation. The efficiency of the third harmonic generation in these experiments reached 0.17 percent. Therefore, at least formally, the third harmonic generation in the filamentation mode was clearly claimed in [70], where the third harmonic was generated in the air during the propagation of a weakly focused pulse (NA $\sim 10^{-3}$) from a Ti-sapphire laser.

It was stated in papers [71–73], that for a collimated laser beam, the supercontinuum generation efficiency increases with a filament elongation. A similar dependence on the filament length was observed for the third harmonics generation [74]. Simultaneous propagation of the third harmonic and supercontinuum along the filament was studied in papers [75–77]. In [75], when considering the simultaneous propagation of the fundamental frequency and the third harmonic beams, the term a two-color filament was introduced. As a result of a numerical study carried out in [75], it was demonstrated that under such a propagation, in addition to the spectrum broadening of the fundamental (laser) frequency, the spectrum broadening of the third harmonic occurred (Figure 7.18).

It was shown in [76] that the mutual influence of the supercontinuum and the fundamental frequency during this propagation along the filament had a strong effect on the third harmonic generation. It should be noted, that in the case of an extended interaction of a laser beam with the medium of propagation, the supercontinuum spectrum can overlap the third harmonic one and even exceed its amplitude. An experimental and numerical study of the two-color filamentation demonstrated that in a weakly focused laser beam, energy and intensity of the third harmonic generated during filamentation were stabilized with an increase in the fundamental frequency pulse energy [77]. It was shown in [78] that the generation of the third harmonic occurred spatially in the form of concentric rings. Studies of the third harmonic generation were also made in papers [79–81]. The conversion

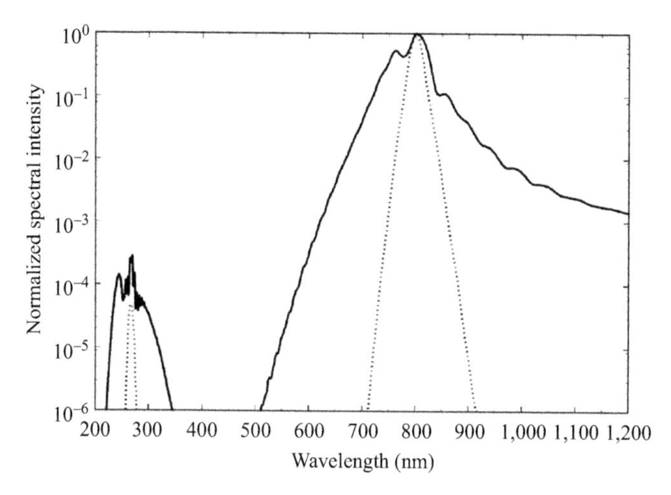

Figure 7.18 Spectra of fundamental frequency and the third harmonic at the beginning (dotted lines) and behind the filamentation region (solid lines) [75]

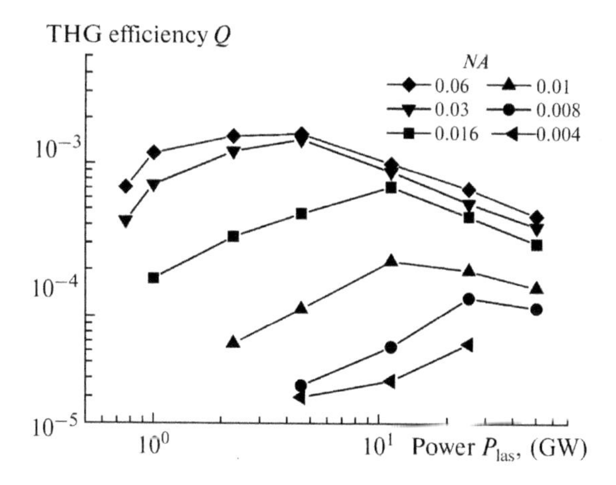

Figure 7.19 The third harmonic conversion efficiency Q versus laser pulse peak power P_{las} for various numerical apertures NA [82]

efficiency of the laser pulse energy into the third harmonic was 10^{-3} [79] and 1.2×10^{-3} [80]. In [81], with relatively tight focusing (NA = 0.01) of high-power (100 GW) laser pulses with a wavelength of 1.54 μm, the energy conversion coefficient to the third harmonic was 2×10^{-3}. In a later paper [82], the influence of focusing tightness on the third harmonic generation was studied. The authors paid attention to a non-monotonic dependence of the conversion efficiency of a laser beam energy into the third harmonic (Figure 7.19). It is interesting to note that the

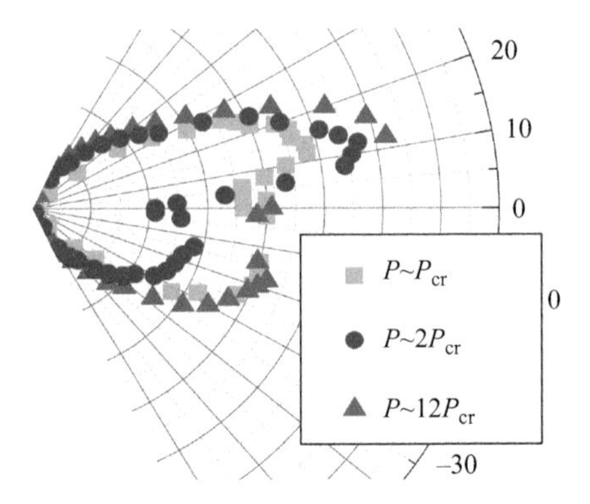

Figure 7.20 Angular distribution of THz emission for different peak powers of a laser pulse [86]

highest efficiency was observed at the maximum numerical aperture (NA = 0.06) and at the subcritical pulse power, i.e., in the absence of filamentation.

As was mentioned earlier, the plasma channel formed during the filamentation of femtosecond laser pulses could serve as a source of terahertz emission [11,83–85]. Typically, THz emission is generated in a so-called two-color filament, where the first and second harmonics of a laser pulse are combined into a single filament. The mechanism of generating THz emission in such a way was studied and understood. There are much fewer publications on THz generation in plasma of a single-color filament, and there is no complete understanding of how it occurs. For instance, in paper [83], the Cherenkov mechanism of THz generation was assumed. In later paper [84], a dipole or quadrupole generation mechanism was proposed. It was shown in [85] that the dipole mechanism made a decisive contribution to the generation of THz emission. It was reported in these papers [83–85] that the angular pattern of THz emission significantly depended on the focusing conditions—i.e., the focal length of a focusing element—by changing of which, the length of the plasma channel length corresponding to the filament of light was varied, the cone angle being determined by the expression $\theta \sim (\lambda_{THz}/L)^{-1/2}$. However, there was demonstrated in a later paper [86] that the change of the plasma channel length by several times with a change in the laser pulse power by more than an order of magnitude had practically no effect on the angular divergence of THz emission (Figure 7.20). As was shown in [87], the focusing conditions—in addition to affecting the angular divergence of THz emission—had a significant effect on its spectral characteristics (Figure 2.21): namely, an increase in the intensity of a filament due to a growth of the numerical aperture allowed one to obtain a higher frequency THz emission.

Thus, for a number of applications, it is quite convenient to use the filamentation of geometrically focused laser pulses, for instance, to obtain efficient third harmonic generation by increasing the focusing numerical aperture despite a decrease in the effective nonlinear interaction length. By the numerical aperture

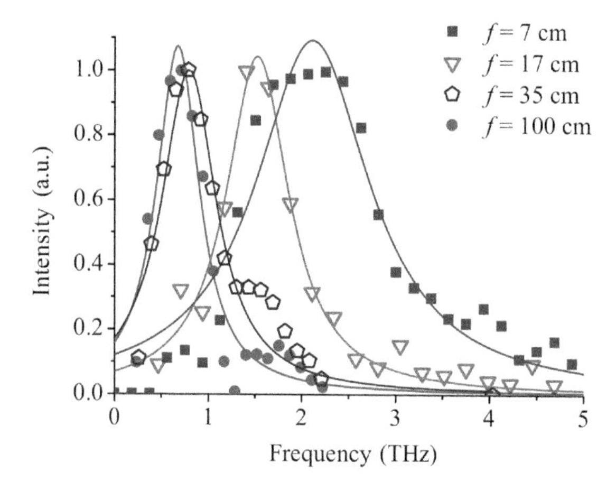

Figure 7.21 Experimental spectra of THz emission from a single-color filament under different focusing conditions (colored markers) and Lorentz fits of these experimental data (curves). Color of the curves corresponds to the color of markers [87]

growth, quite higher frequency THz radiation can be obtained due to the higher intensity of light and plasma density in the filament.

7.7 Conclusions

This chapter presents the results on studying the filamentation that occurs when focusing femtosecond laser pulses. It is shown that even with a fairly tight focusing (numerical aperture up to NA ~ 0.2), an extended (as compared to laser beam waist) plasma channel is formed, the geometric dimensions of which decline with increasing numerical aperture of the beam, and the plasma density at the same time increases. The transverse size of such a channel can be as low as several microns, which is an order of magnitude smaller than the size of the plasma channel formed during the filamentation of collimated laser beams. For laser pulses with peak power several times exceeding the critical one for the Kerr-nonlinearity self-focusing, the formation of several parallel filaments of light and corresponding plasma channels is possible (multiple filamentation). It is demonstrated that the energy reservoir is focused to a size comparable to the size of the filament near the geometric focus, and after passing through the focus, an extended post-filamentation channel is formed, angular divergence of which depends on the focusing tightness and initial laser pulse energy. In addition, an increase in the focusing tightness results in a growth of the laser pulse peak intensity and plasma density. It results in a higher efficiency of various nonlinear processes, such as generation of harmonics and supercontinuum, generation of terahertz emission, which can be useful for different applications.

References

[1] Sugioka K. and Chen Y. 'Ultrafast lasers—reliable tools for advanced materials processing'. *Light: Science & Applications*. 2014; 3(4): e149.

[2] Sugioka K. and Cheng Y. 'Fundamentals of femtosecond laser processing' in *Femtosecond Laser 3D Micromachining for Microfluidic and Optofluidic Applications*. London: Springer; 2014. pp. 19–33.

[3] Sugioka K. 'Progress in ultrafast laser processing and future prospects'. *Nanophotonics*. 2017; 6(2): 393–413.

[4] Pfeifenberger M. J., Mangang M., Wurster S.*et al.* 'The use of femtosecond laser ablation as a novel tool for rapid micro-mechanical sample preparation'. *Materials & Design*. 2017; 121: 109–18.

[5] Balling P. and Schou J. 'Femtosecond-laser ablation dynamics of dielectrics: basics and applications for thin films'. *Reports on Progress in Physics*. 2013; 76(3): 036502.

[6] Dausinger F. and Friedemann L. *Femtosecond Technology for Technical and Medical Applications*. Springer-Verlag Berlin Heidelberg; 2004. Vol. 96.

[7] Kabashin A. V. and Meunier M. 'Synthesis of colloidal nanoparticles during femtosecond laser ablation of gold in water'. *Journal of Applied Physics*. 2003; 94(12): 7941–43.

[8] Eliezer S., Eliaz N., Grossman E. *et al.* 'Synthesis of nanoparticles with femtosecond laser pulses'. *Physics Review B*. 2004; 69(14): 144119.

[9] Barcikowski S., Hahn A., Kabashin A. V., and Chichkov B. N. 'Properties of nanoparticles generated during femtosecond laser machining in air and water'. *Applied Physics A*. 2007; 87(1): 47–55.

[10] Chin S. L., Hosseini S. A., Liu W. *et al.* 'The propagation of powerful femtosecond laser pulses in optical media: physics, applications, and new challenges'. *Canadian Journal of Physics*. 2005; 83(9): 863–905.

[11] Couairon A. and Mysyrowicz A. 'Femtosecond filamentation in transparent media'. *Physics Reports*. 2007; 441(2–4): 47–189.

[12] Kandidov V. P., Shlenov S. A., and Kosareva O. G. 'Filamentation of high-power femtosecond laser radiation'. *Quantum Electronics*. 2009; 39 (5): 205–28.

[13] Marburger J. H. 'Self-focusing: theory'. *Progress in Quantum Electronics*. 1975; 4(1): 35–110.

[14] Talanov V. I. 'Focusing of light in cubic media'. *JETP Letters*. 1970; 11(6): 303–5.

[15] Liu W., Luo Q., and Chin S. L., 'Competition between multiphoton or tunnel ionization and filamentation induced by powerful femtosecond laser pulses in air'. *Chinese Optics Letters*. 2003; 1(1): 56–9.

[16] Liu W., Kosareva O., Golubtsov I. S. *et al.* 'Random deflection of the white light beam during self-focusing and filamentation of a femtosecond laser pulse in water'. *Applied Physics B*. 2002; 75(4–5): 595–99.

[17] Liu J., Duan Z., Zeng Z. *et al.* 'Time-resolved investigation of low-density plasma channels produced by a kilohertz femtosecond laser in air'. *Physics Review E*. 2005; 72: 026412.

[18] Deng Y. P., Zhu J. B., Ji Z. G. *et al.* 'Transverse evolution of a plasma channel in air induced by a femtosecond laser'. *Optics Letters*. 2006; 31(4): 546–8.

[19] Theberge F., Liu W. W., Simard P. T., Becker A., and Chin S. L. 'Plasma density inside a femtosecond laser filament in air: strong dependence on external focusing'. *Physics Review E*. 2006; 74(3): 036406.

[20] Ionin A. A., Kudryashov S. I., Makarov S. V., Seleznev L. V., and Sinitsyn D. V. 'Multiple filamentation of intense femtosecond laser pulses in air'. *JETP Letters*. 2009; 90(6): 423–7.

[21] Geints Y. E. and Zemlyanov A. A. 'Self-focusing of a focused femtosecond laser pulse in air'. *Applied Physics B*. 2010; 101(4): 735–42.

[22] Papazoglou D. G. and Tzortzakis S. 'In-line holography for the characterization of ultrafast laser filamentation in transparent media'. *Applied Physics Letters*. 2008; 93(4): 041120.

[23] Liu X.-L., Lu X., Liu X. *et al.* 'Tightly focused femtosecond laser pulse in air: from filamentation to breakdown'. *Optics Express*. 2010; 18(25): 26007–17.

[24] Liu X.-L., Lu X., Liu X. *et al.* 'Broadband supercontinuum generation in air using tightly focused femtosecond laser pulses'. *Optics Letters*. 2011; 36 (19): 3900–2.

[25] Daigle J.-F., Jaron-Becker A., Hosseini S. *et al.* 'Intensity clamping measurement of laser filaments in air at 400 and 800 nm'. *Physics Review A*. 2010; 82(2): 023405.

[26] Kiran P. P., Bagchi S., Arnold C. L., Krishnan S. R., Kumar G. R., and Couairon A. 'Filamentation without intensity clamping'. *Optics Express*. 2010; 18(20): 21504–10.

[27] Geints Y. E., Zemlyanov A. A., Ionin A. A. *et al.* 'Peculiarities of filamentation of sharply focused ultrashort laser pulses in air'. *Journal of Experimental and Theoretical Physics*. 2010; 111(5): 724–30.

[28] Lim K., Durand M., Baudelet M., and Richardson M. 'Transition from linear- to nonlinear-focusing regime in filamentation'. *Scientific Reports*. 2014; 4: 7217.

[29] Reyes D., Baudelet M., Richardson M., and Fairchild S. R. 'Transition from linear- to nonlinear-focusing regime of laser filament plasma dynamics'. *Journal of Applied Physics*. 2018; 124: 053103.

[30] Chin S. L., Brodeur A., Petit S., Kosareva O. G., and Kandidov V. P. 'Filamentation and supercontinuum generation during the propagation of powerful ultrashort laser pulses in optical media (white light laser)'. *Journal of Nonlinear Optical Physics & Materials*. 1999; 8(1): 121–46.

[31] Brodeur A., Chien C. Y., Ilkov F. A., Chin S. L., Kosareva O. G., and Kandidov V. P. 'Moving focus in the propagation of ultrashort laser pulses in air'. *Optics Letters*. 1997; 22(5): 304–6.

[32] Dubietis A. A., Gaizauskas E., Tamosauskas G., and Trapani P. D.'Light filaments without self-channeling'. *Physics Review Letters*. 2004; 92(25): 253903.
[33] Liu W., Gravel J. F., Théberge F., Becker A., and Chin S. L. 'Background reservoir: its crucial role for long-distance propagation of femtosecond laser pulses in air'. *Applied Physics B*. 2005; 80: 857–60.
[34] Mills M. S., Kolesik M., and Christodoulides D. N. 'Dressed optical filaments'. *Optics Letters*. 2013; 38: 25–7.
[35] Geints Y. E. and Zemlyanov A. A. 'Ring-Gaussian laser pulse filamentation in a self-induced diffraction waveguide'. *Journal of Optics*. 2017; 19(10): 105502.
[36] Walter D., Bürsing H., and Ebert R. 'Emissions from single filaments in air triggered by tailored background energy flows'. *Optics Letters*. 2017; 42(5): 931–4.
[37] Hosseini S. A., Luo Q., Ferland B. *et al.* 'Competition of multiple filaments during the propagation of intense femtosecond laser pulses'. *Physics Review A*. 2004; 70: 033802.
[38] Daigle J.-F., Kosareva O., Panov N. *et al.* 'Formation and evolution of intense, post-filamentation, ionization-free low divergence beams'. *Optics Communications*. 2011; 284(14): 3601–6.
[39] Mlejnek M., Kolesik M., Moloney J. V., and Wright E. M. 'Optically turbulent femtosecond light guide in air'. *Physical Review Letters*. 1999; 83 (15): 2938–41.
[40] Kandidov V. P., Kosareva O. G., and Koltun A. A. 'Nonlinear-optical transformation of a high-power femtosecond laser pulse in air'. *Quantum Electronics*. 2003; 33(1): 69–75.
[41] Dergachev A. A., Ionin A. A., Kandidov V. P.*et al.* 'The influence of the energy reservoir on the plasma channel in focused femtosecond laser beams'. *Laser Physics*. 2015; 25(6): 065402.
[42] Pilipetskii N. F. and Rustamov A. R. 'Observation of self-focusing of light in liquids'. *JETP Letters*. 1965; 2(2): 55–6.
[43] Bespalov V. I. and Talanov V. I. 'Filamentary structure of light beams in nonlinear liquids'. *JETP Letters*. 1966; 3(10): 307–10.
[44] Schillinger H. and Sauerbrey R. 'Electrical conductivity of long plasma channels in air generated by self-guided femtosecond laser pulses'. *Applied Physics B*. 1999; 68(4): 753–6.
[45] Chin S. L., Petit S., Liu W. *et al.* 'Interference of transverse rings in multifilamentation of powerful femtosecond laser pulses in air'. *Optics Communications*. 2002; 210: 329–41.
[46] Chin S. L., Talebpour A., Yang J. *et al.* 'Filamentation of femtosecond laser pulses in turbulent air'. *Applied Physics B*. 2002; 74(1): 67–76.
[47] Liu W., Hosseini S. A., Luo Q. *et al.* 'Experimental observation and simulations of the self-action of white light laser pulse propagating in air'. *New Journal of Physics*. 2004; 6: 6.
[48] Mejean G., D'Amico C., Andre Y.-B. *et al.* 'Range of plasma filaments created in air by a multi-terawatt femtosecond laser'. *Optics Communications*. 2005; 247: 171–80.

[49] Point G., Brelet Y., Houard A. *et al.* 'Superfilamentation in air'. *Physical Review Letters*. 2014; 112(22): 223902.

[50] Xi T.-T., Lu X., and Zhang J. 'Interaction of light filaments generated by femtosecond laser pulses in air'. *Physical Review Letters*. 2006; 96(2): 025003.

[51] Ma Y.-Y., Lu X., Xi T.-T., Gong Q.-H., and Zhang J. 'Filamentation of interacting femtosecond laser pulses in air'. *Applied Physics B*. 2008; 93: 463–8.

[52] Shim B., Schrauth S. E., Hensley C. J. *et al.* 'Controlled interactions of femtosecond light filaments in air'. *Physical Review A*. 2010; 81: 061803.

[53] Kosareva O. G., Liu W., Panov N. A. *et al.* 'Can we reach very high intensity in air with femtosecond PW laser pulses?'. *Laser Physics*. 2009; 19(8): 1776–92.

[54] Xu S., Zheng Y., Liu Y., and Liu W. 'Intensity clamping during dual-beam interference'. *Laser Physics*. 2010; 20(11): 1968–72.

[55] Reyes D., Peña J., Walasik W., Litchinitser N., Fairchild S. R., and Richardson M. 'Filament conductivity enhancement through nonlinear beam interaction'. *Optics Express*. 2020; 28(18): 26764–73.

[56] Shipilo D. E., Panov N. A., Sunchugasheva E. S. *et al.* 'Fusion of regularized femtosecond filaments in air: far field on-axis emission'. *Laser Physics Letters*. 2016; 13(11): 116005.

[57] Pushkarev D. V., Mitina E. V., Uryupina D. S. *et al.* 'Nonlinear increase in the energy input into a medium at the fusion of regularized femtosecond filaments'. *JETP Letters*. 2017; 106(9): 561–4.

[58] Pushkarev D., Mitina E., Shipilo D. *et al.* 'Transverse structure and energy deposition by a subTW femtosecond laser in air: from single filament to superfilament'. *New Journal of Physics*. 2019; 21: 033027.

[59] Pushkarev D., Shipilo D., Lar'kin A. *et al.* 'Effect of phase front modulation on the merging of multiple regularized femtosecond filaments'. *Laser Physics Letters*. 2018; 15: 045402.

[60] Murzanev A., Bodrov S., Samsonova Z., Kartashov D., Bakunov M., and Petrarca M. 'Superfilamentation in air reconstructed by transversal interferometry'. *Physical Review A*. 2019; 100: 063824.

[61] La Fontaine B., Vidal F., Jiang Z. *et al.* 'Filamentation of ultrashort pulse laser beams resulting from their propagation over long distances in air'. *Physics of Plasmas*. 1999; 6(5): 1615–21.

[62] Méchain G., Couairon A., André Y.-B. *et al.* 'Long-range self-channeling of infrared laser pulses in air: a new propagation regime without ionization'. *Applied Physics B*. 2004; 79: 379–82.

[63] Chen Y., Théberge F., Kosareva O., Panov N., Kandidov V. P., and Chin S. L. 'Evolution and termination of a femtosecond laser filament in air'. *Optics Letters*. 2007; 32(24): 3477–9.

[64] Gao H., Liu W., and Chin S. L. 'Post-filamentation multiple light channel formation in air'. *Laser Physics*. 2014; 24: 055301.

[65] Ivanov N. G., Losev V. F., Prokop'ev V. E., and Sitnik K. A. 'Generation of a highly directional supercontinuum in the visible spectrum range'. *Optics Communications*. 2017; 387: 322–7.
[66] Geints Y. E., Ionin A. A., Mokrousova D. V. *et al.* 'High intensive light channel formation in the post-filamentation region of ultrashort laser pulses in air'. *Journal of Optics*. 2016; 18(9): 095503.
[67] Geints Yu. E., Zemlyanov A. A., Ionin A. A., Mokrousova D. V., Seleznev L. V., and Sinitsyn D. V. 'Post-filamentation propagation of high-power laser pulses in air in the regime of narrowly focused light channels'. *Quantum Electronics*. 2016; 46(11): 1009–14.
[68] Geints Yu. E., Ionin A. A., Mokrousova D. V. *et al.* 'Energy, spectral, and angular properties of post-filamentation channels during propagation in air and condensed media'. *Journal of the Optical Society of America B*. 2019; 36(10): G19.
[69] Fedotov A. B., Koroteev N. I., Loy M. M. T., Xiao X., and Zheltikov A. M. 'Saturation of third-harmonic generation in a plasma of self-induced optical breakdown due to the self-action of 80-fs light pulses'. *Optics Communications*. 1997; 133: 587–95.
[70] Akozbek N., Iwasaki A., Becker A., Scalora M., Chin S. L., and Bowden C. M. 'Third-harmonic generation and self-channeling in air using high-power femtosecond laser pulses'. *Physical Review Letters*. 2002; 89(14): 143901.
[71] Gaeta A. L. 'Catastrophic collapse of ultrashort pulses'. *Physical Review Letters*. 2000; 84: 3582–5.
[72] Kandidov V. P., Kosareva O. G., Golubtsov I. S. *et al.* 'Self-transformation of a powerful femtosecond laser pulse into a white-light laser pulse in bulk optical media (or supercontinuum generation)'. *Applied Physics B*. 2003; 77: 149–65.
[73] Skupin S. and Bergé L. 'Supercontinuum generation of ultrashort laser pulses in air at different central wavelengths'. *Optics Communications*. 2007; 280(1): 173–82.
[74] Theberge F., Luo Q., Liu W., Hosseini S. A., Sharifi M., and Chin S. L. 'Long-range third-harmonic generation in air using ultrashort intense laser pulses'. *Applied Physics Letters*. 2005; 87(8): 081108.
[75] Akozbek N., Becker A., Scalora M., Chin S. L., and Bowden C. M. 'Continuum generation of the third-harmonic pulse generated by an intense femtosecond IR pulse in air'. *Applied Physics B*. 2003; 77: 177–83.
[76] Kolesik M., Wright E. M., Becker A., and Moloney J. V. 'Simulation of third-harmonic and supercontinuum generation for femtosecond pulses in air'. *Applied Physics B*. 2006; 85: 531–8.
[77] Theberge F., Filion J., Akozbek N., Chen Y., Becker A., and Chin S. L. 'Self-stabilization of third-harmonic pulse during two-color filamentation in gases'. *Applied Physics B*. 2007; 87: 207–10.
[78] Theberge F., Akozbek N., Liu W., Gravel J.-F., and Chin S. L. 'Third harmonic beam profile generated in atmospheric air using femtosecond laser pulses'. *Optics Communications*. 2005; 245: 399–405.

[79] Yang H., Zhang J., Zhao L. *et al.* 'Third-order harmonic generation by self-guided femtosecond pulses in air'. *Physical Review E*. 2003; 67: 015401.

[80] Ganeev R. A., Suzuki M., Baba M., Kuroda H., and Kulagin I. A. 'Third-harmonic generation in air by use of femtosecond radiation in tight-focusing conditions'. *Applied Optics*. 2006; 45(4): 748–55.

[81] Naudeau M. L., Law R. J., Luk T. S., Nelson T. R., Cameron S. M., and Rudd J. V. 'Observation of nonlinear optical phenomena in air and fused silica using a 100 GW, 1.54 μm source'. *Optics Express*. 2006; 14(13): 6194–200.

[82] Ionin A. A., Kudryashov S. I., Seleznev L. V., Sinitsyn D. V., Sunchugasheva E. S., and Fedorov V. Y. 'Third harmonic generation by ultrashort laser pulses tightly focused in air'. *Laser Physics*. 2011; 21(3): 500–4.

[83] D'Amico C., Houard A., Franco M. *et al.* 'Conical forward THz emission from femtosecond-laser-beam filamentation in air'. *Physical Review Letters*. 2007; 98(23): 235002.

[84] Panov N. A., Kosareva O. G., Andreeva V. A. *et al.* 'Angular distribution of the terahertz radiation intensity from the plasma channel of a femtosecond filament'. *JETP Letters*. 2011; 93(11): 638–41.

[85] Shkurinov A. P., Sinko A. S., Solyankin P. M. *et al.* 'Impact of the dipole contribution on the terahertz emission of air-based plasma induced by tightly focused femtosecond laser pulses'. *Physical Review E*. 2017; 95(4): 043209.

[86] Koribut A. V., Rizaev G. E., Mokrousova D. V. *et al.* 'Similarity of angular distribution for THz radiation emitted by laser filament plasma channels of different lengths'. *Optics Letters*. 2020; 45(14): 4009–11.

[87]. Mokrousova D. V., Savinov S. A., Seleznev L. V. *et al.* 'Tracing air-breakdown plasma characteristics from single-color filament terahertz spectra'. *International Journal of Infrared and Millimeter Waves*. 2020; 41: 1105–13.

Chapter 8

Linear and nonlinear exotic light wave packets, physics and applications

Dimitris G. Papazoglou[1,2] *and Stelios Tzortzakis*[1–3]

8.1 Airy beams

In 2007 Siviloglou *et al.* [1,2] demonstrated a new kind of non-spreading optical waves, the Airy beams. Their name originates from their envelope that is described by an Airy function [3]. In contrast to other non-spreading optical waves, for example the Bessel beams [4], the Airy distribution is the only dispersion-free solution of the paraxial wave equation in one dimension [2,5]. Furthermore, Airy wave packets present an exciting property, their main intensity lobes shift transversely in a quadratic fashion during propagation [1,2,6]. For this, they are also referred to as "accelerating" since their trajectory is similar to those of a projectile moving under the action of gravity [6]. Their unique properties, such as propagating in curved trajectories [2,6,7] and resisting to diffraction or dispersion [8,9], have ignited the research interest of various research communities working in linear and nonlinear optics [7,9–13]. Landmarks in this field include, among others, the use of intense Airy beams to generate curved plasma strings in air [7] and the generation of stable linear light bullets by combining non-dispersing Airy pulses with non-diffracting beams, first as Airy–Bessel [9], and then pure Airy–Airy–Airy (or Airy^3) light bullets [13].

The nonlinear propagation of Airy beams has also attracted a lot of attention [14–16]. As the peak intensity of the Airy beam increases, it either breaks up [17] radiating a series of tangential emissions or exhibits shrinking and modification of the Airy profile [12,13]. Kaminer *et al.* [14] and Lotti *et al.* [15] demonstrated the existence of stationary accelerating Airy wave packets in the presence of third-order (Kerr) nonlinearity and nonlinear losses. Furthermore, a surprising property was recently revealed for the family of optical wave packets whose distribution is described by the Airy function: their harmonics preserve the phase distribution of

[1]Institute of Electronic Structure and Laser, Foundation for Research and Technology Hellas, Heraklion, Greece
[2]Materials Science and Technology Department, University of Crete, Heraklion, Greece
[3]Science Program, Texas A&M University at Qatar, Doha, Qatar

the fundamental beam [18]. Consequently this "phase memory" imparts to the harmonics the propagation properties of the fundamental.

The idea of a beam that propagates in a curved trajectory has been further exploited by Efremidis and Christodoulides in 2010 [19] by introducing a cylindrically symmetric variation, the ring-Airy beam. The radially symmetric intensity lobes in this case shrink down in a nonlinear fashion as the beam propagates resulting in an abrupt autofocus [19,20]. For their abrupt autofocusing behavior, they are often referred to as CABs an acronym for circularly symmetric abruptly autofocusing beams. Interestingly, in the nonlinear propagation regime, they preserve their abrupt autofocus, but they also experience a minor nonlinear focus shift [21], while they are transformed to nonlinear intense light bullets. These properties make them ideal for applications that involve accurate deposition of energy in space, such as multiphoton polymerization [22] and intense THz fields [23]. Furthermore, through the "phase memory" effect, which imparts to the harmonics the abrupt autofocusing behavior [18], it is possible, under certain conditions, for the foci of the harmonics to coincide in space with the ones of the fundamental.

8.1.1 Properties

The exotic properties of the Airy distribution were first theoretically introduced in 1979 in the context of quantum mechanics, by Berry and Balazs [5]. As they have shown, by solving the Schrödinger equation for a particle in free space, when its wave packet distribution is described by an Airy function, the probability density propagates in free space without distortion and with constant acceleration.

Furthermore, in the one-dimensional (1D) domain the Airy distribution, along with the trivial plane wave, are the only non-spreading solutions [1,2,5], i.e. an Airy pulse does not spread in time.

8.1.1.1 Linear propagation

The mathematical equivalence of the Schrödinger equation in quantum mechanics with the paraxial wave propagation equation in optics was the basis for the first realization of optical Airy wave packets [1]. The paraxial wave equation in a homogenous medium takes the form:

$$i\frac{\partial U}{\partial z}+\frac{1}{2k}\nabla_{\perp}^{2}U=0 \tag{8.1}$$

where U is the wave amplitude and k the wavenumber. In this case [1,5] the solution for the propagation of an ideal 1D Airy beam is given by

$$U(x,z)=U_{o}Ai\left(x-\frac{z^{2}}{4k^{2}w^{4}}\right)e^{i\left(\left(xz/2kw^{3}\right)-\left(z^{3}/12k^{3}w^{6}\right)\right)} \tag{8.2}$$

where U_o is the amplitude and w is the width parameter of the Airy distribution. Clearly from (8.2) the transverse intensity profile does not change in a curvilinear coordinate system under free propagation. The intensity maxima, as shown in Figure 8.1, follow a

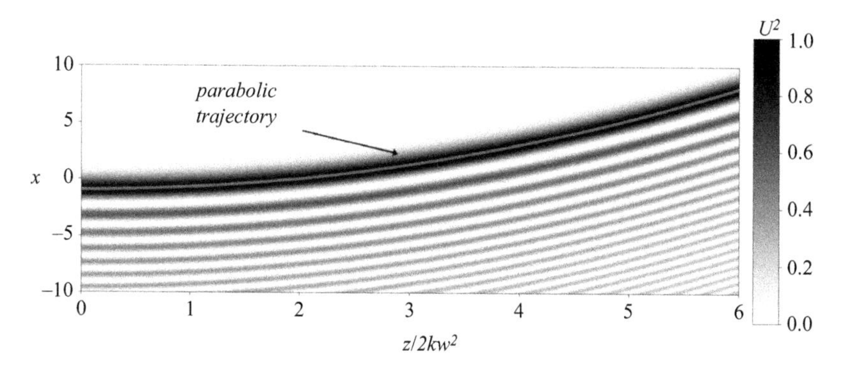

Figure 8.1 Intensity profile of a propagating 1D Airy beam (normalized values)

parabolic trajectory described by $x_{\max}(z) = x_{\max}(0) = z^2/4k^2w^4$. This trajectory is similar to that of a projectile moving under the action of gravity [6]; thus these wave packets are referred to as "accelerating."

Airy wave packets, like any other pure non-diffracting solution of the wave equation, carry infinite power [2]. Experimental implementation of any non-diffracting beam is correlated with some kind of apodization or truncation [2,4,16]. In the case of Airy wave packets, as Siviloglou *et al.* [2] demonstrated, it is beneficial to perform this apodization using an exponential mask. In this case, the field at $z = 0$ is described as

$$U(x,0) = U_o Ai(x) e^{ax} \tag{8.3}$$

where $0 < a \ll 1$ is the truncation factor. Although other truncation functions can be used, this, as we describe in more detail later on, has the interesting property that its spatial Fourier transform (FT) is a Gaussian distribution with a cubic chirp in the spatial phase. The propagation of the exponentially truncated Airy beam is similar to that of an ideal Airy and can be described [2] by

$$U(x,z) = U_o Ai\left(\frac{x_o + x}{w} - \frac{z^2}{4k^2w^4} + i\frac{a\,z}{k\,w^2}\right) e^{a((x_o+x)/w)-\left(az^2/2k^2w^4\right)+i\left(\left(a^2z/2kw^2\right)+\left(x_o+x/2kw^3\right)z-\left(z^3/12k^3w^6\right)\right)} \tag{8.4}$$

where x_o controls the peak position of its primary lobe. Since the Airy distribution is 1D, it can be independently applied to both x and y spatial dimensions to generate a two-dimensional (2D) Airy beam [2,7], shown in Figure 8.2(a). In this case, the field of an exponentially truncated 2D Airy beam at $z = 0$ is described as

$$U(x,y,0) = U_o Ai(x) Ai(y) e^{a(x+y)} \tag{8.5}$$

This approach can be extended even further to generate three-dimensional (3D) Airy wave packets [13], depicted in Figure 8.2(b), referred to as Airy^3 or Airy–Airy–Airy. Since in the linear regime the space and time are decoupled [9],

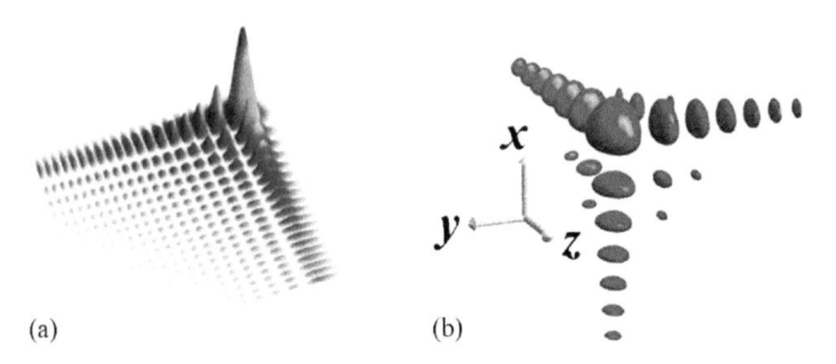

Figure 8.2 *(a) Intensity profile of a 2D Airy (normalized values) and (b) iso-intensity profile of an Airy^3 wave packet*

the spatiotemporal distribution of the intensity of an Airy^3 wave packet, assuming for simplicity an apodization factor $a = 0$, can be written as [13]

$$I(x,y,z,t) = U_o^2 Ai^2\left(\frac{x_o + x}{w_x} - \frac{z^2}{4k^2 w_x^4}\right) Ai^2\left(\frac{y_o + y}{w_y} - \frac{z^2}{4k^2 w_y^4}\right) Ai^2\left(\frac{\tau_o + \tau}{\tau_w} - \frac{k_o'' 2z^2}{4\tau_w^4}\right) \tag{8.6}$$

where $\tau = t - z/v_g$ is the reduced time, v_g is the group velocity of the wave packet envelope, w_x, w_y, τ_w are the wave packet spatiotemporal width parameters, x_o, y_o, τ_o control the peak position of its primary lobe in space and time at $z, \tau = 0$, and $k''_0 = \partial^2 k/\partial\omega^2$ is the dispersion coefficient of the medium at the central frequency. By combining the non-diffracting properties of the 2D Airy with the non-dispersing properties of an Airy pulse, Airy^3 wave packets propagate as light bullets [9,13].

8.1.1.2 Generation techniques

There are several approaches that can be used to generate accelerating Airy wave packets [2,9,10,13,24]. The first observation of such beams by Siviloglou *et al.* [2] relied on the fact that the angular Fourier spectrum [25] of an exponentially truncated Airy function is a Gaussian distribution modulated with a cubic phase [1,2]:

$$F\{Ai(s)e^{a\cdot s}\} = e^{-a\cdot k_\perp^2} e^{i(k_\perp^3/3)} e^{-(a^2(a-ik_\perp)/3)} \tag{8.7}$$

where $k_\perp = \sqrt{k_x^2 + k_y^2}/k$ is the normalized transverse wavenumber. In this context, an optical Airy beam can be generated by imprinting a cubic phase modulation on a Gaussian beam and then spatially Fourier transforming [25] it using a converging lens. Likewise, in the temporal domain, Airy pulses can be generated by imprinting a cubic phase modulation in the spectral components of a Gaussian pulse [13].

The most common experimental approach to impose a spatial cubic phase in a Gaussian beam is using a phase spatial light modulator (SLM) [2,7]. As shown in Figure 8.3, a Gaussian beam illuminates an SLM positioned at the first focal plane of a lens of focal length *f*. The SLM imprints a cubic spatial phase on a Gaussian beam and the phase modulated beam is then spatially Fourier transformed by the lens. The Airy distribution is then formed at the second focal plane of the lens. The

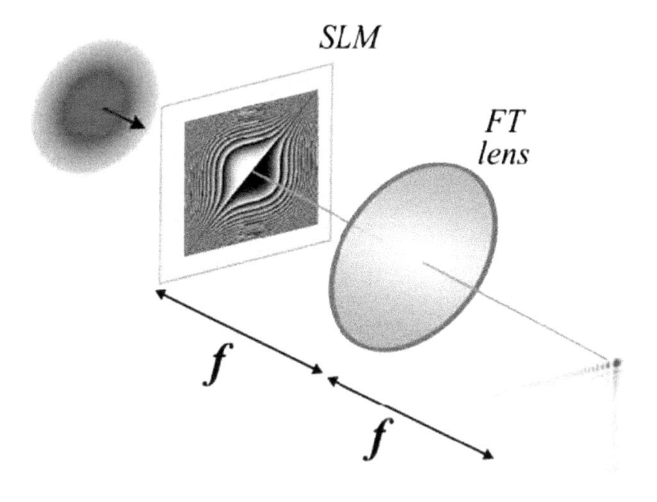

Figure 8.3 Generation of a 2D Airy beam using a phase SLM and a Fourier transforming lens

distribution parameters, such as the width parameter w and the truncation factor a, are controlled by the magnitude of the cubic phase that the SLM imprints, the waist size of the Gaussian beam and the focal length of the FT lens.

Although using an SLM to imprint the spatial cubic phase is straightforward and relatively easy to implement, it has several drawbacks. For example, due to physical limitations SLMs cannot actually impose a phase variation larger than 2π and rely on phase wrapping to achieve larger than 2π spatial phase gradients. This limitation in combination to the fact that they are discrete devices, comprising an array of pixels, sets an upper limit to the spatial phase gradient. Another problem is that SLM modulators have a relatively low damage threshold [2,7,10] and cannot be used for high power applications. On the other hand, high-intensity-accelerating Airy beams were generated by Polynkin *et al.* [7] using phase masks. Although not versatile as SLMs, phase masks have higher damage thresholds enabling them to be used in high power applications. One drawback is that phase masks are designed for a specific wavelength. This, in combination to the fact that they also rely on phase wrapping for achieving large phase variations, makes them unsuitable for broadband beams.

An alternative approach, presented in 2010 by Papazoglou *et al.* [10], that relies on the use of optical aberrations as continuous phase masks can overcome the abovementioned drawbacks. In more detail, coma aberration is related to a cubic phase modulation as revealed by the general formulation of the Seidel wave aberrations for cylindrical lenses [10,26]:

$$\Phi_{cyl}^{(4)} = -\frac{1}{4}Bx^4 - \frac{1}{2}(2C + D)x_o^2x^2 + E\,x_o^3x + Fx_ox^3 \tag{8.8}$$

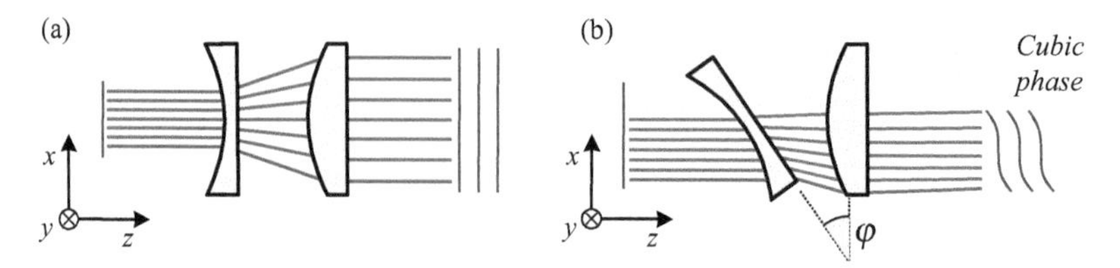

Figure 8.4 *Using coma aberration to induce cubic phase: (a) cylindrical telescope, (b) tilting the first diverging cylindrical lens by an angle φ induces optical aberrations. In order to compensate all the undesired aberrations, except the coma, the second converging lens is displaced in the longitudinal and transverse direction*

where $\Phi_{cyl}^{(4)}$ is the wavefront aberration, B, C, D, E, F are the spherical aberration, astigmatism, field curvature, distortion and coma aberration coefficients, respectively, x_o is the object height and x is the distance, measured from the lens center, along the normal cylindrical axis of the lens. Unfortunately, when using a single lens, the coma-induced cubic term in (8.8) cannot, in general, be isolated from the other terms (quadratic, quartic). However, this is possible using a two-lens optical system [10,27]. By combining lenses with opposite signs of aberration coefficients, and following well-known optical design strategies [26], the optical system can be tuned so that only a single optical aberration will dominate [10,27].

As shown in Figure 8.4, this can be achieved by using a tilted variant of a cylindrical beam expander design [10]. This technique is extended to two dimensions by cascading two cylindrical telescopic systems in a perpendicular orientation, as shown in Figure 8.5(a). A typical example of the resulting 2D phase modulation, as estimated by raytracing, is shown in Figure 8.5(b). The contour curves are in good agreement with a 2D cubic phase $(x^3 + y^3)$ fit for a ~6×6 mm^2 beam. A nice example of an experimentally generated 2D Airy beam is shown in Figure 8.5(c). Compared to other generation techniques, this approach has several advantages. It results in a continuous phase variation that is scalable, both in beam size and in wavelength, and is suitable for high power. For example, by replacing the cylindrical lenses with cylindrical mirrors [27], the same concept can be used to generate Airy beams, throughout the electromagnetic spectrum, like for UV or even THz radiation.

Other generation techniques rely on approximate descriptions of the Airy distribution. For example, the $x^{3/2}$ phase term in the approximate description of the Airy function $Ai(-x) \approx \sin\left[2x^{3/2}/3 + \pi/4\right]/\left(\sqrt{\pi}x^{1/4}\right)$ [3] makes it possible to generate accelerating Airy beams using a 3/2 phase-only pattern [24]. Furthermore, accelerating beams in general can be generated using caustic engineering [28,29]. In this case, the desired caustic is pre-engineered in 3D space enabling the generation of paraxial and non-paraxial accelerating beams. Technically, in all of these approaches SLMs are used to achieve the required phase modulation, so they suffer from the drawbacks, like power limitations, discretization and aliasing mentioned earlier.

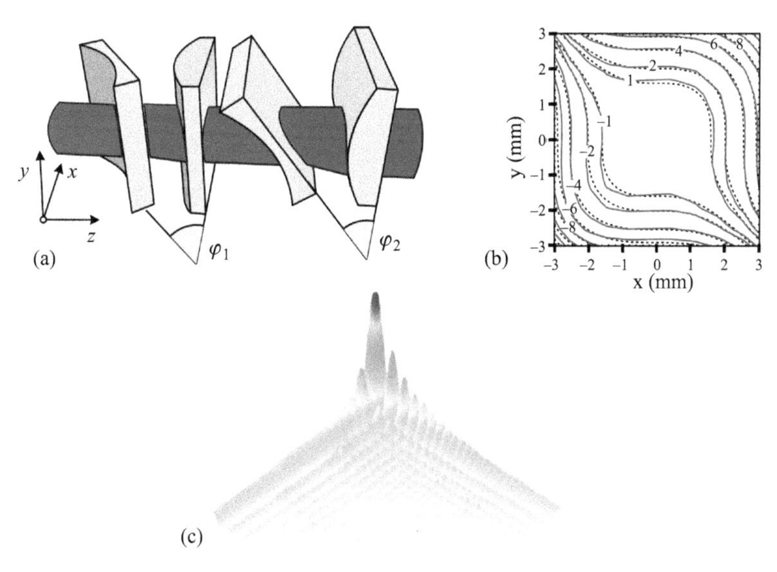

Figure 8.5 *(a) Cascade of two cylindrical telescopes to generate a 2D cubic phase. (b) 2D phase contour curves (in red solid lines) (dotted black lines: 2D cubic phase fit). (c) Intensity distribution (in false colors) of a 2D Airy generated using such system*

8.1.2 Intense Airy wave packets

The need for generating intense and tunable Airy wave packets is continually growing owing to their numerous applications in filamentation [17,21,30], electric discharge triggering and guiding [31], high-power THz generation [23,32] and nonlinear wave mixing [33]. The first intense Airy wave packet propagation in air was demonstrated in 2009 by Polynkin *et al.* [7], followed by a plethora of publications covering from tunable intense 2D Airy wave packets [10], to light bullets [9,13,34] and other nonlinear applications [11,15,16,33,35–37].

The non-spreading linear propagation of Airy wave packets results in unique properties like self-healing and bypassing obstacles. This behavior is common to other non-spreading wave packets and is due to a strong linear energy flux. The accelerating intensity features of the Airy wave packet are generated through interference from this linear energy flux. This is advantageous for their propagation in the nonlinear regime since nonlinear effects are affecting mainly the high intensity features of the wave packet. The nonlinear propagation of Airy beams has attracted a lot of attention [14–16]. As the peak intensity of the Airy beam is increased, it either breaks up [17] and radiates a series of tangential emissions or exhibits shrinking and modification of the Airy profile [12,13]. Furthermore, Kaminer *et al.* [14] and Lotti *et al.* [15] have demonstrated the existence of stationary accelerating Airy wave packets in the presence of third-order (Kerr) nonlinearity and nonlinear losses.

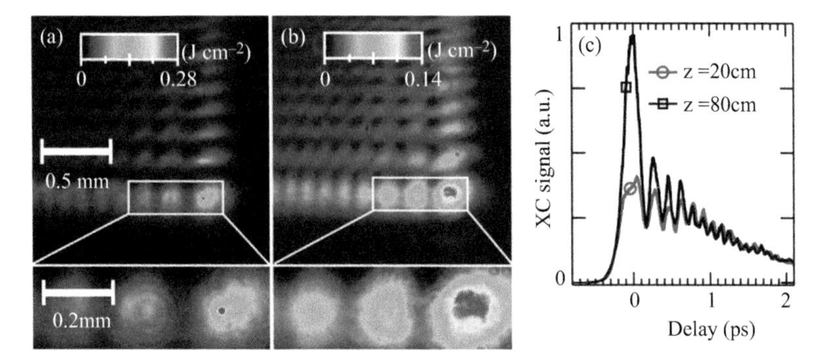

Figure 8.6 Spatial profiles, representing the fluence distribution, of the Airy^3 light bullet with total energy of 0.4 mJ: (a) at the exit surface of the sample and (b) after z=9 cm of propagation in air. (c) Respective temporal profiles at z=20 cm (○) and z=80 cm (□) after the exit surface of the sample. Reprinted, with permission, from Reference [13], APS © 2010

8.1.2.1 Light bullets

Light bullets are optical wave packets that do not spread in space or time as they propagate. They were first introduced by Silberberg [38] and relied on nonlinear effects such as spatial self-focusing and self-phase modulation to balance out the diffraction and dispersion effects. This delicate balance is critical for the stability of the light bullets that break down when passing from one medium to the other [39].

An exciting application exploiting the non-dispersing character of the Airy distribution [1,2,5] is the use of Airy shaped pulses in combination to non-diffracting beams to generate light bullets. These wave packets that do not spread in time and space even in the linear propagation regime were fist demonstrated by Chong *et al.* in 2010 [9] by combining an Airy pulse with a Bessel beam. These linear light bullets are robust and do not break down when the propagation medium is perturbed. A pure Airy light bullet (Figure 8.6), the so-called Airy–Airy–Airy or Airy^3, where the wave packet is described by an Airy distribution in all dimensions, was demonstrated by Abdollahpour *et al.* in 2010 [13]. The experimentally measured iso-intensity profile of such an accelerating light bullet is shown in Figure 8.7. Interestingly the ability of these wave packets to self-heal is advantageous as we increase the intensity and enter the nonlinear propagation regime. As demonstrated in [13], higher order nonlinear effects like filamentation can severely disturb the wave packet's spatiotemporal profile. On the other hand, as shown in Figure 8.6, when nonlinear effects are reduced the wave packet self-heals and the Airy^3 light bullet profile recovers both in space and time.

8.2 Ring-Airy beams

Many applications, like laser micromachining and microprinting [40–42], multiphoton polymerization [22,43–45] and intense THz fields [23], involve the deposition of

Figure 8.7 Iso-intensity profile of an experimentally generated Airy[3] light bullet

energy in confined volumes. Typically, this is performed by using imaging optics that imprint a quadratic spatial phase on an incident Gaussian wave packet. This quadratic phase gradient directs the energy of the beam to a focus, with position and size that are controlled by the phase amplitude and beam size. This approach, although effective, has several drawbacks, with the most profound being the gradual increase in the peak intensity as we approach the focus. Increasing the initial transverse size of the beam, and consequently the dimensions of the focusing optical elements, is a viable solution to this problem, though not so practical to biomedical applications where long focal length optical systems are commonly used.

A cylindrically symmetric variation of the Airy beam theoretically proposed by Efremidis and Christodoulides in 2010 [19], and then experimentally demonstrated by Papazoglou *et al.* in 2011 [20], has opened up the way to solve this problem. Ring-Airy wave packets [18,20,21,46,47], or CABs [48–51] as they are also often referred, exhibit an asymmetric longitudinal intensity profile, so that the energy is abruptly delivered to the focus while their intensity is low up to that point.

8.2.1 Properties

The radial distribution of the spatial profile of a ring-Airy wave packet is described by an Airy function

$$u(r) = u_o Ai(\rho) e^{a\rho}, \tag{8.9}$$

where u_o is the amplitude, Ai is the Airy function, $\rho \equiv (r_o - r)/w$, r is the radial coordinate, and r_o, w and α are, respectively, the primary ring radius, width and apodization parameters. The primary ring of the beam has its peak intensity at a radius of $R_o \equiv r_o - w \cdot g(\alpha) \cong r_o + w, a \ll 1$, where $g(\alpha)$ denotes the first zero of the function $Ai'(x) + \alpha \cdot Ai(x)$, while its full width at half maximum is ~$2.28w$ [20–22]. A typical transverse and radial intensity distribution of a ring-Airy beam is shown in Figure 8.8.

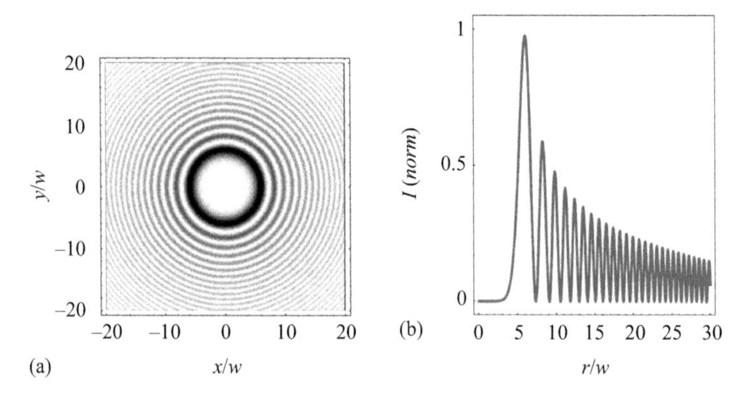

Figure 8.8 Ring-Airy (a) transverse intensity profile and (b) radial intensity distribution (normalized)

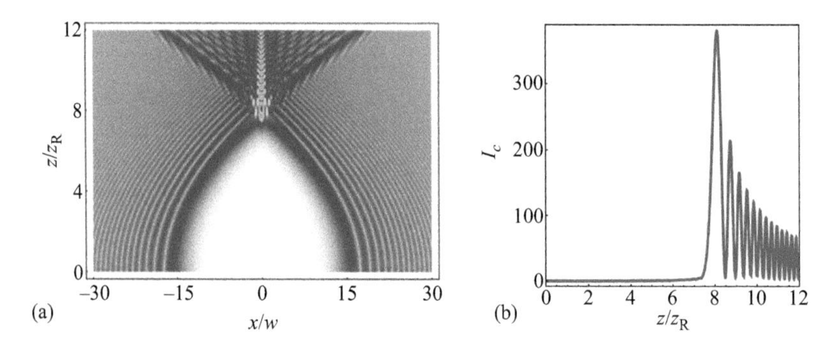

Figure 8.9 Intensity of an abruptly focusing of a ring-Airy: (a) I(x,z) profile and (b) $I_c(z)$ profile

8.2.1.1 Linear propagation

The Airy distribution imparts to the ring–Airy the accelerating behavior of 1D and 2D Airy wave packets [2], causing the caustic of their intensity maxima to converge in a nonlinear fashion toward the focus [19,20]. In contrast to typical wave packet propagation, where diffraction results in the spatial spreading of the wave packet, a ring-Airy exhibits an abrupt autofocus. This unique behavior is shown in Figure 8.9 where from the x–z cross section of the intensity profile of the propagation of a ring-Airy, we can observe the characteristic parabolic trajectory toward the focus. A useful metric of the efficiency of a wave packet to autofocus is the intensity contrast defined as $I_c(z) \equiv I_{\max}(x, y; z)/I_{\max}(x, y; 0)$ where $I_{\max}(x, y; z)$ is the peak intensity at a plane located at position z along the propagation. Large intensity contrast values indicate a significant contrast between the initial intensity and the intensity at the focus. The high- I_c values reached by an abrupt autofocusing ring-Airy beam and the asymmetric nature of the longitudinal intensity distribution are shown in Figure 8.9(b). The intensity contrast achieved is much higher than that of a Gaussian beam making these wave packets ideal for a variety of applications, as in biomedicine and laser nano-surgery. The trajectory of the intensity maxima caustics and consequently the position f_{Ai} of the abrupt autofocus can be predicted by the following 1D Airy analytical formulas [2,20–22,47]:

$$f_{Ai} = 4z_R\sqrt{\frac{r_o + w}{w}} \tag{8.10}$$

where $z_R \equiv \pi w^2/\lambda$ is a scaling parameter similar to the Rayleigh length of Gaussian beams. Since the position of the abrupt autofocus depends only on the r_o, w parameters we can tailor these wave packets so that they abruptly autofocus over an extended range of working distances while keeping almost invariant their voxel shape and dimensions [22,47].

8.2.1.2 Generation techniques

There are a few approaches that can be used to generate a radially symmetric Airy distribution [20,23,49,52]. Despite their functional similarities, ring-Airy wave packets cannot be generated in a simple way as in the case of Airy wave packets [2,10,13]. The FT of a ring-Airy is a Bessel-like distribution with a central peak surrounded by decaying amplitude rings and cannot be described by a simple analytical form [19,20], so a combined phase and amplitude modulation are required.

The first demonstration of ring-Airy beams [20] was performed using an FT technique proposed by Davis *et al.* [53]. In brief, by properly modulating only the phase of an incident beam, and then Fourier transforming by a lens, the desired complex amplitude and phase distribution are generated at the focal plane of the lens. In more detail, the procedure followed to generate the ring-Airy wave packets is as follows. First, the FT $\mathscr{F}_T\{u(r)\} \equiv A(r)\exp[i\Phi(r)]$ of the ring-Airy spatial distribution is calculated numerically. In order to compensate distortions in the encoding process, the amplitude $A(r)$ is then slightly distorted using a lookup table [20,53]. The distorted amplitude is then multiplied with the calculated phase $\Phi(r)$ of the FT; thus the desired phase distribution $\Psi(r) = \exp[i\,A'(r)\cdot\Phi(r)]$ is derived as shown in Figure 8.10.

A typical experimental setup used to generate and study the propagation of ring-Airy wave packets using this method is shown in Figure 8.11(a). A phase only reflecting SLM modulates the phase of an incident Gaussian beam. The ring-Airy distribution is then generated by using a lens to FT the modulated beam. The intensity contrast of the generated ring-Airy wave packet as a function of the propagation distance is depicted in Figure 8.11(b). The main advantage of the FT technique is the capability of generating ring-Airy wave packets of excellent quality, for a large parameters range [18,20–22,46]. On the other hand, its low efficiency in respect of energy conversion is a disadvantage since the generated ring-Airy gets only a small fraction (~15%) of the incident beam power, with the rest being delivered to the zero order.

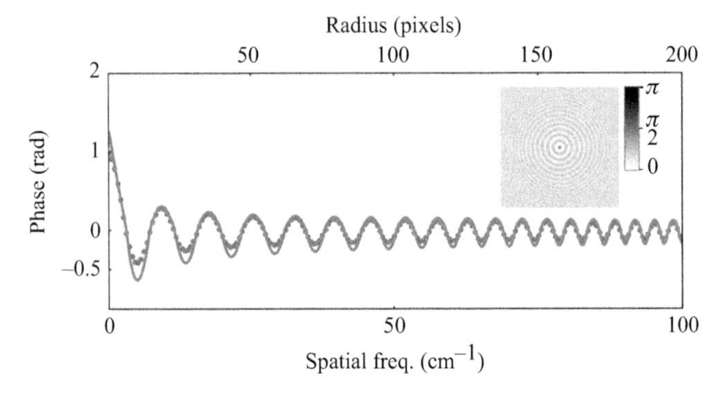

Figure 8.10 Radial profile of a phase mask used to generate a ring Airy (dots: calculated phase, solid line: distorted phase; inset: phase mask)

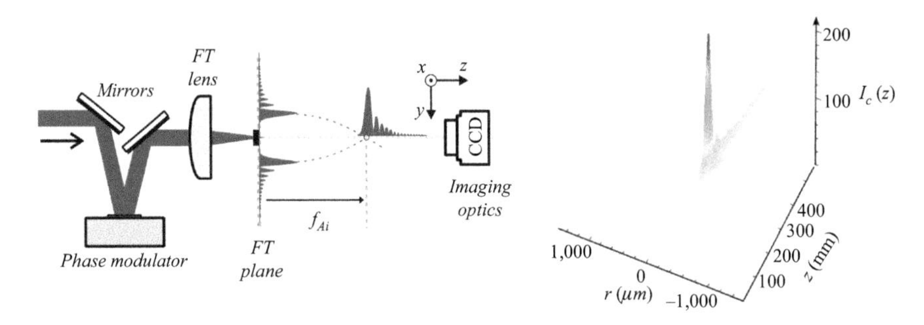

Figure 8.11 (a) Experimental setup to generate and characterize ring-Airy wave packets and (b) intensity contrast of an experimentally generated ring-Airy

Alternatively, the ring-Airy distribution can be generated using holographic techniques [52,54]. By computing the interference between a ring-Airy wave and a tilted plane wave, an off-axis hologram is generated. This off-axis hologram is then imprinted to an amplitude-modulating SLM device that is illuminated by a tilted plane reference wave. The result is the holographic reproduction of the designed ring-Airy. This technique is complementary to the FT generation method, with a lower boundary to the ring-Airy parameters range that can be reproduced, imposed by the discrete nature of the SLM devices.

Other approaches to generate abruptly autofocusing beams are based on the use of approximate representations of the Airy distribution [23,49], or caustic design, where an incident wave is phase modulated in such a way that the trajectory of the intensity maxima (caustics) is a polynomial curve [48,55,56]. For example, accelerating wave packets can be generated by chirping [23,49] the phase $\Phi(r) = -\varphi_o(r - r_o)^\beta$ of an annularly shaped beam $A(r)\exp[i\Phi(r)]$, where $A(r)$ describes the shaped amplitude, and β is the chirp parameter $(1 < \beta < 2)$, and φ_o is a constant.

8.2.2 Intense ring-Airy wave packets

The propagation of intense optical wave packets in transparent media is usually strongly influenced by low and high-order nonlinear effects, like the optical Kerr effect, multiphoton absorption and ionization [30]. These effects, as the peak power is increased, lead to an uncontrolled complex reshaping of the optical wave packet that involves pulse splitting, refocusing cycles in space and significant variations of the focus. In contrast to this behavior, ring-Airy wave packets are able to reshape into intense nonlinear light bullets, propagating over extended distances, while their positioning in space is extremely well defined [21].

The comparative dynamics between a Gaussian and an autofocusing ring-Airy wave packet are depicted in Figure 8.12. One can clearly see that the Gaussian wave packet is spatiotemporally evolving in a complex way, as expected during the filamentation of Gaussian beams [30]. On the other hand, the spatiotemporal dynamics of the ring-Airy beam are very different. The wave packet has the form of a ring with an intense core in its center and no pulse splitting is observed in the temporal domain. This nonlinear light bullet structure propagates without significant changes over 15 cm. Only after 18 cm of propagation, because of nonlinear losses, the nonlinear dynamical balance is perturbed and the ring-Airy wave packet starts to spread due to the action of diffraction and dispersion. The annular structure that surrounds the intense peak of both wave packets is generated by different mechanisms. In the case of the nonlinear ring-Airy beam, the ring is sustained by the combination of the pre-organized energy flux from the tail to the main lobe of the Airy profile and light coming from the nonlinear focus, diffracted by the generated plasma [21,57]. In the case of the Gaussian beam, the ring is generated by a spontaneous transformation of the Gaussian beam into a Bessel-like beam induced by nonlinear effects and featured by conical emission [30,58].

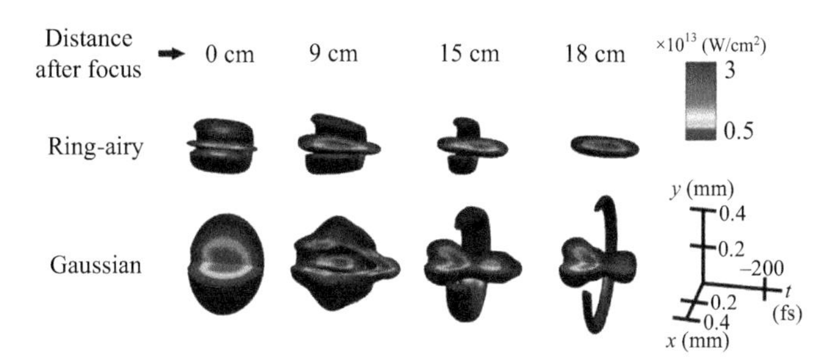

Figure 8.12 Spatiotemporal dynamics in the nonlinear propagation regime (numerical simulation results). Intensity iso-surfaces of an autofocusing ring-Airy wave packet (first row) in comparison to a focused Gaussian wave packet (second row) at various positions along z after the nonlinear focus. Both wave packets carry $10P_{cr}$. Beam propagation direction is from left to right

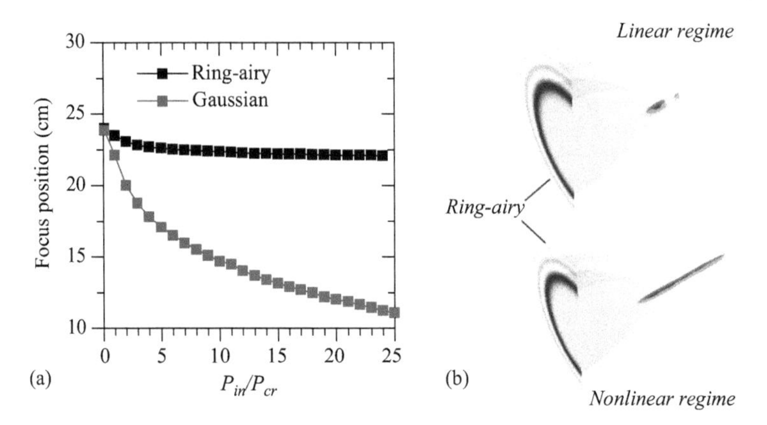

Figure 8.13 (a) Focus position as a function of the input power for the ring-Airy (black squares) and a focused Gaussian wave packet (red dots); (b) graphical representation of the linear and nonlinear propagation of a ring-Airy wave packet

Another interesting feature of intense ring-Airy wave packets is the minor nonlinear focal shift that is observed compared to Gaussian wave packets of similar power. This behavior is shown in Figure 8.13(a) where, as expected, the focus is shifted toward the laser source as the input power is increased. Clearly, in the nonlinear regime, the ring-Airy wave packet focus shift is significantly smaller compared to the Gaussian as the beam power increases. For the Gaussian wave packet, the focus position moves by 54% while for the ring-Airy, the focus shift is only 8%. Furthermore, this small focus shift appears while the input power is increased up to $10P_{cr}$. Further increase in power, up to $24P_{cr}$, seems to have negligible effect on the focus position for the ring-Airy wave packet.

This small focus shift cannot be predicted by directly applying the well-known Marburger formula [30]. As demonstrated by Panagiotopoulos *et al.* [21], it is possible by describing the collapse of a ring-Airy beam in two stages. The first stage involves quasi-linear propagation during which the ring structure shrinks over a propagation distance $f_{\min}$ and follows a parabolic trajectory. In the second stage from $f_{\min}$ to the collapse distance f_{rAi}, the intensity of the ring-Airy beam increases more abruptly since, in analogy with the mutual attraction and merging of two individual filaments induced by cross Kerr effects, the primary ring reaches a sufficiently small diameter to induce significant mutual attraction between opposite parts of the ring. Assuming that the previous is true in the case of ring-Airy beams as well, nonlinear effects will take over when the diameter of the primary ring is ~3 times that of the diameter of a typical filament nonlinear focus shift as a function of the ring-Airy input power P_{in} is given by

$$f_{rAi} = f - (f - f_{\min})(1 - e^{-\sqrt{\gamma P_{in}/P_{cr}}})^2, \tag{8.11}$$

where P_{cr} refers to the critical power of a Gaussian beam with the same central wavelength, $\gamma \equiv P_{Ring}/P_{in}$ is a scaling parameter referring to the fraction of power contained in the primary ring of the ring-Airy distribution, $f_{\min} = f\sqrt{1 - 3w_{\alpha}/2R_0}$ where $w_{\alpha} \cong 90$ μm is the typical filament diameter in air and R_0 is the primary ring radius. The quantity $f - f_{min}$ corresponds to the maximum shift that can be reached at very large powers, where the optical Kerr effect abruptly focuses the primary ring.

8.2.2.1 Applications

As shown in the graphical representation of the linear and nonlinear propagation ring-Airy of Figure 8.13(b), ring-Airy wave packets can be used in delivering large amounts on energy in specified remote locations in a precise and controlled manner. These unique characteristics give a significant advantage in numerous fields like the generation of high harmonics [18] and attosecond physics or the precise micro-engineering of materials [22,59].

For example as shown by Papazoglou *et al.* [20], ring-Airy wave packets represent a very efficient way to deliver a large amount of power in a transparent material especially when operating at long working distances and low *NA*. This is graphically depicted in Figure 8.14(a) where the abrupt increase of the ring-Airy wave intensity near the focus results in a more confined focal volume compared to Gaussian wave packets.

Another interesting application of ring-Airy wave packets is their use in multiphoton polymerization. As shown in Figure 8.14(b), structures of very long aspect ratio (>300) can be generated [22] using ring-Airy wave packets. By properly tuning the initial parameters, like ring width and radius, Manousidaki *et al.* [22] have demonstrated that ring-Airy wave packets result in high aspect ratio voxels that can be positioned at different working distances using solely computer-controlled optical means. As shown in Figure 8.15, ring-Airy beams keep the voxel aspect ratio almost invariant irrespectively of the approach used to manipulate the

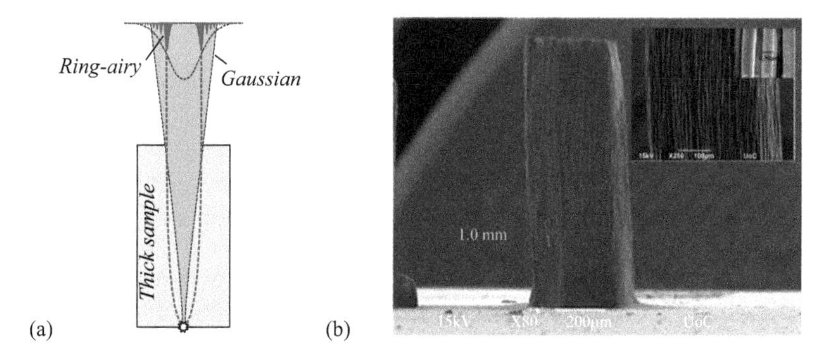

Figure 8.14 (a) Remote processing of a transparent sample using ring-Airy wave packets compared to Gaussian. (b) Scanning electron microscope (SEM) images of 3D structures fabricated using ring-Airy wave packets. Reprinted, with permission, from Reference [22], OPS © 2016

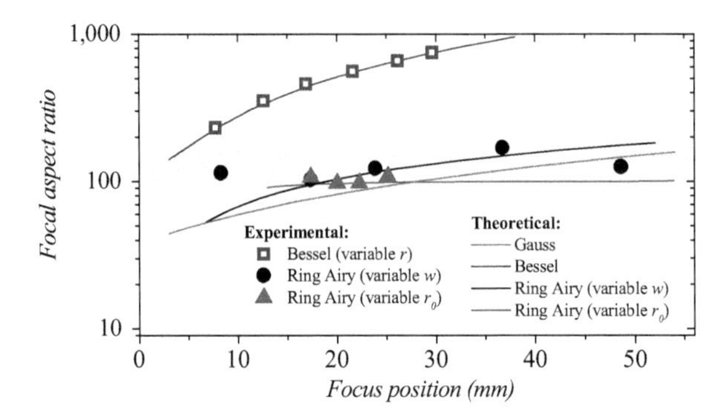

Figure 8.15 Aspect ratio of focal voxel (ratio of voxel longitudinal size over its diameter): comparative curves as a function of the effective focal length for various experimental and simulated ring-Airy, Bessel and Gaussian beams

focus position. In contrast to this behavior, the voxel aspect ratio for Bessel and Gaussian beams is increasing as the beam focus is pushed further away.

On the other hand, in many applications besides the voxel dimensions, a critical parameter is the peak intensity at the focus. The focal intensity can be increased by decreasing the focal voxel dimensions, and in the case of ring-Airy wave packets this can be achieved by simply decreasing the spatial dimensions of the wave packet. One would normally expect that there is a physical lower limit in the size, below which, non-paraxial propagation would degrade the abrupt autofocus of these paraxial beams. In this context, the focal volume would gradually increase, while the peak intensity would decrease in a monotonic fashion. Recently Manousidaki *et al.* [59] have demonstrated that paraxial ring-Airy beams can approach the wavelength limit, while observing a counterintuitive, strong enhancement of their focal peak intensity. This behavior is a result of the coherent constructive action of paraxial and non-paraxial energy flow that occurs in a very narrow parameter window described [59] by the following equation:

$$r_o = NA_{np}^2 k^2 w^3 - 2NA_{np} k\, w^2, \tag{8.12}$$

where r_o, w are the ring-Airy radius and width parameters, $NA_{np} \cong 0.4$ is the typical non-paraxiality limit [60] and k is the wavenumber.

Abruptly autofocusing wave packets are also excellent candidates for replacing Gaussian wave packets in nonlinear optics applications. In 2016 Liu *et al.* [23] demonstrated that ring-Airy wave packets can be used for the remote generation of THz radiation from laser-induced two-color plasma strings. As shown in Figure 8.16, the ring-Airy plasma-generated THz pulse energy is about 5.3 times larger than that from the Gaussian wave packet, under the same experimental conditions. This enhancement is attributed to the elongated plasma length of the ring-Airy-generated plasma. Furthermore, the weaker plasma density leads to a slight narrowing of the THz radiation spectrum.

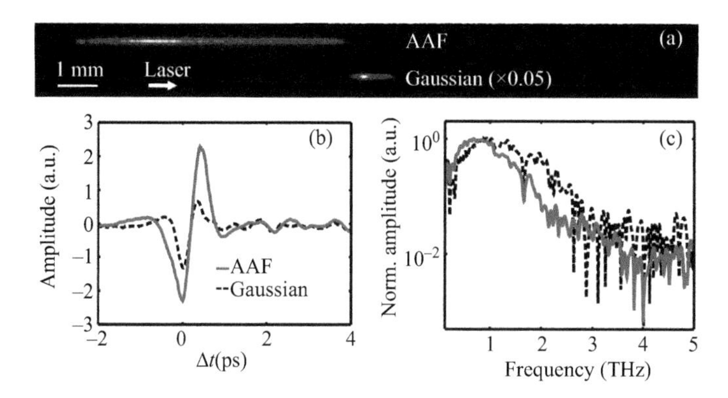

Figure 8.16 (a) Fluorescence image comparison between an AAF beam plasma and a Gaussian plasma (intensity reduced by 20 times), both generated with pump pulse energy of 0.5 mJ. (b) Measured THz waveforms and (c) their corresponding normalized spectra generated by the two plasmas in (a). Red solid curves, emission from AAF plasma; black dashed curves, emission from Gaussian plasma. Reprinted, with permission, from Reference [23], OPS © 2016

Likewise, harmonic generation using intense abruptly autofocusing wave packets presents an interesting property: as demonstrated by Koulouklidis *et al.* [18] they preserve the phase of the fundamental. This unique property, referred to as "phase memory," is an inherent property [18] of a family of optical wave packets whose distribution is described by the Airy function, for example the accelerating Airy and the abruptly autofocusing ring-Airy wave packets. In more detail, an intense pulsed beam of fundamental frequency ω with electric field amplitude $E(\boldsymbol{r}, t)$, propagating in a nonlinear optical medium, modulates the dielectric polarization density $P(\boldsymbol{r}, t) = \varepsilon_o \chi^{(1)} E_\omega(\boldsymbol{r}, t) + \varepsilon_o \chi^{(2)} E_\omega^2(\boldsymbol{r}, t) + \dots$ [61], where ε_o is the vacuum permittivity, and $\chi^{(1)}, \chi^{(p)}, (p = 2, 3, \dots)$ are, respectively, the linear and the nonlinear optical susceptibilities of order p. Each one of the nonlinear polarization terms can be correlated to a harmonic of order p, with phase and amplitude that are in general different to that of the fundamental. Phase memory is then expressed as a preservation of the spatial phase in the powers of the field distribution.

For example, in the case of ring-Airy wave packets, the spatial distribution is described by $u_\omega(r) = u_o Ai(\rho) e^{\alpha\rho}$ where $\rho \equiv (r_o - r)/w$, u_o is the amplitude, $Ai()$ denotes the Airy function and r_o, w and α are, respectively, the primary ring radius, width and apodization parameters. The amplitude of the second harmonic generated in a thin nonlinear crystal is $u_{2\omega}(r) \propto \chi^{(2)} u_\omega(r)^2 = \chi^{(2)} u_o^2 Ai(\rho)^2 e^{2\alpha\rho}$; thus its spatial distribution is proportional to the square of the Airy function term that can be analytically approximated by [18]

$$Ai(\rho)^2 \cong 2E(\rho)^2 + E(\rho) Ai(\widehat{S} \cdot \rho) \equiv Ai_2^a(\rho), \tag{8.13}$$

where $E(\rho) = 1/(2\sqrt{\pi}f(\rho)^{1/4})$, $\widehat{S}\cdot\rho \equiv 2^{2/3}\rho + (\pi/8 \times 2^{1/3})$ is a linear scaling operator, and f is an apodization function. Using the analytic approximation of $Ai(\rho)^2$ we can rewrite the second harmonic amplitude to

$$u_{2\omega}(r)\alpha\,\chi^{(2)}u_{\omega}(r)^2 = 2\chi^{(2)}u_o^2E(\rho)^2e^{2\alpha\rho} + \chi^{(2)}u_o^2E(\rho)Ai(\rho')e^{2\alpha\rho}, \tag{8.14}$$

where $\rho' = \widehat{S}\cdot\rho \equiv (r_o' - r/w'), r_o' = r_o + \pi w/16, w' = 2^{-2/3}w$. Clearly, the second harmonic is composed by two terms, a smooth pedestal that does not lead to any autofocusing behavior and a ring-Airy term that is strongly apodized, carrying ~25% of the total energy. The second term is a clear demonstration of phase memory, which will be resulting in abrupt autofocusing of the second harmonic. We can estimate the focus position of the fundamental and the autofocusing component of the second harmonic, using the analytical formulas for 1D Airy propagation [2,18,20,21]:

$$\begin{aligned} f_{Ai}^{\omega} &= 4\pi\frac{w^2}{\lambda}\sqrt{\frac{r_o + w}{w}} \simeq 4\pi\frac{w^{3/2}}{\lambda}\sqrt{r_{o'}}(r_o \gg w) \\ f_{Ai}^{2\omega} &= 4\pi\frac{w'2}{\lambda/2}\sqrt{\frac{+w'r_o'}{w'}} \simeq 4\pi\frac{w^{3/2}}{\lambda}\sqrt{r_o} = f_{Ai} \end{aligned} \tag{8.15}$$

Interestingly, not only the second harmonic imparts the abruptly autofocusing properties of the fundamental, but also their foci coincide. This behavior is clearly demonstrated in Figure 8.17 where the experimental results of the propagation of the fundamental and the second harmonic of a ring-Airy beam are shown near the abrupt autofocus.

This analysis can be generalized for the generation of ring-Airy harmonics of any higher order [18]. In this case (8.13) becomes

$$\begin{aligned} Ai(\rho)^{2m} &\cong Ai_2^a(\rho)^m = \sum_{n=0}^{m}\frac{2^{m-n}\cdot m!}{n!(m-n)!}E(\rho)^{2m-n}Ai(\widehat{S}\cdot\rho)^n \\ Ai(\rho)^{2m+1} &\cong Ai_2^a(\rho)^m\cdot Ai(\rho) = \sum_{n=0}^{m}\frac{2^{m-n}\cdot m!}{n!(m-n)!}E(\rho)^{2m-n}Ai(\widehat{S}\cdot\rho)^n\cdot Ai(\rho) \end{aligned} \tag{8.16}$$

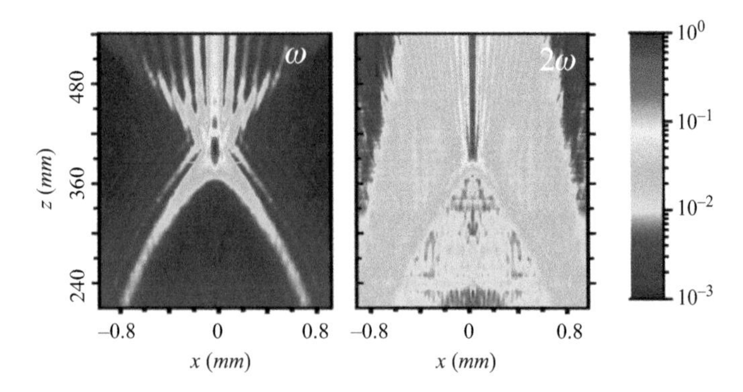

Figure 8.17 Propagation of the fundamental and second harmonic of a ring-Airy in air: experimental results and normalized values

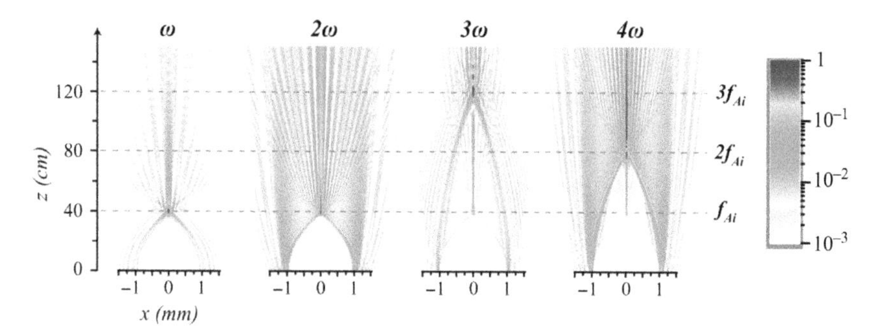

Figure 8.18 Propagation of ring-Airy fundamental and harmonics (2nd, 3rd, 4th): simulation results and normalized values

In the case of even orders, $Ai(\rho)^{2m}$, whenever $n = 2^l, (l = 0, 1, \ldots)$ the corresponding power term $Ai\left(\widehat{S} \cdot \rho\right)^n$ in (8.16) will result in a sum of "pure" ring-Airy terms [18]. These terms represent parts of the beam that will exhibit abrupt autofocusing as they propagate at positions [18]:

$$f_{Ai(q)}^{(2m\cdot\omega)} \simeq m \cdot 2^{1-q} \cdot f_{Ai}^{\omega},\ q = 1, \ldots, [\log_2 m] + 1, (r_o \gg w) \tag{8.17}$$

where $\lambda_{(2m\cdot\omega)} = \lambda/2m$ is the $2m^{\text{th}}$ harmonic wavelength, and f_{Ai}^{ω} refers to the autofocus position of the first harmonic. Interestingly, for all the even harmonics that are a power of 2 (i.e., 2nd, 4th, 8th, ...), there will exist a term that will be autofocusing at f_{Ai}^{ω} [18]. On the other hand, for odd orders $Ai(\rho)^{2m+1}$, there exists only one "pure" ring-Airy term $(\ldots)Ai(\rho)$ (for $n=m$) in (8.16) that will autofocus at

$$f_{Ai}^{(2m+1)\cdot\omega} \simeq (2m+1) \cdot f_{Ai}^{\omega}, \quad (r_o \gg w) \tag{8.18}$$

Figure 8.18 shows simulation results on the propagation of the fundamental (ω) and the corresponding ring-Airy harmonics (2ω, 3ω, 4ω). The fundamental abruptly autofocuses at a distance f_{Ai} (~40 cm), while all harmonics are also abruptly autofocusing exhibiting multiple foci. The positions of the foci are located at multiples of f_{Ai}, as predicted by (8.17) and (8.18) while the foci of the second harmonic coincides with the fundamental.

8.3 Janus waves

Exotic wave packets, like the accelerating Airy [1,2,10,13], ring-Airy [19–22,59] and higher order accelerating [62,63] wave packets, share a common property affecting the way they propagate and focus. As shown by Papazoglou *et al.* [46] these waves belong to a broad class of waves that are referred to as Janus waves (JWs). Like the god Janus from the Roman mythology, who was depicted with two faces looking in opposite directions, JWs can be decomposed to the propagation of two waves that are conjugate to each other under inversion of the propagation direction. The formation of twin waves is a well-known effect in holography [64] where the amplitude and phase of the original wavefront is spatially modulated by the hologram. This modulated wavefront can be decomposed into two waves [25,64], one of which leads to the formation of the "real" image of the original object at some distance z_o and the other to the formation of the "virtual" image at distance $-z_o$. In contrast to classical holograms, and due to their spatial distribution, in JWs the part of the wave that is related to the "virtual" image is diffracting out with minor interference with the part of the wave that is related to the "real" image. The presence of the twin "virtual" image clearly manifests when these beams are focused by a lens [46]. In this case both the converging and the diverging parts of the beam are brought to focus and form a pair of opposite facing focal distributions.

8.3.1 Properties

In order to provide a formal definition of JWs let us assume that the field $U(\mathbf{r}, z)$, of a harmonic wave, can be described as the superposition of two waves, $\Psi_o(\mathbf{r}, z)$ and a "conjugate" $\Psi_o^*(\mathbf{r}, -z)$:

$$U(\mathbf{r}, z) = \Psi_o(\mathbf{r}, z) + \Psi_o^*(\mathbf{r}, -z) \tag{8.19}$$

where $\mathbf{r}$ is the transverse position vector and z is the position along the propagation axis. Clearly, this wave is conjugate symmetric under inversion of the propagation direction, $U(r, -z) = U^*(r, z)$. Furthermore, the distribution at $z=0$, which defines the plane of symmetry, is always a real-valued function:

$$U(\mathbf{r}, 0) = \Psi_o(\mathbf{r}, 0) + \Psi_o^*(\mathbf{r}, 0) = 2\mathrm{Re}[\Psi_o(\mathbf{r}, 0)] \tag{8.20}$$

Based on (8.20) we can define a simple mathematical criterion for CWs: *"A wave is Conjugate if its field $U(\mathbf{r}, z)$, is real valued at a transverse plane along its propagation."* By real valued we mean that the phase can take only $m \cdot \pi$ values (where $m = 0, \pm 1, \pm 2, \ldots$). Based on this criterion the propagation of a wave that is real valued, at a transverse plane along the propagation, can always be described [46] as the superposition of two waves, $\Psi_o(\mathbf{r}, z)$ and a "conjugate" $\Psi_o^*(\mathbf{r}, -z)$. The CW criterion is fulfilled by a large variety of wave packets, ranging from accelerating beams such as Airy [1,13] and ring-Airy beams [19–22,59], but also the widely used Gaussian beams and Bessel beams [4,65]. On the other hand, Helical beams [66] are waves that do not fulfill this criterion due to their helical phase gradient.

From this broad category of CWs, we now concentrate on those waves that exhibit a discrete focus away from their symmetry plane. We define such waves as JWs. These include, for example, all accelerating beams [1,2,10,13,19–22,59,62,63] but not Gaussian and Bessel beams [4,65]. Since JWs are also CW for each discrete focus, another "virtual" one, located symmetrically to the symmetry plane, will exist.

This effect is depicted in Figure 8.19, where in equivalence to the formation of a virtual and real image in holography [25,64] we observe the existence of a real and a virtual focus in a JW. In this graphical example, the focal point lies at a distance z_o while a "virtual" focal point lies on the opposite side (at $-z_o$) and is not accessible.

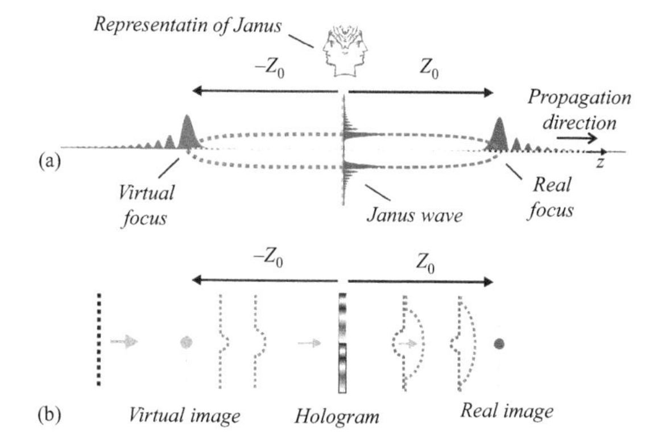

Figure 8.19 *Equivalence between Janus waves and holography. (a) Janus wave (example of ring-Airy beam) and real and virtual foci. The two-faced roman mythology god Janus is also represented. (b) Holographic formation of twin images*

8.3.1.1 Linear propagation

The effect of the twin "real" and "virtual" waves on the wave packet propagation can be visualized by numerically simulating the propagation of ring-Airy wave packets [19,20,46] that represent a typical example of JWs. Ring-Airy wave packets fulfill the CW criterion and exhibit a discrete focus away from their symmetry plane. More specifically, they are real valued at their generation plane since they are described [19,20,46] by $Ai(\rho)e^{a\rho}$, where $\rho = (r - r_o)/w$, r is the radius, r_o and w are the ring radius and width parameters, respectively, and a is the apodization coefficient. Furthermore, they also exhibit a discrete focus, away from their symmetry plane, since they abruptly autofocus [19,20] at predefined distance z_o [18,20,21,59].

As shown in Figure 8.20, the characteristic parabolic trajectory toward the real focus is clearly visible both in the intensity, and the phase of the beam observing the $\pm\pi$ iso-phase curves that are uncovered by the phase wrap. On the other hand, the virtual wave can be seen only in the phase map. The intensity, although in logarithmic scale, shows no trace of the component that diffracts out. The "virtual" wave manifests as a diffracting out wave and is usually ignored [46,48]. In this perspective, one could argue that the JW description of abruptly wave packets does not offer a significant advantage on understanding their behavior. What makes JWs interesting is their peculiar behavior when they are focused by a lens. As we will show in the next session, two foci, instead of one, can be observed.

8.3.2 "Manipulating" Janus waves

In applications, for example laser direct laser writing [22,40–45], a critical parameter is the peak intensity at the focus and the size of the voxel. Abruptly autofocusing wave packets rely on diffraction to generate a focus. A straightforward way to decrease the voxel volume, and thus increase the focal intensity, is to decrease the physical dimensions of the beam [55,59]. This can be performed either

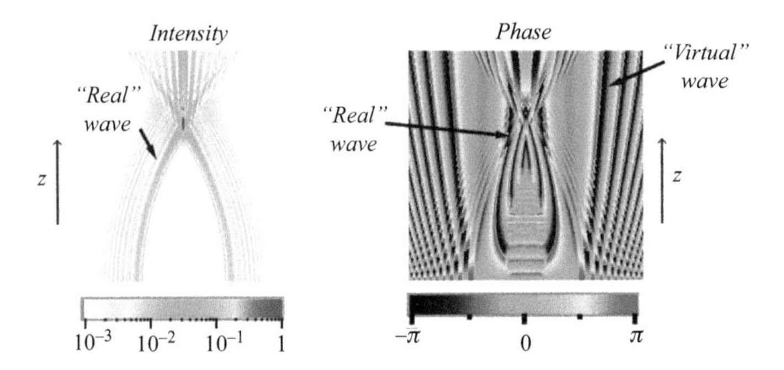

Figure 8.20 Simulation results of ring-Airy propagation (cross-sectional r,z). The parabolic trajectory toward the focus is clearly visible both in the intensity (logarithmic scale) and the phase (wrapped phase). The virtual wave is clearly visible in the phase but practically invisible in the intensity map

by tuning the generating phase mask design [20,55,59], or by imaging using telescopic (or 4*f*) optical systems [21,59]. On the other hand, in the case of Gaussian wave packets, this can be achieved by simply using a lens. The lens reshapes the spatial phase of the wave packet by adding a quadratic phase that leads to focusing. By decreasing the focal length of the lens, the focal voxel is also decreased.

Using a focusing element, like a lens, to further focus, an abruptly autofocusing wave packet is an appealing alternative way to increase the focal intensity. The JW character of these waves leads to a counterintuitive behavior in this case. The twin real and virtual waves that compose the JW are independently reshaped by the focusing element and lead to two foci instead of one.

8.3.2.1 The action of a thin lens

If we insert a thin lens on the plane of symmetry ($z = 0$), as shown in Figure 8.21, then the wave distribution after the lens is described by [46,67]

$$U'(r,z) = \frac{1}{\zeta} e^{\left(-ik\left(x^2+y^2\right)/2f\zeta\right)} \left\{ \Psi_0\left[\frac{\mathbf{r}}{\zeta}, \frac{z}{\zeta}\right] + \Psi_0^*\left[\frac{\mathbf{r}}{\zeta}, \frac{-z}{\zeta}\right] \right\} \tag{8.21}$$

where $\zeta \equiv 1 - z/f$. The intensity distribution after the lens can then be estimated by [46]

$$\begin{aligned} I'(\mathbf{r},z)\alpha\, U'(\mathbf{r},z) \cdot U'*(\mathbf{r},z) = \frac{1}{\zeta^2} \Bigg\{ \left| \Psi_0\left[\frac{\mathbf{r}}{\zeta}, \frac{z}{\zeta}\right] \right|^2 + \left| \Psi_0\left[\frac{\mathbf{r}}{\zeta}, \frac{-z}{\zeta}\right] \right|^2 \\ +2\mathrm{Re}\left[\psi_0\left[\frac{\mathbf{r}}{\zeta}, \frac{z}{\zeta}\right] \psi_0\left[\frac{\mathbf{r}}{\zeta}, \frac{-z}{\zeta}\right] \right] \Bigg\} \end{aligned} \tag{8.22}$$

As expected, the resulting intensity distribution is the result of the interference of the two twin waves. Interestingly, a local maximum of the wave distribution

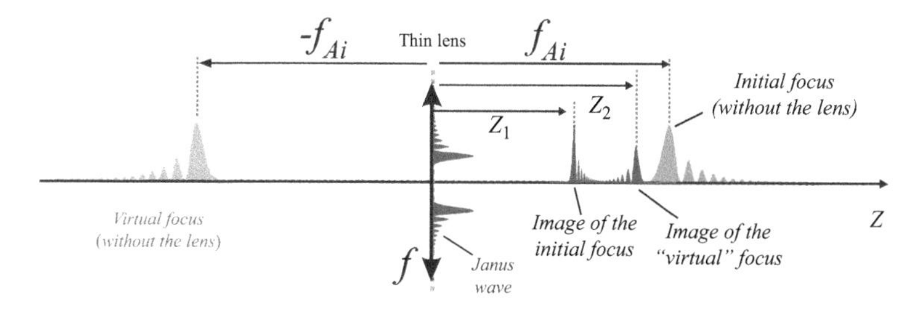

Figure 8.21 Using a thin lens to focus a ring-Airy Janus wave packet. Without the lens the abrupt autofocus intensity peak is accompanied by a symmetrically positioned "virtual" intensity peak. With the lens, two foci with opposite facing focal distributions are created

$\Psi_o(\mathbf{r}, z)$ at position $(\mathbf{r}_o, f_{Ai})$ will be imaged after the lens at two positions $(\mathbf{r}_1, z_1), (\mathbf{r}_2, z_2)$ [46]:

$$-\frac{1}{f_{Ai}} + \frac{1}{z_1} = \frac{1}{f}, \mathbf{r}_1 = M_T(-f_{Ai}) \cdot \mathbf{r}_o, \quad \frac{1}{f_{Ai}} + \frac{1}{z_2} = \frac{1}{f}, \mathbf{r}_2 = M_T(f_{Ai}) \cdot \mathbf{r}_o \tag{8.23}$$

where $M_T(z) = (1 - z/f)^{-1}$ is the transverse magnification of the lens. Although the accurate location of the intensity maxima from (8.22) involves a complex analysis, the positions of the twin images, as given by (8.23), are correct as long as $f_{Ai} \neq 0, |1 - z_i/f| > 0.1$. Counter intuitively, classical imaging formulas can be used to predict the maxima positions if one takes into account that the peak intensity of the $\Psi_o(\mathbf{r}, z)$ wave can be treated as an imaginary object to the lens while the peak of the "conjugate" $\Psi_o^*(\mathbf{r}, z)$ wave can be treated as a real object. Likewise, by taking into account the transverse magnification M_T for each position of the object, the transverse position of the peaks can be predicted. The two foci will exhibit opposite signs of M_T for $f_{Ai} > f$.

As shown schematically in Figure 8.21 without the presence of the lens, each initial intensity peak (focus) is accompanied by a symmetrically positioned virtual focus. The focusing action of the lens leads to the creation of two opposite facing focal distributions located at positions z_1 and z_2 that can be estimated using (8.23). The two foci are distinct as long as the twin waves, $\Psi_o(\mathbf{r}, z)$ and $\Psi_o^*(\mathbf{r}, z)$ do not have overlapping maxima along z. It is easy to show that overlapping maxima can occur only on the symmetry plane $f_{Ai} = 0$, a case already excluded from the definition of JWs. Thus, a JW, in contrast to the broader category of CWs, will always exhibit two foci under the action of a lens.

The x–z cross-sectional intensity profile of a ring-Airy JW is shown in Figure 8.22. As expected the focused JWs form two focal regions. The "real" and "virtual" waves produce two foci at different positions and the positions of the two foci agree well, within 2%, to the theoretical predictions.

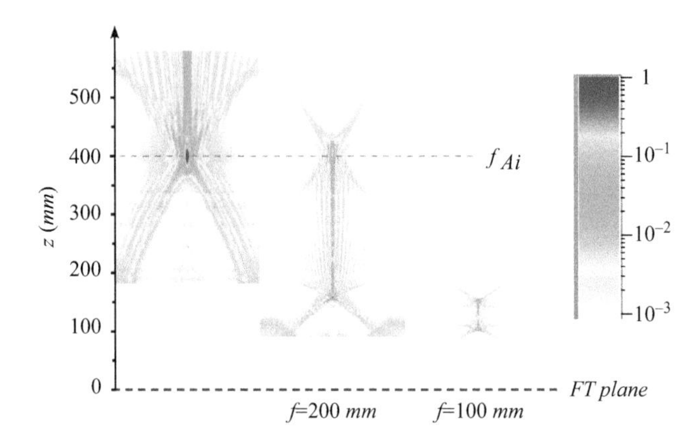

Figure 8.22 Experimental cross-sectional intensity profiles I(x,0,z) of ring-Airy Janus waves. Free propagation (left column), focusing by a 200-mm lens (middle column), focusing by a 100-mm lens (right column)

8.4 Conclusion

In conclusion, the introduction of engineered wave packets with emphasis on accelerating beams and light bullets has opened numerous opportunities in linear and nonlinear optical sciences. The precise deposition of laser energy, even at very high powers, is among the most exciting ones. For instance, using ring-Airy beams the energy can be deposited in space in three different ways; well-localized in small volumes with high contrast, as intense light bullets over extended distances when the power exceeds a critical threshold, or as light strings with precise origin and termination using the JWs property when these beams are Fourier transformed through a focusing lens.

The excellent control over the deposition of energy along well-defined paths, including curved ones, offers unique advantages in laser materials engineering. Beyond the one we reported here in photopolymerization, many more applications are emerging in laser cutting and in particle manipulation.

Finally, in the nonlinear regime, these beams have considerably impacted the field of laser filamentation and allowed for the observation of intense self-healing light bullets, as well as the remote generation of intense THz radiation in the atmosphere.

In the future we expect studies in this field to allow for even more exciting applications that can address technological challenges in many fields, like in materials processing, biomedical imaging, telecommunications at different parts of the spectrum—from the visible all the way to THz—through scattering and lossy media, as well as in the filamentation of intense laser beams in the mid and far-infrared.

References

[1] G. A. Siviloglou and D. N. Christodoulides, *Opt. Lett.* 32, 979 (2007).
[2] G. A. Siviloglou, J. Broky, A. Dogariu, and D. N. Christodoulides, *Phys. Rev. Lett.* 99, 213901 (2007).
[3] O. Vallée and M. Soares, *Airy Functions and Applications to Physics* (Imperial College Press, 2004).
[4] J. Durnin, J. J. Miceli, and J. H. Eberly, *Phys. Rev. Lett.* 58, 1499 (1987).
[5] M. V Berry and N. L. Balazs, *Am. J. Phys.* 47, 264 (1979).
[6] G. A. Siviloglou, J. Broky, A. Dogariu, and D. N. Christodoulides, *Opt. Lett.* 33, 207 (2008).
[7] P. Polynkin, M. Kolesik, J. V Moloney, G. A. Siviloglou, and D. N. Christodoulides, *Science* (80-.). 324, 229 (2009).
[8] J. Broky, G. A. Siviloglou, A. Dogariu, and D. N. Christodoulides, *Opt. Express* 16, 12880 (2008).
[9] A. Chong, W. H. Renninger, D. N. Christodoulides, and F. W. Wise, *Nat. Photonics* 4, 103 (2010).
[10] D. G. Papazoglou, S. Suntsov, D. Abdollahpour, and S. Tzortzakis, *Phys. Rev. A* 81, 61807 (2010).
[11] G. Porat, I. Dolev, O. Barlev, and A. Arie, *Opt. Lett.* 36, 4119 (2011).
[12] J. Kasparian and J. P. Wolf, *J. Eur. Opt. Soc. Publ.* 4, (2009).
[13] D. Abdollahpour, S. Suntsov, D. G. Papazoglou, and S. Tzortzakis, *Phys. Rev. Lett.* 105, 253901 (2010).
[14] I. Kaminer, M. Segev, and D. N. Christodoulides, *Phys. Rev. Lett.* 106, 213903 (2011).
[15] A. Lotti, D. Faccio, A. Couairon, *et al.*, *Phys. Rev. A* 84, 21807 (2011).
[16] P. Panagiotopoulos, D. Abdollahpour, A. Lotti, *et al.*, *Phys. Rev. A* 86, 13842 (2012).
[17] P. Polynkin, M. Kolesik, and J. Moloney, *Phys. Rev. Lett.* 103, 123902 (2009).
[18] A. D. Koulouklidis, D. G. Papazoglou, V. Y. Fedorov, and S. Tzortzakis, *Phys. Rev. Lett.* 119, 223901 (2017).
[19] N. K. Efremidis and D. N. Christodoulides, *Opt. Lett.* 35, 4045 (2010).
[20] D. G. Papazoglou, N. K. Efremidis, D. N. Christodoulides, and S. Tzortzakis, *Opt. Lett.* 36, 1842 (2011).
[21] P. Panagiotopoulos, D. G. Papazoglou, A. Couairon, and S. Tzortzakis, *Nat. Commun.* 4, 2622 (2013).
[22] M. Manousidaki, D. G. Papazoglou, M. Farsari, and S. Tzortzakis, *Optica* 3, 525 (2016).
[23] K. Liu, A. D. Koulouklidis, D. G. Papazoglou, S. Tzortzakis, and X.-C. Zhang, *Optica* 3, 605 (2016).
[24] D. M. Cottrell, J. A. Davis, and T. M. Hazard, *Opt. Lett.* 34, 2634 (2009).
[25] J. W. Goodman, *Introduction to Fourier Optics*, 2nd ed. (McGraw-Hill, New York, 1996).

[26] M. Born and E. Wolf, *Principles of Optics: Electromagnetic Theory of Propagation, Interference and Diffraction of Light*, 7th ed. (Cambridge University Press, 1999).
[27] D. Mansour and D. G. Papazoglou, *Opt. Lett.* 43, 5480 (2018).
[28] F. Courvoisier, A. Mathis, L. Froehly, *et al.*, *Opt. Lett.* 37, 1736 (2012).
[29] L. Froehly, F. Courvoisier, A. Mathis, *et al.*, *Opt. Express* 19, 16455 (2011).
[30] A. Couairon and A. Mysyrowicz, *Phys. Rep.* 441, 47 (2007).
[31] S. Tzortzakis, B. Prade, M. Franco, A. Mysyrowicz, S. Hüller, and P. Mora, *Phys. Rev. E* 64, 057401 (2001).
[32] S. Moradi, A. Ganjovi, F. Shojaei, and M. Saeed, *Phys. Plasmas* 22, 043108 (2015).
[33] I. Dolev and A. Arie, *Appl. Phys. Lett.* 97, 171102 (2010).
[34] P. Piksarv, A. Valdmann, H. Valtna-Lukner, and P. Saari, *Laser Phys.* 24, (2014).
[35] L. Li, T. Li, S. M. Wang, C. Zhang, and S. N. Zhu, *Phys. Rev. Lett.* 107, 126804 (2011).
[36] Y. Zhang, M. Beli, Z. Wu, *et al.*, *Opt. Lett.* 38, 4585 (2013).
[37] Y. Hu, G. Siviloglou, P. Zhang, N. Efremidis, D. Christodoulides, and Z. Chen, in *Nonlinear Photonics Nov. Opt. Phenom.*, edited by Z. Chen and R. Morandotti (Springer, 2012), pp. 1–46.
[38] Y. Silberberg, *Opt. Lett.* 15, 1282 (1990).
[39] D. Majus, G. Tamosauskas, I. Grazuleviciute, *et al.*, *Phys. Rev. Lett.* 112, (2014).
[40] D. G. Papazoglou and M. J. Loulakis, *Opt. Lett.* 31, 1441 (2006).
[41] D. G. Papazoglou, I. Zergioti, and S. Tzortzakis, *Opt. Lett.* 32, 2055 (2007).
[42] I. Zergioti, A. Karaiskou, D. G. Papazoglou, C. Fotakis, M. Kapsetaki, and D. Kafetzopoulos, *Appl. Phys. Lett.* 86, 3 (2005).
[43] M. Farsari, M. Vamvakaki, and B. N. Chichkov, *J. Opt.* 12, 124001 (2010).
[44] M. Beresna, M. Gecevičius, and P. G. Kazansky, *Adv. Opt. Photonics* 6, 293 (2014).
[45] M. Malinauskas, A. Žukauskas, S. Hasegawa, *et al.*, *Light Sci. Appl.* 5, e16133 (2016).
[46] D. G. Papazoglou, V. Y. Fedorov, and S. Tzortzakis, *Opt. Lett.* 41, 4656 (2016).
[47] D. Mansour and D. G. Papazoglou, *OSA Continuum* 1, 104 (2018).
[48] I. D. Chremmos, Z. Chen, D. N. Christodoulides, and N. K. Efremidis, *Phys. Rev. A* 85, 23828 (2012).
[49] I. Chremmos, N. K. Efremidis, and D. N. Christodoulides, *Opt. Lett.* 36, 1890 (2011).
[50] Y. Jiang, K. Huang, and X. Lu, *Opt. Express* 20, 18579 (2012).
[51] H. Deng, Y. Yuan and L. Yuan, Annular Arrayed-Waveguide Fiber for Autofocusing Airy-like Beams, *Opt. Lett.* 41, 824 (2016).
[52] I. Chremmos, P. Zhang, J. Prakash, N. K. Efremidis, D. N. Christodoulides, and Z. G. Chen, *Opt. Lett.* 36, 3675 (2011).
[53] J. A. Davis, D. M. Cottrell, J. Campos, M. J. Yzuel, and I. Moreno, *Appl. Opt.* 38, 5004 (1999).

[54] P. Zhang, J. Prakash, Z. Zhang, *et al.*, *Opt. Lett.* 36, 2883 (2011).
[55] Z. Zhao, C. Xie, D. Ni, *et al.*, *Opt. Express* 25, 30598 (2017).
[56] R.-S. Penciu, K. G. Makris, and N. K. Efremidis, *Opt. Lett.* 41, 1042 (2016).
[57] P. Panagiotopoulos, A. Couairon, M. Kolesik, D. G. Papazoglou, J. V. Moloney, and S. Tzortzakis, *Phys. Rev. A* 93, 033808 (2016).
[58] D. Faccio, A. Averchi, A. Lotti, *et al.*, *Opt. Express* 16, 1565 (2008).
[59] M. Manousidaki, V. Y. Fedorov, D. G. Papazoglou, M. Farsari, and S. Tzortzakis, *Opt. Lett.* 43, 1063 (2018).
[60] A. E. Siegman, *Lasers*. (University Science Books, Mill Valley, 1986).
[61] R. W. Boyd, *Nonlinear Optics*, 2nd ed. (Academic Press, 2003).
[62] I. Kaminer, R. Bekenstein, J. Nemirovsky, and M. Segev, *Phys. Rev. Lett.* 108, 163901 (2012).
[63] E. Greenfield, M. Segev, W. Walasik, and O. Raz, *Phys. Rev. Lett.* 106, 213902 (2011).
[64] D. Gabor, *Nature* 161, 777 (1948).
[65] J. Durnin, *J. Opt. Soc. Am. A* 4, 651 (1987).
[66] M. W. Beijersbergen, R. P. C. Coerwinkel, M. Kristensen, and J. P. Woerdman, *Opt. Commun.* 112, 321 (1994).
[67] M. A. Bandres and M. Guizar-Sicairos, *Opt. Lett.* 34, 13 (2009).

Chapter 9

Dimensionless numbers for numerical simulations and scaling of ultrashort laser pulse filamentation

Arnaud Couairon,[1] Christoph M. Heyl[2–4] and Cord L. Arnold[5]

Dimensionless numbers are of key importance in understanding the similarity among problems belonging to the same broad class. They are also extremely useful for parametric analysis of engineering problems. Dimensionless numbers are scarcely used in optics, in striking contrast to their common usage in fluid dynamics. However, *without dimensionless numbers, experimental progress in fluid mechanics would have been almost nil; it would have been swamped by masses of accumulated data* [1]. For instance, two widely used dimensionless numbers in fluid dynamics are the Reynolds and the Mach numbers, quantifying the ratio of inertial forces to viscous forces and compressibility effects, respectively.

To cite only a few of their numerous advantages, dimensionless numbers not only allow equations governing fluid flows to be rewritten in dimensionless forms, with the additional benefit of a compact writing, but also permit flows with different scales that share a similar behavior to be classified into a group characterized by a set of dimensionless numbers. Dimensionless numbers indicate the importance of the different source terms in the system of equations governing the dynamics. From the computational point of view, the dimensionless form of the system of discretized equations provides a physically based scaling that is helpful to minimize the computational cost and to prevent ill-conditioning of the system of equations.

In optics, there is no widespread dimensionless number, with the possible exceptions of the f-number ($f\#$) or N, the numerical aperture NA of an optical system and the Fresnel number F.

[1]Centre de Physique Theorique, Ecole polytechnique, CNRS, Institut Polytechnique de Paris, Palaiseau, France
[2]Deutsches Elektronen-Synchrotron, DESY, Hamburg, Germany
[3]Helmholtz-Institute Jena, Jena, Germany
[4]GSI Helmholtzzentrum für Schwerionenforschung GmbH, Darmstadt, Germany
[5]Department of Physics, Lund University, Lund, Sweden

The f-number of an optical system (e.g., a lens) is the ratio of the focal length f to the diameter D of the entrance pupil (effective aperture) [29].

$$N = \frac{f}{D}. \tag{9.1}$$

Similarly to the $f\#$, the numerical aperture describes convergence or divergence in terms of a range of angles over which an optical system can accept or emit light. The Fresnel number F is used in scalar diffraction theory to distinguish between near- and far-field approximations. For an electromagnetic wave of wavelength λ passing through an aperture of characteristic size (e.g. radius) a and hitting a screen a distance L away from the aperture, the Fresnel number F is defined as

$$F = \frac{a^2}{L\lambda}. \tag{9.2}$$

Conceptually, the Fresnel number represents the path difference expressed in a number of wavelengths, as seen from the center of the imaging screen between rays passing either through the edge or through the center of the aperture. The Fresnel number is useful to determine whether the far-field (Fraunhofer) diffraction theory can be applied (for $F \ll 1$) or the near-field (Fresnel) diffraction theory must be applied (for $F > 1$).

Recently, scaling properties for the nonlinear propagation of ultrashort laser pulses in transparent media were highlighted and experimentally confirmed in dedicated experiments [2,3]. Numerous physical processes accompanying the nonlinear propagation and laser–matter interaction support scalability. For instance, frequency conversion processes like high harmonic generation are fully compatible with the scaling framework derived from the propagation of the driving pulse [2]. These scaling properties are essential due to the universal form taken by the governing equations for the propagation of the laser pulse coupled with the matter response and can be naturally expressed in the language of dimensionless numbers.

The goal of this chapter is to introduce dimensionless numbers for unidirectional propagation equations used for numerical simulations of nonlinear propagation of ultrashort laser pulses and their interaction with matter. After naming a set of useful dimensionless numbers suited for ultrafast nonlinear optics, we derive and interpret the limits of the scaling behaviors presented in [2] in light of a classification of propagation regimes based on these numbers.

9.1 Dimensionless numbers for unidirectional propagation equations

9.1.1 Nonlinear Schrödinger equation

A standard model describing nonlinear propagation of laser pulses in Kerr media was proposed, and the laser pulse dynamics was investigated extensively by means of multidimensional numerical simulations since the 1990s [4–6]. The model is based on the nonlinear Schrödinger equation for the electric field amplitude $\mathcal{E}$:

$$\frac{\partial \mathcal{E}}{\partial z} = \frac{i}{2k_0}\nabla_\perp^2 \mathcal{E} - i\frac{k_0''}{2}\frac{\partial^2 \mathcal{E}}{\partial t^2} + i\frac{\omega_0}{c}n_2|E|^2\mathcal{E}, \tag{9.3}$$

where z denotes the propagation distance, $\nabla_\perp^2 \equiv \partial_{x^2}^2 + \partial_{y^2}^2$ the transverse Laplacian operator, $k_0 = n_0\omega_0/c$ the propagation constant along the z-axis for a wave of angular frequency ω_0, propagating in a medium with linear refractive index n_0, k_0'' the second-order dispersion parameter and n_2 the nonlinear index coefficient. The three terms on the rhs of (9.3) describe diffraction, group velocity dispersion and self-phase modulation, respectively.

A numerical simulation of (9.3) requires the knowledge of the initial condition, i.e., the space and time distribution of the amplitude and phase of the input pulse. As an example, we use an input Gaussian pulse of duration t_p, beam width w_0, carrying a quadratic spatial phase of curvature radius f, and chirp C:

$$\mathcal{E}(z=0,x,y,t) = \mathcal{E}_0 \exp\left(-\frac{r^2}{w_0^2} - i\frac{k_0 r^2}{2f}\right)\exp\left(-\frac{t^2}{t_p^2} + i\frac{Ct^2}{t_p^2}\right), \tag{9.4}$$

where $r^2 \equiv x^2 + y^2$. Without loss of generality, we consider the case of cylindrical symmetry. The remaining of this chapter thus assumes a dependence of electric fields $\mathcal{E}(z,r,t)$.

Dimensionless transverse coordinates and time are built as $\widetilde{r} = r/w_s$ and $\widetilde{t} = t/t_s$, where the transverse scale w_s and the timescale t_s do not need to coincide with w_0 and t_p. A dimensionless propagation coordinate $\widetilde{z} = z/z_s$ is built by choosing a reference length scale z_s. The input field amplitude $\mathcal{E}_0$ is a natural scale to make the field amplitude dimensionless, but again, we may make a different choice and more generally write the dimensionless amplitude as $\widetilde{A} = \mathcal{E}/\mathcal{E}_s$. Incidentally, we note that the writing of (9.3) assumes that $n_2|\mathcal{E}|^2$ must be dimensionless, setting the units for $\mathcal{E}$ as well as for $\mathcal{E}_s$. In the following, we will follow this convention and use the notation for the pulse intensity $I \equiv |\mathcal{E}|^2$. Hence, the intensity scale $I_s \equiv \mathcal{E}_s^2$ is used for the normalization and the calculation of derived quantities such as beam power and pulse energy. Using these scales, (9.3) can be rewritten in dimensionless form as

$$\frac{\partial \widetilde{A}}{\partial \widetilde{z}} = \frac{i}{4}\nabla_{\widetilde{\perp}}^2 \widetilde{A} - i\frac{S_e}{4}\frac{\partial^2 \widetilde{A}}{\partial \widetilde{t}^2} + iK_e|\widetilde{A}|^2\widetilde{A}, \tag{9.5}$$

where the coefficient in front of the transverse Laplacian is normalized to $1/4$ by choosing the length scale z_s to be equal to the Rayleigh length calculated for a collimated beam with a waist w_s, i.e., $z_s = k_0 w_s^2/2$.

With these scales, (9.5) only involves two dimensionless numbers, the Sellmeier number S_e and the Kerr number K_e, which characterize dispersion [7] and the optical Kerr effect [8], respectively, and can be expressed as a ratio of the reference Rayleigh length z_s to a typical length characterizing each effect, namely

L_{S_e} for group velocity dispersion and L_{K_e} for the optical Kerr effect:

$$\text{Sellmeier \#}: \quad S_e = \frac{z_s}{L_{S_e}} = \frac{k_0'' k_0 w_s^2}{t_s^2} \quad \text{with} \quad L_{S_e} = \frac{t_s^2}{2k_0''},$$

$$\text{Kerr \#}: \quad K_e = \frac{z_s}{L_{K_e}} = \frac{\omega_0^2 w_s^2}{2c^2} n_0 n_2 I_s \quad \text{with} \quad L_{K_e} = \frac{c}{\omega_0 n_2 I_s}.$$

Since the power for a Gaussian beam of waist w_s and intensity I_s is expressed as $P_s = \pi w_s^2 I_s/2$, the Kerr number can simply be expressed as

$$K_e = 2\frac{P_s}{P_{\rm cr}} \quad \text{with} \quad P_{\rm cr} = \frac{\lambda_0^2}{2\pi n_0 n_2}, \tag{9.6}$$

where the expression for $P_{\rm cr}$ is a close approximation of the critical power threshold for a beam to undergo catastrophic collapse in a Kerr medium [9].

The input pulse defined by (9.4) can also be rewritten using these scales as

$$\widetilde{A}(\widetilde{z}=0,\widetilde{r},\widetilde{t}) = \sqrt{\frac{I_0}{I_s}} \exp\left(-\frac{w_s^2}{w_0^2}(1+iF)\widetilde{r}^2\right) \exp\left(-\frac{t_s^2}{t_p^2}(1+iC)\widetilde{t}^2\right), \tag{9.7}$$

where the chirp C is the same quantity as in (9.4) and we introduced a modified Fresnel number F in which the beam width w_0 and radius of curvature f ($\sim$ focal distance) play the role of the aperture diameter and distance to the screen, respectively:

$$\text{Fresnel \#}: \quad F = \frac{k_0 w_0^2}{2f}. \tag{9.8}$$

As can be seen from (9.5) and (9.7), the propagation problem (Equations (9.3) and (9.4)) with originally eight parameters $(\lambda_0, w_0, t_p, f, C, I_0, k_0'', n_2)$ has been reduced to a problem with only four dimensionless parameters (C, F, S_e, K_e).[1] If we choose the input values as the reference intensity, radial and temporal scales, respectively, namely, $I_s = I_0$, $w_s = w_0$ and $t_s = t_p$, and the Rayleigh length $z_s = k_0 w_s^2/2$ as longitudinal scale, (9.7) will obviously be parametrized by only two dimensionless numbers, F and C. Together with S_e and K_e in (9.5), we then have a four-dimensional parameter space, and this property is valid for any other choice of the four scales chosen earlier. In other words, the choice of four typical scales is free and always subtracts four dimensions from the initial parameter space.

Marburger's works on self-focusing provide a simple illustration of the advantages in using dimensionless numbers for classifying similar propagation regimes: by numerical simulations of (9.3) and (9.4) in the case of beam self-focusing ($k_0'' = 0$, $t_p \to \infty$), the collapse distance z_c at which a Gaussian beam (initially at waist) with

[1]We assume a given wavelength λ_0, which does not appear explicitly in (9.3) or (9.4). The central laser frequency $\omega_0 = 2\pi c/\lambda_0$, the linear index of refraction n_0 and the propagation constant $k_0 = n_0\omega_0/c$ are not free parameters but can be calculated from λ_0 and thus count for a single one.

power P undergoes a catastrophic singularity was found to follow the law:

$$\widetilde{z}_c = \frac{z_c}{L_R} = \frac{0.367}{\sqrt{\left(\sqrt{\frac{2K_e}{3.77}} - 0.852\right)^2 - (1 - 0.852)^2}}, \tag{9.9}$$

where $L_R = k_0 w_0^2/2$ denotes the Rayleigh length of the collimated input beam, and the values of the coefficients were determined from a fit of numerical simulation results [9]. Marburger's law (9.9) shows that the normalized collapse distance for a collimated Gaussian beam depends on a single dimensionless parameter, the Kerr number K_e. The actual threshold for a catastrophic collapse at distance z_c is $K_e^* = 3.77/2 = 1.885$, that is, $P/P_{\rm cr} = 3.77/4 \sim 0.94$, i.e., the threshold was found by Marburger 6% below the reference critical power as given by (9.6), obtained from a purely dimensional analysis.

This example shows that using dimensionless numbers reduces the dimension of the parameter space to be investigated and facilitates the interpretation of the competition between physical effects. For instance, a comparison of dimensionless numbers appearing in the propagation equation tells us that the optical Kerr effect is stronger than diffraction when $L_{K_e} \ll z_s$ or equivalently $K_e \gg 1$. However, the threshold parameter values that separate different propagation regimes may not be obtained by the exact equality between dimensionless numbers: neither $K_e = 1$ (equivalent to $L_{K_e} = z_s$) nor $K_e = 1/4$ (equivalent to balancing diffraction and the Kerr effect) determines an accurate threshold, as illustrated by the threshold Kerr number K_e^* above which self-focusing prevails over diffraction and leads to a catastrophic collapse: $K_e^* \approx 1.885$. The choice of typical scales to construct dimensionless numbers has an impact on their numerical values and thus on thresholds between propagation regimes. However, it is always possible to choose typical scales so as to set a particular dimensionless number to unity. This will in general introduce another dimensionless quantity to replace the one set to unity.

9.1.2 Extended nonlinear Schrödinger equation coupled with ionization

Carrying on with standard models in the field of ultrashort laser pulse propagation, we consider Kerr media such as air, which can undergo ionization. As originally proposed in [4,5], the system of coupled equations governing pulse propagation and medium response becomes

$$\frac{\partial \mathcal{E}}{\partial z} = \frac{i}{2k_0}\nabla_\perp^2 \mathcal{E} - i\frac{k_0''}{2}\frac{\partial^2 \mathcal{E}}{\partial t^2} + i\frac{\omega_0}{c}\left(n_2 \mathcal{R}(I) - \frac{f_{\nu_c}}{2n_0}\frac{\rho}{\rho_c}\right)\mathcal{E} - \frac{\beta_K}{2} I^{K-1}\mathcal{E} \tag{9.10}$$

$$\frac{\partial \rho}{\partial t} = \sigma_K I^K (\rho_n - \rho) + \frac{\sigma}{U_i}\rho I, \tag{9.11}$$

$$\mathcal{R}(I) = \int_{-\infty}^{t} R(t-t')I(t')dt' \tag{9.12}$$

$$R(t) = (1-f_R)\delta(t) + f_R \frac{\Lambda^2 + \Gamma^2}{\Lambda} \exp(-\Gamma t) \sin \Lambda t \tag{9.13}$$

where $\mathcal{R}(I)$ denotes the Raman–Kerr response, including a Dirac delta function for the instantaneous (electronic) part and a delayed (molecular) part with a weight f_R and a Raman response $R(t)$ (see (9.13)) of characteristic frequencies Γ and Λ. The quantity ρ denotes the electron plasma density, $\rho_c \equiv \varepsilon_0 m_e \omega_0^2/e^2$ denotes the critical density of electrons beyond which the plasma becomes opaque to the electromagnetic radiation of frequency ω_0 and $f_{\widetilde{\nu}_c}$ denotes the complex-valued Drude function of the phenomenological electron neutral collision frequency ν_c:

$$f_{\widetilde{\nu}_c} = \frac{1 - i\widetilde{\nu}_c}{1 + \widetilde{\nu}_c^2}, \quad \widetilde{\nu}_c \equiv \frac{\nu_c}{\omega_0} \tag{9.14}$$

with $f_{\widetilde{\nu}_c} \to 1$ in the limit of a collisionless model, and β_K denotes the multiphoton absorption coefficient for a process involving K photons.

In (9.11) describing ionization of the medium by multiphoton ionization and avalanche, σ_K denotes the multiphoton ionization coefficient, K denotes the number of photons involved in the multiphoton process, ρ_n the density of neutral atoms or molecules and U_i the ionization potential. Since multiphoton absorption describes the laser energy deposited on the medium to liberate electrons, the multiphoton ionization and absorption coefficients are not independent but linked by

$$\beta_K = \sigma_K U_i \rho_n.$$

In a similar way, the avalanche ionization rate is linked to the cross section for inverse bremsstrahlung:

$$\sigma = \frac{\omega_0}{c n_0 \rho_c} \frac{\widetilde{\nu}_c}{1 + \widetilde{\nu}_c^{\,2}}.$$

To write the dimensionless system, we introduce scales I_s, w_s and t_s, as previously for intensity, radius and time, as well as the Rayleigh length $z_s = k_0 w_s^2/2$ for longitudinal distances. We also introduce a scale ρ_s for the electron density and for easier reading, we omit tildes on dimensionless quantities.[2] Using the normalized variable $n = \rho/\rho_s$ for the electron density,[3] the dimensionless version for the

[2]We preserve tildes only in the Raman–Kerr terms to avoid confusion with the corresponding (9.12) and (9.13). The frequencies of the Raman response become $\widetilde{\Lambda} = \Lambda t_s$ and $\widetilde{\Gamma} = \Gamma t_s$.

[3]The normalized electron density should not be confused with the linear refraction index n_0 or the nonlinear index coefficient n_2; the latter quantities always have subscripts throughout the chapter.

system (9.10)–(9.13) reads

$$\frac{\partial A}{\partial z} = \frac{i}{4}\nabla_{\perp}^{2} A - i\frac{S_e}{4}\frac{\partial^2 A}{\partial t^2} + i\left(K_e\widetilde{\mathcal{R}}(|A|^2) - D_r f_{\widetilde{\nu}_c} n\right)A - K_l|A|^{2K-2}A, \tag{9.15}$$

$$\frac{\partial n}{\partial t} = P_e|A|^{2K}(1 - Z_s n) + T_o n|A|^2, \tag{9.16}$$

$$\mathcal{R}(|A|^2) = \int_{-\infty}^{i} \widetilde{R}(t-t')|A|^2(t')dt', \tag{9.17}$$

$$\widetilde{R}(t) = (1 - f_R)\widetilde{\delta}(t) + f_R\frac{\widetilde{\Lambda}^2 + \widetilde{\Gamma}^2}{\widetilde{\Lambda}}\exp(-\widetilde{\Gamma}t)\sin\widetilde{\Lambda}t. \tag{9.18}$$

New dimensionless numbers were introduced in the system of equations, each corresponding to a specific physical effect. We propose to name these numbers in the following way:

- The Drude number D_r [10,11],

$$D_r = \frac{\omega_0}{c}\frac{z_s}{2n_0}\frac{\rho_s}{\rho_c} = \pi^2\frac{w_s^2}{\lambda_0^2}\frac{\rho_s}{\rho_c} \tag{9.19}$$

  represents the importance of plasma defocusing in the collisionless limit, for a plasma of density ρ_s, compared to diffraction for a beam of waist w_s. The dimensionless function $f_{\widetilde{\nu}_c}$ is a complex-valued function, and for a gas, its imaginary part is usually one or two orders of magnitudes smaller than its real part. The corresponding source term (plasma absorption) is negligible for timescales of the order of ultrashort pulses. Nevertheless, it is included in (9.15) and characterized by the product of dimensionless numbers $D_r|f_{\widetilde{\nu}_c,i}|$, where subscript i denotes the imaginary part.
- The Keldysh number K_l [12] is simply defined as the ratio of the Rayleigh length z_s and the length scale characterizing multiphoton absorption (or more generally nonlinear absorption of energy) at the intensity scale I_s:

$$K_l = \frac{z_s}{L_{K_l}} = \frac{1}{2}\beta_K I_s^{K-1} z_s = \frac{1}{4}k_0 w_s^2 \beta_K I_s^{K-1}. \tag{9.20}$$

- The Poynting number P_o [13] is defined as

$$P_o = \frac{\rho_s U_i z_s}{2I_s t_s}. \tag{9.21}$$

While the Poynting number does not directly appear in (9.15), it appears in a dimensionless version of the Poynting theorem applied to (9.15). Indeed, P_o represents the ratio between scales for the average density of dissipated energy to generate electrons at density ρ_s from atoms of ionization potential U_i, by the average density of laser energy deposited in the medium. As shown next, dimensionless numbers appearing in (9.16) are redundant with combinations of D_r, K_l and P_o.

- The Perelomov number P_e [14] reads

$$P_e = \frac{\sigma_K I_s^K t_s \rho_n}{\rho_s} = \frac{K_l}{P_o}. \tag{9.22}$$

Using the relation $\beta_K = \sigma_K U_i \rho_n$ shows the link with the Keldysh and Poynting numbers. The Perelomov number represents a scaled ionization degree. Indeed, P_e is the ratio of the plasma density generated by a square pulse of constant intensity I_s and duration t_s, by the density scale ρ_s. Defining the scale Z_s for the ionization degree as

$$Z_s = \frac{\rho_s}{\rho_n}$$

completes the interpretation of P_e as a scaled ionization degree.

- The Townsend number T_o [15] represents a normalized avalanche ionization rate in (9.16) but can be expressed in terms of previously defined dimensionless numbers as

$$T_o = \frac{\sigma I_s t_s}{U_i} = \frac{D_r}{P_o} |f_{\widetilde{\nu}_c, i}|. \tag{9.23}$$

Table 9.1 summarizes the definitions of all dimensionless numbers proposed in this chapter and their expressions as functions of the radius, length, time, intensity and electron density scales. These numbers are also relevant in all extensions of the nonlinear propagation equation (9.10) in the form of unidirectional pulse propagation equations using a similar matter response model (Equations (9.11)–(9.13)).

Choosing the scales $w_s = w_0$, $z_s = k_0 w_0^2/2$, $t_s = t_p$, $I_s = I_0$ and $\rho_s = \rho_n$ shows that the parameter space for the model (Equations (9.7), (9.15), (9.16)) is determined by seven dimensionless numbers: F, C, S_e, K_e, D_r, K_e and P_o. Five additional parameters complete this space: the dimensionless collision frequency $\widetilde{\nu}_c$, the Raman frequencies $\widetilde{\Gamma}$ and $\widetilde{\Lambda}$, the Raman fraction f_R and the number of photons K involved in the multiphoton process.

9.2 Numerical simulations

In this section, we present the results of numerical simulations of ultrashort laser pulse filamentation in air for a set of Kerr numbers from 4 to 40 and calculate several characteristics of filaments such as the filament radius or the clamping intensity [16–18]. These typical quantities can be defined in several ways from the complex field envelope; hence, we first provide definitions for a quantitative evaluation of these numerical diagnostics. We then compare the values corresponding to different definitions for the range of Kerr numbers characterizing the pulse propagation.

Table 9.1 Dimensionless numbers for the nonlinear Schrödinger and matter response model equations (9.10)–(9.13)

Physical effect	Dimensionless number	Symbol	Expression	Length ratio	Reference length
Focusing	Fresnel #	F	$k_0 w_0^2/2f$	L_R/f	$L_R = k_0 w_0^2/2$
Dispersion	Sellmeier #	S_e	$k_0'' k_0 w_s^2/t_s^2$	z_s/L_d	$L_{S_e} = t_s^2/2k_0''$
OKE	Kerr #	$K_e = 2(P_s/P_{\mathrm{cr}})$	$(w_s^2/\lambda_0^2)2\pi^2 n_0 n_2 I_s$	z_s/L_{K_e}	$L_{K_e} = \lambda_0/2\pi n_2 I_s$
PLD	Drude #	D_r	$\pi^2(w_s^2/\lambda_0^2)(\rho_s/\rho_c)$	z_s/L_{D_r}	$L_{D_r} = \lambda_0(n_0\rho_c/\pi\rho_s)$
MPA	Keldysh #	K_l	$k_0 w_s^2 \beta_K I_s^{K-1}/4$	z_s/L_{K_l}	$L_{K_l} = 2/\beta_K I_s^{K-1}$
	Poynting #	P_o	$\rho_s U_i k_0 w_s^2/4I_s t_s$		
MPI	Perelomov #	$P_e = K_l/P_o$	$\sigma_K I_s^K t_s(\rho_n/\rho_s)$	t_s/T_{P_e}	$T_{P_e} = \rho_s/\sigma_K I_s^K \rho_n$
AVA	Townsend #	$T_o = D_r/P_o\lvert f_{\nu_c, i}\rvert$	$\sigma I_s t_s/U_i$	t_s/T_{T_o}	$T_{T_o} = U_i/\sigma I_s$

The first five dimensionless numbers (F, S_e, K_e, D_r, K_l) enter in the input pulse (9.7) and propagation equation (9.15). The last two (P_e and T_o) enter in the rate equation (9.16) and are expressed as a timescale ratio rather than a length ratio. OKE, optical Kerr effect; PLD, plasma defocusing; MPA, multiphoton absorption; MPI, multiphoton ionization; AVA, avalanche ionization. See text for the definition of symbols.

Table 9.2 Table of dimensionless numbers used in simulations of filamentation in air for the physical parameters listed in Section 9.2.1

Equation (9.7)	I_0/I_s	w_0^2/w_s^2	t_p/t_s	ρ_s/ρ_b	(F, C)
Input pulse	1	1	1	1	$(0, 0)$
Filament	0.125	80.3	1	1.5×10^{-3}	$(0, 0)$
Equation (9.15)	K_e	D_r	K_l	P_o	S_e
Input pulse	$2\widetilde{P}$	1.1×10^5	$8.1 \times 10^{-12}\widetilde{P}^{K-1}$	$1.2 \times 10^5\widetilde{P}^{-1}$	2.8×10^{-2}
$\widetilde{P} = 10$	20	1.1×10^5	7.9×10^{-6}	1.2×10^4	2.8×10^{-2}
Filament	2	2	2.7×10^{-2}	2.7×10^{-2}	3.4×10^{-4}
Equation (9.16)	P_e	T_o	K	$\widetilde{\nu}_c \sim f_{\widetilde{\nu}_c, i}$	Z_s
Input pulse	6.8×10^{-10}	1.1×10^{-2}	7	1.2×10^{-3}	1
Filament	1	8.9×10^{-2}	7	1.2×10^{-3}	1.5×10^{-3}
Equation (9.18)	f_R	$\widetilde{\Gamma}$	$\widetilde{\Lambda}$		
	0.5	2.2	2.7		

Each row corresponds to one of the dimensionless equations of the model. The choice of typical scales can be based on the input pulse or on the filament characteristics. The corresponding line in each row gives the list of dimensionless numbers identifying a simulation.

9.2.1 Input pulse vs filament-based scales

The parameters used in all simulations for the medium (air at the pressure of 1 bar) and for the pulse were taken from [4], allowing us to retrieve all results of this publication and use them as reference cases.[4] Namely, $\lambda_0 = 775$ nm, $n_2 = 5.57 \times 10^{-19}$ cm^2/W, $k'' = 2$ fs^2/cm, $\beta_K = 6.5 \times 10^{-82}$ cm^{11}/W^6, $\nu_c = 2.86$ THz, $\rho_n = 2.5 \times 10^{19}$ cm^{-3} (at 1 bar), $U_i = 11$ eV, $K = 7$, $\sigma = 5.1 \times 10^{-20}$ cm^2, $t_p = 170$ fs (Full Width at Half Maximum (FWHM) of 200 fs). $f_R = 1/2$, $\Gamma = 13$ THz, $\Lambda = 16$ THz. $w_0 = 700$ µm and the input power is varied up to $20P_{\text{cr}}$. In this list, n_2, k'', β_K and ν_c implicitly depend on the medium density ρ_n and thus on air pressure. We discuss implications of this dependency on dimensionless numbers in Section 9.3.

Table 9.2 presents the list of all the dimensionless parameters and numbers necessary to perform a simulation, namely, (i) the five input pulse parameters: I_0/I_s, w_0/w_s, t_p/t_s, C and F; (ii) the five parameters in the propagation equation (9.15): K_e, D_r, K_l, S_e and $\widetilde{\nu}_c$; (iii) the two dimensionless numbers P_e, T_o in the rate equation (9.16) as well as the density scale Z_s and the number of photons K needed for multiphoton ionization; (iv) the three Raman parameters for (9.18). Once typical scales are chosen, for instance input pulse–based scales I_0, w_0, t_p and ρ_n as indicated in the first row, five dimensionless numbers (K_e, D_r, K_l, S_e and P_o) are sufficient to cover the entire

[4]In this aim, the overestimated value $k'' = 2$ fs^2/cm for the GVD coefficient in air from [4] has not been corrected. All results from [4] can be retrieved using an FWHM pulse duration of 200 fs rather than 100 fs (a possible typo in [4]).

[5]The Raman parameters and the collision frequency actually constitute four additional dimensions of the parameter space, which we do not explore here, since these are determined by the gas and not free to change.

parameter space.[5] The Fresnel number and the chirp coefficient are equal to zero in all of our simulations because we begin at the waist. Only the input pulse intensity I_0 is varied (equivalent to changing the normalized pulse power $\widetilde{P}$ from 4 to 40, or the input pulse power from 2 to 20 P_{cr}), keeping constant all other parameters. Three dimensionless parameters depend on the intensity scale I_s, namely, K_e, K_l and P_o, and thus vary as indicated in the second row when the power $\widetilde{P}$ increases.

The numerical values of these numbers may differ significantly as exemplified in Table 9.2 for $P = 10P_{cr}$ ($K_e = 20$). This indicates that another choice for the typical scales may better reflect the physical effects playing a role within a filament. For instance, if the typical scales represent the filament scales, pairs of dimensionless numbers with similar order of magnitude reveal a competition between physical effects. In addition, in the dimensionless numbers, it could be interesting to keep separate the medium and the input pulse parameters. Both of these properties can be achieved by choosing an intensity scale $I_s = I_f$ that is only medium dependent and reflects the high intensities reached in the filament. To define this scale I_f, we can write the balance between the plasma defocusing term and the self-focusing term, i.e., we require $D_r = K_e$. Eliminating the scale w_s^2 on both sides and selecting a plasma density scale $\rho_s = \sigma_K I_s^K t_s \rho_n$ (which amounts to requiring $P_e = 1$ or $K_l = P_o$), we find

$$I_f = \left(\frac{2n_0 n_2 \rho_c}{\sigma_K \rho_n t_s}\right)^{1/(K-1)}. \tag{9.24}$$

This intensity scale is fully defined from the medium parameters and a scale for t_s, for instance $t_s = t_p$, and coincides with the clamping intensity.[6] To complete the choice of typical scales, we choose w_s to be close to the radius of a filament by requiring that the power scale $P_s = \pi w_s^2 I_s/2$ for a Gaussian beam of width w_s and intensity I_s be equal to P_{cr}. This leads to a Kerr value and a Drude value both equal to 2 and corresponding scales w_f and ρ_f are calculated as

$$w_f = \sqrt{\frac{2P_{cr}}{\pi I_f}}, \tag{2.25}$$

$$z_f = \frac{k_0 w_f^2}{2}, \rho_f = \sigma_K I_f^K t_s \rho_n. \tag{2.26}$$

Their values calculated from the physical parameters of the medium are in-line with typical filamentation characteristics: $I_f = 1.8 \times 10^{13}$ W/cm^2, $w_f = 78$ μm, $z_f \sim 2.5$ cm and $\rho_f = 3.7 \times 10^{16}$ cm^{-3}. With the filament-based scale choice $I_s = I_f$, $w_s = w_s$, $\rho_s = \rho_f$ and $t_s = t_p$, the remaining dimensionless numbers are expressed as

$$K_l = P_o = \frac{n_0 \beta_K P_{cr} I_f^{K-2}}{\lambda_0}. \tag{2.27}$$

[6] Equation (9.24) corrects the typo of (9.18) in [19]. An alternative choice would be (9.20) in [19]. See also [18].

The list of dimensionless parameters in Table 9.2 therefore shows that all the simulations can be performed equivalently by selecting (i) either the input pulse–based scales, keeping the input pulse (9.7) frozen and changing only K_l in the propagation equation (9.15) when K_e (or $\widetilde{P}$) is increased;[7] (ii) or the filament-based scales, for which some dimensionless parameters are frozen ($K_e = D_r = 2$, $K_l = P_o$, $P_e = 1$) and only the amplitude of the input pulse changes with $\widetilde{P}$. Summarizing, the model (9.7) and (9.15)–(9.18) is simulated by selecting a single set of scales and dimensionless parameters among *input pulse* and *filament*, in each row of Table 9.2. Simulation results of a given simulation can then be entirely rescaled for any combination of physical parameters that provide exactly the same minimal set of five dimensionless numbers. This is one of the most powerful properties of dimensionless numbers based on the fact that only the scales change but the five dimensionless numbers keep their values; the dimensionless model is preserved and a single dimensionless simulation result can finally be rescaled using several physical scale sets.

9.2.2 Filament characteristics

The simulation results provide the electric field envelope $\mathcal{E}(r,t,z)$ within a range of propagation distances $0 \leq z \leq z_{\max}$. This quantity is not directly accessible to measurements; therefore, the complex field envelope is post-processed to obtain the intensity $I(r,t,z) = |\mathcal{E}(r,t,z)|^2$ and the fluence distribution $F(r,z) = \int_{-\infty}^{+\infty} I(r,t,z)dt$, which is the closest quantity to that recorded by a CCD camera when it is possible to insert one in the beam.

An example of fluence distribution is shown in Figure 9.1.

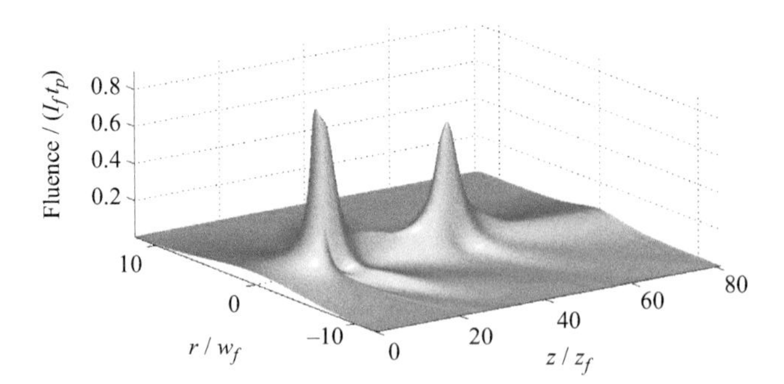

Figure 9.1 Typical fluence distribution obtained in the simulation of a single filament in air for $K_e = 4$. The filament scales are used for normalization: $w_f \sim 78$ μm, $z_f \sim 2.5$ cm *and* $I_f t_p \sim 3$ J/cm^2

[7] The variation of P_o as a function of $\widetilde{P}$ allows us to calculate the dimensionless parameters of the rate equation P_e and T_o.

The fluence distribution exhibits two peaks, separated by a channel of relatively high fluence. The radius of this channel is calculated as the radial distance at which the radial profile of the fluence reaches half the maximum (axial) value for each propagation distance, z, and is plotted as a function of z in Figure 9.2(a). This diagnostics clearly shows the initial self-focusing of the beam down to a radius R_{sf} at position z_{sf}, followed by a region where the radius oscillates around an average value of 50 μm (red part of the curve), and by a final diffraction stage. This leads to a first formal definition of the filament length as the length L_{fch} of the fluence channel, i.e., of the region between the positions of the local minimum radii, z_{sf} that terminates the self-focusing stage and z_{df} that starts the diffraction stage from a radius R_{df}. The filament length can be evaluated from this definition by adding a Rayleigh length on each side:

$$L_{fch} = z_{df} - z_{sf} + \frac{k_0 R_{sf}^2}{2} + \frac{k_0 R_{df}^2}{2}. \tag{9.28}$$

However, fluence is a time-integrated quantity, which smoothes the longitudinal intensity variation inside the filament region. For instance, the two peaks in Figure 9.1 indicate the presence of two zones with high intensity within the filament, but localized high-intensity peaks can be reached over a larger region, which is hardly visible in Figure 9.2 alone. Fortunately, numerical simulations allow us to easily access the maximum intensity and electron density as functions of the propagation distances, as plotted in Figure 9.2(b) and (c).

In the self-focusing region, the maximum intensity grows with increasing propagation distance. Beyond the nonlinear focus, it exhibits plateau regions characterized by the so-called intensity clamping at a few tens of TW/cm^2. It is common to obtain a single plateau followed by the diffraction stage. In this case, the filament can be defined by the region of clamped intensity and its length L_{fil} can be defined as the full width at half maximum of the profile $I_{max}(z)$. Several plateau regions may be obtained, usually when the beam is defocused by the plasma generated in a plateau region and refocused further, leading to another plateau region. Focusing and defocusing events may also occur without a gap between the two events, leading to a single plateau. In all cases, the filament length L_{fil} can be defined similarly by merely adding the lengths of the plateau regions (length of straight lines in Figure 9.2(b).

The third definition of a filament, often found in the literature, corresponds to the region of large plasma density, i.e., the filament refers to the plasma channel. Since the ionization rate is a nonlinear function of the intensity, these regions are closely connected to the high-intensity regions (see Figure 9.2(b) and (c)). However, due to the high-order nonlinearity of ionization rates, the length of the region over which the plasma density exceeds the half maximum is shorter than L_{fil}, the length of the region over which the intensity exceeds the half maximum of the intensity profile.[8] To be closer to L_{fil}, the definition for the length L_{pch} of the plasma

[8]If the maximum intensity profile in Figure 9.2(b) were Gaussian, a ratio of $\sqrt{K}$ would be found between the two lengths, K, corresponding to the number of photons involved in the ionization process.

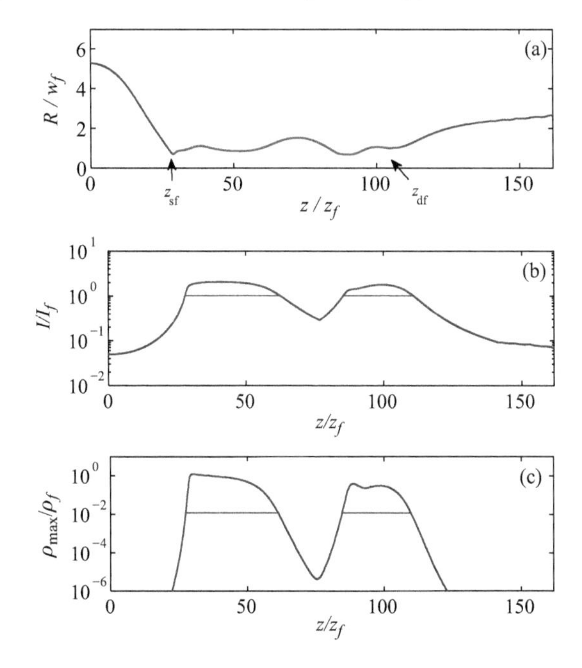

Figure 9.2 *Three different definitions of the filament length. Results are obtained for $K_e = 8$ and normalized to the filament scales $w_f \sim 78$ μm, $z_f \sim 2.5$ cm, $I_f \sim 18$ TW/cm^2 and $\rho_f \sim 3.7 \times 10^{16}$ cm^{-3}. (a) Beam radius as a function of the propagation distance z/z_f. The red part of length L_{fch} corresponds to the filament (fluence channel). (b) Axial peak intensity as a function of z_f. The straight lines mark the length L_{fil} of the high-intensity channel. (c) Maximum electron density on axis as a function of z/z_f. The straight lines mark the length L_{pch} of the plasma channel*

channel must then be adapted: L_{pch} is thus defined as the total length of the regions where the plasma density exceeds 1% of the highest electron density (length of straight lines in Figure 9.2(c)).

Figure 9.3(a) shows a comparison of the lengths of the fluence channel, intensity channel (filament) and plasma channel for values of the Kerr number from 4 to 40. The three lengths have the same order of magnitude and reach several tens of diffraction lengths z_f calculated from the filament radial scale w_f. Figure 9.3(b) represents these three lengths as a function of the Kerr number, highlighting the presence of discontinuity in the clamping intensity (for $K_e = 8$, 10, 28–34, 38). When this occurs, the fluence channel covers also the gaps in intensity clamping, and its length is thus found to exceed those of the intensity and plasma channels. A clear trend of the global increase of the filament lengths with the Kerr number is obtained, albeit it is non-monotonic and overall slow. Above the Kerr number of 40, it is likely that the beam would break up into multiple filaments, thereby decreasing the total length of individual filaments found at this power.

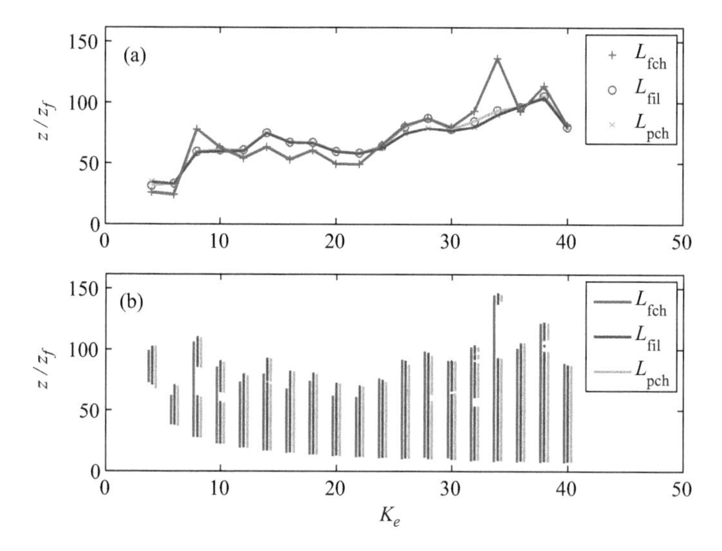

Figure 9.3 Filamentation length as a function of the Kerr number. The filament scale $z_f = 2.5$ cm *is used for normalization. (a) The three curves correspond to the three different definitions.* L_{fch}*: length of the fluence channel (red curve with +markers).* L_{fil}*: length of the light (intensity) channel (blue curve with circle markers).* L_{pch}*: length of the plasma channel (green curve with ×markers). (b) The filament lengths are shown as lines indicating regions of discontinuity*

We finally monitored the highest intensity $I_{cl} = \max_z(I_{max}(z))$ in each filament, which slowly increases from 30 to 40 TW/cm^2 for increasing Kerr numbers between 4 and 40 (see Figure 9.4(a)) and is found slightly above the average intensity in the filament:

$$I_{ave} = \frac{1}{L_{fil}} \int_{I_{max}(z) > \frac{1}{2} I_{cl}} I_{max}(z) dz$$

It is the quasi-constant value (very slow increase) of I_{cl} as a function of the Kerr number that justifies calling this quantity the *clamping* intensity, together with the flatness of the curve $I_{max}(z)$. Figure 9.4(b) shows a comparison of the highest and the average plasma densities in each filament as a function of the Kerr number. Both quantities follow the trend of the clamping and average intensities in the filaments, except for the range of Kerr numbers between 24 and 32, for which a significant increase of the clamping plasma density is observed. This increase actually reflects the enhancement of the slight increase of the clamping intensity visible in Figure 9.4(b) by the high-order nonlinearity featuring multiphoton ionization.[9]

[9]Since the clamping plasma density is a time-integrated quantity, changes in the pulse profile during filamentation are also responsible for fluctuations in the highest plasma density.

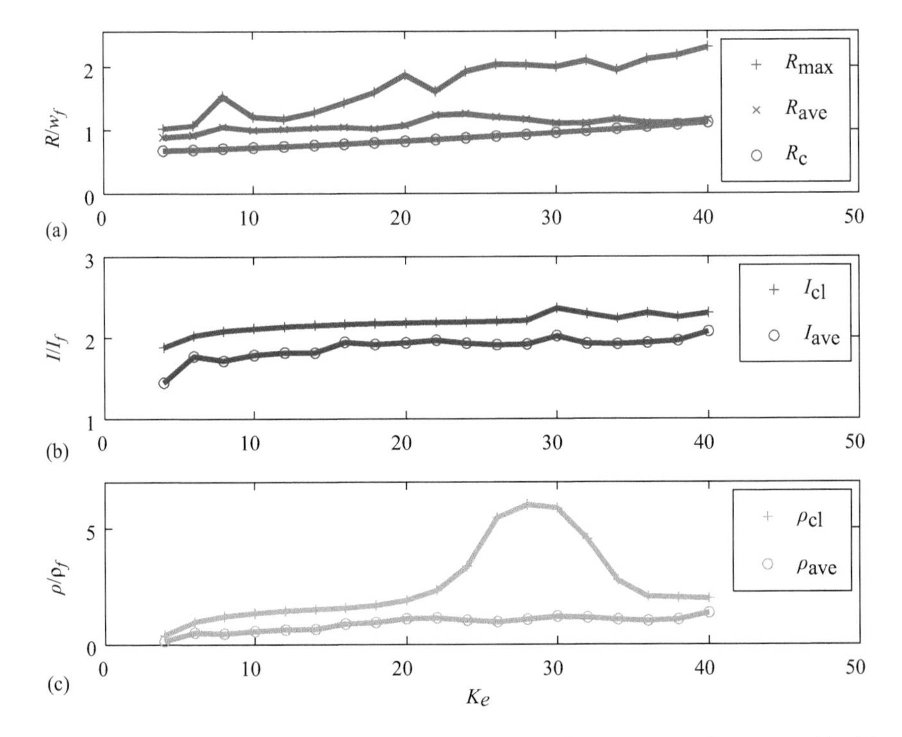

Figure 9.4 Filament characteristics for Kerr numbers ranging from 4 to 40. (a) R_{max}: Maximum radius (+ marker); R_{ave}: average radius (× marker) and R_c (○ marker): radius of the filament at the end of the self-focusing stage (near collapse point). (b) Clamping intensity I_{cl} (+ marker) and average intensity I_{cl} (○ marker). (c) Highest (clamping) plasma density ρ_{cl} (+ marker) and average plasma density ρ_{cl} (○ marker)

9.2.3 Pulse shortening

Few-cycle pulses were shown experimentally and from numerical simulations to be generated by pulse self-shortening during filamentation [20,21]. Several papers presented schemes to extract the few-cycle pulse from the filament at the point of best compression [22,23] or scenarios for optimizing the energy throughput [24]. In this section, we show that in all simulations presented in this chapter, we found several propagation distances within the filament, for which the pulse exhibits self-shortening down to a few-cycle duration. The next section will discuss how to upscale the energy in these few-cycle structures generated by filamentation.

Figure 9.5 shows an example of pulse shortening occurring during filamentation in air. The region of high intensity bounded by red curves is obtained by calculating the full width at half maximum duration for the axial temporal profiles at each propagation distance. A *tongue*-shaped region appears when the pulse undergoes self-shortening, as a result of defocusing of its trailing part (positive times) due to the

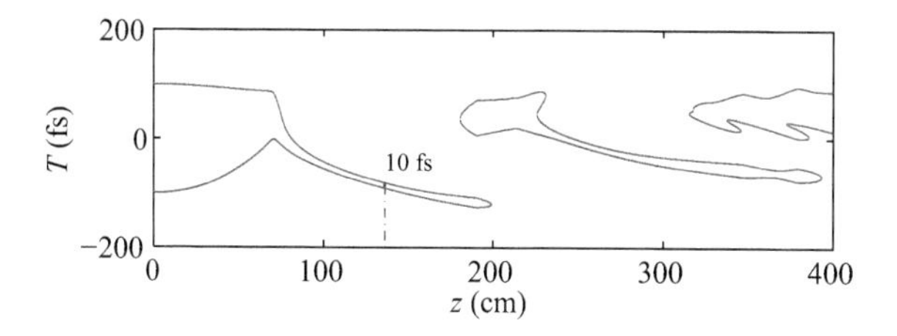

Figure 9.5 Full-width-at-half-maximum pulse duration as a function of propagation distance for a Kerr number of 8. The successive tongues indicate pulse-shortening events. The shortest pulse (10 fs) is indicated by the dashed line (at z = 130 cm*)*

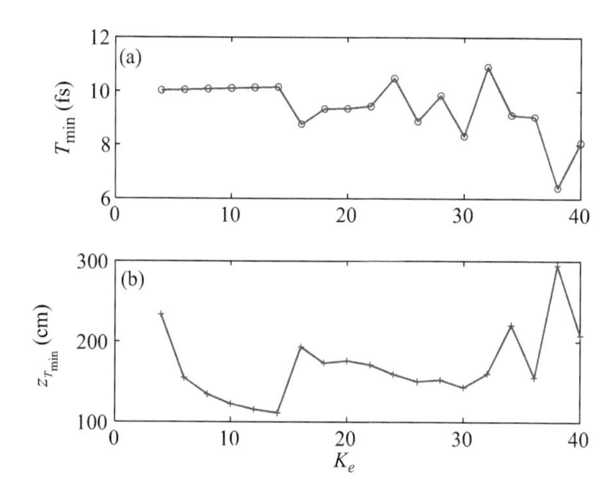

Figure 9.6 (a) Shortest pulse duration (FWHM) in the filament as a function of the Kerr number and (b) propagation distance corresponding to the shortest pulse duration

plasma-induced negative index change. A 20-fold decrease of the pulse duration is commonly obtained, leading to a 10-fs pulse as indicated in Figure 9.5. Successive tongues may be obtained and usually indicate refocusing events in the trailing part of the pulse, further evolving toward the leading part as the pulse propagates. Multiple refocusing events lead to a complex pulse temporal structure; thus, the shortest pulse with the best contrast is usually found in the first tongue.

Figure 9.6 shows the shortest pulse duration obtained in the filament for each Kerr number and the propagation distance at which it is obtained. For Kerr numbers between 4 and 14, a self-shortened pulse with remarkably constant duration is obtained systematically in the first tongue. For higher Kerr numbers, the self-shortening process that occurs after different refocusing events may result in even shorter pulses. However, refocusing events are detrimental to the overall pulse temporal structure, and the short pulses that are obtained in the successive tongues (from the second one) exhibit lower contrast in comparison with those appearing in the first tongue. The energy content of the self-shortened pulse does not exceed a few tens of μJ. However, several strategies were proposed for upscaling this level both in the near-infrared and in the mid-infrared regions of the electromagnetic spectrum[10] [2,25]. We discuss later how possible strategies for this upscaling can be inferred from the analysis of dimensionless numbers used in the propagation equation.

9.3 Discussion and conclusion

In this chapter, we have proposed to define and name dimensionless numbers that span the parameter space of the most standard model used for simulations of ultrashort laser pulse filamentation and laser–matter interaction, namely, the extended nonlinear Schrödinger equation. We have shown that apart from the normalized Raman response parameters and the collision frequency,[11] five dimensionless numbers and the number of photons involved in multiphoton ionization are sufficient to span the entire parameter space of this model. A single point in this five-dimensional space, for instance, a set of five values corresponding to the Sellmeier, the Kerr, the Drude, the Keldysh and the Poynting numbers, allows us to perform a single simulation of model (9.7) (9.15)–(9.18), the results of which can be rescaled, thanks to the associated input pulse scales $I_s = I_0$, $w_s = w_0$, $t_s = t_p$ and $\rho_s = \rho_n$, to a multiplicity of physical results, provided the physical scales yield the same values for the set of dimensionless numbers.

This property can be illustrated with the scaling laws for pulse propagation equations proposed in [2,3], which state that by playing on the gas pressure, it is possible to obtain a similar pulse propagation for a small beam with characteristic width w_0 and typical Rayleigh length $L_R = k_0 w_0^2/2$ and for a large beam with characteristic width ηw_0 and typical Rayleigh length $\eta^2 L_R$, where η is the scaling parameter. This property is justified by the facts that (i) the diffraction term in (9.3) is not affected by this scaling. (ii) Other terms can be scaled similarly by decreasing the density of the gas (or pressure) by a factor η^2. Since the dispersive coefficients and the nonlinear polarization are proportional to the density of neutral atoms ρ_n, (9.3) is quasi-invariant for the scaling $z \to \eta^2 z$, $r \to \eta r$, $t \to t$ and $\rho_n \to \rho_n/\eta^2$.

This scaling simply follows from the dependence of the dimensionless numbers upon pressure. The definitions of dimensionless numbers in Table 9.1 show that all values of S_e, K_e, D_r K_l and P_o are proportional both to w_s^2 and to a single parameter that is itself proportional to the gas pressure p: $k'' \propto p$, $n_2 \propto p$, $\beta_K \propto p$

[10]This region corresponds to anomalous dispersion, hence to a negative Sellmeier number.

[11]For gases, and zero or small Fresnel numbers, the collision frequency has an almost negligible impact on results.

and $\rho_n \propto p$. Thus, for the dimensionless parameters to be preserved by a change of gas pressure, one must ensure that the product $w_s^2 p$ be preserved, which is exactly the case for the proposed scaling where an increase of the input beam width by a factor η, and thus an increase of $w_s^2 = w_0^2$ by a factor η^2, compensates for a decrease of the pressure by the same factor η^2. This is associated with a larger propagation distance as it is increased by a factor η^2 corresponding to the rescaling of the Rayleigh length. Note that this scaling would work as well in the case of a focused geometry since the Fresnel number is preserved, provided the increase of w_0 by a factor η is compensated by an increase of the spatial quadratic phase f by factor η^2. The only limitation of this scaling appears in the expression for the Townsend number: if T_o were only the ratio of the preserved Drude and Poynting numbers, it would be preserved as well. However, it is also proportional to the normalized collision frequency (via $|f_{\nu_c,i}|$), and hence to the pressure. Strictly speaking, the avalanche term is thus a source of departure from the proposed scaling property, but applying the scaling property in a situation where the gas pressure is decreased makes avalanche ionization more and more negligible, thereby leaving open numerous possible applications of the scaling properties.

By exploiting the proposed scaling properties with the aim of improving the energy content of few-cycle pulses generated during filamentation, we can infer a design for an air plasma–based pulse compressor from the results of Figure 9.6. In the range of Kerr numbers between 4 and 14, we obtained for $p = 1$ bar, a range of propagation distances (up to 200 cm) where a few-cycle pulse of $\sim$10 fs duration is formed, with a clamping intensity of 30 TW/cm^2 and a filament radius of 80 μm. This represents a relatively low energy of 38 μJ (lower bound estimation), but considering the proposed scaling with a factor $\eta = 4$ leads to numerous possibilities for designing a compressor delivering 600 μJ few-cycle pulses within a low pressure air-filled tube. The energy throughput is indeed calculated from the peak intensity, shortest pulse duration and filament radius that are almost constant quantities when the Kerr number varies, leaving some freedom for the length of the tube. If principle, this length should be large enough to include η^2 times the propagation distance of the first tongue in Figure 9.5. However, larger Kerr numbers (e.g. $K_e = 14$ in Figure 9.6) lead to a shift of this tongue toward shorter distances, and almost constant few-cycle pulse duration obtained in the first tongue beyond the nonlinear focus allows us to select an extraction point close to the beginning of the filament. A focusing geometry constitutes another option to minimize the tube length required to increase the few-cycle pulse energy produce in such a compressor design.

To conclude, the dimensionless numbers proposed in this chapter allow for an easy exploration of the parameter space of one of the most widespread propagation and laser–matter interaction model in the field of ultrashort laser pulse filamentation. We hope that their use in optics will help identifying interesting laser–matter interaction regimes for numerous applications, from the development of sources of secondary radiation [26] to material ablation or processing with femtosecond laser pulses [27,28].

References

[1] Olson RM. *Essentials of Engineering Fluid Mechanics*. New York, NY: International Textbook Company; 1966.
[2] Heyl CM, Coudert-Alteirac H, Miranda M, *et al.* Scale-invariant nonlinear optics in gases. *Optica*. 2016; 10: 1364.
[3] Zhokhov PA and Zheltikov AM. Scaling laws for laser-induced filamentation. *Phys Rev A*. 2014; 89: 043816.
[4] Mlejnek M, Wright EM, and Moloney JV. Dynamic spatial replenishment of femtosecond pulses propagating in air. *Opt Lett*. 1998; 23 (5): 382–384.
[5] Kandidov VP, Kosareva OG, and Shlenov SA. Influence of transient self-defocusing on the propagation of high-power femtosecond laser pulses in gases under ionisation conditions. *Quantum Electron*. 1994; 24 (10): 905–911.
[6] Chernev P and Petrov V. Self-focusing of light pulses in the presence of normal group-velocity dispersion. *Opt Lett*. 1992; 17 (3): 172–174.
[7] Sellmeier W. Zur Erklärung der abnormen Farbenfolge im Spectrum einiger Substanzen. *Ann Phys Chem*. 1871; 219: 272–282.
[8] Kerr J. A new relation between electricity and light: Dielectrified media birefringent. *Philos Mag*. 1875; 4 (50): 337–348.
[9] Marburger JH. Self-focusing: Theory. *Prog Quantum Electron*. 1975; 4: 35–110.
[10] Drude P. Zur Elektronentheorie der Metalle. *Ann Phys*. 1900; 306 (3): 566–613.
[11] Drude P. Zur Elektronentheorie der Metalle; II. Teil. Galvanomagnetische und thermomagnetische Effecte. *Ann Phys*. 1900; 308(11): 369–402.
[12] Keldysh LV. Ionization in the field of a strong electromagnetic wave. *Sov Phys JETP*. 1965; 20 (5): 1307–1314.
[13] Poynting JH. On the transfer of energy in the electromagnetic field. *Phil Trans R Soc London*. 1884; 175: 343.
[14] Perelomov AM, Popov VS, and Terent'ev MV. Ionization of atoms in an alternating electric field. *Sov Phys JETP*. 1966; 23 (5): 924–934.
[15] Townsend JS. VIII. Ionization by collisions of positive ions. *Lond, Edinb, Dublin Philos Mag J Sci*. 1939; 28 (186): 111–117.
[16] Kasparian J, Sauerbrey R, and Chin SL. The critical laser intensity of self-guided light filaments in air. *Appl Phys B*. 2000; 71: 877.
[17] Becker A, Aközbek N, Vijayalakshmi K, *et al.* Intensity Clamping and re-focusing of intense femtosecond laser pulses in nitrogen molecular gas. *Appl Phys B*. 2001; 73: 287–290.
[18] Kandidov VP, Fedorov VY, Tverskoi OV, *et al.* Intensity clamping in the filament of femtosecond laser radiation. *Quantum Electron*. 2011; 41 (4): 382–386.
[19] Couairon A and Mysyrowicz A. Femtosecond filamentation in transparent media. *Phys Rep*. 2007; 441: 47–189.
[20] Hauri CP, Kornelis W, Helbing FW, *et al.* Generation of intense, carrier-envelope phase-locked few-cycle laser pulses through filamentation. *Appl Phys B*. 2004; 79: 673–677.

[21] Couairon A, Biegert J, Hauri CP, *et al.* Self-compression of ultrashort laser pulses down to one optical cycle by filamentation. *J Mod Opt*. 2006; 53 (1−2): 75–85.
[22] Couairon A, Franco M, Mysyrowicz A, *et al.* Pulse-compression to the single cycle limit by filamentation in a gas with a pressure gradient. *Opt Lett*. 2005; 30 (19): 2657–2659.
[23] Steingrube DS, Schulz E, Binhammer T, *et al.* Generation of high-order harmonics with ultra-short pulses from filamentation. *Opt Express*. 2009; 17 (18): 16177–16182.
[24] Kosareva OG, Panov NA, Uryupina DS, *et al.* Optimization of a femtosecond pulse self-compression region along a filament in air. *Appl Phys B*. 2008; 91: 35–43.
[25] Voronin AA and Zheltikov AM. Power-scalable subcycle pulses from laser filaments. *Sci Rep*. 2016; 6: 36263.
[26] Heyl CM, Arnold CL, Couairon A, *et al.* Introduction to macroscopic power scaling principles for high-order harmonic generation. *J Phys B: At Mol Opt Phys*. 2016; 50 (1): 013001.
[27] Gattass RR and Mazur E. Femtosecond laser micromachining in transparent materials. *Nat Photonics*. 2008; 2 (4): 219–225.
[28] Courvoisier F, Stoian R, and Couairon A. Ultrafast laser micro and nano processing with nondiffracting and curved beams. *J Opt Laser Technol*. 2016; 80: 125–137.
[29] W. J. Smith, *Modern Optical Engineering*, 4th ed. (SPIE Press, 2008).

Chapter 10

Kerr instability amplification

Michael Nesrallah,[1] T.J. Hammond,[2] Ashwaq Hakami,[1] Graeme Bart,[1] Chris McDonald,[1] Thomas Brabec,[1] and Giulio Vampa[3]

10.1 Introduction

In this chapter, the theory of an amplification scheme based on Kerr nonlinearity is developed. It is well known that Kerr nonlinearity exhibits both temporal (modulation) and spatial (filamentation) instability. Typically, a single, intense beam propagating in a Kerr nonlinear material has small perturbations in the form of noise that become exponentially amplified. In turn, the noise drastically modifies the beam given a long enough propagation distance, and the beam can form filaments. Subsequently, conical emission ensues—the emission of broadband radiation at a frequency-dependent angle to the filament [1].

Recently, it was found that a different regime of conical emission occurs when the Kerr instability is seeded by a second pulse at a noncollinear angle [2–4]. In this regime, the seed pulse experiences substantial growth through amplification long before filamentation occurs, see Figure 10.1. As the seed beam is amplified via the Kerr instability, this process is called Kerr instability amplification (KIA). The pump and seed beams are aligned noncollinearly at the angle of maximum amplification that depends on the seed pulse frequency. In Appendix A, we discuss the relation between KIA and four wave mixing (FWM). The initial equations of KIA and FWM are shown to be identical. The solution technique used for KIA is more general, yielding the most general analytical expression for the FWM gain that contains commonly used FWM expressions as a limit. Thus, the theories of FWM, conical emission, and KIA are unified.

Theoretical and experimental analyses have shown that the Kerr instability has promising properties as an amplification mechanism of ultrashort infrared pulses. The amplification bandwidth extends over a wide frequency range between the second harmonic of the pump pulse and the mid-infrared wavelength regime.

[1]Department of Physics, University of Ottawa, Ottawa, Canada
[2]Department of Physics, University of Windsor, Windsor, Ontario, Canada
[3]Joint Attosecond Science Laboratory, National Research Council of Canada and University of Ottawa, Ottawa, Canada

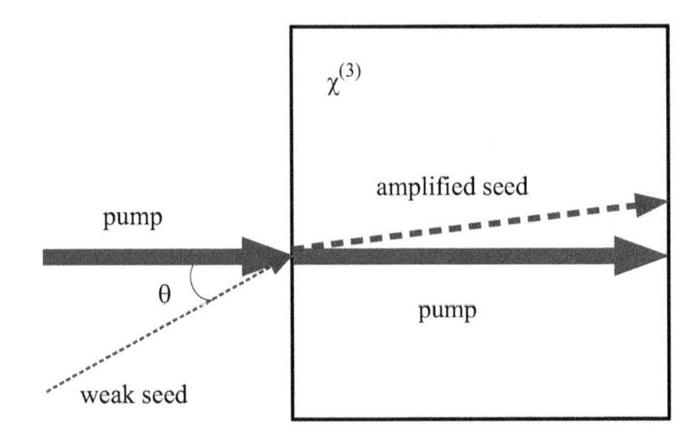

Figure 10.1 Schematic of seeded Kerr instability amplification. A strong, monochromatic pump beam (thick solid blue arrow) interacts noncollinearly at an angle θ with a weak monochromatic seed beam (thin dashed red arrow) in a $\chi^{(3)}$ nonlinear medium. Through Kerr nonlinear instability, the seed beam is amplified (thicker dashed red arrow). The pump beam remains approximately undistorted

Experimentally, the amplification of more than a factor of 1,000 was demonstrated in a yttrium aluminum garnet (YAG) crystal, with a pump frequency of 800 nm and a seed frequency ranging from 600 nm to 2 μm [5]. Theoretically, it was predicted that, by optimizing pump wavelength and materials, the amplification of mid-infrared one- to two-cycle pulses up to sub-mJ energies is possible.

Development of high intensity, ultrashort laser sources in the mid-infrared is crucial for advancements in strong-field physics and attosecond science [6,7]. Currently, the most common generation and amplification methods are based on the second-order nonlinearity, such as chirped pulse optical parametric amplifiers (OPAs) with periodically poled optical phase matching gratings [8–10]. Recently, the potential for single-cycle infrared pulse generation by difference frequency generation has been demonstrated [11]. State-of-the-art OPAs supply amplification factors of up to three orders of magnitude with amplified pulses in the 10-μJ energy range depending on seed wavelength, corresponding to 5%–10% of the pump energy.

Although OPAs are currently the leading technology for ultrashort mid-infrared pulse amplification, their development is challenging. For their efficient operation, a series of stringent conditions must be met, which are intrinsically linked to the nonlinear crystal properties. Amplification of single-cycle pulses either requires thin crystal—reducing the efficiency—or low dispersion across a spectrum covering the frequencies of the three interacting waves. Moreover, many second-order nonlinear crystals absorb light in the mid-infrared, and moderate damage thresholds also present a limitation.

The KIA process is described schematically in Figure 10.2. In a Kerr nonlinear material, parametric FWM processes of the type $\omega_p + \omega_p - (\omega_p \pm \Omega_s) = \omega_p \mp \Omega_s$ occur. Two photons of the pump frequency, ω_p, are converted into fields $\varepsilon_x(\Omega_s)$ and $\varepsilon_x^*(-\Omega_s)$ with photon energies shifted to the red and blue side of ω_p by $\Omega_s = \omega_s - \omega_p$. Coupling between the red and blue sides of the pulse produces instability. For a wide range of seed frequencies in the interval, $-\omega_p < \Omega_s < \omega_p$,

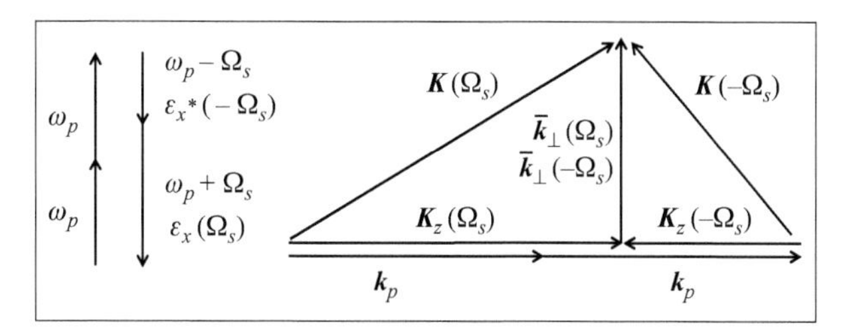

Figure 10.2 *Schematic of KIA. Left: parametric four wave mixing of the type $2\omega_p - (\omega_p \pm \Omega_s) = \omega_p \mp \Omega_s$, where ω_p is the pump frequency and $-\omega_p < \Omega_s < \omega_p$ is the seed frequency. Right: transverse wavevectors exist for which unstable behavior occurs. All parameters shown will be described in the chapter. Gain is maximum for the transverse wavevector $\bar{k}_\perp(\Omega_s)$, see (10.45). The instability evolves as $\varepsilon_x(\Omega_s) = exp\left[i(\omega_p + \Omega_s)t - i\mathbf{K}(\Omega_s)\cdot\mathbf{x}\right]$ and $\varepsilon_x^*(-\Omega_s) = exp\left[-i(\omega_p - \Omega_s)t + i\mathbf{K}(-\Omega_s)\cdot\mathbf{x}\right]$, where $\mathbf{K}(\Omega_s) = \left(0, \bar{k}_\perp(\Omega_s), K_z(\Omega_s)\right)$, see (10.50)*

there exist transverse wavevectors $k_\perp$ for which exponential growth occurs. The transverse wavevector for maximum amplification, $\bar{k}_\perp(\Omega_s)$, (see (10.45)) is finite, so interaction and hence emission with respect to the pump pulse should be non-collinear for optimal amplification. Phase matching and momentum conservation are automatically fulfilled at maximum gain, since the instability wavevector, $\mathbf{K}(\Omega_s)$, fulfills the relation $\mathbf{K}(\Omega_s) = 2\mathbf{k}_p + \mathbf{K}(-\Omega_s)$, see Figure 10.2.

In this chapter, we discuss the potential of KIA for the amplification of ultra-short mid-infrared pulses. In Section 10.2, we develop the vectorial perturbation equations beginning from Maxwell's equations for a Kerr nonlinear medium and solve the resulting equations of motion. In Section 10.3, we review the theory of plane wave KIA. First, a brief description of the optical properties of two materials used for the analysis is given. Then, it is shown that for most materials' vectorial effects are negligible and the scalar theory gives an accurate account of KIA. Subsequently, we discuss KIA for the two example materials in the plane wave limit. Finally, effects arising from finite pulse KIA are discussed.

10.2 Vectorial perturbation equations

In the following section, the equations of motion of the vector Kerr instability are derived and then solved by means of linear stability analysis. The wavevector solution we obtain exhibits an instability regime that will be explored.

10.2.1 Derivation of the equation of motion of the vector Kerr instability

The nonlinear polarization vector for a Kerr nonlinear medium is given by

$$\mathbf{P}^{(3)} = c^{-2}\mu_0^{-1}\chi^{(3)}\mathbf{E}^3, \tag{10.1}$$

where c is the vacuum speed of light, μ_0 is the vacuum permeability, $\chi^{(3)}$ is the third-order nonlinear susceptibility, and $\mathbf{E}$ is the electric field vector. Here, we assume $\chi^{(3)}$ to be instantaneous. The finite response of the Kerr nonlinearity becomes important when part of $\chi^{(3)}$ comes from the Raman nonlinearity. In that case, the theory needs to be extended to account for the delayed Raman response. To avoid undue complexity here, we derive the vectorial theory without the Raman nonlinearity, which is reasonable approximation as long as pump and seed wavelengths are far away from the material bandgap [12]. When the Raman nonlinearity becomes important, it can be incorporated into the scalar theory of KIA [13]. Further, the light is assumed to be linearly polarized so that $\chi^{(3)}$ is reduced from a tensor to a scalar [14].

The pump electric field can be perturbed weakly to have the following form

$$\mathbf{E} = \hat{\mathbf{x}} E_p e^{i\omega_p t -} ik_p z + \boldsymbol{\varepsilon} + \text{c.c.}, \tag{10.2}$$

where E_p, ω_p, and k_p are the pump electric field amplitude, angular frequency, and wavevector, respectively, and $\boldsymbol{\varepsilon} = (\varepsilon_x, \varepsilon_y, \varepsilon_z)$ is a weak perturbation, $|\boldsymbol{\varepsilon}| \ll E_p$. We assume a plane wave pump so that E_p is constant; k_p will be defined later. The goal is to perform a stability analysis on the pump beam and to show that $\boldsymbol{\varepsilon}$ exhibits instability. Thus, controlled amplification is possible if the perturbation is taken to be a weak seed beam.

We insert (10.2) into (10.1) and linearize to order $\mathcal{O}(\boldsymbol{\varepsilon})$ and neglect higher harmonic terms to obtain

$$\begin{aligned}\mathbf{P}^{(3)} &= c^{-2}\mu_0^{-1}\chi^{(3)}\left(\hat{\mathbf{x}} E_p e^{i\omega_p t - ik_p z} + \boldsymbol{\varepsilon} + \text{c.c.}\right)^3 \\ &\approx \frac{1}{3} c^{-2}\mu_0^{-1} n_n \left[3\hat{\mathbf{x}} E_p e^{i\omega_p t - ik_p z} + 2\boldsymbol{\varepsilon} + 4\hat{\mathbf{x}}\varepsilon_x + \left(\boldsymbol{\varepsilon}^* + 2\hat{\mathbf{x}}\varepsilon_x^*\right) e^{2i\omega_p t - 2ik_p z}\right],\end{aligned} \tag{10.3}$$

where $n_n = 3\chi^{(3)} E_p^2 = n_2 I_p$, n_2 is the nonlinear coefficient, typically in units of (cm^2/W), and $I_p = n_p E_p^2/(2Z_0)$ is the constant pump intensity, where n_p is the linear refractive index evaluated at frequency, ω_p, and Z_0 is the vacuum impedance, and * denotes the complex conjugate. We insert (10.3) into the general vectorial wave equation derived from Maxwell's equations

$$\left[\nabla^2 - \nabla(\nabla\cdot) - \frac{n^2 *}{c^2}\frac{\partial^2}{\partial t^2}\right]\mathbf{E} = \mu_0 \frac{\partial^2 \mathbf{P}^{(3)}}{\partial t^2}, \tag{10.4}$$

where $n^2 = 1 + \chi^{(1)}$ is the linear refractive index in the time domain, $\chi^{(1)}$ is the first-order electric susceptibility response function, and $*$ denotes convolution. After inserting (10.3), the wave equation for the pump field yields the wavevector solution:

$$k_p = \frac{\omega_p}{c}\sqrt{n_p^2 + n_n}. \tag{10.5}$$

The remaining wave equation for the perturbation reads

$$\left[\nabla^2 - \nabla(\nabla\cdot) - \frac{n^2 *}{c^2}\frac{\partial^2}{\partial t^2}\right]\varepsilon = \frac{1}{3}\frac{n_n}{c^2}\frac{\partial^2}{\partial t^2}\left[2\varepsilon + 4\hat{\mathbf{x}}\varepsilon_x + \left(\varepsilon^* + 2\hat{\mathbf{x}}\varepsilon_x^*\right) e^{2i\omega_p t - 2ik_p z}\right]. \tag{10.6}$$

To proceed, we let $\boldsymbol{\varepsilon} = \mathbf{u}e^{i\omega_p t}$ and Fourier transform (10.6) over transverse coordinates and time, $(x, y, z, t) \to \left(k_x, k_y, z, \omega\right)$. The Fourier transform of the field is given by

$$\begin{aligned}\widetilde{\varepsilon}\left(k_x, k_y, z, \omega\right) &= \mathscr{F}[\varepsilon] = \mathscr{F}\left[\mathbf{u}e^{i\omega_p t}\right] = \frac{1}{(2\pi)^{3/2}} \int_{-\infty}^{\infty} \mathbf{u}e^{-i\left(\omega-\omega_p\right)t} e^{-ik_x x - ik_y y} \mathrm{d}t\mathrm{d}x\mathrm{d}y \\ &= \widetilde{\mathbf{u}}\left(k_x, k_y, z, \Omega\right),\end{aligned} \tag{10.7}$$

where $\Omega = \omega - \omega_p$ is the frequency shift from ω_p. For the complex conjugate terms on the right-hand side of (10.6), we have

$$\widetilde{\mathbf{u}}^*\left(k_x, k_y, z, \Omega\right) = \widetilde{\mathbf{u}}^*\left(-k_x, -k_y, z, -\Omega\right) \equiv \widetilde{\mathbf{u}}^*_{(-)}, \tag{10.8}$$

which will be important when decoupling $\widetilde{\mathbf{u}}$ and $\widetilde{\mathbf{u}}^*_{(-)}$ next. Throughout the text, the subscript $(-)$ means evaluated at $\left(-k_x, -k_y, z, -\Omega\right)$. Next, we turn our attention to taking the Fourier transform of the $\nabla(\nabla\cdot)$ operator in (10.6), which can be expressed as the symmetric matrix:

$$\widehat{G} = \begin{pmatrix} -k_x^2 & -k_x k_y & ik_x \dfrac{\partial}{\partial_z} \\ -k_x k_y & -k_y^2 & ik_y \dfrac{\partial}{\partial_z} \\ ik_x \dfrac{\partial}{\partial_z} & ik_y \dfrac{\partial}{\partial_z} & \dfrac{\partial^2}{\partial_z^2} \end{pmatrix}. \tag{10.9}$$

Using (10.7)–(10.9), we obtain the vectorial wave equation in Fourier space:

$$\left[\frac{\partial^2}{\partial z^2} - k_\perp^2 - \widehat{G} + k_v^2\right] \widetilde{\mathbf{u}} = -\frac{1}{3}k_n^2\left[-4\left(\widehat{\mathbf{y}}\, \widetilde{u}_y + \widehat{\mathbf{z}}\, \widetilde{u}_z\right) + \left(\widetilde{u}^*_{(-)} + 2\widehat{\mathbf{x}}\widetilde{u}^*_{x(-)}\right)e^{-2ik_p z}\right], \tag{10.10}$$

where $k_\perp^2 = k_x^2 + k_y^2$, $k_v^2 = c^{-2}\omega^2\left[n^2(\omega) + 2n_n\right]$, and $k_n^2 = c^{-2}\omega^2 n_n$ have been defined. The z-equation of (10.10) can be used to eliminate the $\widetilde{u}_z$-dependence in the x and y equations. It is found to be approximately

$$k_z^2\, \widetilde{u}_z \approx i\frac{\partial}{\partial_z}\left(k_x\, \widetilde{u}_x + k_y\, \widetilde{u}_y\right), \tag{10.11}$$

where $k_z^2 = k_v^2 - k_\perp^2$. The fact that $n_n/n^2 \ll 1$ was used in obtaining (10.11). This approximation is only questionable for metamaterials at frequencies for which $n \to 0$. The $\widetilde{u}_z$-dependence can now be eliminated in the other two equations of (10.10) to obtain the set of coupled equations for $\widetilde{u}_x$, $\widetilde{u}_y$, and their complex conjugates. We insert (10.11) into (10.10) and obtain

$$\left[1 + \frac{1}{k_z^2}\frac{\partial^2}{\partial_z^2}\right]\left[\left(k_v^2 - k_y^2\right)\widetilde{u}_x + k_x k_y \widetilde{u}_y\right] = -k_n^2 \widetilde{u}^*_{x(-)} e^{-2ik_p z} \tag{10.12a}$$

$$\left[1 + \frac{1}{k_z^2}\frac{\partial^2}{\partial_z^2}\right]\left[\left(k_v^2 - k_x^2\right)\widetilde{u}_y + k_x k_y \widetilde{u}_x\right] = -\frac{1}{3}k_n^2\left[-4\widetilde{u}_y + \widetilde{u}^*_{y(-)} e^{-2ik_p z}\right]. \tag{10.12b}$$

It is convenient to express (10.12a) and (10.12b) as a 2×2-matrix equation for the transversely polarized vector $\widetilde{\mathbf{u}}_\perp = \left(\widetilde{u}_x, \widetilde{u}_y\right)$ as

$$\widehat{A}\left(\frac{\partial^2}{\partial_z^2} + k_z^2\right)\widetilde{\mathbf{u}}_\perp = -k_n^2 k_z^2 \begin{pmatrix} \widetilde{u}^*_{x(-)} \\ \frac{1}{3}\widetilde{u}^*_{y(-)} \end{pmatrix} e^{-2ik_p z}, \tag{10.13}$$

where

$$\widehat{A} \equiv \begin{pmatrix} k_v^2 - k_y^2 & k_x k_y \\ k_x k_y & k_v^2 - k_x^2 \end{pmatrix}. \tag{10.14}$$

We have neglected the first term on the right-hand side of (10.12b). This approximation is justified since $k_n^2 \ll k_v^2$. As a result, the analysis is much simpler by increasing symmetry between the transversely polarized fields.By multiplying (10.13) on the left-hand side by $\widehat{A}^{-1} = \mathrm{adj}\left(\widehat{A}\right)/\det\left(\widehat{A}\right)$, we obtain

$$\left(\frac{\partial^2}{\partial_z^2} + k_z^2\right)\widetilde{\mathbf{u}} \;=\; -\frac{k_n^2}{k_v^2}\widehat{M}\,\widetilde{\mathbf{u}}^*_{\perp(-)}\,e^{-2ik_p z}, \tag{10.15}$$

where

$$\widehat{M} \equiv \begin{pmatrix} k_v^2 - k_x^2 & -\frac{1}{3}k_x k_y \\ -k_x k_y & \frac{1}{3}\left(k_v^2 - k_y^2\right) \end{pmatrix}. \tag{10.16}$$

We eliminate the exponential terms on the right-hand side of (10.15) by making the transformation $\widetilde{\mathbf{u}}_\perp = \widetilde{\mathbf{v}}_\perp e^{-ik_p z}$, with the result

$$\left[\left(\frac{\partial}{\partial_z} - ik_p\right)^2 + k_z^2\right]\widetilde{\mathbf{v}}_\perp = -\frac{k_n^2}{k_v^2}\widehat{M}\,\widetilde{\mathbf{v}}^*_{\perp(-)}. \tag{10.17}$$

At this point, we obtain an equation for $\widetilde{\mathbf{v}}^*_{\perp(-)}$ by taking the complex conjugate of (10.17) and by the replacement $\Omega \to -\Omega$. The resulting equation reads

$$\left[\left(\frac{\partial}{\partial_z} + ik_p\right)^2 + k_{z(-)}^2\right]\widetilde{\mathbf{v}}^*_{\perp(-)} = -\frac{k_{n(-)}^2}{k_{v(-)}^2}\widehat{M}_{(-)}\widetilde{\mathbf{v}}_\perp. \tag{10.18}$$

By using (10.17) in (10.18), we obtain

$$\left[\left(\frac{\partial}{\partial_z} - ik_p\right)^2 + k_z^2\right]\left[\left(\frac{\partial}{\partial_z} + ik_p\right)^2 + k_{z(-)}^2\right]\widetilde{\mathbf{v}}_\perp = \frac{k_n^2 k_{n(-)}^2}{k_v^2 k_{v(-)}^2}\widehat{L}\widetilde{\mathbf{v}}_\perp, \tag{10.19}$$

where $\widehat{L} \equiv \widehat{M}\widehat{M}_{(-)}$. Explicitly, $\widehat{L}$ is written as

$$\widehat{L} = \begin{pmatrix} \left(k_v^2 - k_x^2\right)\left(k_{v(-)}^2 - k_x^2\right) + \frac{1}{3}k_x^2 k_y^2 & -\frac{1}{9}k_x k_y\left(3k_v^2 + k_{v(-)}^2 - 3k_x^2 - k_y^2\right) \\ -\frac{1}{3}k_x k_y\left(3k_{v(-)}^2 + k_v^2 - 3k_x^2 - k_y^2\right) & \frac{1}{9}\left(k_v^2 - k_x^2\right)\left(k_{v(-)}^2 - k_x^2\right) + \frac{1}{3}k_x^2 k_y^2 \end{pmatrix}.$$
$$\equiv \begin{pmatrix} l_{11} & l_{12} \\ l_{21} & l_{22} \end{pmatrix} \tag{10.20}$$

Equation (10.19) is formally solved by diagonalizing $\widehat{L}$ since the matrix elements contain wavevectors only and are independent of z. Thus, we make the following transformation:

$$\widetilde{\mathbf{v}}_\perp = \widehat{U}\,\widetilde{\mathbf{w}}_\perp, \tag{10.21}$$

where $\widehat{U}$ is a unitary matrix that shifts the coordinate axes such that $\widehat{U}^{-1}\widehat{L}\widehat{U} = \widehat{D}$ is a diagonal matrix with nonzero elements, given by the eigenvalues of $\widehat{L}$. With this transformation, (10.19) becomes

$$\left[\left(\frac{\partial}{\partial_z} - ik_p\right)^2 + k_z^2\right]\left[\left(\frac{\partial}{\partial_z} + ik_p\right)^2 + k_{z(-)}^2\right]\widetilde{\mathbf{w}}_\perp = \frac{k_n^2 k_{n(-)}^2}{k_v^2 k_{v(-)}^2}\widehat{D}\,\widetilde{\mathbf{w}}_\perp. \tag{10.22}$$

The solution of (10.22) will be in the form of plane waves:

$$\widetilde{\mathbf{w}}_\perp(z) = \begin{pmatrix} e^{iK_x z} & 0 \\ 0 & e^{iK_x z} \end{pmatrix}\widetilde{\mathbf{w}}_\perp(0) \equiv e^{i\widehat{K}z}\widetilde{\mathbf{w}}_\perp(0), \tag{10.23}$$

where K_x and K_y are the complex wavevectors for $\widetilde{w}_x$ and $\widetilde{w}_y$, respectively, and $\widetilde{\mathbf{w}}_\perp(0)$ denotes the initial condition of the field $\widetilde{\mathbf{w}}_\perp(z)$ at $z = 0$. Notice from (10.21) that

$$\widetilde{\mathbf{w}}_\perp(0) = \widehat{U}^{-1}\widetilde{\mathbf{v}}_\perp(0). \tag{10.24}$$

Thus, by means of (10.21), (10.23), and (10.24), we have the following formal solution:

$$\widetilde{\mathbf{v}}_\perp(z) = \widehat{U}e^{i\widehat{K}z}\widehat{U}^{-1}\widetilde{\mathbf{v}}_\perp(0) \equiv \widehat{P}(z)\widetilde{\mathbf{v}}_\perp(0), \tag{10.25}$$

where $\widehat{P}(z)$ is the z-dependent evolution matrix that propagates $\widetilde{\mathbf{v}}_\perp$ from its initial state at $z = 0$ to z. The procedure for the solution of (10.19) is given by (10.20)–(10.25). In order to proceed, the eigenvalues, $\lambda_{x,y}$, of $\widehat{L}$ must be found. These will be used to obtain $\widehat{D}$, and the eigenvectors will yield $\widehat{U}$. We set

$$\det\left(\widehat{L} - \lambda_{x,y}\widehat{I}\right) = 0, \tag{10.26}$$

where $\widehat{I}$ is the identity matrix. From (10.20), we see that the off-diagonal elements, $l_{12,21}$, are much weaker than the diagonal elements, $l_{11,22}$. To order $\mathcal{O}(l_{12}l_{21})$, we obtain

$$\lambda_x = l_{11} + \frac{l_{12}l_{21}}{l_{11} - l_{22}} \tag{10.27a}$$

$$\lambda_y = l_{22} - \frac{l_{12}l_{21}}{l_{11} - l_{22}}. \tag{10.27b}$$

By means of (10.27a) and (10.27b), we obtain the matrices $\widehat{D}$, $\widehat{U}$, and $\widehat{U}^{-1}$ as

$$\widehat{D} = \begin{pmatrix} l_{11} + \dfrac{l_{12}l_{21}}{l_{11} - l_{22}} & 0 \\ 0 & l_{11} - \dfrac{l_{12}l_{21}}{l_{11} - l_{22}} \end{pmatrix} \tag{10.28a}$$

$$\widehat{U} = \frac{1}{l_{11} - l_{22}} \begin{pmatrix} l_{11} - l_{22} & -l_{12} \\ l_{21} & l_{11} - l_{22} \end{pmatrix} \tag{10.28b}$$

$$\widehat{U}^{-1} = \frac{1}{l_{11} - l_{22}} \begin{pmatrix} l_{11} - l_{22} & l_{12} \\ -l_{21} & l_{11} - l_{22} \end{pmatrix} \tag{10.28c}$$

Using (10.28a)–(10.28c) in (10.25), we find the propagator matrix to be

$$\widehat{P}(z) = \begin{pmatrix} e^{iK_x z} + \dfrac{l_{12}l_{21}}{(l_{11} - l_{22})^2} e^{iK_y z} & \dfrac{l_{12}}{l_{11} - l_{22}} \left(e^{iK_x z} - e^{iK_y z}\right) \\ \dfrac{l_{21}}{l_{11} - l_{22}} \left(e^{iK_x z} - e^{iK_y z}\right) & e^{iK_y z} + \dfrac{l_{12}l_{21}}{(l_{11} - l_{22})^2} e^{iK_x z} \end{pmatrix}. \tag{10.29}$$

The propagator $\widehat{P}$ together with (10.22) and (10.25) governs the evolution of the instability.

10.2.2 Solution of the equations of motion of the vector Kerr instability

To complete the solution, we must solve the differential equation (10.22). We do this by means of the ansatz of (10.23) to obtain wavevectors of the plane wave equation. The resulting equations are quartic in the complex wavevectors $K_{x,y}$ and depend on the reduced eigenvalues $\Lambda_{x,y} \equiv \lambda_{x,y}/k_v^2 k_{v(-)}^2$:

$$\left[\left(K_{x,y} - k_p\right)^2 + k_z^2\right] \left[\left(K_{x,y} + k_p\right)^2 + k_{z(-)}^2\right] = \Lambda_{x,y} k_n^2 k_{n(-)}^2. \tag{10.30}$$

To make progress, the sign flipped functions $k_{z(-)}$, $k_{v(-)}$, and $k_{n(-)}$ need to be determined. To this end, these functions are split into even components that remain unchanged and odd components that change sign when $\Omega \to -\Omega$. First, we write the general linear refractive index, $\eta(\omega) = \sqrt{n^2(\omega) + 2n_n}$ as

$$\begin{aligned} \eta(\omega) &= \eta_p + \Delta\eta(\Omega) \\ &= \eta_p + \eta_g(\Omega) + \eta_u(\Omega), \end{aligned} \tag{10.31}$$

where $\eta_p = \eta(\omega_p)$, and

$$\eta_g(\Omega) = \frac{1}{2}[\Delta\eta(\Omega) + \Delta\eta(-\Omega)] \tag{10.32a}$$

$$\eta_u(\Omega) = \frac{1}{2}[\Delta\eta(\Omega) - \Delta\eta(-\Omega)], \tag{10.32b}$$

are the even and odd functions, respectively, such that $\eta_g(-\Omega) = \eta_g(\Omega)$, and $\eta_u(-\Omega) = -\eta_u(\Omega)$. With these definitions, we are equipped to write similar splitting for k_v,

$$k_v = k_v(\omega_p) + D_u(\Omega) + D_g(\Omega), \tag{10.33}$$

where

$$D_u(\Omega) = \frac{1}{c}\left[\eta_p\Omega + \eta_g(\Omega)\Omega + \eta_u(\Omega)\omega_p\right] \tag{10.34a}$$

$$D_g(\Omega) = \frac{1}{c}\left[\eta_g(\Omega)\omega_p + \eta_u(\Omega)\Omega\right], \tag{10.34b}$$

where $D_u(\Omega)$ and $D_g(\Omega)$ are odd and even dispersion functions, respectively, such that $D_u(-\Omega) = -D_u(\Omega)$ and $D_g(-\Omega) = D_g(\Omega)$. Note that these functions are exact and no approximations, such as a Taylor expansion, have been made. Even when summed to all orders, a Taylor expansion need not necessarily converge. This is particularly the case for far infrared frequencies $\omega = \omega_p + \Omega$ with $\Omega \approx< \omega_p$. In the limit of small detuning, where $\Omega \ll \omega_p$, and since $n^2 \gg 2n_n$, $\eta_u \approx n'_p\Omega$ and $\eta_g \approx n''_p\Omega^2/2$. Here, the primes on n_p denote differentiation of $n(\omega)$ with respect to ω evaluated at ω_p. Thus, to lowest order we obtain for (10.34a) and (10.34b), $D_u \approx \beta_1\Omega$ and $D_g \approx \beta_2\Omega^2/2$, where $\beta_1 = [dk/d\omega](\omega_p) = (n_p + n'_p\omega_p)/c$ and $\beta_2 = [d^2k/d\omega^2](\omega_p) = (2n'_p + n''_p\omega_p)/c$ are the group velocity, and group velocity dispersion terms, respectively.

Finally, the sign-flipped functions are determined as

$$k_{v(-)} = k_v(\omega_p) - D_u(\Omega) + D_g(\Omega), \tag{10.35}$$

and, treating n_n as constant,

$$k^2_{n(-)} = \frac{n_n}{c}(\omega_p - \Omega)^2. \tag{10.36}$$

Now, we can readily use (10.33) and (10.35) in (10.30) to obtain

$$\left(K^2_{x,y} - \sigma^2 D^2_u + k^2_\perp - \kappa^2_\perp\right)^2 - 4k^2_p\left(K_{x,y} + \sigma D_u\right)^2 - \Lambda_{x,y}k^2_n k^2_{n(-)} = 0, \tag{10.37}$$

where we have defined

$$\kappa^2_\perp = \left(k^2_p - D^2_u\right)\left(\sigma^2 - 1\right), \tag{10.38}$$

and

$$\sigma = \frac{k_v(\omega_p) + D_g}{k_p}. \tag{10.39}$$

The dominant solution for $K_{x,y}$ is given by the second term of (10.37), $K_{x,y} \approx -\sigma D_u$, which was verified numerically [13]. With this approximation we can then use

$$\left(K_{x,y}^2 - \sigma^2 D_u^2\right) \approx -2\sigma D_u \left(K_{x,y} + \sigma D_u\right). \tag{10.40}$$

in (10.37). Using this approximation significantly simplifies the equations by reducing them to quadratic equations for $K_{x,y}$:

$$\left(k_p^2 - \sigma^2 D_u^2\right)\left(K_{x,y} - \sigma D_u\right)^2 - \sigma D_u \Delta k_\perp^2 \left(K_{x,y} - \sigma D_u\right) - \Delta k_\perp^4 = -\Lambda_{x,y} k_n^2 k_{n(-)}^2, \tag{10.41}$$

where $\Delta k_\perp^2 = \kappa_\perp^2 - k_\perp^2$ has been used to be concise. Note that when $k_p^2 = \sigma^2 D_u^2$, which occurs at $\omega \approx 2\omega_p$ and $\omega \approx 0$, (10.41) reduces to a linear equation that exhibits no unstable solutions. As we are seeking such unstable solutions, we assume $k_p^2 \neq \sigma^2 D_u^2$ and must avoid frequencies that are double that of the pump and higher. As we are interested in frequencies on the red side of the pump in the near- to mid-infrared regime, it poses no issue for the purposes of this work, provided the pump frequency is chosen appropriately relative to ω such that $\omega \approx 0$ is not realized. In Section 10.3.4, this will become clear when two example materials are discussed.

Equation (10.41) exhibits the solutions $K_{x,y} = K_u(\Omega) \pm K_g(\Omega; \Lambda_{x,y})$

$$K_u(\Omega) = -\sigma D_u \left(1 - \frac{1}{2} \frac{\Delta k_\perp^2}{k_p^2 - \sigma^2 D_u^2}\right) \tag{10.42a}$$

$$K_g(\Omega; \Lambda_{x,y}) = \frac{1}{2} \frac{k_p \sqrt{\Delta k_\perp^4 - \Lambda_{x,y} \delta_\perp^4}}{k_p^2 - \sigma^2 D_u^2}, \tag{10.42b}$$

where we have defined

$$\delta_\perp^2 = \frac{k_n k_{n(-)}}{k_p} \sqrt{k_p^2 - \sigma^2 D_u^2}, \tag{10.43}$$

which is a transverse width whose significance will be seen later. The wavevector solution exhibits instability when the argument under the square root in $K_g(\Omega; \Lambda_{x,y})$ becomes negative. As a result, exponential growth is possible with intensity gain

$$g(\Omega; \Lambda_{x,y}) = -2\mathrm{Im}\left[K_g(\Omega; \Lambda_{x,y})\right]. \tag{10.44}$$

Neglecting vectorial effects, (10.42a) and (10.42b) go over into filamentation instability for $\Omega = 0$ [15]; further in the 1D limit where transverse effects are ignored and looking at weak detuning ($\Omega \ll \omega_p$) we obtain the modulation instability [16]; see [13] for more details.

Now that the full wavevector solutions for $K_{x,y}$ have been obtained, and the propagator, $\widehat{P}(z)$, given by (10.29) can be specified. The propagator describes how $\widetilde{\mathbf{v}}_\perp(z)$ evolves from its arbitrary initial state of $\widetilde{\mathbf{v}}_\perp(0)$, given by (10.25). The difference between scalar instability theory and the vectorial instability theory here lies within the transverse asymmetries of $\Lambda_{x,y}$ contained in $K_{x,y}$ and $\widehat{P}(z)$, which couples the x- and y-polarizations of the initial fields. We can ignore the K_y solution since to leading order $\Lambda_x \approx 9\Lambda_y$, which renders $g(\Omega;\Lambda_y)$ negligible compared to $g(\Omega;\Lambda_x)$. We consider then only the gain $g(\Omega;\Lambda_x) \equiv g(\Omega)$. From (10.42b), it is clear that the gain is maximized for the transverse wavevector $k_\perp = \bar{k}_\perp(\Omega)$:

$$\bar{k}_\perp(\Omega) = \begin{cases} \kappa_\perp(\Omega) & \text{if} \quad \kappa_\perp^2 \geq 0 \\ 0 & \text{if} \quad \kappa_\perp^2 < 0 \end{cases}. \tag{10.45}$$

From (10.38), we see that $\kappa_\perp^2 \propto \sigma^2 - 1 \approx 2\left(\eta_g + \eta_u\Omega/\omega_p\right)/n_p$. Thus, depending on the material and the pump wavelength, η_g and $\eta_u\Omega/\omega_p$ may have equal or opposite signs. As a result, $\kappa_\perp^2 < 0$ is possible, which terminates the instability regime, which is why $\bar{k}_\perp$ must be piece-wise defined. In-line with this, the maximum gain $g\left(\Omega, k_\perp = \bar{k}_\perp(\Omega)\right) = \bar{g}(\Omega)$ is written as

$$\bar{g}(\Omega) = \begin{cases} k_p \dfrac{\left[(1+r_0)\delta_\perp^4 - \left(\kappa_\perp^2 - \bar{k}_\perp^2\right)^2\right]^{1/2}}{k_p^2 - \sigma^2 D_u^2} & \text{otherwise} \\ 0 & \text{if } \kappa_\perp^2 < 0, \quad \kappa_\perp^4 > \delta_\perp^4(1+r_0), \end{cases} \tag{10.46}$$

where $r_0 \approx \Lambda_x - 1$; r_0 is the leading-order asymmetry term arising from the vectorial nature of KIA. Since $g\left(k_\perp^2 = \bar{k}_\perp^2 \pm \sqrt{1+r_0}\delta_\perp^2\right) = 0$, the (squared) transverse wavevector half-width over which instability gain occurs as a function of Ω is given by $\sqrt{1+r_0}\delta_\perp^2(\Omega)$, see (10.43). The spatiospectral coupling given by $k_\perp^2 = \bar{k}_\perp^2 \pm \sqrt{1+r_0}\delta_\perp^2(\Omega)$ defines the range of frequencies over which the instability exists. Note that r_0 weakly disrupts the transverse symmetry, where otherwise this relation would be completely transversally symmetric.

Note that even if r_0 were nonnegligible, it is always possible to eliminate the transverse asymmetry by seeding along the line $\mathbf{k}_\perp = \left(k_x = 0, k_y\right)$ such that $r_0 = 0$ automatically. In essence, vectorial effects reduce the noncollinear interaction from being conically symmetric to symmetric along the line $k_x = 0$ only.

10.3 Amplification of plane waves

In this section, the theory of plane wave KIA for both vectorial and scalar theories is analyzed. At the end of the section, finite pulse KIA is discussed. First, we define the optical properties of the two example materials chosen for the analysis.

10.3.1 Optical properties of CaF_2 and KBr

The materials chosen are two dielectric bulk crystals, calcium fluoride (CaF_2), and potassium bromide (KBr). Two different pump wavelengths $\lambda_p = 0.85, 2.1$ μm are chosen in each case. The CaF_2 and KBr crystals have transmission windows from 0.3 to 8 and 0.25 to 25 μm, respectively [17]. In all figures, the linear and nonlinear indices for CaF_2 are given by Malitson [18] and $n_2 = 2 \times 10^{-16}$ cm^2/W [19], respectively. The linear and nonlinear indices for KBr are given by Li [20] and $n_2 = 6 \times 10^{-16}$ cm^2/W [21], respectively. Finally, the peak pump intensities used are $I_p = 50$ TW/cm^2 and $I_p = 8$ TW/cm^2 for CaF_2 and KBr, respectively. These intensities are chosen carefully and are valid for the damage threshold of each material in question, see [13] for more detail. Next, the role of vectorial effects on KIA is discussed in CaF_2.

10.3.2 Vectorial effects in Kerr instability amplification

Here, it is demonstrated that vectorial effects are negligible in the regime where instability gain is possible. The material chosen here is CaF_2; KBr exhibits the exact same qualitative features of vectorial effects as in CaF_2.

Vectorial effects are negligible as long as $1 + r_0 = \Lambda_x \approx 1$. This approximation is valid as long as the leading-order asymmetric term from (10.27a), $\Lambda_x \approx l_{11}/k_v^2 k_{v(-)}^2$, is small, namely if

$$|r_0| = \left| \frac{k_x^2 \left(k_v^2 + k_{v(-)}^2 \right)}{k_v^2 k_{v(-)}^2} \right| \ll 1. \tag{10.47}$$

This condition is valid for most materials, as the instability regime will have ceased before the transverse asymmetry significantly affects the instability gain, see Figure 10.3. Clearly, the frequencies at which maximum gain occurs are far away from where $|r_0|$ is significant compared to unity (see the dashed white line), even for maximized asymmetry, where $k_\perp = k_x$. The instability regime ceases before vectorial effects alter the scalar theory. In the next section, general plane wave KIA theory is discussed.

10.3.3 Plane wave Kerr instability amplification

Consider the initial condition a seed plane wave, inclined along k_y, $k_x = 0$, and with the assumption that $K_y \approx 0$ as compared to K_x, that is, $\widetilde{\mathbf{v}}_\perp(0) = (\widetilde{v}_x(0), 0)$, with

$$\widetilde{v}_x(0) = (2\pi)^{3/2} E_s \delta(k_x) \delta(k_y - \bar{k}_{\perp s}) \delta(\Omega - \Omega_s), \tag{10.48}$$

where E_s is the constant seed electric field amplitude, $\Omega_s = \omega_s - \omega_p$ is the seed frequency, and $\bar{k}_{\perp s} = \bar{k}_\perp(\Omega_s)$. After propagating a material length $z = l$, we have from (10.29) and (10.25)

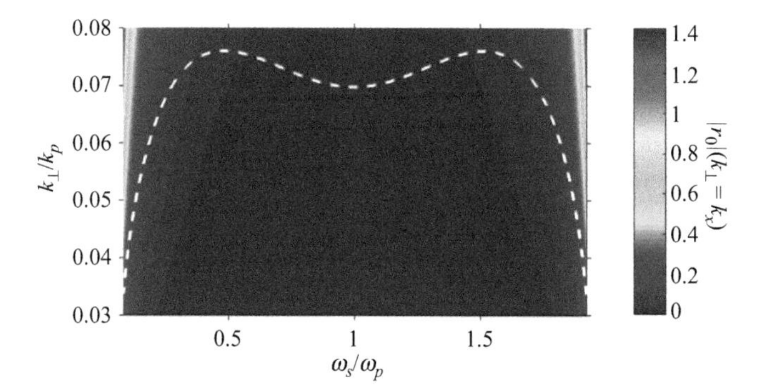

Figure 10.3 The asymmetric vectorial term, $|r_0|$ (see (10.47)), is plotted in CaF_2 with peak pump intensity $I_p = 50$ TW/cm^2. For other parameters used, see Section 10.3.4. We plot along the line $k_y = 0$ where the vectorial asymmetry is maximized at $k_\perp = k_x$. The dashed white line indicates $\bar{k}_\perp$ at which maximum gain $\bar{g} = g(\bar{k}_\perp)$ occurs, see (10.46)

$$\widetilde{\mathbf{v}}_\perp(l) = \begin{pmatrix} 1 \\ \dfrac{l_{21}}{l_{11} - l_{22}} \end{pmatrix} \widetilde{v}_x(0) e^{iK_x l}. \tag{10.49}$$

As we are seeding at $k_x = 0$ and $l_{21} \propto k_x$, $\widetilde{v}_y(l)$ will vanish upon inverse Fourier transforming back to $\varepsilon_y(l)$. After inverse Fourier transforming, what remains is $\boldsymbol{\varepsilon}_\perp = (\varepsilon_x(\mathbf{x}, t), 0)$, where

$$\varepsilon_x(\mathbf{x}, t) = E_s e^{\bar{g}(\Omega_s) l/2} e^{-i\mathbf{K}_s \cdot \mathbf{x}} e^{i\omega_s t}, \tag{10.50}$$

and $\mathbf{K}_s = K(\Omega_s) = (0, \bar{k}_{\perp s}, K_{zs})$ is the seed wavevector, where $K_{zs} = K_z(\Omega_s) = k_p + \sigma(\Omega_s) D_u(\Omega_s)$, and $\mathbf{x} = (x, y, z = l)$. Optimum amplification occurs when the seed propagation axis lies along the line $k_y = \bar{k}_{\perp s}$ inclined with respect to the pump wavevector with half-angle $\theta_s = \theta(\Omega_s)$:

$$\theta_s = \arctan\left(\frac{\bar{k}_{\perp s}}{K_{zs}}\right). \tag{10.51}$$

It should be mentioned that in the scalar limit where $r_0 \approx 0$, optimum amplification can take place along a cone around the pump wavevector as $k_x = 0$ need not be enforced. This is related to but, in general, not the same as the conical emission angle. Conical emission grows out of noise and occurs after filamentation has drastically modified the pump pulse. Seeded amplification assumes the pump pulse to be approximately undistorted and occurs over distances significantly shorter than the self-focusing distance necessary for filamentation [1].

As mentioned in Section 10.1, the KIA process is phase-matched automatically, unlike conventional three- or four-wave mixing processes, see

Figure 10.2. We have $\mathbf{v}_\perp \propto e^{-i\mathbf{K}(\Omega_s)\cdot\mathbf{x}}$ and $\mathbf{v}^*_{\perp(-)} \propto e^{i\mathbf{K}(-\Omega_s)\cdot\mathbf{x}}$. Since $\bar{k}_\perp(-\Omega_s) = \bar{k}_\perp(\Omega_s)$ and $K_z(-\Omega_s) = -K_z(\Omega_s)$, the left-hand- and right-hand side of (10.17) are phase-matched. Outside the instability regime, in the conventional regime of four-wave mixing, $K_g(\Omega_s)$ becomes real and $K_g(-\Omega_s) = K_g(\Omega_s)$; thus, phase-matching is no longer automatic. In the following, the scalar theory of plane wave KIA in CaF_2 and KBr is quantitatively analyzed.

10.3.4 Scalar plane wave KIA in CaF_2 and KBr

Here, we assume $r_0 \approx 0$, as justified in Section 10.3.2. In Figure 10.4(a) the intensity gain profile, g, from (10.44) is plotted versus ω_s/ω_p and $k_\perp/k_p$. The solid white line represents $\bar{k}_\perp$. The pump wavelength is $\lambda_p = 2\pi c/\omega_p = 0.85$ μm and pump intensity is $I_p = 50$ TW/cm^2. Amplification occurs over a wide spectral range from 0.45 to 15 μm. The gain ceases along two curves that are defined by the relation discussed in (10.47). In Figure 10.4(b) the maximum gain $\bar{g}$ is shown on the infrared side versus seed frequency ν_s (bottom axis) and seed wavelength λ_s (top axis). The two pump wavelengths $\lambda_p = 0.85, 2.1$ μm correspond to the solid

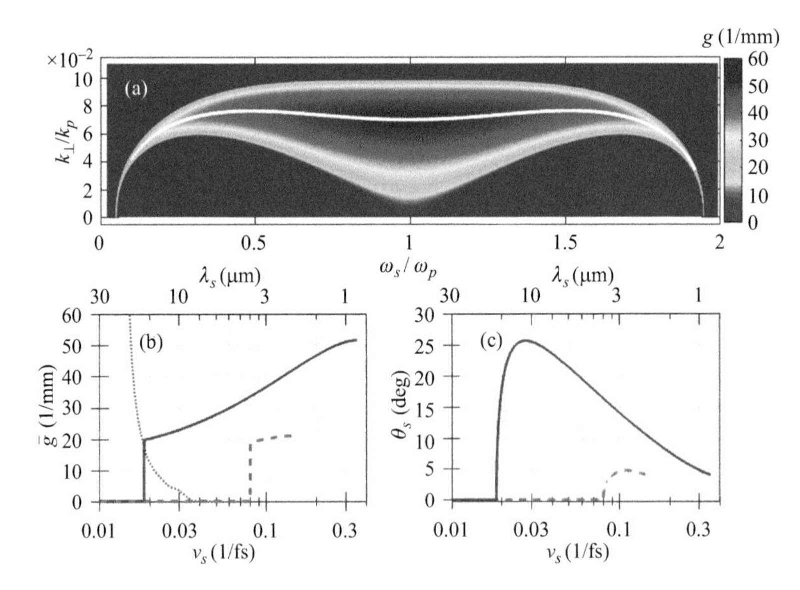

Figure 10.4 Plane wave amplification in bulk CaF_2 crystal. (a) Kerr instability gain, g, versus ω_s/ω_p and $k_\perp/k_p$ (transverse over pump wavevector); pump wavelength $\lambda_p = 0.85$ μm. The white line indicates $\bar{k}_\perp$ at which maximum gain $\bar{g} = g(\bar{k}_\perp)$ occurs, see (10.46). (b) $\bar{g}$ versus seed frequency $\nu_s = \omega_s/(2\pi)$ (bottom) and seed wavelength λ_s (top); red dotted line represents absorption. (b) and (c) $\lambda_p = 0.85, 2.1$ μm corresponds to blue full, green dashed curves, respectively; (c) angle of inclination between pump and seed beam θ_s at which maximum amplification takes place versus ν_s and λ_s, see (10.51)

blue and dashed green curves, respectively, in Figure 10.4(b) and (c). Maximum gain reaches a global maximum when pump and seed frequencies are equal and decreases with longer wavelengths. Further, $\bar{g}$ increases with pump frequency. For $\lambda_p = 0.85$ μm the gain is still significant at $\lambda_s = 15$ μm. Amplification, $e^{\bar{g}l}$, by more than four orders of magnitude can be obtained in a $l = 0.5$-mm long crystal. Note that gain and absorption balance each other at $\lambda_s = 20$ μm. At this wavelength, the medium becomes transparent in the presence of the pump beam.

For $\lambda_p = 2.1$ μm, the gain exists over a narrow spectral range. This is evident from Figure 10.4(c), where the angle for maximum amplification, Equation (10.51), is plotted for the same two pump wavelengths. At $\lambda_p = 2.1$ μm, θ_s reaches a maximum near-the-pump wavelength and then decreases to zero rapidly. This property arises from the functional form of the angle θ_s, due to linear dispersion of $n(\omega)$. We observe that $tan^2\theta_s \propto \kappa_\perp^2 \propto \sigma^2 - 1 \approx 2(n_g + n_u\Omega_s/\omega_p)/\eta_p$. The two terms, η_g and $\eta_u\Omega_s/\omega_p$, can have opposite or equal signs, depending on the material and pump wavelength, as mentioned in (10.45). In this particular case, they are of opposite sign and comparable magnitude, so that for decreasing ν_s, $\kappa_\perp^2$ becomes negative. From (10.45) and (10.46), we see that $\bar{k}_\perp = \bar{g} = 0$ so that both gain and θ_s become zero. Similar behavior can be seen for $\lambda_p = 0.85$ μm, but it is stretched out over a wider spectral range.

In Figure 10.5, the results for KBr crystals are shown for a pump intensity $I_p = 8$ TW/cm^2. The same line styles as in Figure 10.4 are used. The gain in

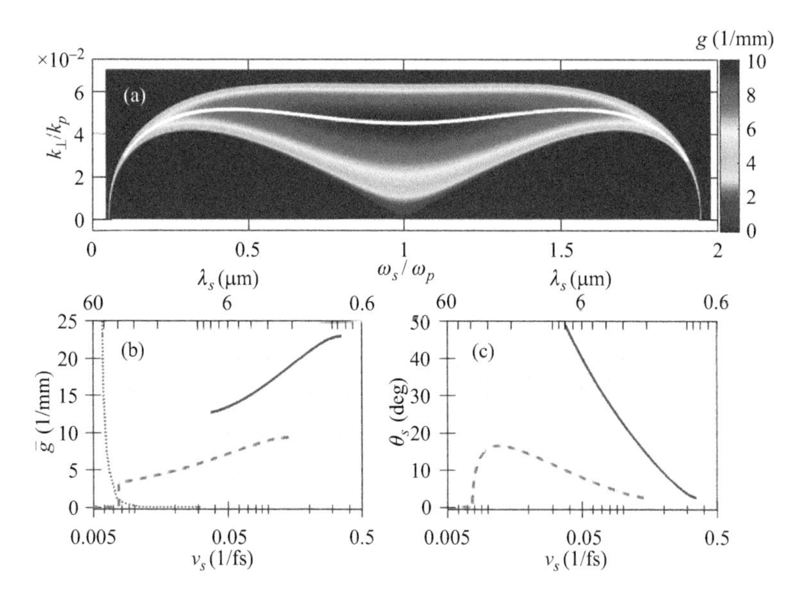

Figure 10.5 Plane wave amplification in bulk KBr crystal. Panels (a)–(c) correspond to those in Figure 10.4. In (a) the pump wavelength is $\lambda_p = 2.1$ μm; all other parameters and definitions in panels (a)–(c) are the same as given in the caption of Figure 10.4

Figure 10.5(a) is plotted for $\lambda_p = 2.1$ μm. Unlike in CaF_2, using $\lambda_p = 2.1$ μm works well in KBr. Gain exists over a transmission window twice as long as in CaF_2, see also Figure 10.5(b). The maximum gain is still substantial near the edge of the transmission window, see the dotted red line. Amplification of more than four orders of magnitude can be achieved over a crystal length $l = 2$ mm. For $\lambda_p = 0.85$ μm, the gain exists in a narrower spectral range, up to 8 μm. The reason is clear from Figure 10.5(c). For $\lambda_p = 0.85$ μm, the angle rapidly increases for increasing λ_s. As mentioned previously for CaF_2, this comes from the fact that both terms in $\sigma^2 - 1$ have the same sign. Here, the gain ceases when the denominator in (10.41) approaches zero for $k_p = \sigma D_u$, as described in (10.42a) and (10.42b). By contrast for $\lambda_p = 2.1$ μm, the signs are different and we see similar behavior as in Figure 10.4(c). Linear dispersion strongly influences KIA and is a crucially important design parameter. The ultrawide frequency range and the large noncollinear angles encountered in Figures 10.4 and 10.5 emphasize the necessity of our theoretical framework that does not rely on dispersion expansion and paraxial approximation, as discussed in (10.34a) and (10.34b).

10.3.5 Discussion of finite pulse KIA

Our analysis of plane wave KIA has been extended to finite pulses [13]. Although the results are important for designing KIA, here we summarize only the results. For a pulse with finite spectral width, each frequency has a slightly different angle θ_s for optimum amplification. As a result, KIA results in an angular chirp that needs to be corrected. The chirp is dominantly linear and can be compensated by standard optical elements, such as a prism. Gain narrowing and pulse widening through spectral chirp imprinted during KIA set a limit to the amplification of ultrashort pulses. These effects can be minimized by choosing peak pump intensities close to the damage threshold minimizing amplification length. Our theoretical analysis suggests that the amplification of one to two cycle pulses of mid-infrared pulses is feasible in KBr. For pulsed amplification the wavelength range is more limited than seen in Figure 10.5. In going toward longer seed wavelengths, longer pump pulse durations and widths are required. This comes from the fact that the seed pulse is required to remain close to the pump peak during the whole amplification length. As a result, the required pump pulse energies go up. Furthermore, the damage intensity for longer pump pulses is reduced limiting the maximum useable pump peak intensities. In addition, the decreasing gain for longer wavelengths sets a limit to the longest obtainable wavelengths. In a detailed analysis of pulsed KIA in KBr for the same parameters as used in Figure 10.5, we find that amplification is limited to seed pulse wavelengths of 15 μm. At this wavelength, we find amplifications factors of about 1,000 and maximum pulse energies after amplification of about 50 μJ.

10.4 Conclusion

KIA is a promising approach for the amplification of ultrashort mid-infrared pulses. In this chapter, the general vectorial theory of KIA is developed. It is shown that vectorial effects are negligible over the transverse wavevector range of the instability regime, proving the validity of the scalar theory. Plane wave KIA theory is discussed and quantitatively analyzed for two example materials, CaF_2 and KBr. The plane wave theory demonstrates the key features of KIA, namely, the importance of linear dispersion properties of the material, and the pump wavelength chosen. In practice, plane wave KIA theory must be extended to allow for finite seed pulses, which introduces limitations to the theory, see Section 10.3.5. An experiment of KIA has shown promising first results [5]. Our analysis suggests that amplification factors of 1,000 of one to two cycle pulses of mid-infrared 50 μJ pulses are feasible in KBr. Future work must be done to explore the vast parameter space of KIA theory in order to optimize the amplification process. Specifically, the materials with optimal dispersive properties—whether natural or man-made—and the optimal profile of the initial seed beam should be explored.

References

[1] A. Couairon and A. Mysyrowicz, "Femtosecond filamentation in transparent media", *Phys. Rep.* 441(47), 47–189 (2007).

[2] E. Rubino, J. Darginavičius, D. Faccio, P. Di Trapani, A. Piskarskas, and A. Dubietis, "Generation of broadly tunable sub-30-fs infrared pulses by four-wave optical parametric amplification", *Opt. Lett.* 36, 382–384 (2011).

[3] M. I. M. Abdul Khudus, F. De Lucia, C. Corbari, *et al.*, "Phase matched parametric amplification via four-wave mixing in optical microfibers", *Opt. Lett.* 41, 761–764 (2016).

[4] M. Nesrallah, G. Vampa, G. Bart, P. B. Corkum, C. R. McDonald and T. Brabec, "Theory of Kerr instability amplification", *Optica* 5(3), 271–278 (2018).

[5] G. Vampa, T. J. Hammond, M. Nesrallah, A. Yu Naumov, P. B. Corkum and T. Brabec, "Light amplification by seeded Kerr instability", *Science* 359, 673–675 (2018).

[6] R. Gattass and E. Mazur, "Femtosecond laser micromachining in transparent materials", *Nat. Photonics* 2, 219 (2008).

[7] S. Ghimire, A. D. Dichiara, E. Sistrunk, P. Agostini, L. F. Dimauro and D. A. Reis, "Observation of high-order harmonic generation in a bulk crystal", *Nat. Phys.* 7, 138 (2011).

[8] P. Malevich, G. Andriukaitis, T. Flöry, *et al.*, "High energy and average power femtosecond laser for driving mid-infrared optical parametric amplifiers", *Opt. Lett.* 38, 2746 (2013).

[9] B. E. Schmidt, N. Thiré, M. Boivin, *et al.*, "Frequency domain optical parametric amplification", *Nat. Commun.* 5, 3643 (2014).

[10] C. Manzoni and G. Cerullo, "Design criteria for ultrafast optical parametric amplifiers", *J. Opt.* 18, 103501 (2016).

[11] P. Krogen, H. Suchowski, H. Liang, *et al.*, “Generation and multi-octave shaping of mid-infrared intense single-cycle pulses”, *Nat. Photonics* 11, 222 (2017).
[12] M. Sheik-Bahae, D. C. Hutchings, D. J. Hagan, and E. W. Van Stryland, “Dispersion of bound electronic nonlinear refraction in solids”, *IEEE J. Quantum Electron.* 27, 1269 (1991).
[13] M. Nesrallah, A. Hakami, G. Bart, C. R. McDonald, C. Varin, and T. Brabec, “Measuring the Kerr nonlinearity via seeded Kerr instability amplification: conceptual analysis”, *Opt. Express* 26(6), 7646–7654 (2018).
[14] R. W. Boyd, *Nonlinear optics*, 3rd edition, Academic Press, Amsterdam (2008).
[15] V. I. Bespalov and V. I. Talanov, “Filamentary structure of light beams in nonlinear liquids”, *JETP Lett.* 3, 307 (1966).
[16] G. Agrawal, *Nonlinear fiber optics*, 5th edition, Academic Press, Oxford (2012).
[17] E. D. Palik, *Handbook of optical constants of solids II*, Academic Press, Boston, MA (1991).
[18] I. H. Malitson, “A redetermination of some optical properties of calcium fluoride”, *Appl. Opt.* 2, 1103 (1963).
[19] D. Milam, M. J. Weber, and A. J. Glass, “Nonlinear refractive index of fluoride crystals”, *Appl. Phys. Lett.* 31, 822 (1977).
[20] H. H. Li, “Refractive index of alkali halides and its wavelength and temperature derivatives”, *J. Phys. Chem. Ref. Data* 5, 329 (1976).
[21] R. DeSalvo, A. A. Said, D. J. Hagan, A. W. Van Stryland, and M. Sheik Bahae, “Infrared to ultraviolet measurement of two-photon absorption and n_2 in wide bandgap solids”, *IEEE J. Quantum Electron.* 32, 1324 (1996).

Appendix A

The goal of this appendix is to show that KIA and FWM are mathematically equivalent and are thus two representations of the same physical process; one can say, two sides of the same coin. The KIA approach is more general, as it does not rely on splitting the electric field of the instability into carrier and envelope. As a result, it yields a general expression for the FWM gain over the whole spectral and transverse wavevector range. This gain can be applied to any beam form, such as noncollinear Gaussian beams or Bessel–Gaussian beams. The theory of FWM emerges from the theory of KIA by splitting the electric field into envelope and carrier. Once this is done, in order to proceed, certain approximations are made, such as the paraxial approximation and a Taylor expansion of the wavevector in a sum over dispersive terms.

By starting from degenerate FWM theory and by avoiding commonly used approximations, it will be shown that the resulting coupled wave equations developed are equivalent to the scalar wave equations of KIA theory. We begin with the

FWM ansatz for the electric field

$$E = E_p e^{i\omega_p t - ik_p z} + E_s(\mathbf{x}, t)e^{i\omega_s t} + E_i(\mathbf{x}, t)e^{i\omega_i t} + \text{c.c.}, \tag{A.1}$$

where subscripts p, s, i stand for the strong plane wave pump, weak signal, and weak idler beams, respectively. Notice we keep E_s and E_i as general functions, consistent with ε in KIA theory; no signal and idler center wavevectors have been introduced. We insert (A.1) into the third-order nonlinear polarization function ((10.1) in the scalar limit) to obtain

$$P^{(3)} \approx \frac{n_n}{\mu_0 c^2}\left[E_p e^{i\omega_p t - ik_p z} + 2\left(E_s e^{i\omega_s t} + E_i e^{i\omega_i t}\right) + e^{2i\omega_p t - 2ik_p z}\left(E_s^* e^{-i\omega_s t} + E_i^* e^{-i\omega_i t}\right)\right] \\ + \text{c.c.}, \tag{A.2}$$

where we have neglected third harmonic generation and assume $E_p \gg E_{s,i}(z = 0)$. We insert (A.2) into the scalar version of the nonlinear wave equation, (A.3),

$$\left[\nabla^2 - \frac{\partial^2}{\partial t^2}\frac{n^2 *}{c^2}\right]E = \mu_0 \frac{\partial^2 P^{(3)}}{\partial t^2}; \tag{A.3}$$

the first term on the right-hand side of (A.2) results in the wavevector solution $k_p = c^{-1}\omega_p\sqrt{n_p^2 + n_n}$ as in KIA theory.

For the remaining terms on the right-hand side of (A.2), we take the Fourier transform of (A.3) of the coordinates (x, y, t) to obtain the coupled equations for the signal and idler fields

$$\left[\frac{\partial^2}{\partial_z^2} + k_z^2\right]\widetilde{E}_s(\omega - \omega_s) = -k_n^2\widetilde{E_i^*}(\omega - \omega_s)e^{-2ik_p z} \tag{A.4a}$$

$$\left[\frac{\partial^2}{\partial_z^2} + k_z^2\right]\widetilde{E}_i(\omega - \omega_i) = -k_n^2\widetilde{E_s^*}(\omega - \omega_i)e^{-2ik_p z}, \tag{A.4b}$$

where $k_z^2 = k_v^2 - k_\perp^2$ and $k_n^2 = c^{-2}\omega^2 n_n$ are defined as before (see (10.10)). As FWM is a parametric process, we have used the energy conservation relation $2\omega_p = \omega_s + \omega_i$ in (A.2), which couples the signal and idler fields with the corresponding conjugates. To proceed, we let $\widetilde{E_{s,i}}(\omega - \omega_{s,i}) = \widetilde{A_{s,i}}(\omega - \omega_{s,i})e^{-ik_p z}$ to eliminate the exponential factors of (A.4a) and (A.4b). Further, $\widetilde{E_{s,i}^*}(\omega - \omega_{s,i}) = \widetilde{E_{s,i}^*}(\omega_{s,i} - \omega)$ is used. In addition, $\omega - \omega_s = \omega - \omega_p - (\omega_s - \omega_p) \equiv \Omega - \Omega_s$, and $\omega - \omega_i = \omega - 2\omega_p + \omega_s = \Omega + \Omega_s$. Using the previous equation, taking the complex conjugate of (A.4a), and letting $\Omega \to -\Omega$ results in the coupled equations

$$\left[\left(\frac{\partial}{\partial_z} - ik_p\right)^2 + k_z^2\right]\widetilde{A}_s(\Omega - \Omega_s) = -k_n^2\widetilde{A}_i^*(\Omega_s - \Omega) \tag{A.5a}$$

$$\left[\left(\frac{\partial}{\partial_z} + ik_p\right)^2 + k_{z(-)}^2\right]\widetilde{A}_i^*(\Omega_s - \Omega) = -k_{n(-)}^2\widetilde{A}_s(\Omega - \Omega_s). \tag{A.5b}$$

If we let $\widetilde{A}_s(\Omega - \Omega_s) \to \widetilde{v}(\Omega)$ and $\widetilde{A}_i^*(\Omega_s - \Omega) \to \widetilde{v}^*(-\Omega)$, we obtain

$$\left[\left(\frac{\partial}{\partial_z} - ik_p\right)^2 + k_z^2\right]\widetilde{v}(\Omega) = -k_n^2\widetilde{v}^*(-\Omega) \tag{A.6a}$$

$$\left[\left(\frac{\partial}{\partial_z} + ik_p\right)^2 + k_{z(-)}^2\right]\widetilde{v}^*(-\Omega) = -k_{n(-)}^2\widetilde{v}(\Omega), \tag{A.6b}$$

which are precisely the coupled equations, (10.17) and (10.18), in the scalar limit [4]. Thus, it can be concluded that the derivations of FWM and KIA start from the same set of equations. Usually, in FWM theory the signal and idler wavevectors are explicitly introduced in the ansatz requiring the paraxial approximation and Taylor expansion of the wavevectors to proceed. As a result, an explicit wavevector mismatch, $\Delta\mathbf{k} = 2\mathbf{k}_p - \mathbf{k}_s - \mathbf{k}_i$, is obtained. Here, we kept the signal and idler electric fields general. As a result, the wavevector mismatch does not show up explicitly; nevertheless, optimum amplification takes place when the sum over wavevectors is conserved; see also Figure 10.2.

Chapter 11
Plasma studies in filament

Xin Lu,[1,2,5] *Tingting Xi,*[2] *and Jie Zhang*[1,3,4]

When the intense femtosecond laser pulse propagates in air, the laser pulse self-focuses owing to the Kerr effect, and the laser intensity increases rapidly. Then, the air is ionized, and electrons are generated. The dynamic competition of self-focusing, electron defocusing, and diffraction supports the long-distance propagation of the femtosecond laser pulse. This process is called filamentation. During filamentation, the ionization plays an important role in the formation of a plasma channel and the generation of supercontinuum. This chapter starts with the illustration of the optical field ionization models that are commonly used for the simulation of filamentation in different field conditions. Based on the ionization coupled with the nonlinear Schrödinger equation, theoretical studies on plasma filaments under different conditions are illustrated. Then several measurement methods for electron density inside filament are introduced. Based on these, characteristics of electron density under different conditions are studied.

11.1 Generation and decay theory of plasma in filament

11.1.1 Multiphoton ionization and tunnel ionization

The ionization potentials of oxygen and nitrogen are $U_O = 12.1\text{eV}$ and $U_N = 15.6\text{eV}$, respectively. Even if the frequency of the femtosecond laser ω_0 satisfies $\hbar\omega_0 \lll U_O, U_N$, the ionization still occurs when filamentation is triggered with a high intensity. The ionization is called optical field ionization and covers two regimes: multiphoton ionization (MPI) and tunnel ionization. Many theories have been proposed to describe the optical field ionization. The adiabatic parameter γ proposed by Keldysh is usually used to separate the two regimes [1]. Here,

[1]Beijing National Laboratory for Condensed Matter Physics, Institute of Physics, Chinese Academy of Sciences, Beijing, China
[2]School of Physical Sciences, University of Chinese Academy of Sciences, Beijing, China
[3]Key Laboratory for Laser Plasmas (Ministry of Education) and Department of Physics, Shanghai Jiao Tong University, Shanghai, China
[4]IFSA Collaborative Innovation Center, Shanghai Jiao Tong University, Shanghai, China
[5]Songshan Lake Materials Laboratory, Dongguan, China

$\gamma = \left(\omega_0\sqrt{2m_eU_i}/|q_e|E_p\right)$, where E_p is the electric field strength, and U_i is the ionization potential of the atom. When the parameter $\gamma \ll 1$, tunnel ionization prevails, and when $\gamma \gg 1$, MPI occurs.

With the increase of the laser intensity, MPI occurs first. MPI is the process in which several photons are absorbed simultaneously by an atom and a free electron is generated. In the MPI regime, the ionization rate is scaled as I^K, where I is the laser intensity and $K = \langle(U_i/\hbar\omega_0)\rangle + 1$ is the photon number of simultaneous absorption.

When the laser intensity is higher, $\gamma \ll 1$, tunnel ionization dominates. The potential barrier that binds the electrons to nucleus is suppressed by the strong electric field. Then, the electron escapes from the potential, and a free electron is generated.

The ionization of air is crucial to the characteristics of filamentation. The density of electrons generated by ionization is usually calculated using the following equations [2]:

$$\frac{\partial\rho_o}{\partial t} = W_O(t)(\rho_{On} - \rho_o), \tag{11.1}$$

$$\frac{\partial\rho_N}{\partial t} = W_N(t)(\rho_{Nn} - \rho_N), \tag{11.2}$$

$$n_e = \rho_O + \rho_N, \tag{11.3}$$

where $n_e, \rho_O, \rho_N, \rho_{On}$, and ρ_{Nn} are the densities of electrons, ionized oxygen and nitrogen, and neutral oxygen and nitrogen, respectively. $W_O(t)$ and $W_N(t)$ are the ionization rates of oxygen and nitrogen, respectively. When the laser intensity is not higher, only the ionization of oxygen, which occurs first, is considered in the simulation. The ionization rate is usually calculated according to the Perelomov–Popov–Terent'ev (PPT) model [3] and the Ammosov–Delone–Kraĭnov (ADK) model [4].

11.1.2 Perelomov–Popov–Terent'ev model

For the PPT model, the general formula for the ionization rate of any atom or ion with quantum numbers l, m is written as follows [3,5,6]:

$$W = \omega_{\text{a.u.}}\sqrt{\frac{3}{2\pi}}\left|C_{n^*,l^*}\right|^2 f(l,m)\frac{U_i}{U_H}\left(\frac{2F_0}{E_p\sqrt{1+\gamma^2}}\right)^{2n^*-|m|-(3/2)}$$
$$A_m(\omega_0,\gamma)\exp\left[-\frac{2F_0}{3E_p}g(\gamma)\right], \tag{11.4}$$

where $n^* = Z/\sqrt{2U_i}$ is the effective quantum number, Z is the charge state of the resulting ion, and U_i and U_H are the ionization potentials of the atom and hydrogen, respectively. $l^* = n^* - 1$ is the effective orbital quantum number. The constants $\left|C_{n^*,l^*}\right|^2$ and $f(l,m)$ are given as

$$\left|C_{n^*,l^*}\right|^2 = \frac{2^{2n^*}}{n^*\Gamma(n^*+l^*+1)\Gamma(n^*-l^*)}, \tag{11.5}$$

$$f(l,m)=\frac{(2l+1)(l+|m|)!}{2^{|m|}|m|!(l-|m|)!}, \tag{11.6}$$

where Γ is the gamma function. The other parameters are $F_0=(2U_i)^{3/2}$ and $\omega_{\mathrm{a.u.}}=4.1\times10^{16}\mathrm{S}^{-1}$. The involved functions are written as follows:

$$A_m(\omega_0,\gamma)=\frac{4}{\sqrt{3\pi}}\frac{1}{|m|!}\frac{\gamma^2}{1+\gamma^2}\sum_{\mathrm{K}\geq\nu}^{+\infty}\exp[-\alpha(K-\nu)]\Phi_m[\sqrt{\beta(K-\nu)}], \tag{11.7}$$

$$\nu=\frac{U_i}{\hbar\omega_0}\left(1+\frac{1}{2\gamma^2}\right), \tag{11.8}$$

$$\Phi_m\left[\sqrt{\beta(K-\nu)}\right]=\exp(-x^2)\int_0^x(x^2-y^2)^{|m|}\exp(y^2)dy, \tag{11.9}$$

$$\alpha(\gamma)=2\left[\sinh^{-1}\gamma-\frac{\gamma}{\sqrt{1+\gamma^2}}\right], \tag{11.10}$$

$$\beta(\gamma)=\frac{2\gamma}{\sqrt{1+\gamma^2}}, \tag{11.11}$$

$$g(\gamma)=\frac{3}{2\gamma}\left[\left(1+\frac{1}{2\gamma^2}\right)\sinh^{-1}\gamma-\frac{\sqrt{1+\gamma^2}}{2\gamma}\right]. \tag{11.12}$$

The calculated ionization rate of oxygen and nitrogen can be fitted to the experimental measurement in the regime ($10^{13}<I<10^{15}\mathrm{W/cm^2}$) by changing the charge parameter with an effective charge [5].

11.1.3 Ammosov–Delone–Kraĭnov model

In the ADK model [4], the ionization rate depends on the real-time electric field, rather than the electric field strength as in the PPT model. The ionization rate of an atom in an arbitrary state is written as

$$W=\left(\frac{3En^{*3}}{\pi Z^3}\right)^{1/2}\frac{Z^2}{2n^{*2}}\left(\frac{2e}{n^*}\right)^{2n^*}\frac{1}{2\pi n^*}\frac{(2l+1)(l+|m|)!}{2^{|m|}|m|!(l-|m|)!}\left(\frac{2Z^3}{En^{*3}}\right)^{2n^*-|m|-1}\exp\left(-\frac{2Z^3}{3n^{*3}E}\right), \tag{11.13}$$

where E is the real-time electric field of the laser. Using this equation, the tunnel ionization rate from the ground state is given as [7]

$$W=\omega_p|C_{n^*}|^2\left(\frac{4\omega_p}{\omega_t}\right)^{2n^*-1}\exp\left(-\frac{4\omega_p}{3\omega_t}\right), \tag{11.14}$$

where $\omega_p=U_i/\hbar$, $\omega_t=e|E(t)|^2/\sqrt{2m_eU_i}$ and $|C_{n^*}|^2=2^{2n^*}/n^*\Gamma(n^*+1)\Gamma(n^*)$. The ADK model is suitable for the ionization of few-cycle laser pulses, when the phase of real-time electric field significantly affects the ionization process [8].

11.1.4 Second-level ionization

If the intensity of the laser pulse is higher, the air may be fully ionized, and second-level ionization should be considered. The evolution of the density of the charged particles can be described by the following equations [2,9]:

$$\frac{d\rho_1}{dt} = -W''_{air}(I)\rho_1 + W'_{air}(I)(\rho_{air} - \rho_1 - \rho_2), \tag{11.15}$$

$$\frac{d\rho_2}{dt} = W''_{air}(I)\rho_1, \tag{11.16}$$

$$n_e(t) = \rho_1(t) + 2 \times \rho_2(t), \tag{11.17}$$

where $\rho_{air} = 2.7 \times 10^{19}\text{cm}^{-3}$ is the initial density of air, and ρ_1 and ρ_2 are the densities of the first- and second-level ions, respectively. n_e is the total electron density. $W'_{air}(I)$ and $W''_{air}(I)$ are the first- and second-level ionization rates of air, which can be calculated according to the PPT model. The second-level ionization can be used for the simulation under tight focusing conditions [9].

11.1.5 Plasma decay in filament

The plasma channel remaining in the tail of the laser pulse is composed of free electrons, positive ions, negative ions, and neutral molecules. Several processes occur, including the electron–ion recombination, ion–ion recombination, and the attachment of electrons to neutral molecules. The densities of the charged particles as a function of time can be written as follows [10,11]:

$$\frac{dn_e}{dt} = -\eta n_e - \beta_{ep} n_e n_p, \tag{11.18}$$

$$\frac{dn_p}{dt} = -\beta_{ep} n_e n_p - \beta_{np} n_n n_p, \tag{11.19}$$

$$\frac{dn_n}{dt} = -\eta n_e - \beta_{np} n_n n_p, \tag{11.20}$$

where n_e, n_p, and n_n are the densities of free electrons, positive ions, and negative ions, respectively. $\eta = 6.2 \times 10^7 \text{s}^{-1}$ is the attachment coefficient. β_{ep} and β_{np} are the coefficients of electron–ion recombination and ion–ion recombination, respectively, with $\beta_{ev} = \beta_{nv} = \beta = 2.2 \times 10^{-13} \text{m}^3\ \text{s}^{-1}$.

The electron density can be analytically solved according to the principle of charge conservation and the initial condition $n_e(0) = n_p(0) = n_{e0},\ n_n(0) = 0$, which is written as follows:

$$n_e(t) = \frac{n_{e0} e^{-\eta t}}{1 + \beta n_{e0} t}. \tag{11.21}$$

When the time $t < 1/\eta = 16$ ns, the electron density decreases, mainly owing to the electron–ion recombination. The decreasing rate of the electron density depends on the product of the ion and electron densities. Therefore, the electron density decreases faster when the initial electron density is higher. The lifetime of the plasma channel cannot be improved by enhancing the initial density. When the time $t \gg 1/\eta = 16$ ns, the attachment of electrons and neutral molecules dominates. The electron density decreases quickly and almost disappears.

11.2 Simulation studies of plasma filament under different conditions

In this section, we present simulation studies on the nonlinear propagation dynamics of the filament. In simulation, the electron density equation should be coupled with the nonlinear Schrödinger equation [2], which includes the effects of diffraction, group-velocity dispersion, instantaneous Kerr response, delayed Raman effect, electron defocusing, and ionization loss.

11.2.1 Interaction of filaments

When the power of the laser is significantly higher than the self-focusing threshold, multiple filaments are formed. The interaction of the filaments significantly influences the propagation dynamics. Therefore, the studies on the interaction among filaments are necessary. The interaction between two filaments has been simulated [12], as briefly described in this part.

Simulation results show that the phase shift between filaments plays an important role in the interaction between filaments. When the two filaments are in-phase, they disperse first and release energy to the background. The overlap of the two filaments leads to the increase of the intensity at the center of them. Then, the refractive index in this region increases owing to the Kerr effect, and the background energy is attracted toward this region. As a result, the two filaments merge into a new one (Figure 11.1(a) and (b)). During the dispersion and fusion of the two filaments, the peak intensity decreases, reducing the multiphoton absorption. Therefore, the plasma channel is elongated (Figure 11.1(b)). For two out-of-phase filaments, the destructive interference leads to the decrease of the central refractive index. The two out-of-phase filaments appear to repel each other and disperse quickly. During this process, the plasma channel is shortened (Figure 11.1(f)). Therefore, in order to form a long and stable plasma channel, the phase shift between two filaments should be as small as possible.

11.2.2 Third harmonic generation in disturbed filament

During filamentation, the laser intensity is high enough to generate third harmonic (TH) emission that can be applied for remote sensing and the generation of intense coherent ultraviolet (UV) light sources. Simulations of the enhancement of the TH emission, which is beneficial for these applications, are introduced in this part [13,14].

The characteristics of TH emission are similar to those of the fundamental wave (FW) owing to its strong driving source. During filamentation, both the intensities of the FW and the TH are clamped, and the conversion efficiency of the TH is obtained as approximately 0.2%–0.3%. With the termination of filamentation, the intensities

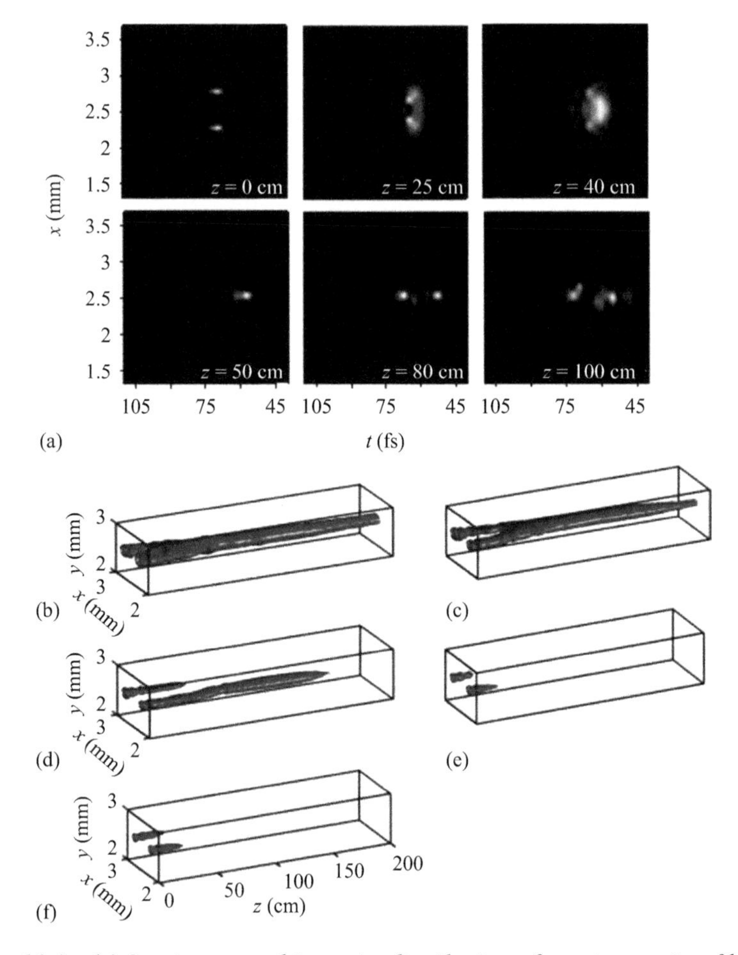

Figure 11.1 (a) Spatiotemporal intensity distribution of two interacting filaments that are parallel and in-phase. (b)–(f) Energy fluence distribution (fluence$_{iso}$ = 1.65) of two parallel filaments with a separation distance of 0.5 mm and relative phase shifts of 0π (b), 0.25π (c), 0.5π (d), 0.75π (e), and π (f) [12]

of both the FW and TH decrease. The energy of TH emission is returned to the FW and almost decreases to zero.

One method for enhancing TH emission is adding a small droplet to the filament center to destroy the destructive interference between the TH and FW [13]. Although the FW and TH emission loses part of the energy owing to the block of the droplet, they both reconstruct. The peak intensity of the FW first decreases owing to the screening of the droplet and then increases owing to the replenishment of its energy reservoir. During this process, the peak intensity of the TH oscillates and is enhanced, as shown in Figure 11.2(b). The TH energy is also enhanced. The

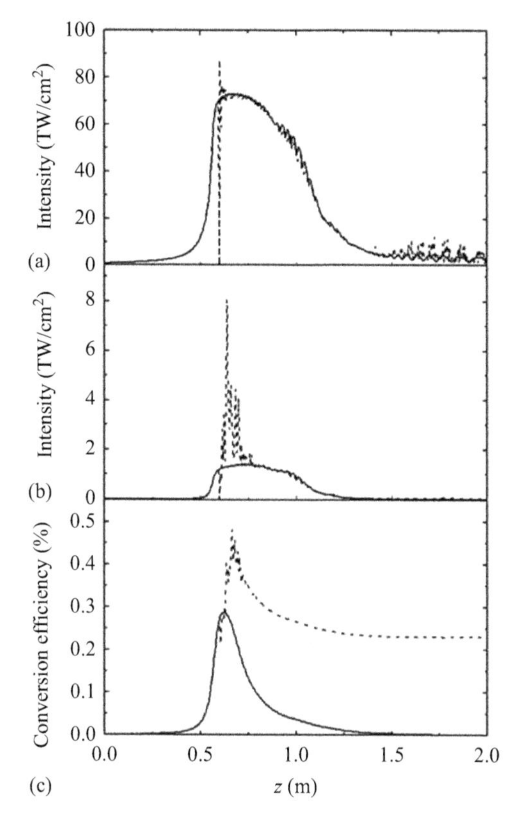

Figure 11.2 *On-axis peak intensities of the fundamental pulse with a peak power of $P_{in} = 4P_{cr}$, a temporal duration (full width at half maximum (FWHM)) of 50 fs, a beam waist (FWHM) of 1 mm (a) and TH emission (b); the TH conversion efficiency (c) for the undisturbed (dashed line) and disturbed filamentation (solid line) [13]*

destructive interference leads to the preservation of part of the TH energy even after the termination of filamentation, as shown in Figure 11.2(c).

Another method is introducing a plasma spot into the filament [14]. The destruction and reconstruction process of filamentation is similar to the case of adding a droplet [13]. The peak intensity of the TH oscillates during this process. Then, the TH emission is enhanced. To distinguish the contribution of the pure plasma effect from the plasma-enhanced third-order susceptibility, the enhancements of the TH with an artificially changed electron density and enhanced $\chi^{(3)}$ are compared (Figure 11.3). In the case of the enhanced $\chi^{(3)}$, the strong spatial reshaping and enhancement of the TH do not occur (Figure 11.3(a) and (c)). After the termination of filamentation, most of the TH energy is returned to the FW. This result suggests that the plasma-enhanced third-order susceptibility cannot be responsible for the enhancement of the TH emission. While for the simulation, both the pure plasma effect and the plasma-enhanced $\chi^{(3)}$ effect are included

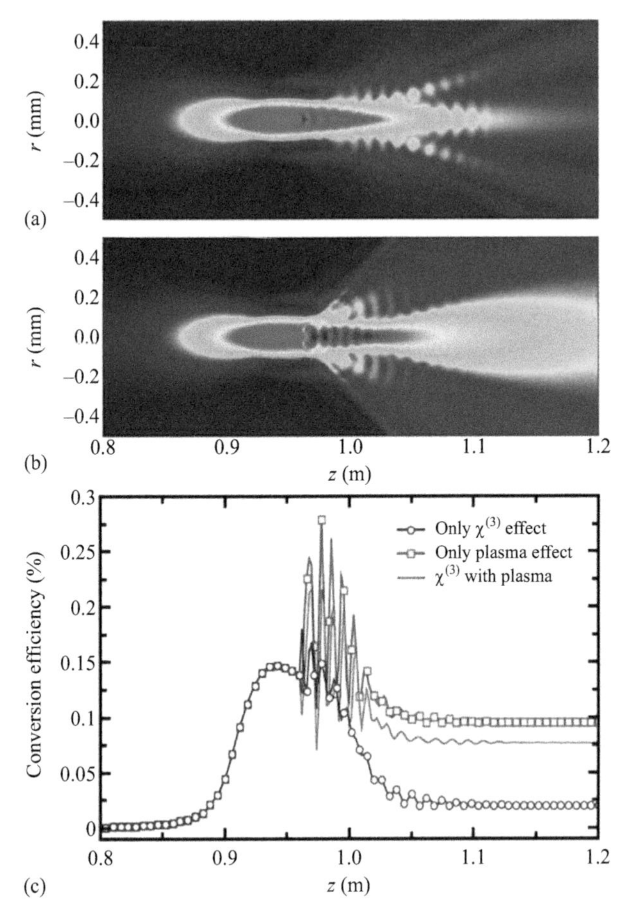

Figure 11.3 Averaged intensity distribution of the TH in the disturbed filament, considering the pure $\chi^{(3)}$ effect (a) and the combination of plasma effect with the $\chi^{(3)}$ effect (b). (c) TH conversion efficiency as a function of the propagation distance for the three models: the pure $\chi^{(3)}$ effect (circles), the pure plasma effect (squares), and the plasma effect with the enhanced $\chi^{(3)}$ (solid line) [14]

(Figure 11.3(b)), and strong spatial reshaping occurs in the TH emission. Moreover, the conversion efficiency of the TH is similar to the case of the pure plasma effect, even after the termination of filamentation. Therefore, the enhancement of the TH mainly arises from the pure plasma effect.

11.2.3 Filamentation in turbulent air

The turbulence of air significantly influences the characteristics of long-distance filamentation. The simulation of the filamentation under different strengths of turbulence is briefly introduced in this part [15].

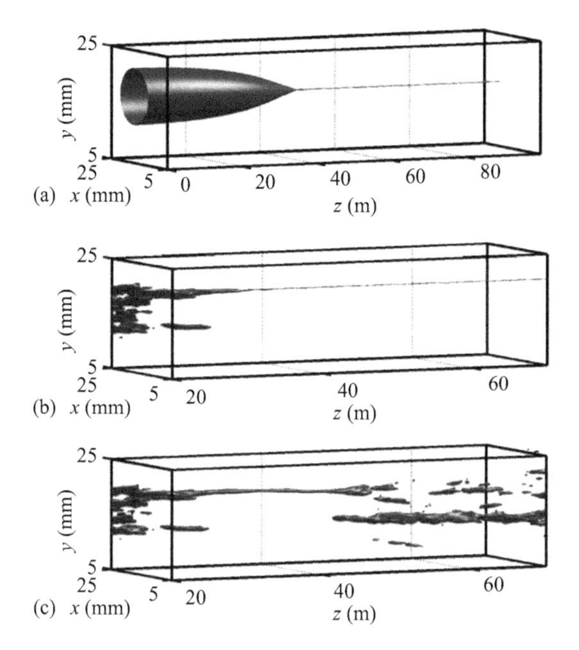

Figure 11.4 Energy fluence distribution of the femtosecond laser pulse in unperturbed air (a), weak turbulent air with a structure constant of $C_n^2 = 2.7 \times 10^{-17} m^{-2/3}$ *(b), and moderate turbulent air with* $C_n^2 = 2.75 \times 10^{-16} m^{-2/3}$ *(c) [15]*

For the unperturbed air (Figure 11.4(a)), the laser pulse undergoes the undisturbed process of self-focusing, and the filament is triggered with a diameter of 146 μm, which is defined by the full width at half maximum (FWHM) of the energy fluence. When the moderate turbulence ($C_n^2 = 2.75 \times 10^{-16}$ m$^{-2/3}$, Figure 11.4(c)) is introduced, a wide filament is formed, and the diameter is approximately 1–2 mm. The widening of the filament results from the suppression of the energy replenishment from the background reservoir to the filament core caused by the distortion of phase of the wave front. With the widening of the filament, the peak intensity is significantly lower than that in the undisturbed case. The generated electron density can be neglected. For the propagation of the laser pulse with a focus lens, the widening of filament can be suppressed owing to the shorter propagation distance before filamentation.

11.2.4 Spatiotemporal moving focus of long filament in air

For long-distance filamentation, the electron density is lower, and the dynamic competition of self-focusing and plasma defocusing cannot support the long-distance propagation. Therefore, it is necessary to study the mechanism in the free propagation of the femtosecond laser pulse. The spatiotemporal moving focus of long filamentation is introduced in this part [16].

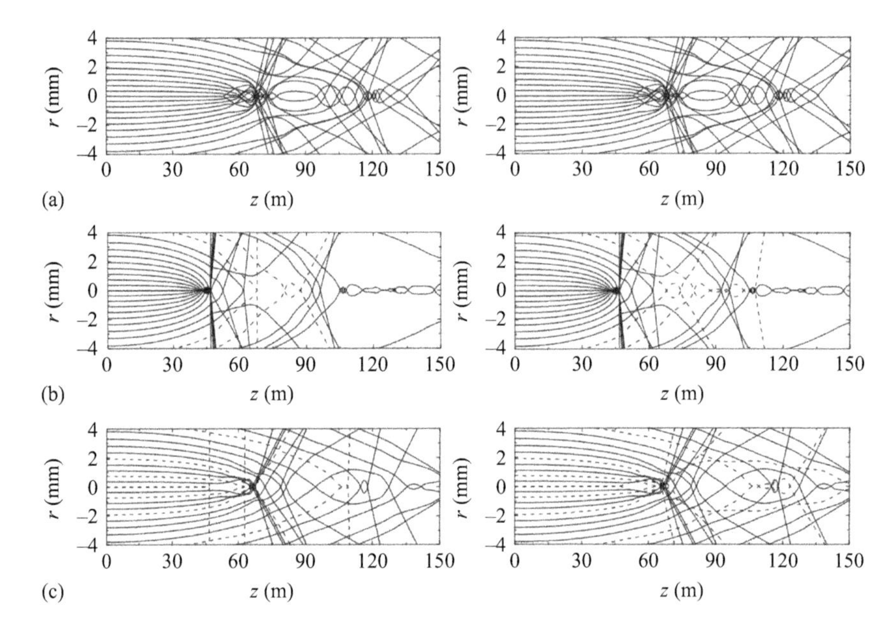

Figure 11.5 Evolution of light rays for time slices of $t = -174$ (a), 0 (b), and 174 fs (c) when the effects of both Kerr nonlinearity and electrons are considered (left) and only the Kerr nonlinearity is considered (right) [16]

In this simulation, the laser beam is considered a bunch of light rays. The left column of Figure 11.5 shows the trajectory of the light rays for three time slices when both the contributions of the Kerr effect and the electrons are considered. The focusing distance of the rays essentially increases with their initial distance from the beam axis owing to their initial intensity distribution. This is called the spatial moving focus effect. The moving focus effect also exists in the temporal region during this long filamentation. Therefore, for the free propagation of the laser pulse, the spatio-temporal moving focus supports this filamentation. In addition, the right column of Figure 11.5 shows the trajectory of the light rays when only the Kerr effect is considered. Comparison between left and right columns reveals that the saturation of electrons does not play a critical role in the formation of the long filament.

11.3 Experimental studies on plasma density in filament

11.3.1 Illustration of measurement methods for electron density inside filament

The electron density is one of the most important parameters for evaluating the quality of the plasma filament and is closely related to many practical applications of filaments. Researchers have developed many effective methods for the diagnosis of the electronic density based on transverse interferometry [17], longitudinal spectral interferometry [18], longitudinal diffraction [19], grazing-incidence interferometry [20], sonographic probing [21,22], stark broadening [23,24], THz scattering [25], microwave probing [26], electrical conductivity (EC) [27–31], electromagnetic induction (EMI) [32], etc. This section is mainly devoted to introducing the transverse interferometry method, EC, and EMI methods.

11.3.2 Transverse interferometry method

Interferometry is a common method for measuring the electron density in plasma physics. Transverse interferometry is a direct and reliable method for measuring the plasma density of the filament, and its physical mechanism is the basis of other methods such as longitudinal interferometry and diffraction. When the laser beam propagates through the plasma, its phase is shifted. If the beam with phase shift interferes with another undisturbed reference beam, the interference fringes are shifted. The shift distance of the interference fringes depends on the electron density of the plasma and the optical path of the laser beam in the plasma.

The transverse interference method was detailed described in the experimental work of Yang *et al.* [17]. The experimental setup is shown in Figure 11.6. A femtosecond laser pulse with energy of 45 mJ and a duration of 30 fs is focused in air by an $f = 2$ m lens to generate a plasma filament. The electron density of the filament is measured by a Nomarski interferometer. The probe pulse passes through a polarizer and the plasma filament from the left side and then converges owing to a lens. After the focusing, the divergent probe beam is divided into two beams with equal amplitude and perpendicular polarization by a Wollaston prism. Then, the two beams passed a polarizer, whose polarization direction is 45° with respect to the first polarizer. Finally, the two beams interfere on a charge-coupled device (CCD) camera and produce interference fringes.

Generally, the plasma density of the filament is significantly lower than the molecular density of air. In this case, the phase shift of the probe pulse induced by

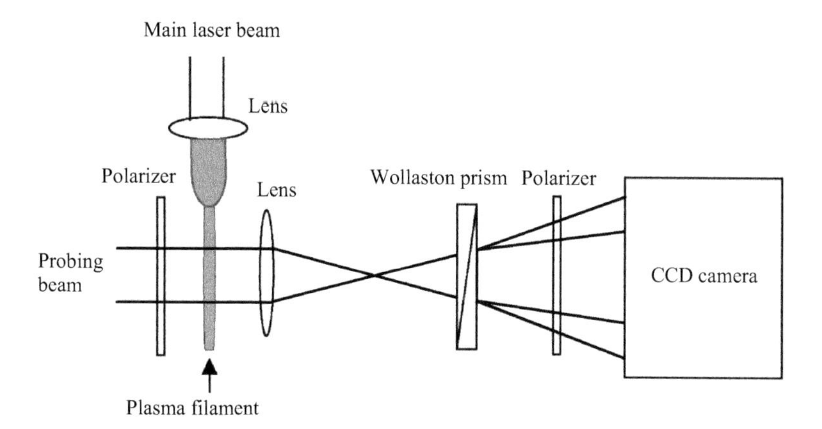

Figure 11.6 Schematic showing the principle of the transverse interferometry method [17]

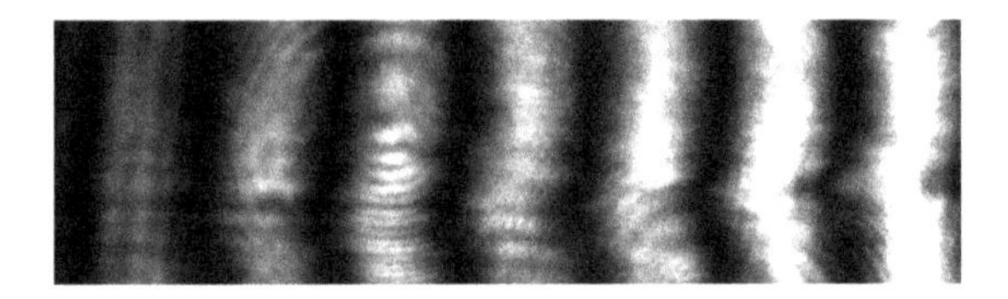

Figure 11.7 Image of the interference fringes recorded on a CCD camera [17]

the filament can be expressed as follows:

$$\Delta\phi = \frac{\omega}{2cn_c}\int n_e dl, \tag{11.22}$$

where ω is the frequency of the probe light, n_e is the electron density of the plasma filament, n_c is the critical plasma density of the probe light, and l is the optical path of the probe light in the plasma. Suppose that the shift number of fringes is D (in units of the distance between two unshifted neighboring fringes); then, $\Delta\phi = 2\pi D$, and

$$\int n_e dl = 2Dn_c\lambda. \tag{11.23}$$

Figure 11.7 shows an image of interference fringes recorded by the CCD camera in the experiment. The maximum shift number is $D = 1/8$. The diameter of the filament is determined to be 120 μm according to the side fluorescence image of filament, and the maximum electron density is determined as $n_e = 2.7 \times 10^{18}$ cm^{-3} from (11.23).

The transverse interference method can observe the obvious interference fringe shift only when the electron density of the filament is sufficiently high (10^{18} cm^{-3} level). This is mainly because when the probe light passes through the filament transversely, the ionization area is determined by the diameter of filament, which is only approximately 100 μm. The phase shift induced by such a short optical path is rather small. For a filament with an electron density lower than 10^{18} cm^{-3}, the shift of the interference fringes can be difficult to be accurately identified. Bodrov *et al.* measured an electron density of 10^{17} cm^{-3} for a filament via the transverse interferometry method but used a complicated procedure to extract a phase shift of only 0.03 rad phase shift at the noise level from 50 phase maps [25]. The electron density of the filament produced by the focusing or free propagation of the femtosecond laser in air is usually at the level of 10^{16} cm^{-3} or even lower [2]. For improving the measurement sensitivity, researchers have developed longitudinal spectral interferometry [18], longitudinal diffraction [19], and grazing-incidence interferometry [20]. These methods used probe light that propagates coaxially with the filament or crosses the filament at a very small angle (0.75°). Thus, the optical path of the probe pulse in the ionization zone becomes significantly longer; thus, a large phase shift can accumulate in the probe pulse. The sensitivity of such methods is two to three orders better than that of the transverse interference method. Moreover, grazing-incidence interferometry can retain an axial spatial resolution of approximately 5 mm.

11.3.3 Electrical conductivity methods

The plasma filament formed by the femtosecond laser pulse in air is electrically conductive. The conductivity of the filament is closely related to the electron density inside it. Therefore, the electron density can be determined by measuring the conductivity of the filament. The basic principle of the conductivity measurement is to bridge an originally disconnected high-voltage circuit loop by using the filament. An instantaneous current can be generated in the circuit owing to the EC of the filament, and the instantaneous current waveform can be recorded by an oscilloscope. The resistance of the filament can easily be obtained by applying Ohm's law to the circuit. Considering the nonmagnetic and steady-state conditions, for a weakly ionized filament, the current density can be obtained as follows:

$$J = \frac{n_e e^2}{m_e \nu_0} E, \tag{11.24}$$

where ν_0 is the collision frequency between free electrons and neutral molecules, and the conductivity of plasma is [29]

$$\sigma = \frac{n_e e^2}{m_e \nu_0}. \tag{11.25}$$

Under atmospheric pressure of 1 atm, the collision frequency ν_0 is approximately 10^{12} s^{-1} [29]. In this way, the electron density of the filament can be calculated using its conductivity.

The conductivity method described earlier measures the longitudinal resistance of the filament. Some researchers have used the transverse conductivity of the filament to determine the electron density [31], and this method has also achieved good results. In the setup of transverse conductivity method, two electrodes are located at two sides of the filament, not contacting with the filament. When the plasma filament is formed, the positive and negative electric charges are displaced between the electrodes under the electric field between the two electrodes, which produces an electric dipole moment, causing a disturbance to the original electric field and the instantaneous drop of the voltage between the electrodes. At this time, the power supply outputs the current to the circuit in order to maintain the voltage between the electrodes. The value of the current is a direct reflection of the electron density of the filament. However, the absolute value of the electron density cannot be obtained directly by transverse conductivity method. The absolute value of the electronic density should be obtained by using other independent methods to calibrate the electric signal.

11.3.4 Electromagnetic induction method

The standard EC measurement described in the previous subsection is easy and sensitive. In ideal conditions where the circuit has no response delay, the single-shot time-dependent electric signal can be considered the temporal evolution of the electron density of the filament. However, each circuit has its own response time,

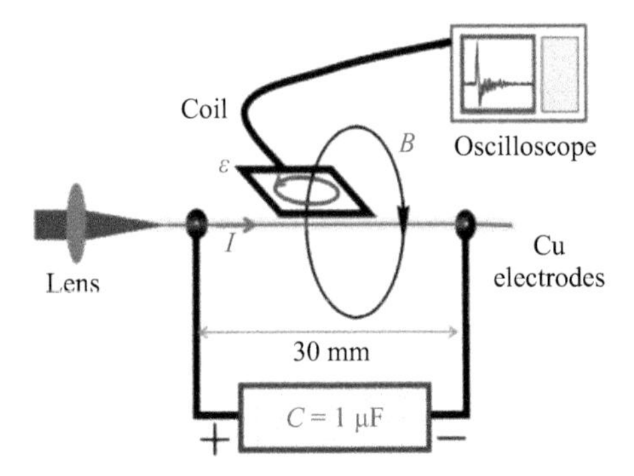

Figure 11.8 Setup for the EMI method [32]

relaxation, and self-oscillation characteristics that are dependent on the parameters of the circuit elements. Therefore, the theory based on the steady-state current may not be fully applicable, especially for the time characteristics of the electron density inside the filament. To obtain an accurate time-dependent electron density of the filament via electrical diagnosis, in 2017, Chen *et al.* reported the EMI method [32]. In the EMI method, the electron density of the filament is measured by non-contact detection of the transient magnetic field around the current in the filament using an induction coil. Owing to the successful decoupling of the self-oscillation in the circuit from the coil, a clean and reliable temporal evolution of the electron density inside the filament is obtained.

The experimental setup of the EMI method is shown in Figure 11.8. Two copper plane electrodes are fixed on the pins of a 1 μF capacitor, which is used as a voltage source. The gap between the two electrodes is 30 mm. Instead of directly measuring the current in the electrical circuit connected by the filament, the transient magnetic field induced by the electrified filament is detected by using a 10×10 mm square single-turn coil. The coil is made of 0.5 mm diameter insulated-coating Cu wires and is located 2.0 mm away from the laser filament, in the same plane. The electromotive force (EMF) signal detected by the coil is recorded by a 25 GS/s high-resolution oscilloscope (Tektronix DPO70804). The voltage loaded on the capacitor is fixed at 100 V.

Figure 11.9(a) presents the EMF signal ε at the strongest ionized part of the filament. Figure 11.9(b) shows the integrated curves of the EMF signal with respect to time, i.e., the transient magnetic flux ($\varphi = \iint BdS = \int \varepsilon dt$), which are proportional to the current and the electron density of the filament. The overall evolution trend of the electron density agrees well with previous results obtained using microwave diagnostics [26]. The lifetime of the filament can be estimated according to the width of the integral signal. When the integral signal decreases to an almost unchanged value (at 5–6 ns in our experiment), the lifetime of the filament is considered to be basically terminated.

The current in the filament can be calculated by using the Faraday EMI law and the Biot–Savart law. The peak plasma current of the strongest ionization part of

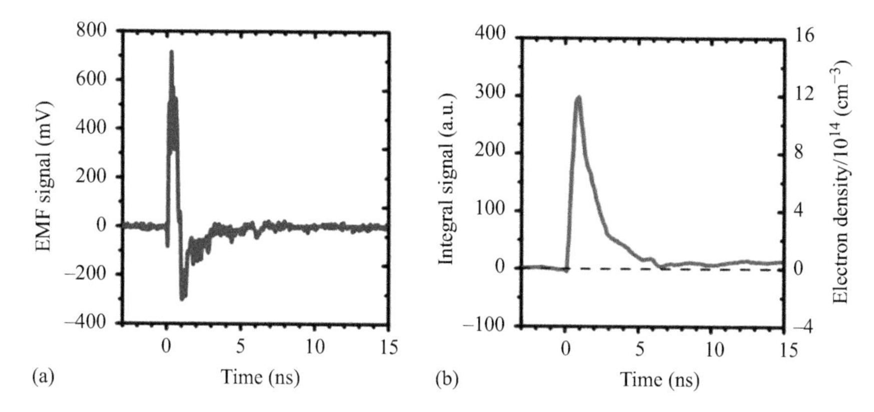

Figure 11.9 (a) EMF signal of the transient magnetic field around the filament; (b) the corresponding integral signal [32]

the filament is calculated to be 94 mA, which corresponds to a resistance of 1.05 kΩ. The typical radius of the filament under similar experimental conditions is approximately 50 μm. Then, the electron density is estimated to be approximately 1.2×10^{17} cm^{-3}. The whole temporal evolution of the electron density at different positions is calibrated by right vertical axis in Figure 11.9(b). The electron density of the filament at the same position is also measured using the standard EC method to be approximately 2.0×10^{16} cm^{-3}, which is less than that measured via the EMI method. In principle, the electron density obtained using the EMI method is closer to the true value because the EMF signal is directly from the filament. The main advantage of the EMI method is the reliable inversion of the time-dependent electron density inside the filament due to the successful decoupling of the self-oscillation and the interference in the circuit. The standard EC method usually gives a lower initial electron density and rough time characteristics of the filament.

11.4 Plasma density of different types of filament

The properties of the optical filament are closely related to the initial parameters of the incident laser pulse. In ref. [33], the characteristics of the optical filaments produced by laser pulses with different initial geometric focusing were systematically studied. The experimental results show that the electron density of the filament increases with the numerical aperture of the laser pulse. In this section, the plasma density in the filament produced by the tightly focused femtosecond pulses, freely propagating femtosecond laser pulses, the carrier envelope phase (CEP)-stabilized few cycle pulses, and the multi-pulse of different combinations are illustrated.

11.4.1 Plasma density of filament by tight focusing

It is known that the plasma defocusing effect limits the enhancement of the light intensity inside the filament. However, Théberge *et al.* found that the electron density

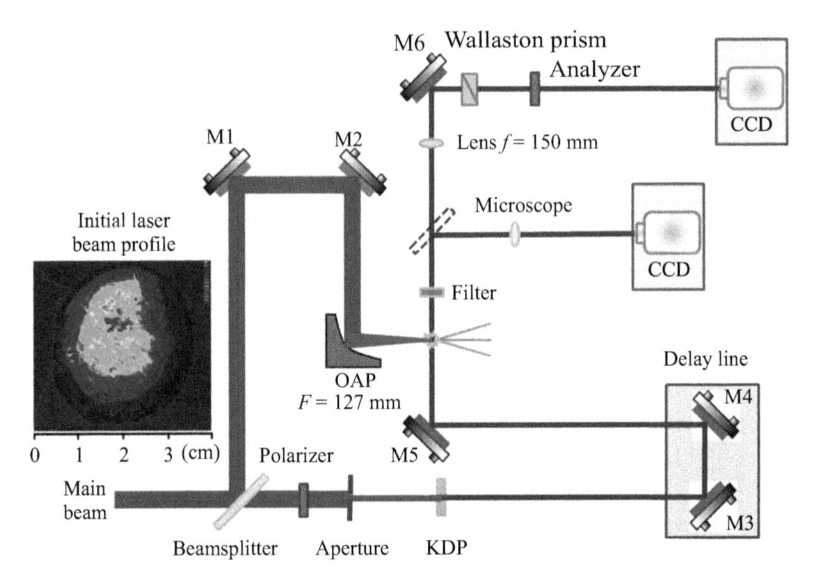

Figure 11.10 Experimental setup for the diagnostics of a tightly focused femtosecond laser pulse in air [9]

in the filament increases significantly with the increase of the focusing numerical aperture, while it is less sensitive to the incident laser power [33]. This raises a question about the filamentation under tight external focusing. Aiming at this problem, Kiran *et al.* studied the filamentation of tightly focused femtosecond lasers in air using side fluorescence imaging [34]. Liu *et al.* have also investigated in-depth the tight focusing propagation of femtosecond laser pulses in air [9]. They have used shadow imaging and transverse interference methods to diagnose the filaments. The detailed filamentation structure was obtained and the electron density of the air plasma was directly measured. The experimental setup used in [9] is shown in Figure 11.10.

The primary laser pulse with an initial beam width of approximately 30 mm and a pulse duration of 60 fs is divided into two parts after passing through the beam splitter. One part (approximately 70% of the total laser energy) is focused in the air through an off-axis parabolic mirror with a focal length of 127 mm. Figure 11.11(a) shows a shadow image of the air plasma produced by femtosecond lasers with different energies (2.4–47 mJ). As shown in Figure 11.11(a), the width of the filament bunch at the filamentation starting position increases with the increase of the laser energy. The light intensity at the filamentation starting point can be estimated according to the laser energy and the initial width of the filament bunch. The calculated laser intensity at the filamentation starting point is 2.5×10^{14} to 5×10^{14} W/cm^2, and the laser intensity at the filamentation starting point tends to be stable when the laser energy increases above 10 mJ. Figure 11.11(b) presents the interference fringes and the corresponding electron density distribution when the laser energy is 38 (above) and 47 mJ (below). As shown, the electron density at the beginning of the filament bunch is approximately 1×10^{19} cm^{-3}.

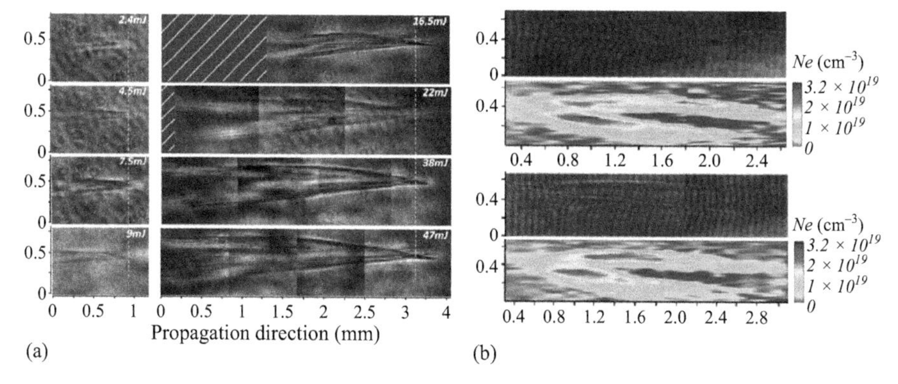

Figure 11.11 *(a) Shadow images of air plasma generated by femtosecond pulses with different energies; (b) interferograms and electron density distributions of air plasma generated by 38 mJ (above) and 47 mJ (below) femtosecond laser pulses [9]*

The electron density increases when the laser pulse approaches the focus. The maximum electron density in the filamentation region is 3×10^{19} to 5×10^{19} cm^{-3}, which is consistent with the numerical simulation results of Kiran *et al.* [34]. However, the maximum electron density and laser intensity did not increase with the increase of the laser energy; rather, they stabilized in a certain range. Therefore, it can be concluded that the intensity clamp effect plays a role under tight focusing.

11.4.2 Plasma density of freely propagating filament

The existing experiments show that if a high-power femtosecond laser pulse propagates freely in air (without initial focusing), a very long filament (up to kilometer range) can be formed, and the properties of such a long filament differ significantly from those of a filament produced by lens focusing. The multifilamentation of femtosecond laser beams in a range of 100 m was studied by Bergé *et al.* [35]. Later, some research groups studied the free filamentation in a significantly larger space. In 2005, Méchain *et al.* observed the filamentation structure of a freely propagating chirped pulse over a distance of 1 km, and bright channels in the beam were extended to 2,350 m [36]. In 2013, Durand *et al.* also studied the free filamentation of a TW femtosecond laser pulse over a 1 km range, and the ablation by the laser filament on a germanium surface was observed at a distance of 1 km [37]. These studies provide good references for the application of filaments over long distances.

Hao *et al.* experimentally compared the properties of the two types of filaments mentioned earlier: thin (100 μm), short (meters) filaments and thick (millimeters), long (tens of meters) filaments. They found that almost all the parameters of the two types of filaments were different and that the freely propagating femtosecond laser multifilament mode exhibited very unique properties [38]. In the experiment, the energy of the prefocused laser pulse was 22 mJ, and its geometric focal length

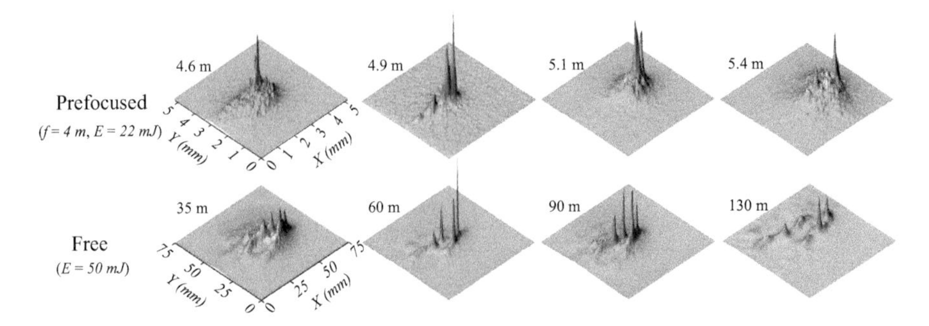

Figure 11.12 Filamentation patterns at different propagation distances. The upper row shows the case of the prefocused laser beam. The lower row shows the free-propagation case, and the pulse duration is 257 fs with a negative chirp. The two cases have different scales for the beam cross section [38]

was approximately 5.05 m. The laser energy used for the freely propagating filamentation was 50 mJ, and the pulse duration was adjusted to 257 fs with a negative chirp. The evolution of the multifilamentation was recorded via cross-sectional scattering spot imaging of the laser beam. Figure 11.12 shows the evolution of the two types of filamentation.

As shown in Figure 11.12, there are significant differences between the prefocused filamentation and the freely propagating filamentation, which can be summarized as follows.

1. Filament diameter (FWHM). The diameter of the prefocused filament is approximately 0.13 mm, while that of the free filament is approximately 1.6 mm.
2. Filament spacing. In prefocused filamentation, the relative distance between the filaments is generally only about 0.3 mm, which is approximately twice the diameter of the filaments. However, for freely propagating filamentation, the filaments are distributed over the entire cross-sectional area, and the distance between the filaments is on the order of centimeters, one order of magnitude larger than the diameter.
3. Length of the optical filament. The length of the prefocused filament is only a few meters. The length of the freely propagating filament can easily reach more than 100 m.
4. Energy background scale. In the case of prefocused filamentation, the diameter of the energy background is only a few millimeters, while in the case of freely propagating filamentation, the energy background is several centimeters. The differences in the energy distribution directly lead to the differences in their various properties.
5. Laser intensity and electron density in the filament. As shown in many references, the typical laser intensity and electron density in the prefocused filament

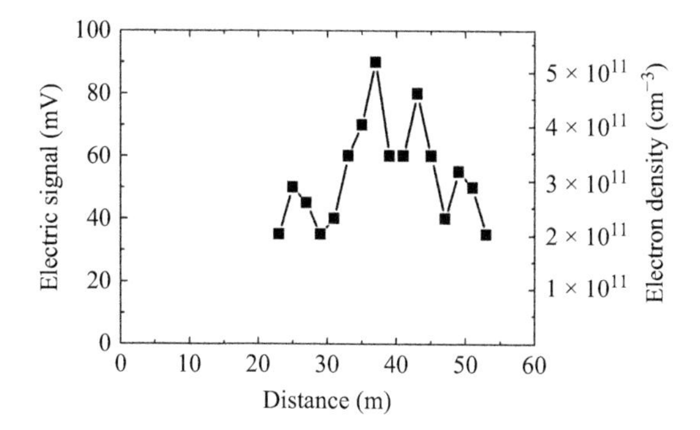

Figure 11.13 Peak electric signal (left vertical axis) and electron density (right vertical axis) of the filament with respect to the propagation distance [40]

are 10^{13}–10^{14} W/cm^2 and 10^{16}–10^{17} cm^{-3}, respectively. The laser energy contained in the core of the freely propagating filament is approximately 1 mJ. Therefore, assuming that the pulse width of the laser pulse in the filament remains unchanged, the laser intensity in the freely propagating filament can be estimated to be approximately 6×10^{11} W/cm^2, which agrees well with the results of Méchain *et al.* [39], where the electron density inside the freely propagating filament was theoretically predicted to be 10^{11}–10^{12} cm^{-3}.

In 2018, Chen *et al.* reported the quantitative experimental measurement of the electron density inside a freely propagating long filament by using the EC method [40]. In the experiment, a 55 mJ negative-chirped femtosecond pulse was employed to form long filaments. The total propagation distance of the laser pulse was approximately 53 m owing to the space limitation. The filamentation started at a propagation distance of 20 m. The electron density of the filament over 33 m was measured via the standard EC method.

Figure 11.13 shows the peak electric signal intensity (left vertical axis) and electron density (right vertical axis) along the plasma filament. The measured electron density inside this freely propagating filament was on the order of 10^{11} cm^{-3}, which agrees well with an earlier theoretical prediction [39]. Notably, the electron density detected via the EC method was slightly undervalued compared with that determined using the newly proposed EMI method [32], although the orders of magnitude were the same.

11.4.3 Dependence of plasma density on carrier envelope phase of few-cycle pulse

When the pulse duration approaches a few optical cycles, the peak electric field of the pulse becomes evidently dependent on the CEP, which may affect the filamentation process and the filament properties. In 2010, Laban *et al.* predicted that the length of the filament generated by 6.3 fs laser pulses in air should change by 790 μm when the CEP shifts by $\pi/2$ rad and performed a preliminary experimental

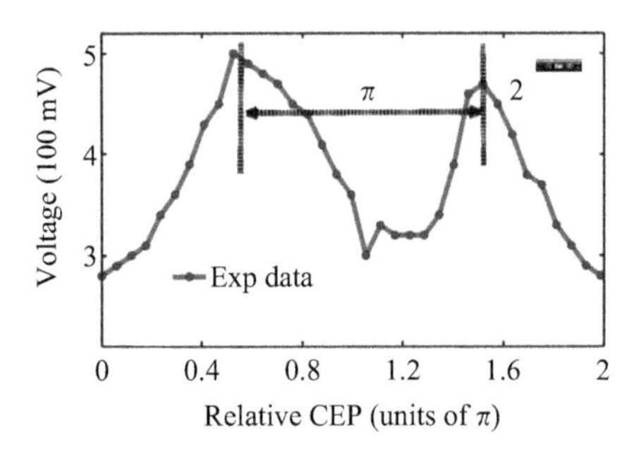

Figure 11.14 Electric signal of the filament as a function of the CEP of the driving laser pulses [8]

test [41]. The first obvious dependence of the plasma density in the filament on the CEP of a few-cycle-driven laser pulse was reported in 2016 by Wang *et al.* [8]. The following is a brief description of this work.

The CEP locked laser system delivers 1 kHz, 0.25 mJ pulses with duration below 7 fs, which include only about three light cycles. The 7 fs laser pulses were focused in air with a 500 mm focal length concave silver mirror to form a filament (plasma channel), and the electron density of the filament was determined using the standard EC method.

Figure 11.14 shows the electric signal of the filament as a function of the CEP of the driving laser pulses. The period of the voltage modulation was π, which is reasonable because the phase shift π leads to exactly the same waveform shape with the reverse direction of the electric field. The modulation depth of the voltage signal was 23.7%. However, when the CEP was not locked, the voltage signal almost remained constant. These experimental results indicate that the CEP can significantly affect the plasma density of the filament when a few-cycle pulse is used.

11.4.4 Prolonging lifetime of plasma filament using supplemental pulse

Some applications of plasma filaments, such as laser lightning [42] and long-distance transmission of electromagnetic energy [43], have higher requirements for the lifetime. However, the lifetime of plasma filaments produced by a single-femtosecond pulse is only nanoseconds [28], which means that the length of the ionized area in the air is only meters. Therefore, prolonging the lifetime of plasma filaments has become a very important research topic. To maintain the electron density of the filament, energy must be continuously injected into the filament to prevent or delay the electron recombination and attachment. Researchers actively attempt this via two main methods. One is to use a single femtosecond laser to produce a filament and then use a long laser pulse to maintain the filament. The other is to use a series of femtosecond pulses to refresh the filament over a long timescale.

The use of a long-pulse laser as a supplemental pulse to prolong the lifetime of filaments was first attempted by Zhao *et al.* in 1995 [10], but the experiment did not measure the lifetime of the filament. Later, Hao *et al.* directly measured the lifetime

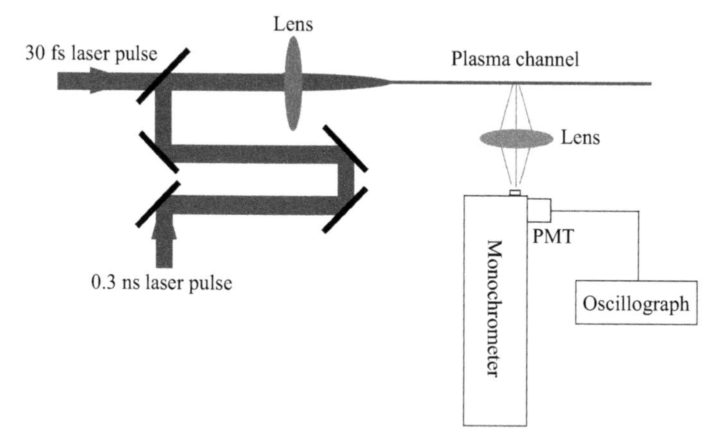

Figure 11.15 Experimental setup of prolonging the filament lifetime by using a sub-nanosecond supplemental pulse [44]

of filament that was produced by an 800 nm femtosecond laser pulse and maintained by a sub-nanosecond laser pulse [44].

The experimental setup of the femtosecond + nanosecond configuration in [44] is presented in Figure 11.15. The main femtosecond laser pulse is focused by a convex lens ($f = 35$ cm) to form a long plasma filament in the air. The femtosecond laser pulse energy is 38 mJ. The subsequent laser is separated from the amplified chirped pulse before the compressor with a duration of approximately 0.3 ns. The subsequent laser passes a delay system to obtain a 10 ns delay relative to the femtosecond pulse. The lifetime of the filaments is reflected by the time evolution of the fluorescence spectral line of nitrogen ions at 504.5 nm (N^+: $3p(^3S)$–$3s(^3P^0)$). The fluorescence signal produced by the filament is collected into a monochromator by a lens. The characteristic spectral line is converted into an electrical signal by a photomultiplier tube and monitored by a digital oscilloscope.

Figure 11.16 shows the time-dependent intensity of the characteristic spectral line in the experiment. The femtosecond laser energy is 38 mJ, and the sub-nanosecond assisted laser pulse energy is 17 mJ (solid line). The dotted line shows the filament lifetime signal without a subsequent sub-nanosecond laser pulse. The lifetime of the single-pulse laser filament is 12 ns at half height and 40 ns at the bottom of the signal. (The signal level at the bottom is 5% of the peak.) When the subsequent sub-nanosecond laser pulse is injected into the filament, the lifetime signal is 26 ns at half height, and the tail of the signal is over 200 ns. The lifetime of the filament is prolonged significantly, and the level of the signal is enhanced by a factor of five compared with the single-femtosecond pulse filament.

11.4.5 Long-lifetime plasma filament produced by femtosecond pulse sequence

Although a long supplemental pulse can prolong the lifetime of filament as mentioned in the previous section, but the effectiveness of this method is very limited

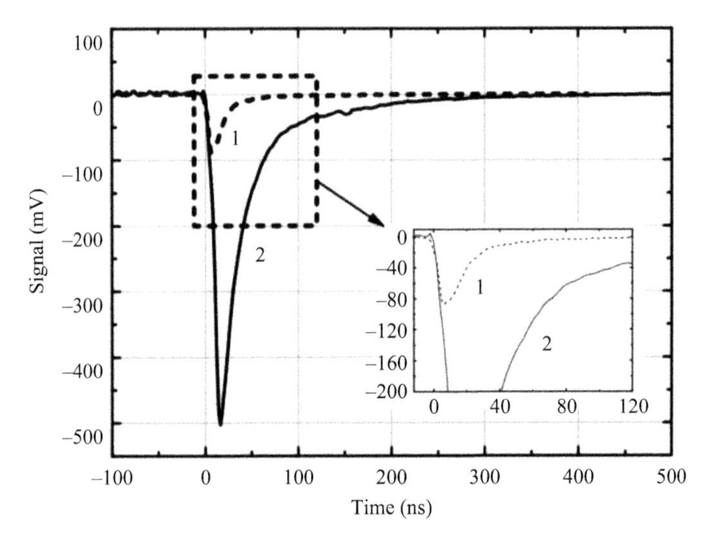

Figure 11.16 Typical fluorescence signals recorded by a digital oscilloscope. The dotted curve represents the free decay of a single- pulse filament; the solid curve represents the temporal evolution of the plasma with a supplemental sub-nanosecond laser pulse after the femtosecond laser pulse [44]

when the filament is very long. Because it is generally difficult to realize the self-guided propagation using a long laser pulse; thus, the filament lifetime cannot be maintained over a long distance. Another way to produce a plasma channel with a long lifetime is using an ultrashort laser pulse sequence to produce temporally separated but spatially overlapping filaments. With this method, the lifetime of the filament was doubled by using dual femtosecond laser pulses [45]. Ji *et al.* used the "leaking" pulses from a regenerative amplifier as a multi-pulse seed and produced a sequence of approximately six pulses, prolonging the lifetime of the plasma channel by a factor of 4.5 [46]. Later, Ionin *et al.* generated a train of picosecond UV pulses overlapped with a long UV pulse by using a hybrid Ti:sapphire-KrF laser facility, where the pulse train seed was obtained via multi-reflection of a UV femtosecond laser pulse in a ring cavity [47,48]. A femtosecond pulse sequence at an 800 nm central wavelength was also produced, via pure multi-pass amplification of the "natural" pulse sequence by a 70 MHz commercial femtosecond oscillator [49]. However, the interval between pulses (14 ns) was significantly longer than the channel lifetime.

To achieve a breakthrough in the quality of the femtosecond pulse sequence, the interval between pulses must be reduced to a few nanoseconds, and the number of amplified pulses needs to be increased further. In 2015, Lu *et al.* used the output from a homemade 350 MHz repetition rate (corresponding to an interval of only 2.9 ns) femtosecond oscillator as the seed and successfully generated an amplified pulse sequence with uniformed distribution of energy [50]. Figure 11.17 shows the

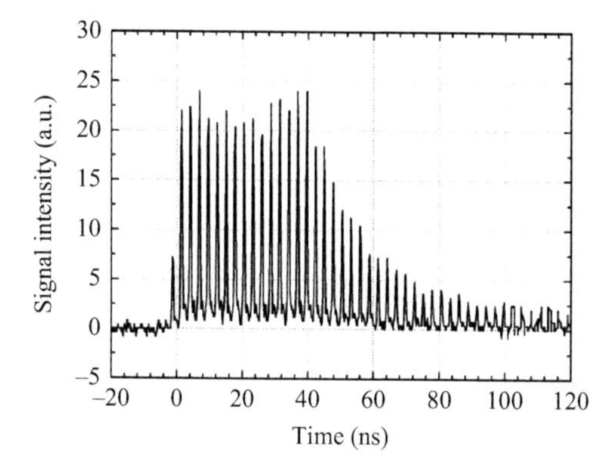

Figure 11.17 Photodiode signal of the femtosecond pulse sequence [50]

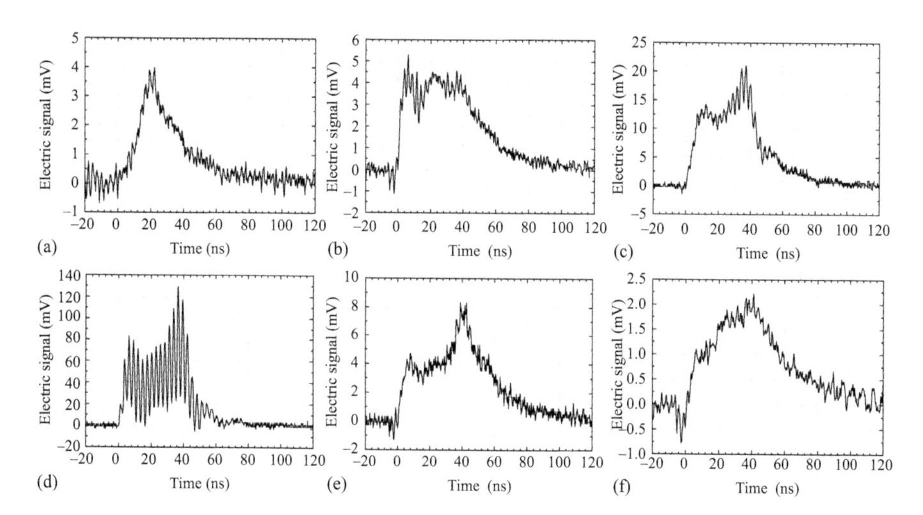

Figure 11.18 Electric signal of the plasma channel under f = 2 m external focusing at different distances: (a) z = 0 cm, (b) z = 5 cm, (c) z = 10 cm, (d) z = 20 cm, (e) z = 25 cm, and (f) z = 35 cm [50]

photodiode signal of the pulse sequence. The total energy of pulse sequence was 70 mJ, and the energy of a single pulse in the plateau was approximately 3 mJ.

The pulse sequence was focused by a lens in air to generate a filament with a long lifetime. The properties of the filament were diagnosed by standard EC method. Figure 11.18 shows typical electric signals of plasma channels generated using the $f = 2$ m convex lens at different positions on the plasma channel: $z = 0$, 5, 10, 20, 25, and 35 cm. The distance of 0 cm corresponds to the starting point of the plasma channel, where the electric signal can hardly be detected. The peak signal intensity varied from several millivolts to more than 100 mV for different propagation distances, but the temporal width of the signal always remained in the 80 ns range (bottom width) owing to the improved uniformity of the pulse sequence.

As shown in Figure 11.18, the lifetime of the plasma channel generated by such a pulse sequence was prolonged up to 60–80 ns (bottom width), while the length of the filament was simultaneously maintained at the scale of meters. Currently, the output energy of femtosecond laser systems can reach joules and even tens of joules, which can support the sequencing of hundreds of pulses with a 100-mJ level single-pulse energy. Therefore, in principle, there is no limitation for generating kilometer-range plasma channels with microsecond-level lifetimes, which are useful for many applications, such as lightning control and remote transportation of radio frequency energy.

References

[1] Keldysh L. V., 'Ionization in the field of a strong electromagnetic wave'. *Soviet Physics Journal of Experimental and Theoretical Physics*. 1965; 20(5): 1307–1314.

[2] Couairon A. and Mysyrowicz A., 'Femtosecond filamentation in transparent media'. *Physics Reports*. 2007; 441(2–4): 47–189.

[3] Perelomov A. M., Popov V. S., and Terent'ev M. V., 'Ionization of atoms in an alternating electric field'. *Soviet Physics Journal of Experimental and Theoretical Physics*. 1966; 23(5): 924–934.

[4] Ammosov M. V., Delone N. B., and Krainov V. P., 'Tunnel ionization of complex atoms and of atomic ions in an alternating electromagnetic field'. *Soviet Physics Journal of Experimental and Theoretical Physics*. 1986; 64(6): 1191–1194.

[5] Talebpour A., Yang J., and Chin S. L., 'Semi-empirical model for the rate of tunnel ionization of N_2 and O_2 molecule in an intense Ti:sapphire laser pulse'. *Optics Communications*. 1999; 163(1–3): 29–32.

[6] Couairon A., Tzortzakis S., Berge L., Franco M., Prade B., and Mysyrowicz A., 'Infrared femtosecond light filaments in air: simulations and experiments'. *Journal of the Optical Society of America B*. 2002; 19(5): 1117–1131.

[7] Priori E., Cerullo G., Nisoli M. *et al.*, 'Nonadiabatic three-dimensional model of high-order harmonic generation in the few-optical-cycle regime'. *Physical Review A*. 2000; 61(6): 063801.

[8] Wang L. F., Lu X., Teng H. *et al.*, 'Carrier-envelope phase-dependent electronic conductivity in an air filament driven by few-cycle laser pulses'. *Physical Review A*. 2016; 94(1): 013827.

[9] Liu X. L., Lu X., Liu X. *et al.*, 'Tightly focused femtosecond laser pulse in air: from filamentation to breakdown'. *Optics Express*. 2010; 18(25): 26007–26017.

[10] Zhao X. M., Diels J. C., Wang C. Y., and Elizondo J. M., 'Femtosecond ultraviolet laser pulse induced lightning discharges in gases'. *IEEE Journal of Quantum Electron*. 1995; 31(3): 599–612.

[11] Lu X., Xi T. T., Li Y. J., and Zhang J., 'Lifetime of the plasma channel produced by ultra-short and ultra-high power laser pulse in the air'. *Acta Physica Sinica*. 2004; 53(10): 3404–3408.

[12] Xi T. T., Lu X., and Zhang J., 'Interaction of light filaments generated by femtosecond laser pulses in air'. *Physical Review Letters*. 2006; 96(2): 025003.

[13] Xi T. T., Lu X., and Zhang J., 'Enhancement of third harmonic emission by interaction of two colored filament with droplet in air'. *Optics Communications*. 2009; 282(15): 3140–3143.

[14] Feng L. B., Lu X., Xi T. T. *et al.*, 'Numerical studies of third-harmonic generation in laser filament in air perturbed by plasma spot'. *Physics of Plasmas*. 2012; 19(7): 072305.

[15] Ma Y. Y., Lu X., Xi T. T., Gong Q. H., and Zhang J., 'Widening of Long-range femtosecond laser filaments in turbulent air'. *Optics Express*. 2008; 16(12): 8332–8341.

[16] Xi T. T., Lu X., and Zhang J., 'Spatiotemporal moving focus of long femtosecond-laser filaments in air'. *Physical Review E*. 2008; 78(5): 055401.

[17] Yang H., Zhang J., Li Y. J. *et al.*, 'Characteristics of self-guided laser plasma channels generated by femtosecond laser pulses in air'. *Physical Review E*. 2002; 66(1): 016406.

[18] Fontaine B. La, Vidal F., Jiang Z. *et al.*, 'Filamentation of ultrashort pulse laser beams resulting from their propagation over long distances in air'. *Physics of Plasmas*. 1999; 6(5): 1615–1621.

[19] Liu J. S., Duan Z. L., Zeng Z. N. *et al.*, 'Time-resolved investigation of low-density plasma channels produced by a kilohertz femtosecond laser in air'. *Physical Review E*. 2005; 72(2): 026412.

[20] Chen Y. H., Varma S., Antonsen T. M., and Milchberg H. M., 'Direct measurement of the electron density of extended femtosecond laser pulse-induced filaments'. *Physical Review Letters*. 2010; 105(2): 215005.

[21] Yu J., Mondelain D., Kasparian J. *et al.*, 'Sonographic probing of laser filaments in air'. *Applied Optics*. 2003; 42(36): 7117–7120.

[22] Yu J., Zhang C. Z., Hao Z. Q. *et al.*, 'Fluorescence measurement and acoustic diagnostics of plasma channels in air'. *Acta Physica Sinica*. 2006; 55(1): 299–303.

[23] Liu W., Bernhardt J., Théberge F., Chin S. L., Châteauneuf M., and Dubois J., 'Spectroscopic characterization of femtosecond laser filament in argon gas'. *Journal of Applied Physics*. 2007; 102(3): 033111.

[24] Bernhardt J., Liu W., Théberge F. *et al.*, 'Spectroscopic analysis of femto-second laser plasma filament in air'. *Optics Communications*. 2008; 281(5): 1268–1274.

[25] Bodrov S., Bukin V., Tsarev M. *et al.*, 'Plasma filament investigation by transverse optical interferometry and terahertz scattering'. *Optics Express*. 2011; 19(7): 6829–6835.

[26] Papeer J., Mitchell C., Penano J., Ehrlich Y., Sprangle P., and Zigler A., 'Microwave diagnostics of femtosecond laser-generated plasma filaments'. *Applied Physics Letters*. 2011; 99(14): 141503.

[27] Schillinger H. and Sauerbrey R., 'Electrical conductivity of long plasma channels in air generated by self-guided femtosecond laser pulses'. *Applied Physics B*. 1999; 68(4): 753–756.
[28] Tzortzakis S., Prade B., Franco M., and Mysyrowicz A., 'Time-evolution of the plasma channel at the trail of a self-guided IR femtosecond laser pulse in air'. *Optics Communications*. 2000; 181(1–3): 123–127.
[29] Ladouceur H. D., Baronavski A. P., Lohrmann D., Grounds P. W., and Girardi P. G., 'Electrical conductivity of a femtosecond laser generated plasma channel in air'. *Optics Communications*. 2001; 189(1–3): 107–111.
[30] Zhang Z., Zhang J., Li Y. T. *et al.*, 'Measurements of electric resistivity of plasma channels in AI'. *Acta Physica Sinica*. 2006; 55(01): 357–361.
[31] Abdollahpour D., Suntsov S., Papazoglou D. G., and Tzortzakis S., 'Measuring easily electron plasma densities in gases produced by ultrashort lasers and filaments'. *Optics Express*. 2011; 19(18): 16866–16871.
[32] Chen S. Y., Liu X. L., Lu X. *et al.*, 'Temporal evolution of femtosecond laser filament detected via magnetic field around plasma current'. *Optics Express*. 2017; 25(26): 32514–32521.
[33] Théberge F., Liu W. W., Simard P. T., Becker A., and Chin S. L., 'Plasma density inside a femtosecond laser filament in air: Strong dependence on external focusing'. *Physical Review E*. 2006; 74(3): 036406.
[34] Kiran P. P., Bagchi S., Krishnan S. R., Arnold C. L., Kumar G. R., and Couairon A., 'Focal dynamics of multiple filaments: Microscopic imaging and reconstruction'. *Physical Review A*. 2010; 82(1): 013805.
[35] Bergé L., Skupin S., Lederer F. *et al.*, 'Multiple filamentation of terawatt laser pulses in air'. *Physical Review Letters*. 2004; 92(22): 225002.
[36] Méchain G., D'Amico C., André Y.-B. *et al.*, 'Range of plasma filaments created in air by a multi-terawatt femtosecond laser'. *Optics Communications*. 2005; 247(1–3): 171–180.
[37] Durand M., Houard A., Prade B. *et al.*, 'Kilometer range filamentation'. *Optics Express*. 2013; 21(22): 26836–26845.
[38] Hao Z. Q., Zhang J., Zhang Z. *et al.*, 'Characteristics of multiple filaments generated by femtosecond laser pulses in air: Prefocused versus free propagation'. *Physical Review E*. 2006; 74(6): 066402.
[39] Méchain G., Couairon A., André Y.-B., D'Amico C., Franco M., and Prade B., 'Long-range self-channeling of infrared laser pulses in air: a new propagation regime without ionization'. *Applied Physics B*. 2004; 79(3): 379–382.
[40] Chen S. Y., Teng H., Lu X. *et al.*, 'Properties of long light filaments in natural environment'. *Chinese Physics B*. 2018; 27(8): 085203.
[41] Laban D. E., Wallace W. C., Glover R. D., Sang R. T., and Kielpinski D., 'Self-focusing in air with phase-stabilized few-cycle light pulses'. *Optics Letters*. 2010; 35(10): 1653–1655.
[42] Ball L. M., 'The laser lightning rod system: thunderstorm domestication'. *Applied Optics*. 1974; 13(10): 2292–2296.

[43] Châteauneuf M., Payeur S., Dubois J., and Kieffer J.-C., 'Microwave guiding in air by a cylindrical filament array waveguide'. *Applied Physics Letters*. 2008; 92(9): 091104.

[44] Hao Z. Q., Zhang J., Li Y. T. *et al.*, 'Prolongation of the fluorescence lifetime of plasma channels in air induced by femtosecond laser pulses'. *Applied Physics B*. 2005; 80(4–5): 627–630.

[45] Zhang Z., Lu X., Liang W. X. *et al.*, 'Triggering and guiding HV discharge in air by filamentation of single and dual fs pulses'. *Optics Express*. 2009; 17(5): 3461–3468.

[46] Ji Z. G, Zhu J. B., Wang Z. X. *et al.*, 'Low resistance and long lifetime plasma channel generated by filamentation of femtosecond laser pulses in air'. *Plasma Science and Technology*. 2010; 12(3): 295–299.

[47] Ionin A. A., 'High-power IR- and UV-laser systems and their applications'. *Physics-Uspekhi*. 2012; 55(7): 721–738.

[48] Ionin A. A., Kudryashov S. I., Levchenko A. O., Seleznev L. V., Shutov A. V., and Sinitsyn D. V., 'Triggering and guiding electric discharge by a train of ultraviolet picosecond pulses combined with a long ultraviolet pulse'. *Applied Physics Letters*. 2012; 100(10): 104105.

[49] Liu X. L., Lu X., Ma J. L. *et al.*, 'Long lifetime air plasma channel generated by femtosecond laser pulse sequence'. *Optics Express*. 2012; 20(6): 5968–5973.

[50] Lu X., Chen S. Y., Ma J. L. *et al.*, 'Quasi-steady-state air plasma channel produced by a femtosecond laser pulse sequence'. *Scientific Reports*. 2015; 5: 15515.

Chapter 12

Filamentation for atmospheric remote sensing and control

Jérôme Kasparian[1] *and Jean-Pierre Wolf*[1]

12.1 Introduction

Laser filamentation is especially appealing for atmospheric applications, due to its unique ability to propagate over long distances in the atmosphere. The high intensity conveyed remotely by filaments is particularly useful to induce non-linear laser–matter interaction for spectroscopy, like light detection and ranging (Lidar) remote sensing of traces in the air. Furthermore, as filaments can remotely trigger multiphoton photochemistry and ionization, they allow one to initiate various physical and chemical processes, including the manipulation of the transmission through fog and clouds, or the control of high-voltage discharges. In this chapter, we review such applications successively.

12.2 Remote sensing using femtosecond filamentation

12.2.1 Introduction

Lidar [1] is a versatile and powerful analysis method not only for three-dimensional range measurements, but also for remotely sensing the atmosphere composition. A pulsed laser is emitted into the atmosphere to be analysed, and the backscattered light is collected, time-resolved, on a telescope. For each time of flight, i.e., for each distance *d*, the collected light depends on both the backscattering efficiency of the atmosphere at distance *d* and its transmission integrated over the whole path between the scattering and detection location. As both depend on the spatially resolved trace gas concentration and aerosol density, size distribution, shape, and composition, inverting the Lidar signal provides rich knowledge on the atmospheric composition.

The main limitation of the Lidar analysis of pollutants stems from the need to tune the excitation or illumination wavelength, as well as the detection system, on the absorption or emission lines of the species under investigation. Such

[1]University of Geneva, Geneva, Switzerland

requirement implies a priori knowledge of this target species. However, in situations like industrial accidents or leaks of complex mixtures of pollutants, the primary need resides in identifying a priori unknown species or mixtures. A similar difficulty arises for the characterization of atmospheric aerosols which display various shapes, sizes, and compositions (hence, refractive indices). All these parameters need to be analysed to reach a full characterization, since they play a key role in the particle effect on public health [2], their atmospheric reactivity [3], or the radiative balance of the atmosphere [4].

Overcoming these limitations requires multichannel Lidar with either fast wavelength switching capability or simultaneous broadband light illumination and/or detection. Furthermore, non-linear interaction with the air mass under scrutiny may also enrich the information contents of the Lidar signal.

As detailed in the previous chapters, and unlike nanosecond-pulsed lasers used in general for Lidars, femtosecond pulses feature several unique properties that are highly beneficial to Lidar. These properties include the following:

- The generation of a broadband supercontinuum, offering a 'white laser' source.
- Peak powers up to the TW level, for an average power of a few watts only, allowing highly non-linear interaction processes with the target air masses and opening unprecedented information channels.
- Femtosecond pulses have spatial extensions in the micrometre range, allowing an unprecedented size resolution in particle analysis.
- Non-linear processes can induce enhanced backward emission, improving the signal collection.
- The broad bandwidth of ultrashort pulses allows pulse shaping by spectral phase and intensity tailoring, offering a new dimension in the wealth of the available information.

Besides, above the critical power, ultrashort laser pulses give rise to filamentation [5–9]. This highly robust propagation mode can withstand the perturbations brought by the atmosphere, such as turbulence [10,11] or temperature and pressure gradients [12]. Filaments self-heal after hitting an obscurant like a cloud particle, allowing them to propagate unaffected even through clouds [13–15] as long as the photon bath surrounding them, and feeding them with energy is not fully attenuated. Furthermore, filaments can be generated at km-scale distances [16,17] and propagate over hundreds of metres [18], i.e., on distances consistent with the atmospheric scales. Together, these properties designate laser filaments as good candidates for atmospheric applications, among which remote sensing holds a key position. In this section, we shall successively review the advantages opened for remote sensing by the various properties of ultrashort laser pulses and the associated filamentation.

12.2.2 Geometry of backward emission and signal enhancement

The self-guiding at the root of filamentation substantially affects the laser beam geometry, as compared with the regularly diverging beam typical of linear propagation. This geometry affects the overlap term in the Lidar equation, mostly in the direction of a more efficient signal collection. Such enhanced efficiency historically belonged to the first motivations to switch to filament-based femtosecond Lidar, with the hope to increase the fraction of light emitted in the backward direction.

In the case of filamentation, the emission geometry and the associated illuminated volume depends on a convolution [16] between the spatial distribution of the filaments in the beam profile and the wavelength-dependent aperture angle of the white-light emission, or conical emission [19–21]. The impact of this complex geometry on the Lidar equation has been investigated in detail in [22]. In short, by reducing the incident beam divergence, the non-linearity tends to concentrate the light within the field of view of the receiving telescope, therefore increasing the overlap term, hence the Lidar signal.

A second impact of the non-linear propagation on the Lidar equation regards the propagation of the backscattered signal to the detector. While most of the elastic processes have more or less isotropic emission, the non-linearity of the filamentary propagation induces an increase of the white light emission in the backward direction, resulting in a factor of ~2 in efficiency as compared with elastic Rayleigh scattering [23]. Obviously, this directional backward emission is favourable to the Lidar detection. The origin of this backward enhancement has not been fully elucidated to date. Among the interpretations under discussion, one may mention a partial reflection of the light on a self-generated Bragg grating due to local variations of the electronic density [23], phase conjugation [24], backward stimulated Raman [25], Brillouin scattering [26], or guiding by the transient waveguide induced by the local heating of the air and the associated shockwave [27].

Finally, even more directional emission as well as chemical selectivity may be achieved by backward amplified stimulated emission (ASE), also known as backward lasing, of atmospheric species in the filament itself. The excited nitrogen molecules in the C $^3\Pi_u$ excited state generated by the recombination of free electrons left behind by the laser pulse with N_2^+ ions give rise to a population inversion, inducing ASE at 357 [28] and 337 nm [29]. A second pulse, or pulse train, heating the plasma and/or increasing its density via avalanche ionization, has been proposed to enhance this effect [30,31] and allowed oxygen ASE at 226 nm [32]. Besides offering a highly directional emission towards the detection system, this process has been proposed as an active diagnostic process in highly polluted atmospheres, although no demonstration at atmospheric relevant concentrations has been provided to date.

12.2.3 White-light Lidar

One of the most spectacular properties of ultrashort pulses is their ability to generate a broadband, or 'white-light' supercontinuum when propagating in air (Figure 12.1(a)). This supercontinuum originates from self-phase modulation [33], i.e., the distortion of the pulse due to the refractive index change associated with high local intensities. It covers the whole spectrum from 230 nm in the ultraviolet [34] up to 4.5 μm [35], or even 14 μm in the infrared [36]. It, therefore,

Figure 12.1 White-light emission from laser filaments: (a) true-colour image of a white-light filament in air [39], (b) visible-infrared spectrum of the white-light continuum produced by laser filaments. The absorption bands of the main atmospheric trace species are superimposed on the spectrum [38]

encompasses the absorption spectrum of most of the atmospheric trace species, including methane (1.7 μm), volatile organic compounds (VOCs) (3–3.5 μm), and water vapour (~820 nm) (Figure 12.1(b)). Similarly, its extension in the UV was measured down to 230 nm, i.e., over the full atmospheric window [34]. It covers atmospheric trace gases like SO_2, NO, NO_2, or ozone (Figure 12.1(b)). The white-light supercontinuum, therefore, provides a versatile source for the remote sensing of many atmospheric trace gases, including the analysis of mixtures and/or unknown gases. In spite of its broadband, the supercontinuum has been shown to be coherent, a property that led to designate it as a white-light laser [37].

Such broadband, coherent light source is extremely attractive to Lidar. It can be combined with a spectrally resolved detection to offer a multispectral source that may allow simultaneous multispecies and/or multiparameter Lidar. It attracted interest as soon as 1997, with the first launch of 110-fs, 220-mJ (2-TW) pulses into the atmosphere. Although emitted it in the near-infrared, it yielded a remarkably bright white-light channel (Figure 12.1(a)) [39] with a spectrum ranging over a 400-nm bandwidth that allowed one to record range-resolved atmospheric spectra [40]. Subsequent tries at even higher incident powers, up to 30 J [41] and 100 TW [42], showed that laser filamentation still occurs in such extreme conditions. Furthermore a white-light Lidar signal could be recorded from up to 20-km distance [41], supported by a 30% conversion efficiency from the laser incident light into the continuum [42,43]. Such achievements illustrate the long-range capability of a femtosecond-based white-light Lidar.

The first demonstration of femtosecond Lidar [39,40] raised a huge interest that motivated the construction of a dedicated system, the *Teramobile* laser [44]. For the first time, a femtosecond-terawatt laser was adapted to a compact design allowing its installation into a standard 20-ft (6 m) freight container equipped as a mobile laboratory. It provided 4-TW pulses in 100 fs, at a repetition rate of 10 Hz, centred at 800 nm. Furthermore, it was equipped with a full Lidar detection system, including a 40-cm Newtonian telescope with a focal length of 1.2 m, limited by the space available for detection in the container. Various detection systems were available from photomultipliers equipped with bandpass filters, spectrometers with photomultipliers or gated intensified CCD, and infrared detectors like specific photomultiplier (Hamamatsu R 5509-72) or InSb diodes. Such a new type of facility opened the way to field experiments based on femtosecond-terawatt lasers on a routine basis.

As such, the *Teramobile* system allowed a variety of atmospheric experiments, like the simultaneous measurement of O_2 and H_2O [46], at an altitude of up to 5 km, using an Echelle spectrograph and a CCD camera on a 2-m Coudé telescope (Figure 12.2). It was then applied to determine the size distribution of atmospheric

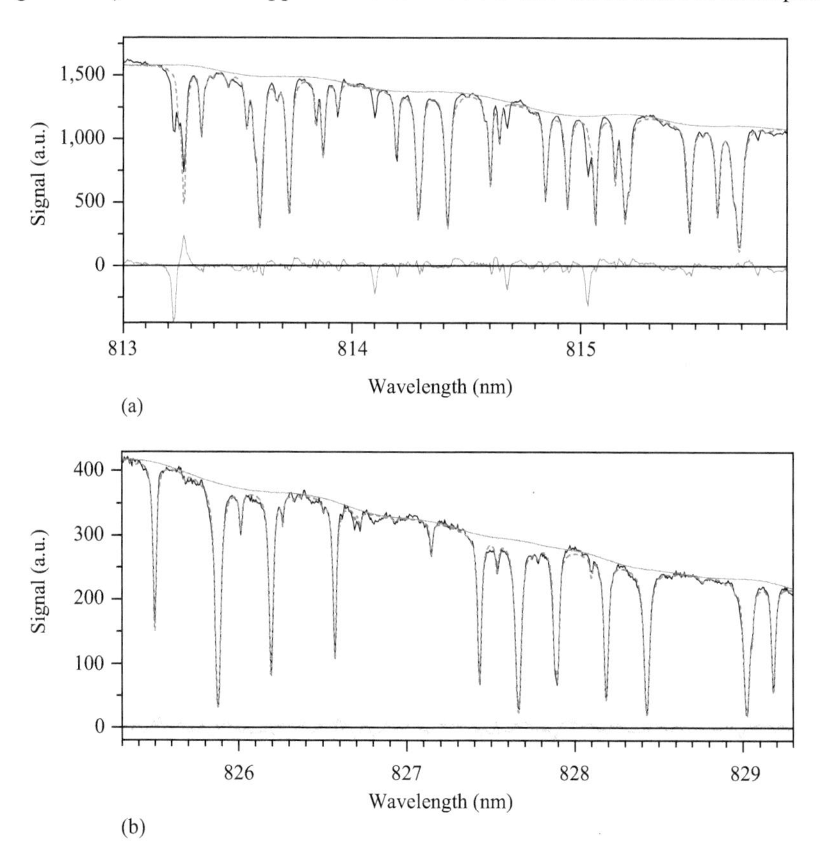

Figure 12.2 Fit of the water absorption spectrum in the 4v-overtone band in the 813–816-nm (a) and 825–829-nm (b) regions. Black solid line: measurement; grey dotted line: fit; grey solid line near zero: residuum of the fit. The smooth line is the estimation of the baseline. Note that the extremely long absorption path allows one to measure weak lines not tabulated in the HITRAN 2000 database [45,46]

aerosols in 14 size classes, by combining multi-FOV and multispectral Lidar measurements [46]. Finally, a Lidar signal in the mid-IR, up to 1.7 μm, was detected, sowing the applicability and providing an estimation of the yield of the white light in this spectral region [47].

In parallel, Galvez *et al.* [48–50] introduced a complementary approach, in which the white light was generated at the laser exit by self-phase modulation in a 9-m long cell filled with 1-bar krypton gas and launched from the ground as a white-light laser. Associated with a multichannel detection set-up, they were able to perform three-channel depolarization Lidar and estimate size distributions of aerosol particles in the atmosphere (Figure 12.3). However, their set-up was limited to three wavelengths, restricting the retrieved size distribution to the same amount of classes. The main advantage of this approach is the simplicity of the signal analysis, since the beam and detection geometry are similar to those of classical Lidar and do not require to know the filamentation altitude, unlike in the case of filament-based Lidar to take the proper geometry into account [22].

12.2.4 High power

In white-light Lidar, non-linearity resides in the generation of the white super-continuum, i.e., in the light source. The interaction of the light with the air mass or the species to be analysed is, however, linear. In this section, we discuss a deeper use of the non-linearity, with techniques where the non-linear interaction occurs between the light and the atmospheric constituents to be analysed themselves.

The first demonstration of such non-linear interaction was based on multiphoton-excited fluorescence (MPEF), which allowed detecting and

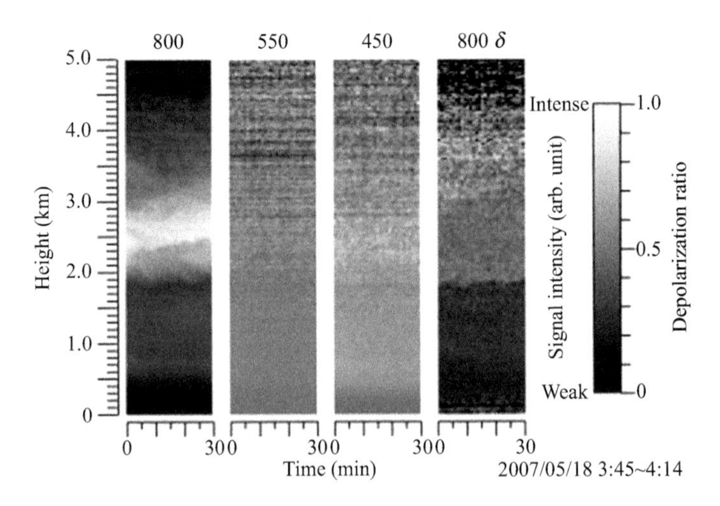

Figure 12.3 Range squared corrected backscattered power (arbitrary units) and depolarization ratio with white-light Lidar system during an Asian dust episode in Osaka [49]

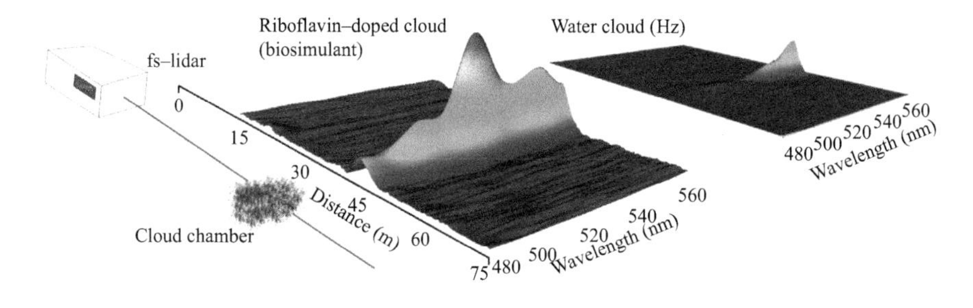

Figure 12.4 Remote detection and identification of bioaerosols: range- and spectrally resolved backward-emitted two-photon-excited fluorescence (2PEF) of riboflavin-doped (simulating bioaerosols, left) and undoped (simulating haze, right) water droplets [51]

identifying, at a distance of 50 m, riboflavin-doped water droplets acting as bacteria simulants (Figure 12.4) [51]. Such approach takes benefit of the highly directional backward MPEF emission from microparticles [52]. It also allows using near-infrared excitation light instead of the ultraviolet light required for single-photon-excited fluorescence. Since Rayleigh scattering decreases with the fourth power of the wavelength, the near-infrared radiation is 16 times less attenuated than 400-nm light, and 81 times less than 270-nm UV radiation, which would furthermore be absorbed by the atmospheric ozone. As a consequence, the MPEF-based Lidar signal decays much more slowly than its linear counterpart [51]. Furthermore, the use of fluorescence and the broadband, spectrally resolved detection allows selectively identifying the species at play.

Short pulses also reach the high intensity required to excite non-linear effects in dilute media at relatively low pulse energies in the millijoule range or even below. Consequently, thermal effects are limited, avoiding sample heating [53], degradation, and parasitic emissions of, e.g., atmospheric constituents [53] (Figure 12.5). The latter property, known as 'clean fluorescence' [54], allows us to detect relatively weak fluorescence lines from atmospheric trace gases, without the need to filter out, either spectrally or temporally, the thermal background typical of the nanosecond counterpart of these measurements. The resulting improvement in sensitivity was demonstrated on the laboratory scale for 25% CF_4 and C_2F_6 halocarbons in air [55], as well as sub% concentrations of methane, acetylene [56], and ethanol [57]. It could even allow one to detect 2% methane in air at a distance of 20 m in field tests [58].

The same property applies to laser-induced breakdown spectroscopy (LIBS). This elemental analysis technique consists in illuminating a sample with a high-intensity laser pulse in order to create plasma on its surface and to analyse the emission lines from the excited surface and the plasma plume. These lines that are mostly atomic also include small radicals and ions [59]. As LIBS requires neither sample preparation nor contact to the sample, it is particularly suited to remote monitoring of inaccessible and/or dangerous samples in contexts like nuclear waste

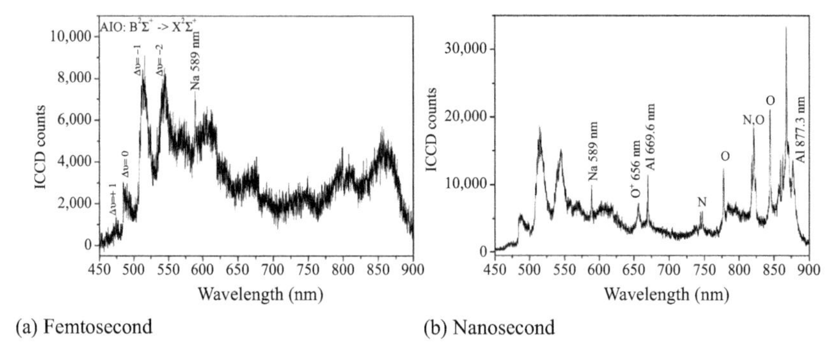

(a) Femtosecond (b) Nanosecond

Figure 12.5 Clean R-FIBS and clean fluorescence in the emission spectrum from AlO in the femtosecond regime (a), only lines from the sample are visible, while atmospheric species significantly contribute to the nanosecond spectrum (b) [53]

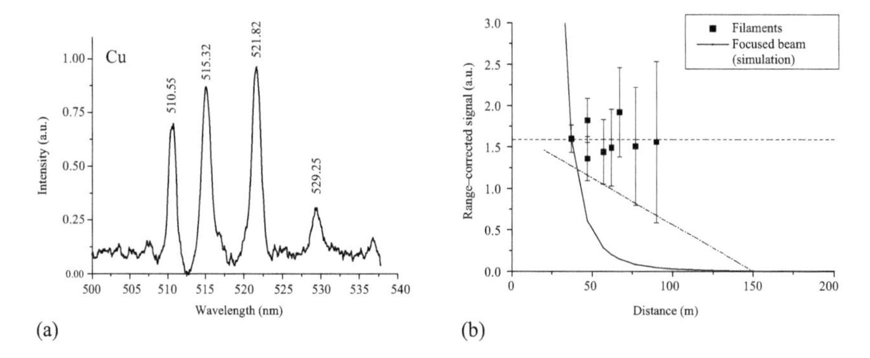

(a) (b)

Figure 12.6 Remote filament-induced breakdown spectroscopy: (a) copper plasma spectrum excited and detected from 90-m distance, and (b) range-dependence of the corresponding signal, normalized through the collection efficiency [61]

management, industrial processes, smart mining, or environmental monitoring. However, remote LIBS with nanosecond pulses is limited by diffraction and the associated size of the focusing optics, restricting the technique to a few tens of metres [60]. In contrast, by allowing an 'extended focus' over hundreds of metres, and a filament onset at a distance of up to several kms [16,17], filamentation allows to circumvent this limitation and deliver the required high intensity at arbitrary distances. As a result, plasma lines from copper, iron, and aluminium targets could be excited and recorded from a distance of 90 m (Figure 12.6), and signal could even be detected from 190-m distance [53,61]. This technique was named R-FIBS, standing for remote filament-induced breakdown spectroscopy. It was subsequently applied to multi-constituent detection [62] and biological material [63] (Figure 12.6(b)).

The non-linear interaction between the ultrashort pulses and the atmosphere can be even more spectacular, when considering, e.g., laser-induced condensation (See Section 12.3 below) [64–66]. As this phenomenon drastically depends on the physical and chemical state of the atmosphere [temperature, relative humidity, availability of condensable and/or hygroscopic species like VOCs, SO_4^{2-}, HNO_3^{-}, and others], the number density and size distribution of the laser-generated particles bear information about the atmospheric composition and status, and, consequently, about its capability of condensing naturally. Probing the laser-induced particles with a second (nanosecond) laser-based Lidar was proposed as a pump–probe technique allowing one to remotely measure the atmospheric condensability and improve short-term local weather forecast [67]. Similarly, in the laboratory, the yield in laser-induced nitrogen monohydride NH radical, measured by its fluorescence, was proposed to characterize the local relative humidity [68]. Here, NH radicals are taken as a tracer of the reactions between water vapour and the nitrogen from the air. In the future, further pump–probe Lidar techniques may emerge based on similar associations and proxies.

12.2.5 Short pulse and extreme size resolution

The ultrashort pulse duration in itself provides an extremely high spatial resolution. A resolution of 100 fs corresponds to 30 μm at the speed of light in air. Of course, range-resolved measurements and mapping do not need such resolution, and current detectors are far from offering such fast reaction time. However, 30 μm is the circumference of a 10-μm diameter droplet, i.e., of the corresponding whispering gallery modes. A shorter pulse will, therefore, be localized within this mode. Conversely, any longer pulse will wrap over the mode and keep it filled with a stationary wave, the lifetime of which is controlled by the quality factor of the cavity, i.e., by the losses within the particle. Such duration can be up to several ns in the case of a realistic quality factor 10^5.

As such, it can excite stimulated processes like stimulated Raman scattering (SRS) with much higher efficiency. As a consequence, the transition between the localized regime and the fully overlapped regime will give rise to a jump in the SRS efficiency [69]. As this relatively sharp transition is directly related to the particle size, it could be used as a remote sensor for particle sizes in the micrometre size range [69].

Furthermore, continuously varying the pulse duration by means of detuning the grating compressor of the laser system allows recording the signal as a function of the pulse duration. Differentiating this signal with respect to pulse duration can provide the particle size distribution.

Even more flexibility for measuring a wider size range, as well as particle sizes beyond reasonable laser chirping capabilities, could be achieved in a pump–probe scheme, where a two-photon process like two-photon-excited fluorescence is excited by two pulses, the delay between which being varied continuously. Only

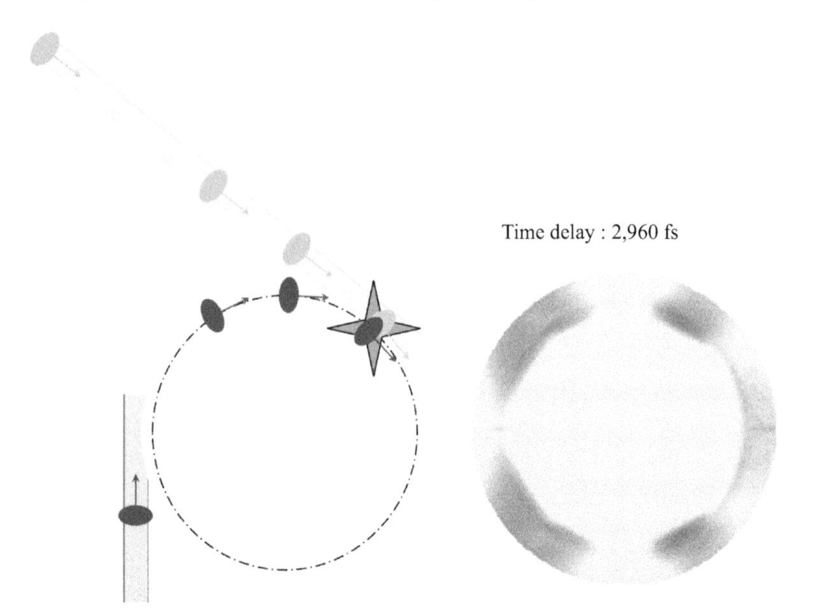

Figure 12.7 (Left) Principle of 2-PEF pump–probe measurement based on localized pulses in a microparticle whispering gallery mode; (right) simulation of the two-photon fluorescence excitation in the equatorial plane of the droplet, integrated over time [71]

the pulse overlap yields to the expected emission. As this overlap is obtained modulo, one round trip within the gallery mode of the particle, the recurrence of the signal while sweeping the delay is directly related to the particle size or the cloud size distribution [70] (Figure 12.7).

12.2.6 Outlook – shaping: selective sensing

Beyond the wealth of available variants of non-linear Lidar based on ultrashort pulses and filaments, the bandwidth associated with ultrashort pulses may allow to further develop remote sensing selectivity and sensitivity by allowing pulse shaping. This technique consists in tailoring the pulse in terms of spectral and/or temporal shape and intensity. As a result, the pulse can be optimized to perform coherent control, via open or closed loop optimization [72]. In spite of the atmospheric turbulence, such predesigned pulse shape can be propagated without damage through the atmosphere over relevant distances for remote sensing [73]. This is due to the fact that the beam diameter can be kept smaller, or at most on the same order of magnitude, than the inner scale of the atmospheric turbulence [74].

However, as the filamentation scrambles the spectral phase of the pulse during its propagation, via pulse reshaping, the desired pulse shape must not be produced at the laser output, as is the case in laboratory experiments. Instead, it should be generated at the measurement location, i.e., at the end of the filaments. Recently, a method was proposed to determine the pulse shape to launch into the air in order to output the desired pulse shape at the end of the filamenting region. This approach relies on the reversibility of the non-linear Schrödinger equation governing the filamentation [75,76]. Numerical reverse calculations of the pulse propagation, therefore, allow to design the pulse shape arbitrarily to obtain the target shape at the filament output, i.e., on the target or in the atmospheric region of interest.

12.2.7 Conclusion

As reviewed in the present section, ultrashort laser pulses offer several original properties of interest to remote sensing: their short spatial extension, their ability to interact non-linearly with the target air masses, the generation of a broadband white-light continuum, and an enhanced backward scattering. These properties enhance the Lidar signal and enrich its information contents to multispecies or multiparameter capability, as well as more selective identification without a priori knowledge of the target.

The practical development of these techniques is likely to be supported in the near future by the development of more reliable and compact ultrashort lasers, with less stringent environmental requirements and affordable price. Indeed, the development of direct diode-pumped, high-energy ytterbium systems is a significant step in this direction [77].

12.3 Light- and laser-induced water condensation

12.3.1 Laboratory-scale experiments

Beyond the sensing of the atmosphere, lasers have very early been applied to tailor it. One of these fields of application is water condensation, following early observations of C.T.R. Wilson in an expansion chamber [78] that ionizing radiation favours droplet concentration. At much lower photon energy, Wilson showed that UV radiation also triggers fog formation and even requires a lower saturation ratio (S~1, i.e., ~100% relative humidity). This stabilization is related to the photo-induced formation of H_2O_2 that once dissolved reduces the Gibbs energy of the droplets, stabilizing them.

Nanosecond UV lasers at 193 [79], 248 [80], and 266 nm [81] also turned out to trigger condensation. However, the low peak power and intensity of these nanosecond lasers (10^7 W/cm^2) did not allow two-photon ionization, so that these results were interpreted in terms of photodissociation of molecular oxygen into the ^3P state resulting in the formation of ozone that was subsequently photodissociated with the formation of a singlet oxygen, forming $^{\bullet}OH$ radicals and finally hydrogen peroxide H_2O_2, i.e., the mechanism evidenced by Wilson and later confirmed by Clark and Noxon [82]. Indeed, CRDS spectroscopy led an estimation of 6×10^{14} cm^{-3} (~20 ppm vol) of H_2O_2 at 263K, as compared to four orders of magnitude less in the open atmosphere [83].

The high fluence and peak intensity provided by ultrashort pulses and filaments are much more favourable to photochemistry. The first demonstration was performed by the *Teramobile* team [5] with a Ti:Sa laser providing 50-fs, 1-mJ pulses at 800 nm, at 1-kHz repetition rate. Filaments produced by this laser propagated through a diffusion chamber where supersaturation was achieved by imposing strong opposing vertical temperature and humidity gradient. The ~50-TW/cm^2 intensity in the laser filaments and the associated photoionization with electrons up to 10^{14}–10^{16} cm^{-3} [6,7] resulted in the spectacular production of fog droplets (Figure 12.8).

Filament-induced condensation turned out not to be restricted to saturated conditions. In laboratory experiments, the *Teramobile* laser (220 mJ, 50 fs, 800 nm [44]) generated a bundle of several tens of filaments, with an individual intensity of 50 TW/cm^2, in a diffusion chamber. Micrometre-sized droplets were massively produced within a few seconds, for relative humidity as low as 75% and remained stable after the laser was switched off (Figure 12.9). This stability owes to the Raoult effect, i.e., chemical stabilization balancing the Kelvin surface tension

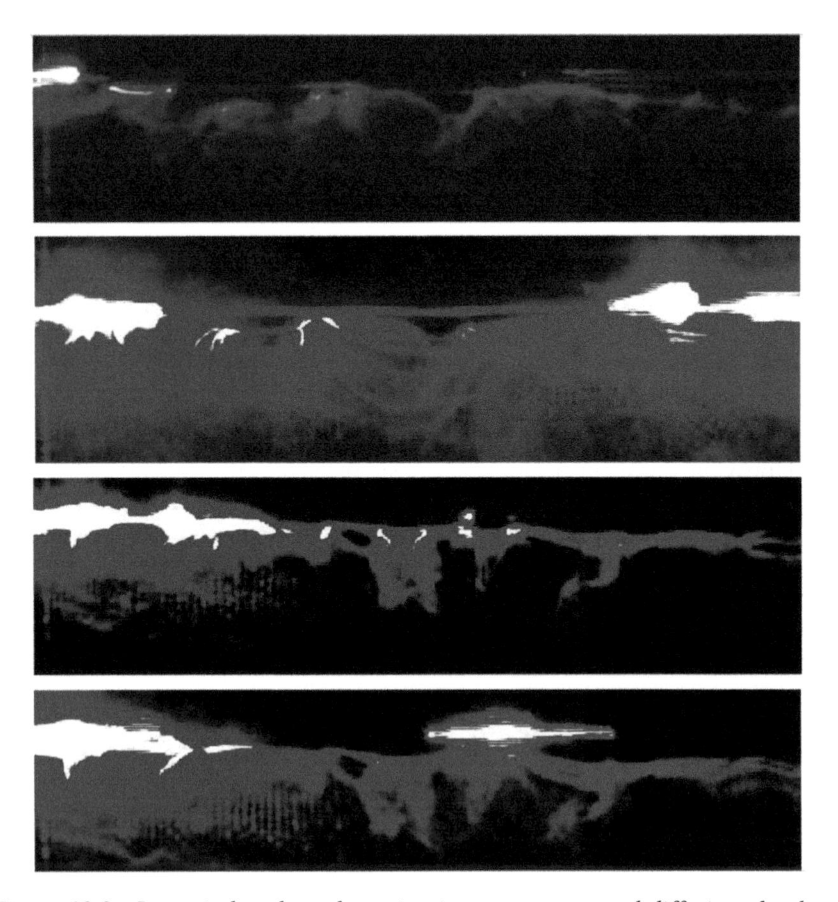

Figure 12.8 Laser-induced condensation in a supersaturated diffusion cloud chamber (Luderer, 2001 [5]). A low-power cw laser illuminates the droplets produced by the filaments. Vortices stem from shockwaves due to laser heating

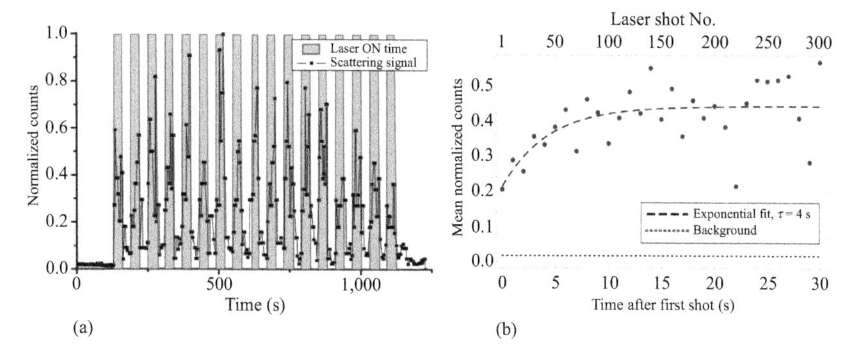

Figure 12.9 Laser filament–induced water condensation in a sub-saturated diffusion cloud chamber (T=60°C, RH=75%–85%). (a) Evolution of the particle concentration over cycles of laser on/off periods; (b) continuation of the droplet density increase after the laser pulses. Reproduced, with permission, from Reference [64]

related to the droplet radius of curvature [84]. Indeed, the photoionization and photodissociation of air components (nitrogen, oxygen, and water vapour) generate various chemically active ions and radicals [85,86]. In particular, in a nitric acid pathway [3,87], oxygen radicals form ozone, and atomic nitrogen ions and radicals react with oxygen to form NO and then NO_2. The latter combines with O_3 to form N_2O_5, which dissolves in water, forming nitric acid. The high abundance of N_2 in the atmosphere favours this pathway as compared with the classical sulphuric acid pathway typical of natural condensation in the atmosphere.

Indeed, filaments can produce up to 200 and 50 ppb of ozone and NO_x, respectively, in a flow chamber [88]. Concentrations of HNO_3 [89] as well as NO_2, O_3, and NO_3 [90] 1,000–10,000 times over natural concentration [91] were also directly measured in the filaments by spectroscopic measurements. These concentrations, although limited to the very small volume of the filaments, contribute to stabilizing particles in sub-saturated conditions, as expected from the Köhler theory [92], as updated by Laaksonen *et al.* [93]. This approach that quantifies the chemical stabilization balancing the surface tension has been applied to laser-produced particles [94,95], assuming that they were nucleated around a dissolved, non-volatile core of ammonium nitrate of several nm diameter [96–98]. The resulting Köhler plots (Figure 12.10) define, for each size, the relative humidity ensuring the particle stability for a given atmospheric HNO_3 concentration. Positive slopes define stable conditions, as the growth of the particle would deplete the relative humidity while increasing the humidity required for its stability. This model shows that both particles below 30 nm and above 2 μm are stable for sub-saturated atmospheres, validating the ternary HNO_3–NH_4NO_3–H_2O particles as plausible candidates for laser-induced condensation.

The possibility of nucleating particles without the initial presence of condensation nuclei was investigated at the AIDA cloud chamber featuring a

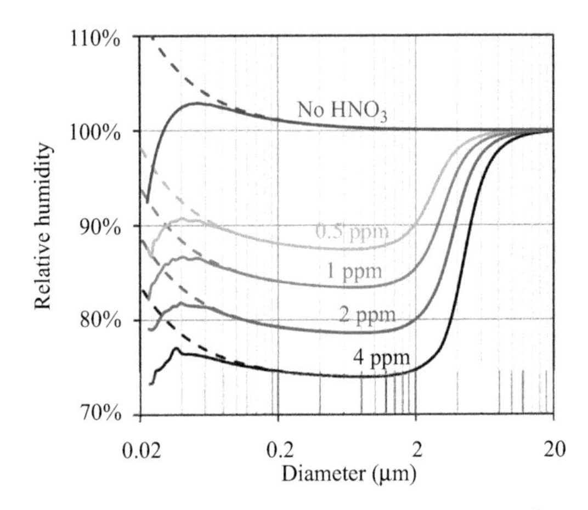

Figure 12.10 *Köhler plots for a droplet density of 1,000 cm^{-3}, at a temperature of 279K and a pressure of 1,013.25, for various initial concentrations of gaseous HNO_3; solid lines: droplets including a 15-nm NH_4NO_3 core; dashed lines: droplets without NH_4NO_3 core. Reproduced, with permission, from Reference [65]*

volume of 85 m^3 [99,100], which limits the walls-related artefacts. Filaments were produced by the *Teramobile* laser. Particle formation within the filament volume was five–six orders of magnitude more efficient in ambient air [101] than in synthetic air [102]. In the latter conditions, the particle formation rate increases exponentially with the water vapour concentration, pointing the key role of water molecule clusters [103] and the associated formation of OH radicals in the laser-generated plasma, which, in turn, can oxidize organic molecules [104]. Adding traces of NH_3, SO_2, toluene, or α-pinene, even at the ppb level, boosted particle formation and growth. The role of organic traces has been further evidenced by the occurrence of condensation in humid argon, i.e., without nitrogen. A similar role of traces of VOCs has been observed under proton synchrotron illumination simulating galactic cosmic rays [105,106].

Besides their photochemical effect, laser filaments are known to locally deposit energy in the air, resulting in shockwaves [107–110] and turbulence. The resulting mixing of warm moist air with clod air favours particle growth in diffusion chambers (Figure 12.11) [111–115], by generating a local supersaturation up to 1.3 and 2.1 relative to water and ice, respectively. The stabilization of the resulting particles and their accumulation as snow on the cold baseplate of the chamber (Figure 12.12) occurs even in nitrogen-free atmospheres such as humid helium, pointing to a role of contamination by organic traces. The shapes of the particles suggest that they have been produced at a temperature between −25°C and −15°C. Particles, therefore, formed in air, before they reached the colder baseplate (−46°C).

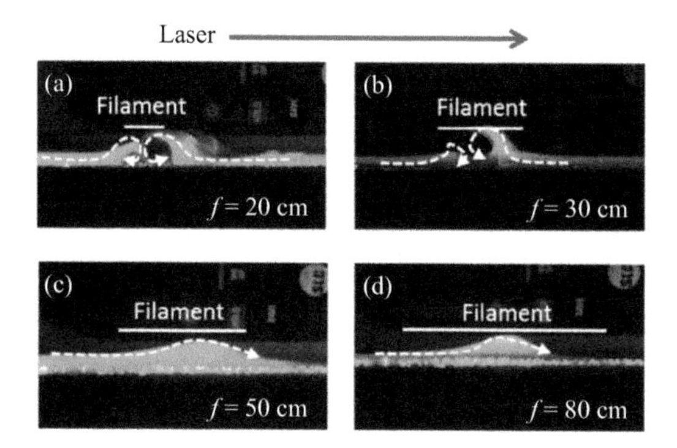

Figure 12.11 *Laser-induced turbulent flow around filaments produced with different focal length lenses (a) f = 20 cm; (b) f = 30 cm; (c) f = 50 cm; (d) f = 80 cm. Reproduced, with permission, from Reference [114]*

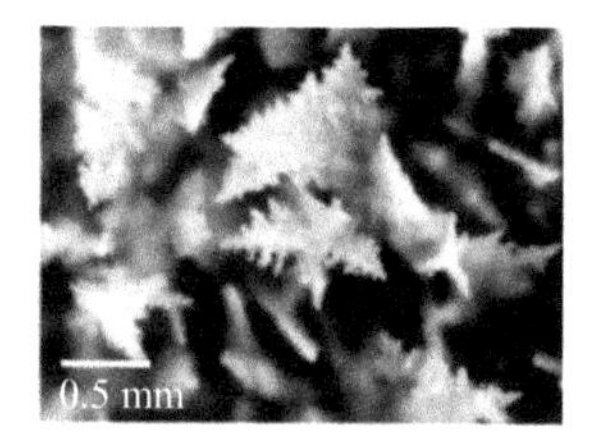

Figure 12.12 *Snow particles formed on the cold bottom plate of the cloud chamber, cooled at −46°C, under laser filament illumination. Reproduced, with permission, from Reference [112]*

12.3.2 Upscaling

While spectacular, the laser-induced condensation remains restricted to limited volumes and, therefore, moderate amounts of particles. Macroscopic effects, in-line with atmospheric scales, definitely require an upscaling of these processes. Prospects in this regard are fuelled by the development of lasers offering 10–100-PW peak powers and multi-kW average powers [116].

Evaluations have been performed at various sub-PW facilities [117–119]. The results show that the upscaling of small-scale experiments is not straightforward. While the number of filaments tends to saturate beyond a photon bath intensity of 0.5 TW/cm^2 (Figure 12.13, red squares), the production of aerosols below 150 nm rises with the fifth power of the photon bath intensity (blue circles), pointing to a substantial contribution of the latter (green triangles). At this peak power level, due to its large volume and high intensity, the photon bath bears 95% of the laser energy, and its intensity is sufficient to efficiently initiate the photodissociation of molecular oxygen and release atomic oxygen radicals, which, in turn, will oxidize organic traces and produce condensable species. This possibility to perform filament-free fs-laser-induced condensation opens a much larger volume to

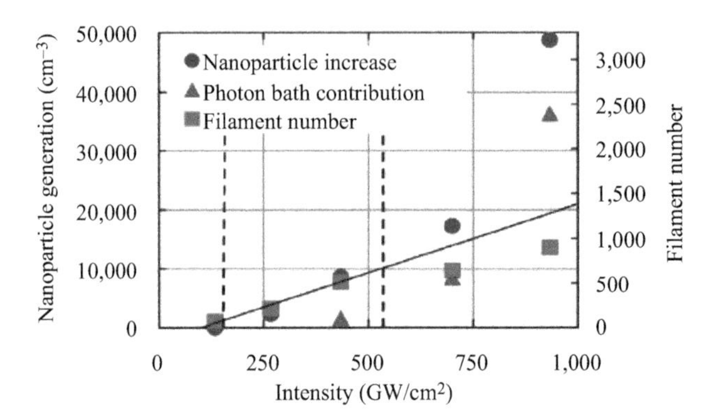

Figure 12.13 Number of filaments and of nanoparticles generated by the laser as a function of incoming beam intensity. Reproduced, with permission, from Reference [117], AIP Publishing

condensation, also overcoming growth limitations related to water vapour diffusion to the growing particles.

Multi-laser, multi-wavelength schemes have also been shown to enhance particle production. The addition of a frequency-quadrupled nanosecond Nd:YAG laser providing 250 mJ at 266 nm doubles the particle production rate of a 100-TW near-infrared laser, although the UV laser produces no particle by itself [118]. Indeed, the UV laser produces singlet oxygen radicals that react with water and produce OH radicals, with an estimated rate of 10^{13} cm^{-3} s^{-1}. This rate lies seven orders beyond the natural atmosphere. It was confirmed experimentally by fluorescence measurements [120].

This result, together with the early UV-triggered condensation experiments, points to investigating the effect of UV laser filaments propagating in air [121–125]. The high photon energy allows one to reduce the order of multiphoton processes. As a consequence, a KrF laser delivering 11-mJ, 110-fs pulses, i.e., only 0.1 TW, produced five to ten times more nanoparticles than the above-mentioned 100-TW near-infrared laser [126]. Again, the efficiency of OH radical production appears to boost HNO_3 production. However, the H_2O_2 pathway evidenced by Yoshihara *et al.* [80] could also contribute.

While very efficient for particle generation, ultrashort UV lasers however suffer limitations for real-atmosphere applications, in particular the stronger Rayleigh scattering and the ozone absorption that restrict its propagation over hundreds of metres to kilometres in the atmosphere.

12.3.3 Field experiments

Beyond the laboratory tests, the mobility of the *Teramobile* laser [5] allowed one to perform field experiments with laser filaments. The first campaign in that purpose was conducted close to the Rhône River in Geneva, Switzerland. The location downstream of the Geneva Lake was chosen to ensure a high relative humidity

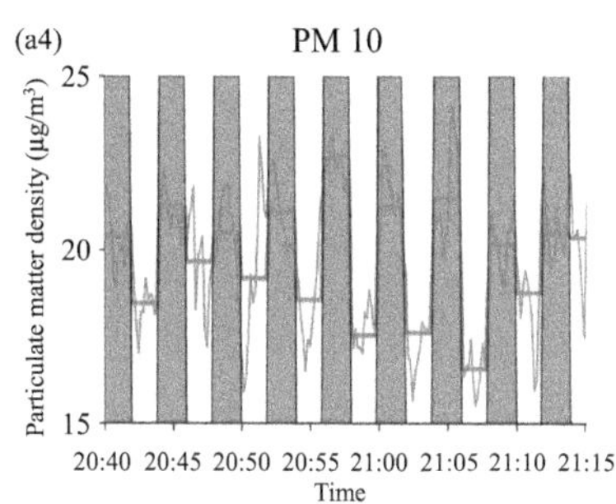

Figure 12.14 Left: The Teramobile system close to Rhône River (Geneva, Switzerland), and the van-based air analyser. Right: particle production in the PM-10 class, in the open atmosphere (shaded: laser on, unshaded: laser off). Reproduced, with permission, from Reference [65]

related to the relatively warm water flow from the lake. The campaign lasted over 6 months in 2009–10, ensuring a wide variety of conditions (relative humidity between 35% and 100%, temperature between 2°C and 36°C). The 160-mJ, 240-fs pulses were slightly focused and generated about ten filaments over 15–20 m (Figure 12.14, left). Condensation was observed over the whole range of measured sizes, up to several μm (see Figure 12.14, right) [65], although the most efficient particle production occurred below 300 nm. Gas-phase measurements and ion chromatography confirmed high local concentrations of ozone (~200 ppb) and NO_2 (25 ppb), as well as the presence of NO_3^- ions in the laser-generated particles, confirming the nitric acid pathway.

Further analysis was performed with an Aerodyne high resolution time of flight aerosol mass spectrometer (AMS) [127,128]. The AMS provided the full composition of the laser-generated particles as a function of size (Figure 12.15). Nitric acid was shown to condense as ammonium nitrate, validating the above-described growth model [65,94]. Oxidized VOCs constitute around one-third of the particle dry mass. The corresponding absolute concentration of laser-condensed organics (<1 μg/m^3) is only a small percentage of the total urban VOC concentration [129], suggesting that the limiting factor for their condensation is the OH radical production by the laser, subsequently oxidizing the trace volatile organics in a highly accelerated 'ageing' producing secondary organic aerosols [130–132], which are efficient cloud condensation nuclei.

The particle concentration generated by the laser at an elevation of 50–100 m over ground is even sufficient to be detected remotely from the ground by a Lidar [133] synchronized with the filaments [64]. As displayed in Figure 12.16, the difference between the Lidar signals 1 and 100 ms after the filaments were emitted into the atmosphere was significant in the filamenting region, in spite of a relative humidity of

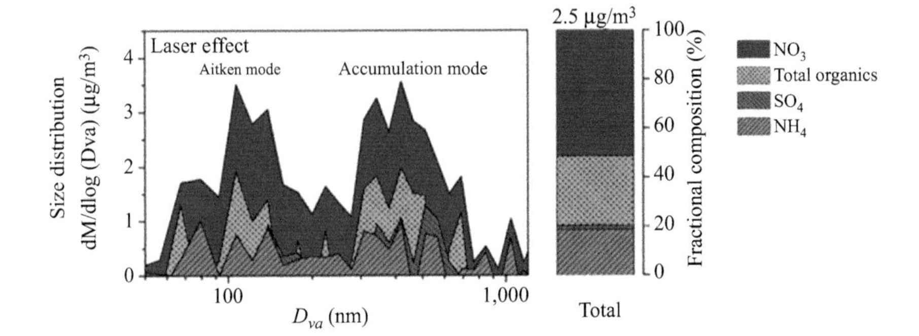

Figure 12.15 Mass fractions of the condensable species within the particles (dry mass): difference between irradiated and non-irradiated particles. Reproduced, with permission, from Reference Mongin et al., Scientific Reports 5, 14978 (2015)

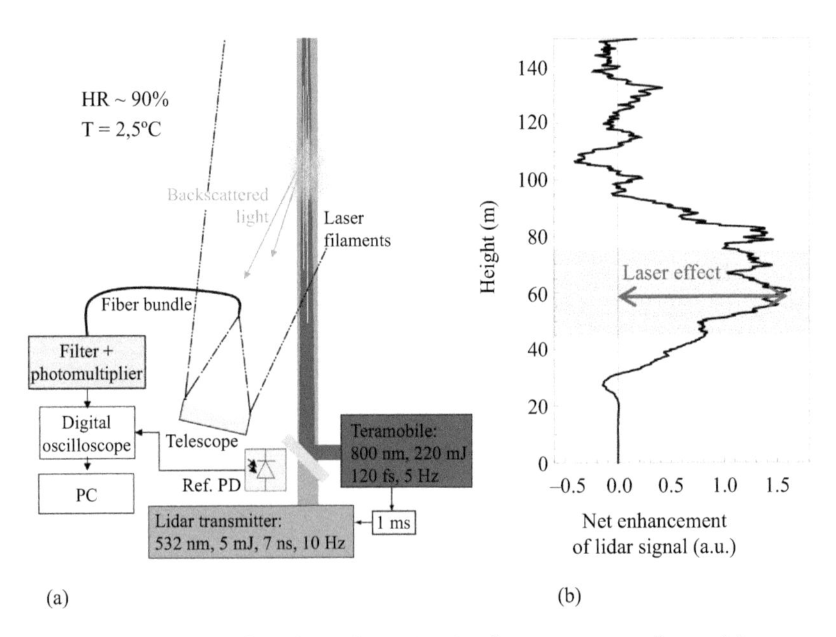

Figure 12.16 Laser-induced condensation in the open atmosphere. (a) Experimental set-up. The Teramobile laser (red) is fired 1 ms before the ns pulse of the Lidar (green). (b) Time-averaged relative increase of the Mie backscattering when firing the terawatt laser. The signal increase is a signature of the rise of the particle density under laser illumination. Reproduced, with permission, from Reference [64]

only 90%. Here, the slight transverse wind clears up the atmosphere between two filamenting pulses, so that the first pulse detects the Mie scattering [134] of the laser-generated particles, while the second one provides a clear-sky background. The Mie scattering increase due to the laser-generated particles amounts to a factor of 20, from $\beta_{Mie}=10^{-6}\ m^{-1}\ sr^{-1}$ to $2\times10^{-5}\ m^{-1}\ sr^{-1}$, although converting these values into particle size distribution and density would need to perform additional assumptions [135].

These atmospheric results raise the prospect of applications to modulate precipitation and/or manipulate the Earth albedo. Obviously, the energies delivered even by the most powerful laser are minimal as compared with the ones involved in large-scale atmospheric processes. The effect of the laser can, therefore, only be a trigger, seeding condensation nuclei or small-sized particles that would subsequently evolve subject to favourable atmospheric conditions. These conditions may be assessed based on Lidar measurements with the same laser [136–138] and/or evaluated from atmospheric modelling. It may be noted that the ability to switch on and off the laser and aim it precisely to specific locations in the atmosphere would drastically facilitate the assessment of the laser efficiency in the context of cloud seeding, unlike classical methods [139–142].

12.4 Modulating the optical transmission of fogs and clouds

Ultrashort, ultra-intense laser pulses can not only trigger the formation of fog droplets, but they have also been considered a tool to clear fogs and clouds. Such clearing would provide a great leap forward for ground-to-satellite or satellite-to-satellite optical communications. First efforts date back to the 1970s and 1980s, using high-energy, high-power CO_2 lasers [143], but it turned out that intensities up to 10-kW/cm^2 continuous wave (cw) [143] or 10–1,000 MW/cm^2 for pulsed lasers [144,145] were required to shatter and evaporate the cloud particles, preventing their application at the atmospheric scale. Shockwaves associated with plasma sparks (Figure 12.17) were also observed for fluences of 200 J/cm^2 and above [145–149] in focused beams. However for nanosecond pulses, avalanche ionization leads to high plasma densities that prevent further propagation of the beam.

In contrast, femtosecond laser filaments can propagate over much longer distances due to their moderate plasma density. Furthermore, filaments can self-heal after hitting obscurants like droplets and, therefore, propagate unaffected through clouds [150,151]. This is due to the refocusing of the photon bath (aka energy reservoir) immediately after the filament light is scattered out by the droplet [14,151,152]. The only remaining limitation is the linear losses due to the Mie scattering of the photon bath [151].

Furthermore, drop clearing by femtosecond pulses do not rely on particle evaporation. Rather, the droplet curvature focuses the beam internally close to the shadow side [153] and to a lesser extent to the illuminated side [154]. The resulting nanometric plasma spot [153,155] triggers a shockwave leading to the drop shattering (Figure 12.18) [156]. The energy cost of this process, ~50 μJ per droplet, is much lower than the energy required to evaporate it as with longer CO_2 laser pulses.

Figure 12.17 *200-J CO_2 laser producing a 15-m long spark in fog [143]*

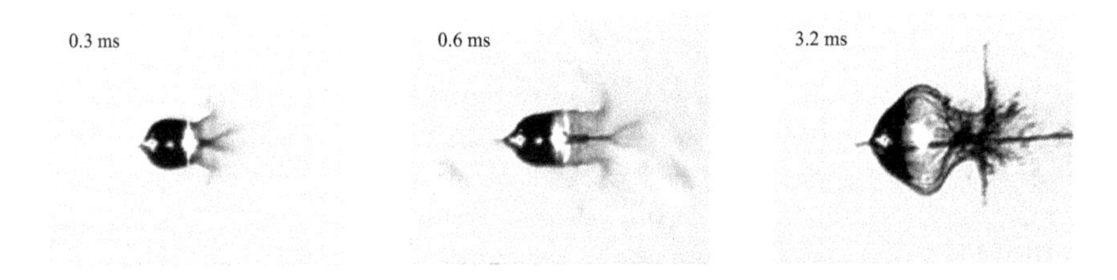

Figure 12.18 *Explosion dynamics of individual water droplets by 60-fs laser pulses at an intensity of 4×10^{11} W/cm^2. Note that times of the frames are in μs and not ms as stated in the original labels. Reproduced, with permission, from Reference [156], the OSA*

A similar shattering was observed in ice particles, representative of cirrus clouds. Cirrus are clouds of 50–100-μm ice particles forming in the upper troposphere below the threshold for supercooled water, −37°C [157]. They usually have a net warming effect on climate, although this effect depends on their altitude and particle size [158].

Ice particles of 90 μm, representative of cirrus clouds and produced in a Paul trap at a temperature of −41°C, were illuminated by laser filaments and their subsequent evolution was observed, together with that of their neighbourhood, at 140,000 frames/s. In the first microseconds after the laser pulse, the energy deposited by the filaments shatters the ice particles as it does for liquid droplets, but it also vaporizes 15%–20% of its volume. However, after ~12 μs, several tens of

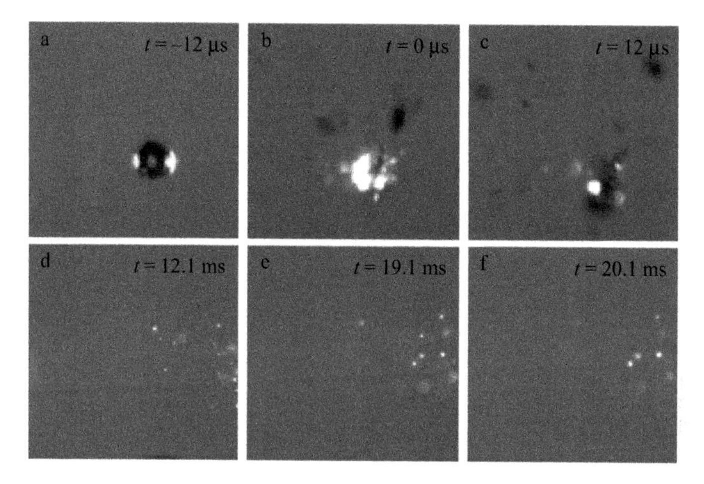

Figure 12.19 Direct observation of FISIM: (a) initial ice particles before the filament illumination, (b) filament illumination, (c) particle shattering. The main fragment is ejected backward and leaves the field of view; smaller particles are also ejected from the shadow face. (d–f) Formation of smaller particles along the trajectory of the large fragment after re-illuminations by the filament. Reproduced, with permission, from Reference [159]

new particles nucleate and reach several microns in diameter (Figure 12.19) [159]. These particles are recondensed from the water vapour released by the partial vaporization of the initial particles that leads to supersaturation beyond the threshold for ion-induced nucleation (S=4) [160], or even for homogeneous nucleation (S=15) [161]. This recondensation, named filament-induced secondary ice multiplication (FISIM), drastically shifts the size particle distribution to small sizes, expectedly turning the warming effect of the cloud to a cooling one.

FISIM was even observed on a large collection of particles at the AIDA cloud chamber. Particles initially nucleated on mineral dust and grown to a few tens of μm were shattered by the action of the filaments produced by the *Teramobile* laser, instantly shifting the size distribution to around 1 μm and multiplying their number density by a factor of 5. These tiny ice particles are indeed secondary particles, as they are not generated if the large particles do not pre-exist [162].

Laser filaments deposit energy not only in the particles, but also in the air along their propagation path, due to the ionization. As a result of this local heating, a low-density channel ten times as large as the filament itself is created in the filament wake [107,108,110,163]. This occurs through the emission of a cylindrical shockwave propagating in air (Figure 12.20, left). The associated drag radially expels particles present in this volume. If the laser repetition rate is faster than the time required for new particles to diffuse or to be advected into the depleted volume, a permanently cleared channel will establish. This permanent clearing was achieved with a picosecond thin-disc ytterbium laser producing 100-mJ pulses at a

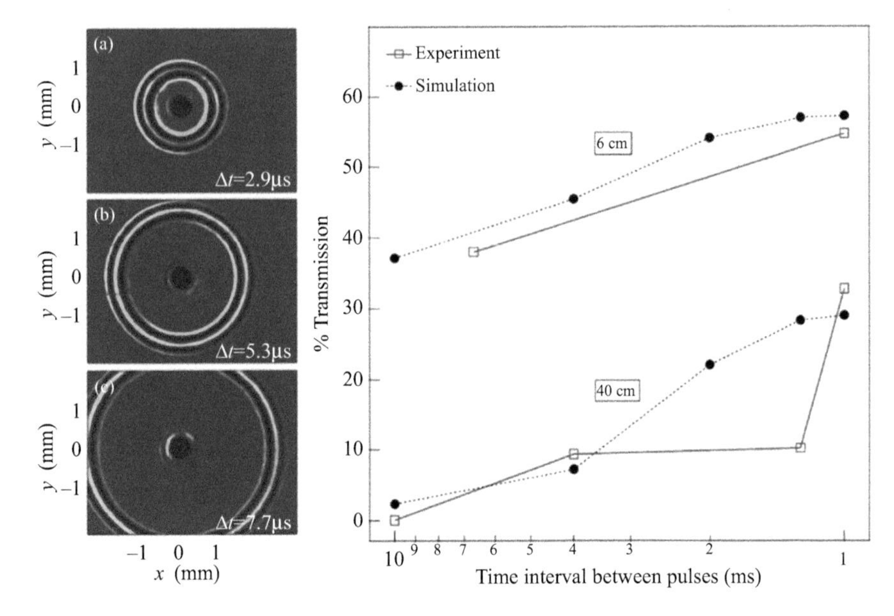

Figure 12.20 Left: filament shockwave propagating in air. © American Physical Society, 2014. Reproduced, with permission, from Reference [108]; right: laser transmission through fog based on droplet optomechanical expulsion by the filament-induced shockwave. Reproduced, with permission, from Reference [164], AIP Publishing

repetition rate of 1 kHz [164]. The experiment was performed with longitudinally integrated optical thickness comparable to typical fogs. It shows a clear transition from an opaque fog for laser repetition rates of 100 Hz and below to much clearer conditions in which information can be transmitted when the repetition rate exceeds several hundred Hz (Figure 12.20, right). As also observed in Monte Carlo simulations, the diameter of the cleared channel rises from 3 mm at 100 Hz to 12 mm at 1 kHz. These results, therefore, open the way to fog-immune optical transmission, in which laser filaments open a depleted channel where a fast-modulated telecom laser could transmit information. Progress in laser peak- and average-power is key to allow the generation of the required cylindrical shockwave over the required distances.

12.5 Control of high-voltage discharges with ultrashort lasers

12.5.1 Triggering and guiding high-voltage discharges with laser filaments

The ability of intense lasers to ionize the air initiated attempts to trigger high-voltage discharges shortly after the onset of lasers [165,166]. However, it was shown that the depletion of the air density also contributed to the discharge guiding

[167]. Further work with nanosecond CO_2 [168–170] and UV [171] lasers extended these results to megavolts discharges over gaps reaching 2 m. Still, the formation of dense, opaque plasma balls [172] prevented the laser propagation, hence the discharge guiding, beyond the metre-scale. In spite of these limitations, Uchida *et al.* [173] attempted to trigger and guide lightning in real scale. Using a complex set-up involving three lasers and a 50-m high tower [174,175], only two events were reported, so that statistical significance could not be established.

In contrast to nanosecond lasers, femtosecond filaments produce long, continuous plasma channels, virtually allowing guiding over unlimited distances. The group of J.C. Diels was the first to use femtosecond laser filaments, in the UV, to trigger discharges over 26 cm, under 100 kV [176,177]. This distance was extended to the multimetre-scale by using a positive polarity and near-infrared lasers producing several hundred mJ pulses at 800 nm, resulting in a bundle of tens of filaments [178,179]. A decay of the leader inception voltage by 50%, associated with a 10-fold acceleration of the leader propagation speed [180], was well reproduced in numerical simulations [181]. Similar results were obtained by the *Teramobile* group with negative discharges (Figure 12.21, left) [182,183]. Laser-triggered space-leader discharges between the electrodes were also reported [183], as well as the persistence of the triggering effect in spite of an 1.4-mm/min artificial rain imposed between the electrodes. Laser filaments were also used to divert electric discharges from their direct path between two electrodes [184].

The most realistic electrode configuration featured a 2-m high lightning rod with an aperture allowing the laser filaments to get through and simulate negative leaders. This lightning rod was installed on a circular 15-m grounded electrode and faced a flat, 5-m diameter negative electrode at a distance of 5 m. The voltage on the negative electrode was increased with rise time similar to that of the progression of a real leader to the ground. Once fired in the gap, the filament immediately triggers a corona and the leader inception translates in a current spike (Figure 12.21 right) [185,186]. The leader then propagates to the ground electrode and the guided final jump occurs.

The MV-range voltages needed for metre-scale discharges require Marx generators that produce pulsed high-voltage discharges. However, experiments have been also performed, at smaller scales, with AC voltages, mostly using Tesla coils [187–190]. The associated mechanism involves no streamer, but only leader propagation [189]. This regime however allows discharges at lower electric field, e.g., 1.8 m for an electric field as low as 2 kV/cm [190].

Applying a DC voltage provides even more surprising behaviours. In this regime, a slow mode, occurring over the millisecond timescale and related to the mobility of the ions, can coexist with the usual fast discharges [191]. Such ionic mobility was also identified as the key ingredient of spark-free unloading of HV capacitors with a kHz-repetition-rate laser [192]. Here, the ions flow quasi-continuously between the electrodes, with a current below the breakdown threshold but sufficient to unload a capacitor or an HV power supply.

Besides spark discharges, filaments also influence corona discharges in two ways. First, they can be deviated (Figure 12.22) and temporally extended by nearby

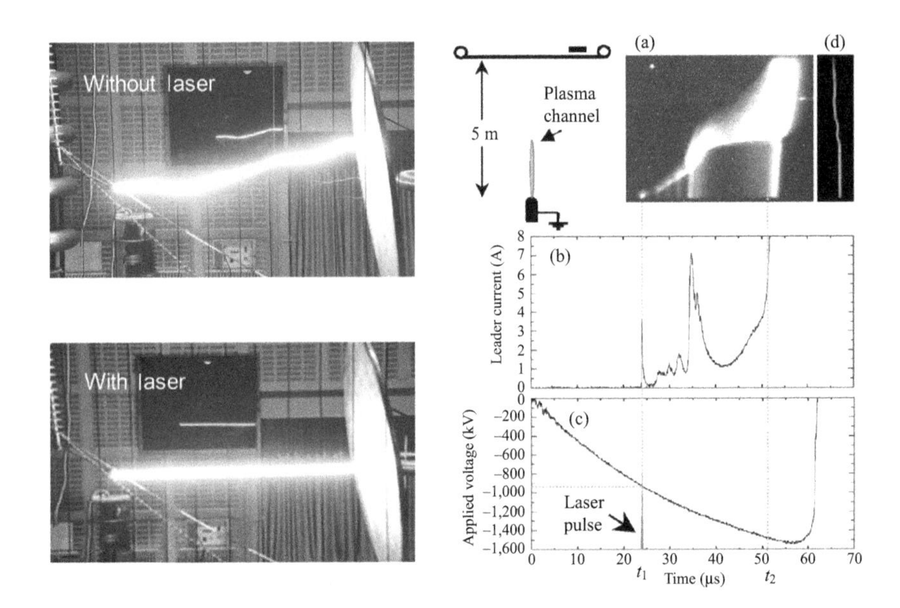

Figure 12.21 Triggering and guiding multi-megavolt discharges with laser filaments. Left: unguided (top) and laser-guided (bottom) discharges over 1.2-m distance. Reproduced, with permission, from Reference [5]. Right: characterization of the laser-guided discharge in a realistic lightning rod configuration. (a) Streak picture of the leader propagation; (b) current intensity; (c) voltage impulse applied to the negative electrode. (d) Time-integrated image of the discharge. Reproduced, with permission, from Reference [185]

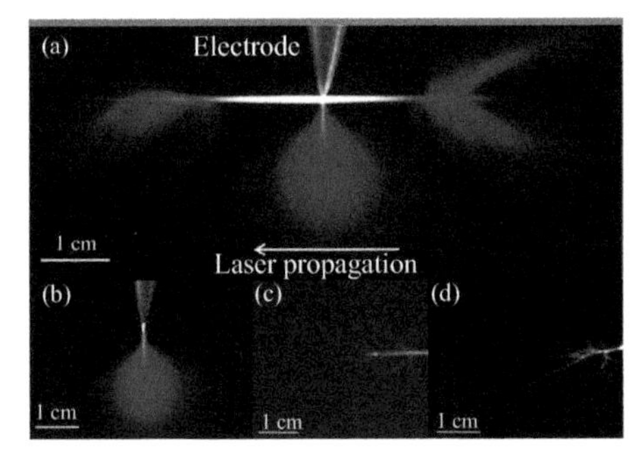

Figure 12.22 (a) True-colour images of a filament-guided corona discharge under 50 kV. The filament from the 7.5-mJ laser appears as a white elongated horizontal region; (b) corona discharge without filament; (c and d) fine structure of the streamers in the direction of laser propagation from (a) and (b), respectively. Reproduced, with permission, from Reference Wang et al., Scientific Reports 5, 18681 (2015)

filaments [175]. Second, runaway electrons, travelling at a speed close to that of light and producing secondary hot electrons via avalanche ionization, X-rays, and γ-rays, [193–195], were found associated with the laser filaments [196,197]. However, the laser beam direction is critical to this observation, as a filament perpendicular to the discharges quenches the runaway electrons [198]. These observations are key to understand the impact of laser filaments on lightning, since runaway electrons are expected to be a key ingredient of lightning strike development [199–201].

12.5.2 Temporally and spatially extending the plasma channel

Although the laser filaments accelerate the leader propagation by a factor of typically 10, its velocity is still finite. Together with the lifetime of the filament-generated plasma, this velocity defines the spatial range on which the laser can be expected to influence discharges. Several time constants are to be considered in the evolution of the plasma left behind by the filaments. Electron–ion recombination occurs in nanoseconds. Electron attachment to O_2 molecules has a typical time constant of hundreds of nanoseconds [202,203]; however, this duration is deeply influenced by the molecular and electronic temperatures [176]. Finally, the acceleration of the free electrons in the electric field imposed between the electrodes heats up the plasma, depleting the molecular density and there reducing or even stopping electron attachment (Figure 12.23) [107,204].

Several groups simulated the decay of carriers for a situation including a laser and external high voltage field, i.e., close to thunderstorm conditions or experiments on metre-long discharges. An updated model of the laser-induced HV-discharge triggering mechanisms, which relies on a compilation of the most recent values of the parameters needed for the simulation, was, for instance, recently

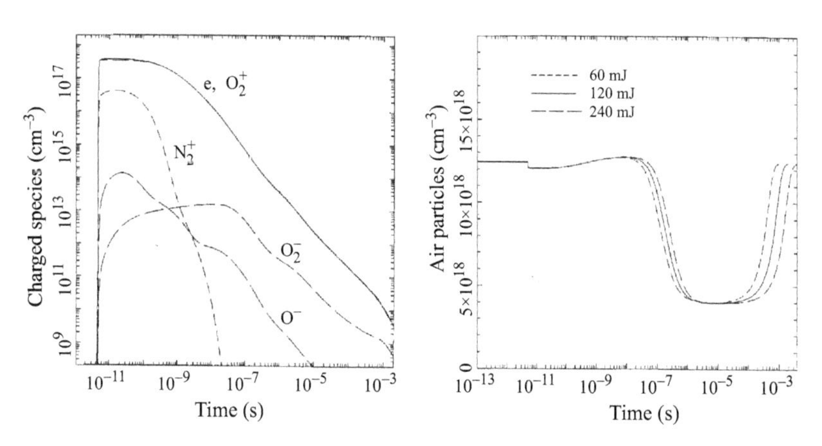

Figure 12.23 Dynamics of charge carrier densities and evolution of the air density after a 120-mJ laser pulse, in a 5-kV/m electric field. Reproduced, with permission, from Reference [107]

presented in Schubert *et al.* [204]. Furthermore, the reduced gas density lessens the breakdown voltage [205] by increasing the mean free path of the free electrons and the associated kinetic energy they acquire in the electric field between two collisions. Within the millisecond lifetime of the air-depleted channel, a leader could propagate over a distance of kilometres. But this requires an electric field of at last 4 kV/cm and a transversely-integrated free-electron density of 10^{11} electrons/cm [107].

A straightforward approach to extend the plasma lifetime would be to regenerate it with a new laser pulse before it decays: in a word using pulse trains separated with a delay comparable with or smaller than the plasma lifetime. The intensity clamping in the filaments, together with the occurrence of multiple filaments when the pulse power rises, decouple to a large extent the plasma density from the incident pulse energy. Therefore, provided each sub-pulse is above the filamenting threshold, partitioning the energy into a train of sub-pulses separated by a few ns increases the plasma lifetime [204,206–208]. However, this does not guarantee an increase of the triggering nor on the guiding of high-voltage discharges [209]. Due to the fast attachment, even pulses separated by only 2.9 ns show no cumulative effect, suggesting that the plasma lifetime is only extended by the successive re-creations of the plasma by the repeated illuminations [210]. This conclusion could however be different if the repetition rate between the first trains reaches the kHz range [211], or in the mid-infrared where heating is more efficient [212].

12.5.2.1 Photo-detachment and reheating with nanosecond lasers

To slow down electron attachment to molecular oxygen, J.C. Diels proposed to heat the femtosecond laser-produced plasma with a subsequent, high-energy (hundreds of mJ), nanosecond pulse [213]. Experimental realizations with ultrashort UV pulses followed by an alexandrite laser [177,214] allowed him to evaluate that maintaining a sufficient plasma density over 10 m would require up to 5 J in 10 μs from the alexandrite laser. Widely available Nd:YAG lasers providing pulses in the Joule range were applied for plasma reheating [215–221]. The plasma lifetime could, however, not be extended well beyond the Nd:YAG pulse duration, while the peak density of free electrons increased by up to 200 times [220], reducing the breakdown voltage by 75% [219]. Simulations [100] suggested that a similar effect can be reached with CO_2 lasers for a 100-time lower incident intensity as with Nd:YAG lasers. However, propagation effects are not considered in this model, although they could cancel the advantage of the mid-infrared pulses (e.g., [171,220]).

Increases of the breakdown probability over 1.2 m have also been observed with a moderate (0.4 J) pulse energy at 532 nm. At this wavelength, heating is inefficient; rather, photo-detachment of the O_2^- ions has been proposed as an interpretation, supported by the lack of effect of the fundamental wavelength at 1,064 nm in the same configuration [222]. In a similar thread, combining the fundamental and frequency-doubled pulses from a Nd:YAG laser was able to revive the plasma produced by a preceding femtosecond igniter [218]. Again, the infrared radiation heats by inverse bremsstrahlung the electron photodetached by the visible light.

Multi-laser approaches require a good spatial overlap of the beams. Bessel beams produced with an axicon can overlap filaments over a fraction of a meter [218–221], but this technique is difficult to scale up beyond. Rather, promising results have been obtained by generating the different pulses from the same laser, e.g., a picosecond pulse trains a nanosecond pulse from the same KrF UV laser system [124], doubling the length of the triggered discharges as compared with the nanosecond pulse alone [124,223].

12.5.2.2 Cumulative heating of a depleted channel

As mentioned earlier, besides ionization, the heating and associated density depletion of the channel left behind by the filaments contributes to the discharge guiding and triggering. Early attempts to enhance this heating consisted in adding 15% of NH_3 in the air, boosting the absorption of CO_2 laser pulses (Saum *et al.*, 1972), and extending the discharge length by one half.

In raw air, absorption in the near-infrared stems from inverse bremsstrahlung. Energy depositions from μJ/cm [224] to ~10 mJ/cm [225] can raise the temperature to 1,000K (Point *et al.*, 2015), similar to a natural leader. As mentioned earlier, such heating can even accumulate from pulse to pulse for laser repetition rates over several hundred Hz. These low-density channels again lower electron attachment to neutral O_2 molecules and increases the free-electron mean free path, reducing the breakdown voltage via Paschen's law, as experimentally demonstrated by a 3-fold drop of the discharge voltage [226].

12.5.2.3 Filament seeding by high voltage boosters

As the discharge itself is known to heat up the plasma [227], connect an additional high-voltage source to the filaments, provides a fraction of the main high voltage supplied between the electrodes (20 kV vs 100 kV). During this voltage pulse, i.e., 130 μs, the guiding remained steady. Using a 500-kV Tesla coil as an AC booster providing 1 J/pulse on top of the 30-kV DC applied to the electrodes reduced the breakdown electric field to 0.13 kV/cm, a value 230 times below its natural value. This was obtained due to a plasma temperature of 30,000 K and an air temperature of 4,800 K (Figure 12.24) [228].

12.5.3 Field experiments

While Uchida *et al.* [173] attempted to extend an existing tower with a set of 3 ns lasers, the first and only published attempt to trigger lightning with a femtosecond laser aimed at triggering lightning from thunderclouds to the ground [229]. It relied on the *Teramobile* laser, installed at the Langmuir Laboratory (South Baldy Peak, NM, USA) at an altitude of 3,200 m. The laser produced a bundle of filaments at a repetition of 10 Hz, with an elevation of 70° above ground, towards the south. The effect of the laser was assessed by recording radiofrequency emission at 63 MHz on a network of antennas (lightning mapping array) [230] and reconstructing the time-resolved tridimensional map of electric activity by triangulation based on GPS time-stamping. The laser was operated continuously and an a posteriori statistical

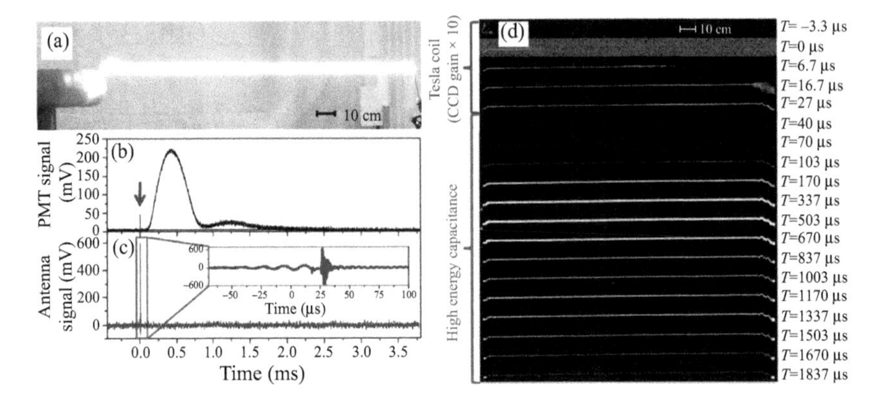

Figure 12.24 (a) Real-colour image of a laser-guided AC–DC discharge over 200-cm in air. (b) Plasma fluorescence, (c) electric field waveform during the discharge. (d) Successive high-speed frames of the discharge. Reproduced, with permission, from Reference [228]

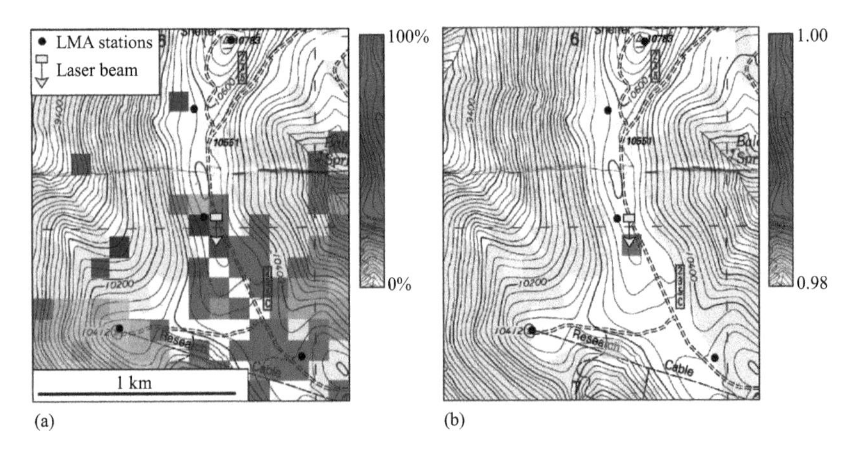

Figure 12.25 (a) Pulses synchronized with the laser repetition rate; (b) corresponding statistical confidence level. The colour scale is set to transparent below 98% (i.e., for error risks above 2%). Topographic background courtesy of US Geological Survey. Reproduced, with permission, from Reference [229], the OSA

analysis allowed one to assess the synchronicity of the electric activity in the vicinity of the laser location with the laser pulses (Figure 12.25). While lightning activity is intense over a wide area around the experimental spot, statistical significance is achieved only at the laser beam location (red square on the left side of Figure 12.25). Furthermore, significant correlations between the laser and thundercloud activity were only observed for ground-level electric fields above 10 kV/m, i.e., when the thunderstorm was most active.

However, no lightning strike was triggered and guided to the ground, due to the limited plasma lifetime. To overcome this limitation, the same team initiated the development of an 1-J, 1-ps, 1-kHz thin-disc laser emitting at 1,030 nm [231], providing both high peak and average power, 1 TW and 1 kW, respectively. Frequency doubling and tripling the laser allows one to emit 68 mJ at 515 nm and 15 mJ at 343 nm [211]. This laser has been installed at the top of the Säntis mountain, at 2,502-m altitude, close to a 120-m tall telecommunication tower equipped with lightning diagnostics [232]. The laser is protected from the harsh environment by a double-shell composed of the telecommunication radome and an air-conditioned and air-tight tent, with a temperature stable within $\pm 2°C$. In a configuration close to Uchida's [173], the laser beam is expanded to a 400-mm diameter and focused in the vicinity of the top of the telecommunication tower, with an angle of 7° with regard to nadir. A pre-chirp ensures that the filaments start at the antenna top and propagate beyond, so as to guide triggered lightning to the earthing system of the tower. Air safety is ensured by a no-fly zone activated before each experiment.

Assessing the effect of the laser in spite of the intrinsically stochastic and fast-evolving character of lightning is based on a firing sequence designed to optimize the statistical analysis. More specifically, measurements are decomposed in periods of 10 min, corresponding to the temporal resolution of the weather data. During each period, the laser is switched ON and OFF alternatively on a regular cycle, with an ON time allowing for cumulative thermal effects and an OFF time sufficient to ensure the renewal of by side wind of the air parcel swept by the laser filaments. Furthermore, radio-frequency antennas provide a precise time- and spatially resolved map of the lightning activity to be compared with the GPS timestamps of the laser pulses [232].

This experimental campaign follows a line of continuous efforts over the last 25 years. It is tailored to provide a conclusive assessment of the ability of the best laser technology available today to effectively control lightning in real scale. It will also contribute to a new understanding of the physics of ascending lightning strikes.

12.6 Quantum control of HV discharges and fog clearing

As already described, filament-induced hydrodynamics play an important role in controlling atmospheric processes like lightning guiding and fog creation/clearing. In the former section, the air density depletion, associated with shockwave generation, was produced by the thermalization of the filament plasma. However, it was recently shown [233] that such fast temperature jump could also be produced in air at lower, non-ionizing, intensity levels by inducing a fast rotation of nitrogen and oxygen molecules via molecular alignment and field-free coherent rotation (called 'quantum molecular wake') [233–238].

Laser-induced alignment of a molecular ensemble is described by solving the following time-dependent Liouville equation:

$$i\frac{\partial\rho(t)}{\partial t} = [H_{Rot} + H_{int}; \rho(t)]$$

where $[A;B]=AB-BA$ is the commutator operator, $\rho(t)$ the density operator, and $H_{Rot}=BJ^2-DJ^4$ the rotational Hamiltonian, with J the angular momentum and B (resp. D) the rotational (resp. centrifugal distortion) constants. For a linear molecule and a linearly polarized radiation,

$$H_{int} = -\frac{1}{4} \in (t)^2 \Delta\alpha \cos^2\theta$$

with $\Delta\alpha$ the molecular polarizability anisotropy and θ the angle between the molecular axis and the laser polarization direction z [239]. The degree of molecular alignment with respect to z can be assessed by

$$\langle\langle\cos^2\theta\rangle\rangle = Tr(\rho\cos^2\theta)$$

As an example, Figure 12.26(a) displays the alignment of nitrogen molecules, calculated for different laser intensities. Molecular alignment is characterized by

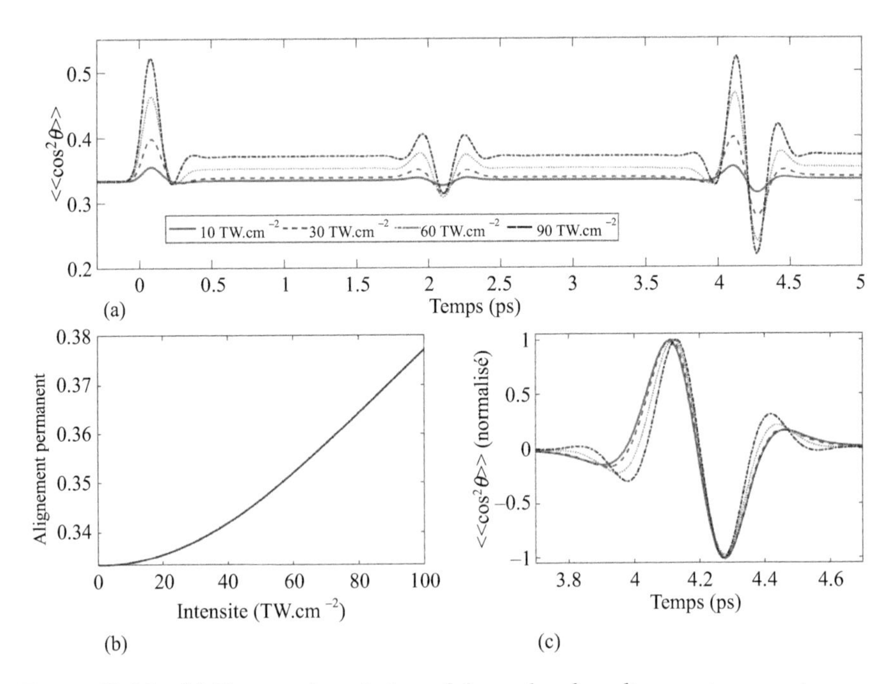

Figure 12.26 (a) Temporal evolution of the molecular alignment parameter induced in N_2 at 300K by a 100-fs pulse with different peak intensities. (b) Permanent alignment as a function of intensity. (c) First revival for different peak intensities

revivals, which appear at each quarter of the rotational period. The revival amplitude is proportional to the laser intensity in the weak-field regime, i.e., for intensities below 60 TW/cm^2 [238]. Beyond these intensities, the structural shape of the revivals is no longer conserved (Figure 12.26(c)). Notice that a permanent alignment is also induced by the interaction, due to the population redistribution among the rotational states.

More precisely, impulsive Raman scattering induces a coherent superposition of the rotational states $|J, m\rangle$, i.e., rotational wave packets:

$$|\Psi_R(t)\rangle = \Sigma_{J,m} c_{J,m} |J, m\rangle e^{-iE_R t/\hbar}$$

where J and m are the quantum numbers of the rotational angular momentum and E_R are the eigenvalues of the free rotor Hamiltonian. In the rigid rotor approximation (i.e., assuming $D=0$), $E_R=hc\ B_e\ J(J+1)$, where B_e is the spectroscopic rotational constant in wavenumbers, h the Planck constant, and c the speed of light. At $t=0$, the non-adiabatic laser excitation locks in phase all the Raman-excited rotational states. After the pulse, the evolution of the wave packet is driven by the field-free Hamiltonian, which generates dephasing and spread (as seen from the equation above). However, as the evolution is coherent, rephasing occurs in the form of revivals at specific times, which leads to a full realignment period of $t=T_R=1/2B_e c$ (8.36 ps in N_2 and 11.2 ps in O_2). Due to the even integer distribution of the rotational frequencies, additional alignment or anti-alignment recurrences also occur at $t=T_R/4$, $t=T_R/2$, and $t=3T_R/4$ [240,241].

The molecular rotation induced by the laser leads to a fast increase of the gas temperature, as shown by Zahedpour *et al.* [233]. In their pioneering work, they also showed that this heating can be controlled by launching a train of pulses separated by the rephasing times. By using a sequence of four pulses separated by T_R, the molecular rotation of N_2 was driven to equivalent temperatures as high as 450 K (30 meV/molecule of additional heating). Conversely, the rotation could be coherently hindered by applying series of pulses resonant at times where molecules are anti-aligned ($T_R/2$). The thermalization of the rotationally heated molecules in the surrounding gas created a shockwave that exceeds the amplitude of the plasma shockwave previously observed in filaments.

In a recent study [242], the coherent control of rotational wave packets in air was applied to produce a *plasma-less* shockwave and open transmission channels in fog and cloud without necessarily requiring filamentation. Larger diameter beams and/or mid-IR lasers could then be used for fog clearing, for which long-distance propagation is more easily achievable and controllable.

For the experimental demonstration of the concept, a train consisting of eight ultrashort pulses at 800 nm was produced with a Michelson interferometer [243], repeated at a repetition rate of 1 kHz.

A cw counter-propagating laser at 1.55 μm, superimposed to the TiSA beam, was used for measuring the gain in transmission and for investigating the dynamics of the droplets expulsion by the shockwave. The size distribution of the droplets in the cloud chamber was centred about 5 μm with a number density variable from 0

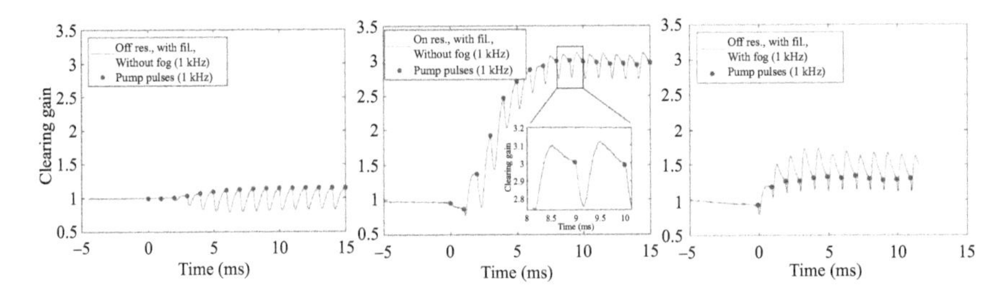

Figure 12.27 Fog clearing induced by the rotational quantum wake. Transmission gain in the case of (a) no fog with the eight-pulse train pump laser, (b) fog (1.3×10^5 cm^{-3} droplet concentration) with the eight-pulse train tuned on-resonance with the quantum wake rephasing time, and (c) fog (0.7×10^5 cm^{-3} droplet concentration) with the eight-pulse train tuned off-resonance with the quantum wake rephasing time. Total energy of the pump: 3.8 mJ. Transmission is normalized to one in each case, corresponding to the situation where no pump is present ($t<0$) [242]

to 1.5×10^5 cm^{-3} (1,000 times more than in a cumulus cloud). This variation corresponds to attenuation values ranging from 0 (no fog) to -10 dB.

Figure 12.27(a) shows the effect of the eight-pulse train pump on the telecom laser transmission in the absence of fog. The air density depletion due to rotational heating causes an initial defocusing of the telecom laser due to the induced refractive index gradient. This leads to the observed modulation of the signal at 1 kHz, due to the finite aperture of the detection system. In this geometry, the envelope of the modulation requires 2–3 ms to reach a steady state. In the presence of fog (Nd$\sim1.3\times10^5$ cm^{-3}), corresponding to 9-dB attenuation, the eight-pulse train increases the telecom laser transmission by a factor 3 (i.e., 4.8 dB) but only if the pulse interval in the train is tuned on-resonance with the full revival time of 8.36 ps (Figure 12.27(b)). When the eight-pulse interval is tuned slightly off-resonance, the gain drops to a factor of only $\sim$1.5 (1.7 dB, Figure 12.27(c)). This remaining gain is caused by partial rotational heating.

This discovery may open new perspectives for real, atmospheric scale operation like FSO. For instance, intense mid-IR lasers, which provide only moderate ionization as they propagate in air [244–247], are now potentially very attractive candidates for fog clearing applications. It was recently reported in particular that TW peak powers CO_2 lasers can provide wide-diameter self-guiding channels over kilometric distances [248], fulfilling the requirements for FSO between earth and satellites.

Another quantum-controlled atmospheric process is high-voltage breakdown and lightning control. As for the fog clearing case, the air depletion can be induced by rotational heating and thus increases the mean free path of the available charge carriers in the external field and thus their kinetic energy before collision. A recent study of this process was reported in the case of short spark gaps [243]. A detailed comparison of ionizing laser pulses versus non-ionizing pulse trains confirmed that the main mechanism at the origin of triggering breakdown in air is the air depletion due to heating, whatever its origin.

Interestingly, the peak current for a single intense pulse increases versus gap field more than in the eight-pulse case (Figure 12.28), owing to its higher initial electron density and increased electrode-driven air heating and channel depth. Even though the eight-pulse train generates less than ~1% of the electron density of the single pulse, the peak current is roughly 50% of the single-pulse case until they converge when the gap voltage is close to the breakdown threshold.

In the case of plasma and density–hole generation by a single filamenting pulse, the electrons are initially driven by the gap field and further heat the air and deepen the depression channel, with subsequent current provided by the electrodes and channelled through the density hole. In the case of little plasma but comparable density–hole generation from an eight-pulse sequence, the lower current initially provided by the electrodes is preferentially channelled through the density–hole, heating it further. As gap voltage is increased, impact ionization increases the gap current to the point where the current and the density–hole volume in both cases are comparable. This leads to similar breakdown thresholds. Below the breakdown threshold, laser excitation of the gap triggers an electrode-supplied current transient through the reduced density channel. As the channel widens and decreases in depth due to transverse thermal diffusion, the current decays on the thermal diffusion timescale of ~100 μs and the air gap returns to a non-conducting equilibrium. These results also show, therefore, that ionization is not necessarily required for triggering discharges over short distances, but that a heated air column with reduced density is.

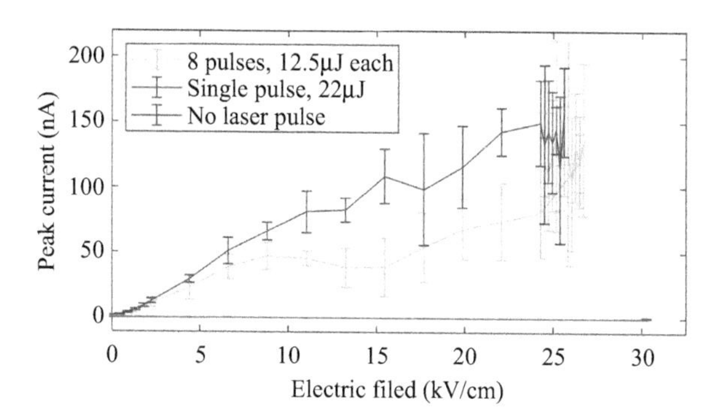

Figure 12.28 Peak gap current versus gap field below the breakdown threshold for the single-pulse (blue, plasma) and eight-pulse (green, no plasma) cases. The blue and green curves converge and terminate where breakdown occurs at ~25 kV/cm. The no-laser case (red) is shown for comparison; breakdown occurs near ~30 kV/cm [243]

12.7 Conclusion

As reviewed in this chapter, ultrashort and ultra-intense laser filamentation opens promising application perspectives in several fields of atmospheric sciences, including remote sensing, cloud transmission manipulation, and lightning control. While the upscaling of these processes from the laboratory to the atmospheric scale is still under investigation, the fast increase of the available laser peak and average power, repetition rate, and pulse energy and the corresponding decrease of the costs permanently renews the prospects to effectively develop the current results to the real world.

References

[1] R. M. Measures, *Laser Remote Sensing*, Krieger Publishing Company, Malabar, Florida (1992).

[2] U. Pöschl. 'Atmospheric aerosols: Composition, transformation, climate and health effects', *Angewandte Chemie* 44, 7520–7540 (2005).

[3] J. H. Seinfeld and S. N. Pandis, *Atmospheric Chemistry and Physics: From Air Pollution to Climate Change*, 2nd Edition, Wiley (2006).

[4] International Panel on Climate Change, Climate Change 2013: The Physical Science Basis. *The Physical Science Basis. Contribution of Working Group I to the Fifth Assessment Report of the Intergovernmental Panel on Climate Change*, eds. T. F. Stocker, D. Qin, G.-K. Plattner, *et al.* Cambridge University Press, Cambridge, United Kingdom and New York, NY, USA (2013).

[5] J. Kasparian, M. Rodriguez, G. Méjean, *et al.* 'White-light filaments for atmospheric analysis', *Science* 301, 61 (2003).

[6] A. Couairon and A. Mysyrowicz. 'Self-focusing and filamentation of femtosecond pulses in air and condensed matter: Simulations and experiments', *Physics Reports* 44, 47 (2007).

[7] L. Bergé, S. Skupin, R. Nuter, J. Kasparian, and J.-P. Wolf. 'Ultrashort filaments of light in weakly-ionized, optically-transparent media', *Reports on Progress in Physics* 70, 1633 (2007).

[8] J. Kasparian and J.-P. Wolf. 'Physics and applications of atmospheric nonlinear optics and filamentation', *Optics Express* 16, 466, (2008).

[9] S. L. Chin, S. A. Hosseini, W. Liu, *et al.* 'The propagation of powerful femtosecond laser pulses in optical media: Physics, applications, and new challenges', *Canadian Journal of Physics* 83, 863 (2005).

[10] R. Ackermann, G. Méjean, J. Kasparian, J. Yu, E. Salmon, and J.-P. Wolf. 'Laser filaments generated and transmitted in highly turbulent air', *Optics Letters* 31, 86–88 (2006).

[11] R. Salamé, N. Lascoux, E. Salmon, J. Kasparian, and J. P. Wolf. 'Propagation of laser filaments through an extended turbulent region', *Applied Physics Letters* 91, 171106 (2007).

[12] G. Méchain, G. Méjean, R. Ackermann, *et al.* 'Propagation of fs-TW laser filaments in adverse atmospheric conditions', *Applied Physics B* 80, 785–789 (2005).

[13] F. Courvoisier, V. Boutou, J. Kasparian, *et al.* 'Light filaments transmitted through clouds', *Applied Physics Letters* 83, 213–215 (2003).

[14] M. Kolesik and J. V. Moloney. 'Self-healing femtosecond light filaments', *Optics Letters* 29, 590–592 (2004).

[15] S. Skupin, L. Bergé, U. Peschel, and F. Luderer. 'Interaction of femtosecond light filaments with obscurants in aerosols', *Physical Review Letters* 93, 023901–023901 (2004).

[16] M. Rodriguez, R. Bourayou, G. Méjean, *et al.* 'Kilometer-range non-linear propagation of femtosecond laser pulses', *Physical Review E* 69, 036607–036607 (2004).

[17] G. Méchain, A. Couairon, Y.-B. André, *et al.* 'Long-range self-channeling of infrared laser pulses in air: A new propagation regime without ionization', *Applied Physics B* 79, 379 (2004).

[18] B. La Fontaine, F. Vidal, Z. Jiang, *et al.* 'Filamentation of ultrashort pulse laser beams resulting from their propagation over long distances in air', *Physics of Plasmas* 6, 1615–1621 (1999).

[19] G. G. Luther, A. C. Newell, J. V. Moloney, and E. M. Wright. 'Short-pulse conical emission and spectral broadening in normally dispersive media', *Optics Letters* 19, 789–791 (1994).

[20] E. T. J. Nibbering, P. F. Curley, G. Grillon, *et al.*'Conical emission from self-guided femtosecond pulses in air', *Optics Letters* 21, 62–64 (1996).

[21] O. G. Kosareva, V. P. Kandidov, A. Brodeur, C. Y. Chen, and S. L. Chin. 'Conical emission from laser-plasma interactions in the filamentation of powerful ultrashort laser pulses in air', *Optics Letters* 22, 1332–1334 (1997).

[22] G. Faye, J. Kasparian, and R. Sauerbrey. 'Modifications to the lidar equation due to nonlinear propagation in air', *Applied Physics B* 73, 157–163 (2001).

[23] J. Yu, D. Mondelain, G. Ange, *et al.* 'Backward supercontinuum emission from a filament generated by ultrashort laser pulses in air', *Optics Letters* 26, 533–535 (2001).

[24] Y. V. Rostovtsev, Z. E. Sariyanni, and M. O. Scully. 'Electromagnetically induced coherent backscattering', *Physical Review Letters* 97, 113001 (2006).

[25] L. Yuan, K. E. Dorfman, A. M. Zheltikov, and M. O. Scully. 'Plasma-assisted coherent backscattering for standoff spectroscopy', *Optics Letters* 37, 987–989 (2012).

[26] S. Weber, C. Riconda, L. Lancia, J.-R. Marques, G. A. Mourou, and J. Fuchs. 'Amplification of ultrashort laser pulses by Brillouin backscattering in plasmas', *Physical Review Letters* 111, 055004 (2013).

[27] N. Jhajj, E. W. Rosenthal, R. Birnbaum, J. K. Wahlstrand, and H. M. Milchberg. 'Demonstration of long-lived high-power optical waveguides in air', *Physical Review X* 4, 011027 (2014).

[28] Q. Luo, W. Liu, and S. L. Chin. 'Lasing action in air induced by ultra-fast laser filamentation', *Applied Physics B* 76, 337 (2003).

[29] D. Kartashov, S. Ališauskas, G. Andriukaitis, *et al.* 'Free-space nitrogen gas laser driven by a femtosecond filament', *Physical Review A* 86, 033831 (2012).
[30] P. Sprangle, J. Peñano, B. Hafizi, D. Gordon, and M. Scully. 'Remotely induced atmospheric lasing', *Applied Physics Letters* 98, 211102 (2011).
[31] P. R. Hemrner, R. B. Miles, P. Polynkin, *et al.* 'Standoff spectroscopy via remote generation of a backward-propagating laser beam', *Proceedings of the National Academy of Sciences of the United States of America* 108, 3130–4 (2011).
[32] A. Dogariu, J. B. Michael, M. O. Scully, and R. B. Miles. 'High-gain backward lasing in air', *Science* 331, 442 (2011).
[33] R. W. Boyd, *Nonlinear Optics*, Cambridge: Academic Press (1992).
[34] P. Maioli, R. Salamé, N. Lascoux, *et al.* 'Ultraviolet-visible conical emission by multiple laser filaments', *Optics Express* 17, 4726–4731 (2009).
[35] J. Kasparian, R. Sauerbrey, D. Mondelain, *et al.* 'Infrared extension of the supercontinuum generated by fs-TW-laser pulses propagating in the atmosphere', *Optics Letters* 25, 1397–1399 (2000).
[36] F. Théberge, M. Châteauneuf, V. Ross, P. Mathieu, and J. Dubois. 'Ultrabroadband conical emission generated from the ultraviolet up to the far-infrared during the optical filamentation in air', *Optics Letters* 33, 2515–2517 (2008).
[37] S. L. Chin, S. Petit, F. Borne, and K. Miyazaki. 'The white light supercontinuum is indeed an ultrafast white light laser', *Japanese Journal of Applied Physics* 38, L126-L128–L126-L128 (1999).
[38] J. Kasparian. 'Some properties of femtosecond laser filamentation relevant to atmospheric applications II. Large-scale filamentation', in *Progress in Ultrafast Intense Laser Science II*, eds. K. Yamanouchi, S. L. Chin, P. Agostini, G. Ferrante, Heidelberg, Germany: Springer (2007). pp. 301–318.
[39] L. Wöste, C. Wedekind, H. Wille, *et al.* 'Femtosecond atmospheric lamp', *Laser und Optoelektronik* 29, 51–51 (1997).
[40] P. Rairoux, H. Schillinger, S. Niedermeier, *et al.* 'Remote sensing of the atmosphere using ultrashort laser pulses', *Applied Physics B* 71, 573–580 (2000).
[41] P. Béjot, L. Bonacina, J. Extermann, *et al.* '32 Terawatt Atmospheric White-Light Laser', *Applied Physics Letters* 90, 151106–151106 (2007).
[42] M. Petrarca, S. Henin, N. Berti, *et al.* 'White-light femtosecond Lidar at 100 TW power level', *Applied Physics B* 114, 319–325 (2014).
[43] Y. Petit, S. Henin, W. M. Nakaema, *et al.* '1-J white-light continuum from 100-TW laser pulses', *Physical Review A* 83, 013805 (2011).
[44] H. Wille, M. Rodriguez, J. Kasparian, *et al.* 'Teramobile: A mobile femtosecond-terawatt laser and detection system', *European Physical Journal – Applied Physics* 20, 183–190 (2002).
[45] L. S. Rothman, A. Barbe, D. C. Benner, *et al.* 'The HITRAN molecular spectroscopic database: Edition of 2000 including updates of 2001', *Journal of Quantitative Spectroscopy and Radiation Transfer* 82, 5 (2003).

[46] R. Bourayou, G. Méjean, J. Kasparian, *et al.* 'White-light filaments for multiparameter analysis of cloud microphysics', *Journal of the Optical Society of America B* 22, 369 (2005).
[47] G. Méjean, J. Kasparian, E. Salmon, *et al.* 'Towards a supercontinuum-based infrared Lidar', *Applied Physics B* 77, 357–359 (2003).
[48] M. C. Galvez, M. Fujita, N. Inoue, R. Moriki, Y. Izawa, and C. Yamanaka. 'Three-wavelength backscatter measurement of clouds and aerosols using a white light lidar system', *Japanese Journal of Applied Physics* 41, L284-L286–L284-L286 (2002).
[49] T. Somekawa, C. Yamanaka, M. Fujita, and M. C. Galvez. 'Observation of Asian dust aerosols with depolarization lidar using a coherent white light continuum', *Japanese Journal of Applied Physics* 47, 2155–2157 (2008).
[50] T. Somekawa, C. Yamanaka, M. Fujita, and M. C. Galvez. 'Development of white light depolarization lidar system', *Review of Laser Engineering* 37, 758–62 (2009).
[51] G. Méjean, J. Kasparian, J. Yu, S. Frey, E. Salmon, and J.-P. Wolf. 'Remote detection and identification of biological aerosols using a femtosecond terawatt Lidar system', *Applied Physics B* 78, 535–537 (2004).
[52] S. C. Hill, V. Boutou, J. Yu, *et al.* 'Enhanced backward-directed multi-photon-excited fluorescence from dielectric microspheres', *Physical Review Letters* 85, 54–57 (2000).
[53] P. Rohwetter, J. Yu, G. Méjean, *et al.* 'Remote LIBS with ultra-short pulses: Characteristics in picosecond and femtosecond regimes', *Journal of Analytical Atomic Spectrometry* 19, 437–437 (2004).
[54] H. L. Xu, J. Bernhardt, P. Mathieu, G. Roy, and S. L. Chin. 'Understanding the advantage of remote femtosecond laser-induced breakdown spectroscopy of metallic targets', *Journal of Applied Physics* 101, 033124 (2007).
[55] J.-F. Gravel, Q. Luo, D. Boudreau, X. P. Tang, and S. L. Chin. 'Sensing of halocarbons using femtosecond laser-induced fluorescence', *Analytical Chemistry* 76, 4799 (2004).
[56] H. L. Xu, G. Mejean, W. Liu, *et al.* 'Remote detection of similar biological materials using femtosecond filament-induced breakdown spectroscopy', *Applied Physics B-Lasers and Optics* 87, 151, 156 (2007).
[57] Q. Luo, H. L. Xu, S. A. Hosseini, *et al.* 'Remote sensing of pollutants using femtosecond laser pulse fluorescence spectroscopy', *Applied Physics B* 82, 105–109 (2006).
[58] Y. Kamali, J.-F. Daigle, F. Théberge, *et al.* 'Remote sensing of trace methane using mobile femtosecond laser system of T&T Lab', *Optics Communications* 282, 2062–2065 (2009).
[59] L. J. Radziemski and D. A. Cremers, *Spectrochemical Analysis Using Laser Plasma Excitation, in Laser-Induced Plasma: Physical, Chemical and Biological Applications*, eds. L. J. Radziemski and D. A. Cremers, Marcel Dekker, New York, NY (1989).

[60] C. C. García, M. Corral, J. M. Vadillo, and J. J. Laserna. 'Angle-resolved laser-induced breakdown spectroscopy for depth profiling of coated materials', *Applied Spectroscopy* 54, 1027–1031 (2000).
[61] K. Stelmaszczyk, P. Rohwetter, G. Méjean, *et al.* 'Long-distance remote laser-induced breakdown spectroscopy using filamentation in air', *Applied Physics Letters* 85, 3977 (2004).
[62] J.-F. Daigle, P. Mathieu, G. Roy, J.-R. Simard, and S. L. Chin. 'Multi-constituents detection in contaminated aerosol clouds using remote-filament-induced breakdown spectroscopy', *Optics Communications* 278, 147–152 (2007).
[63] H. L. Xu, W. Liu, and S. L. Chin. 'Remote time-resolved filament-induced breakdown spectroscopy of biological materials', *Optics Letters* 31, 1540–1542 (2006).
[64] P. Rohwetter, J. Kasparian, K. Stelmaszczyk, *et al.* 'Laser-induced water condensation in air', *Nature Photonics* 4, 451–456 (2010).
[65] S. Henin, Y. Petit, P. Rohwetter, *et al.* 'Field measurements suggest the mechanism of laser-assisted water condensation', *Nature Communications* 2, 456 (2011).
[66] J. Ju, J. Liu, C. Wang, *et al.* 'Laser-filamentation-induced condensation and snow formation in a cloud chamber', *Optics Letters* 37, 1214–1216 (2012).
[67] J. C. S. Chagas, T. Leisner, and J. Kasparian. 'Pump-probe differential Lidar to quantify atmospheric supersaturation and particle forming trace gases', *Applied Physics B* 117, 667 (2014).
[68] S. Yuan, T. Wang, P. Lu, S. L. Chin, and H. Zeng. 'Humidity measurement in air using filament-induced nitrogen monohydride fluorescence spectroscopy', *Applied Physics Letters* 104, 091113 (2014).
[69] J.Kasparian and J.-P. Wolf. 'A new transient SRS analysis method of aerosols and application to a non-linear femtosecond Lidar', *Optics Communications* 152, 355–360 (1998).
[70] J.-P. Wolf, Y.-L. Pan, G. M. Turner, *et al.* 'Ballistic trajectories of optical wave packets within microcavities', *Physical Review A* 64, 023808–023808 (2001).
[71] L. Méès, J.-P. Wolf, G. Gouesbet, and G. Gréhana. 'Two-photon absorption and fluorescence in a spherical micro-cavity illuminated by using two laser pulses: Numerical simulations', *Optics Communications* 208, 371–375 (2002).
[72] R. S. Judson and H. Rabitz. 'Teaching lasers to control molecules', *Physical Review Letters* 68, 1500–1503 (1992).
[73] J. Extermann, P. Béjot, L. Bonacina, P. Billaud, J. Kasparian, and J.-P. Wolf. 'Effects of atmospheric turbulence on remote optimal control experiments', *Applied Physics Letters* 92, 041103 (2008).
[74] J. L. Meyzonnette and T. Lépine, *Bases de radiométrie optique 2ème édition*, Toulouse, France: Cépaduès-Editions (1999).
[75] N. Berti, W. Ettoumi, J. Kasparian, and J.-P. Wolf. 'Time-reversibility of laser filamentation', *Optics Express* 22, 21061–21068 (2014).

[76] N. Berti, W. Ettoumi, S. Hermelin, J. Kasparian, and J.-P. Wolf. 'Nonlinear synthesis of complex laser waveforms at remote distances', *Physical Review A* 91, 063833 (2015).
[77] S. Ricaud, P. Georges, P. Camy, *et al.* 'Diode-pumped regenerative Yb: SrF_2 amplifier', *Applied Physics B-Lasers and Optics* 106, 823–827 (2012).
[78] C. T. R. Wilson, Phil Trans. 'On the condensation nuclei produced in gases by the action of Röntgen rays. Uranium rays, ultra-violet light, and other agents', *Royal Society of London. A* 192, 403–453 (1899).
[79] K. Yoshihara. 'Laser-induced mist and particle formation from ambient air: A possible new cloud seeding method', *Chemistry Letters* 34, 1370 (2005).
[80] K. Yoshihara, Y. Takatori, and Y. Kajii. 'Water aerosol formation upon irradiation of air using KrF laser at 248 nm', *Bulletin of the Chemical Society of Japan* 85, 11551159 (2012).
[81] K. Yoshihara, Y. Sakamoto, M. Kawasaki, Y. Takatori, S. Kato, and Y. Kajii. 'BCSJ Award article', *Bulletin of the Chemical Society of Japan* 87, 593602 (2014).
[82] I. D. Clark and J. F. Noxon. 'Particle formation during water-vapor photolysis', *Science* 174, 941–944 (1971).
[83] W. Hua, Z. M. Chen, C. Y. Jie, *et al.* 'Atmospheric hydrogen peroxide and organic hydroperoxides during PRIDE-PRD'06, China: Their concentration, formation mechanism and contribution to secondary aerosols', *Atmospheric Chemistry and Physics* 8, 6755–6773 (2008).
[84] H. R. Pruppacher and J. D. Klett, *Microphysics of Clouds and Precipitation*, Kluwer Academic Publishers, Dordrecht, Boston (1997).
[85] I. A. Kossyi, A. Kostinsky, A. A. Matveyev, and V. P. Silakov. 'Kinetic scheme of the non-equilibrium discharge in nitrogen-oxygen mixtures', *Plasma Sources Science and Technology* 1, 207 (1992).
[86] H. L. Xu, A. Azarm, J. Bernhardt, Y. Kamali, and S. L. Chin. 'The mechanism of nitrogen fluorescence inside a femtosecond laser filament in air', *Chemical Physics* 360, 171 (2009).
[87] M. Kulmala, A. Laaksonen, P. Korhonen, T. Vesala, T. Ahonen, and J. C. Barrett. 'The effect of atmospheric nitric acid vapor on cloud condensation nucleus activation', *Journal of Geophysical Research* 98, 22949–22958 (1993).
[88] Y. Petit, S. Henin, J. Kasparian, and J. P. Wolf. 'Production of ozone and nitrogen oxides by laser filamentation', *Applied Physics Letters* 97, 021108 (2010).
[89] F. Valle Brozas, C. Salgado, J. I. Apiñaniz, *et al.* 'Determination of the species generated in atmospheric-pressure laser-induced plasmas by mass spectrometry techniques', *Laser Physics* 26, 055602 (2016).
[90] A. Camino, S. Li, Z. Hao, and J. Lin. 'Spectroscopic determination of NO_2, NO_3, and O_3 temporal evolution induced by femtosecond filamentation in air', *Applied Physics Letters* 106, 021105 (2015).

[91] M. Chiwa, H. Kondo, N. Ebihara, and H. Sakugawa. 'Atmospheric concentrations of nitric acid, sulfur dioxide, particulate nitrate and particulate sulfate, and estimation of their dry deposition on the urban- and mountain-facing sides of Mt. Gokurakuji, Western Japan', *Environmental Monitoring and Assessment* 140, 349 (2008).

[92] H. Kohler. 'The nucleus in and the growth of hygroscopic droplets', *Transactions of the Faraday Society* 32, 1152–1161 (1936).

[93] A. Laaksonen, P. Korhonen, M. Kulmala, and R. Charlson. 'Modification of the Köhler equation to include soluble trace gases and slightly soluble substances', *Journal of the Atmospheric Sciences* 55, 853 (1998).

[94] P. Rohwetter, J. Kasparian, L. Woste, and J. P. Wolf. 'Modelling of HNO_3-mediated laser-induced condensation: A parametric study', *Journal of Chemical Physics* 135, 134703 (2011).

[95] J. Kasparian, P. Rohwetter, L. Woeste, and J. P. Wolf. 'Laser-assisted water condensation in the atmosphere: A step towards modulating precipitation?', *Journal of Physics D* 45, 293001 (2012).

[96] S. L. Clegg and A. S. Wexler. 'Densities and Apparent Molar Volumes of Atmospherically Important Electrolyte Solutions. 1. The Solutes H_2SO_4, HNO_3, HCl, Na_2SO_4, $NaNO_3$, NaCl, $(NH_4)_2SO_4$, NH_4NO_3, and NH_4Cl from 0 to 50°C, Including Extrapolations to Very Low Temperature and to the Pure Liquid State, and $NaHSO_4$, NaOH, and NH_3 at 25°C', *Journal of Physical Chemistry* 115, 3393–3460 (2011).

[97] S. L. Clegg, P. Brimblecombe, and A. S. Wexler. 'Thermodynamic Model of the System $H^+-NH_4^+-SO_4^{2-}-NO_3^--H_2O$ at Tropospheric Temperatures', *Journal of Physical Chemistry A* 102, 2137–2154 (1998).

[98] C. S. Dutcher, A. S. Wexler, and S. L. Clegg. 'Surface tensions of inorganic multicomponent aqueous electrolyte solutions and melts', *Journal of Physical Chemistry A* 114, 12216–12230 (2010).

[99] O. Möhler, O. Stetzer, S. Schaefers, *et al.* 'Experimental investigation of homogeneous freezing of sulphuric acid particles in the aerosol chamber AIDA', *Atmospheric Chemistry and Physics* 3, 211–223 (2003).

[100] M. Schnaiter, S. Buttner, O. Mohler, J. Skrotzki, M. Vragel, and R. Wagner. 'Influence of particle size and shape on the backscattering linear depolarisation ratio of small ice crystals – Cloud chamber measurements in the context of contrail and cirrus microphysics', *Atmospheric Chemistry and Physics* 12, 10465–10484 (2012).

[101] H. Saathoff, S. Henin, K. Stelmaszczyk, *et al.* 'Laser filament-induced aerosol formation', *Atmospheric Chemistry and Physics* 13, 4593–4604 (2013).

[102] M. Kulmala, H. Vehkamaki, T. Petaja, *et al.* 'Formation and growth rates of ultrafine atmospheric particles: A review of observations', *Journal of Aerosol Sciences* 35, 143–176 (2004).

[103] T. Kurten, L. Torpo, C. G. Ding, *et al.* 'A density functional study on water-sulfuric acid-ammonia clusters and implications for atmospheric cluster formation', *Journal of Geophysical Research D* 112, 04210 (2007).

[104] F. Fresnet, G. Baravian, L. Magne, *et al.* 'Influence of water on NO removal by pulsed discharge in $N_2/H_2O/NO$ mixtures', *Plasma Sources Science and Technology* 11, 152–160 (2002).
[105] W. Tröstl, K. Chuang, H. Gordon, *et al.* 'The role of low-volatility organic compounds in initial particle growth in the atmosphere', *Nature* 533, 527–531 (2016).
[106] J. Kirkby, K. Duplissy, Sengupta, *et al.* 'Ion-induced nucleation of pure biogenic particles', *Nature* 533, 521 (2016).
[107] F. Vidal, D. Comtois, C.-Y. Chien, *et al.* 'Modeling the triggering of streamers in air by ultrashort laser pulses', *IEEE Transactions of Plasma Science* 28, 418 (2000).
[108] O. Lahav, L. Levi, I. Orr, *et al.* 'Long-lived waveguides and sound-wave generation by laser filamentation', *Physical Review A* 90, 021801 (2014).
[109] N. Jhajj, E. W. Rosenthal, R. Birnbaum, J. K. Wahlstrand, and H. M. Milchberg. 'Demonstration of long-lived high-power optical waveguides in air', *Physical Review X* 4, 011027 (2014).
[110] Y. H. Cheng, J. K. Wahlstrand, N. Jhajj, and H. M. Milchberg. 'The effect of long timescale gas dynamics on femtosecond filamentation', *Optics Express* 21, 4740 (2013).
[111] J. Ju, J. Liu, C. Wang, *et al.* 'Laser-filamentation-induced condensation and snow formation in a cloud chamber', *Optics Letters* 37, 1214–1216 (2012).
[112] J. Ju, J. Liu, C. Wang, *et al.* 'Effects of initial humidity and temperature on laser-filamentation-induced condensation and snow formation', *Applied Physics B* 110, 375–380 (2013).
[113] J. Ju, H. Sun, A. Sridharan, *et al.* 'Laser-filament-induced snow formation in a subsaturated zone in a cloud chamber: Experimental and theoretical study', *Physical Review E* 88, 062803 (2013).
[114] J. Ju, T. Leisner, H. Y. Sun, *et al.* 'Laser-induced supersaturation and snow formation in a sub-saturated cloud chamber', *Applied Physics B* 117, 1001–1007 (2014).
[115] J. Ju, J. S. Liu, H. Liang, *et al.* 'Corona discharge induced snow formation in a cloud chamber', *Scientific Reports* 6, 25417 (2017).
[116] A. V. Bashinov, A. A. Gonoskov, A. V. Kim, G. Mourou, and A. M. Sergeev. 'New horizons for extreme light physics with mega-science project XCELS', *European Physical Journal* 223, 1105–1112 (2014).
[117] M. Petrarca, S. Henin, K. Stelmaszczyk, *et al.* 'Multijoule scaling of laser-induced condensation in air', *Applied Physics Letters* 99, 141103 (2011).
[118] M. Matthews, S. Henin, F. Pomel, *et al.* 'Cooperative effect of ultraviolet and near-infrared beams in laser-induced condensation', *Applied Physics Letters* 103, 264103 (2013).
[119] M. Petrarca, S. Henin, N. Berti, *et al.* 'White-light femtosecond Lidar at 100 TW power level', *Applied Physics B* 114, 319–325 (2014).
[120] S. Yuan, T. Wang, Y. Teranishi, *et al.* 'Lasing action in water vapor induced by ultrashort laser filamentation', *Applied Physics Letters* 102, 224102 (2013).

[121] T. Nagy and P. Simon. 'Single-shot TG FROG for the characterization of ultrashort DUV pulses', *Optics Express* 17, 8144 (2009).

[122] A. Rastegari, E. Schubert, C. Feng, *et al.* 'Beam control through nonlinear propagation in air and laser induced discharges', *Proc. SPIE 9727, Laser Resonators, Microresonators, and Beam Control XVIII*, 97271H (April 22, 2016).

[123] C. Y. Feng, X. Z. Xu, and J. C. Diels. 'Generation of 300 ps laser pulse with 1.2 J energy by stimulated Brillouin scattering in water at 532 nm', *Optics Letters* 39, 3367–3370 (2014).

[124] V. D. Zvorykin, A. A. Ionina, A. O. Levchenkoa, *et al.* 'Extended plasma channels created by UV laser in air and their application to control electric discharges', *Plasma Physics Reports* 41, 112–146 (2015).

[125] V. D. Zvorykin, A. A. Ionin, A. O. Levchenko, *et al.* 'Effects of picosecond terawatt UV laser beam filamentation and a repetitive pulse train on creation of prolonged plasma channels in atmospheric air', *Nuclear Instruments and Methods in Physics Research Section B: Beam Interactions with Materials and Atoms* 309, 2218–222 (2013).

[126] P. Joly, M. Petrarca, A. Vogel, *et al.* 'Laser-induced condensation by ultrashort laser pulses at 248 nm', *Applied Physics Letters* 102, 091112 (2013).

[127] P. DeCarlo, J. Kimmel, A. Trimborn, *et al.* 'Field-deployable, high-resolution, time-of-flight aerosol mass spectrometer', *Analytical Chemistry* 78, 8281–8289 (2006).

[128] M. R. Canagaratna, J. T. Jayne, J. L. Jimenez, *et al.* 'Chemical and microphysical characterization of ambient aerosols with the aerodyne aerosol mass spectrometer', *Mass Spectrometry Review* 26, 185–222 (2007).

[129] R. G. Derwent, J. I. R. Dernie, G. J. Dollard, *et al.* 'Twenty years of continuous high time resolution volatile organic compound monitoring in the United Kingdom from 1993 to 2012', *Atmospheric Environment* 99, 239–247 (2014).

[130] J. E. Shilling, Q. Chen, S. M. King, *et al.* 'Particle mass yield in secondary organic aerosol formed by the dark ozonolysis of α-pinene', *Atmospheric Chemistry and Physics* 8, 2073–2088 (2008).

[131] J. Knoll, M. Canagaratna, Q. Zhang, J. Jimenez, and J. E. A. Tian. 'Organic aerosol components observed in Northern Hemispheric datasets from Aerosol Mass Spectrometry', *Atmospheric Chemistry and Physics* 10, 4625–4641 (2010).

[132] J. Knoll, M. R. Canagaratna, J. L. Jimenez, P. S. Chhabra, J. H. Seinfeld, D. R. Worsnop. 'Changes in organic aerosol composition with aging inferred from aerosol mass spectra', *Atmospheric Chemistry and Physics* 11, 6465–6474 (2011).

[133] C. Weitkamp, *Lidar: Range-Resolved Optical Remote Sensing of the Atmosphere*, Heidelberg, Germany: Springer Verlag (2005).

[134] C. R. Bohren and D. R. Huffman, *Absorption and Scattering of Light by Small Particles*, New York: Wiley and Sons, 1998, ISBN: 978-0-471-29340-8.

[135] J. Kasparian, E. Frejafon, P. Rambaldi, *et al.* 'Characterization of urban aerosols using SEM-microscopy, X-ray analysis and Lidar measurements', *Atmospheric Environment* 32, 2957–2967 (1998).

[136] P. Rairoux, H. Schillinger, S. Niedermeier, *et al.* 'Remote sensing of the atmosphere using ultrashort laser pulses', *Applied Physics B* 71, 573–580 (2000).

[137] R. Bourayou, G. Mejean, J. Kasparian, *et al.* 'White-light filaments for multiparameter analysis of cloud microphysics', *Journal of the Optical Society of America B* 22, 369–377 (2005).

[138] H. L. Xu and S. L. Chin. 'Femtosecond laser filamentation for atmospheric sensing', *Sensors* 11, 32–53 (2011).

[139] National Research Council, *Critical Issues in Weather Modification Research*, Washington, DC: The National Academies Press, 2003.

[140] J. Qiu and D. Cressey. 'Taming the sky: Is it really possible to stop rain, invoke lightning from the heavens or otherwise manipulate the weather?', *Nature* 453, 970–974 (2008).

[141] Q. Miao and B. Geerts. 'Airborne measurements of the impact of ground-based glaciogenic cloud seeding on orographic precipitation', *Advances in Atmospheric Sciences* 30, 1025–1038 (2013).

[142] B. Geerts, Q. Miao, Y. Yang, R. Rasmussen, and D. Breed. 'An airborne profiling radar study of the impact of glaciogenic cloud seeding on snowfall from winter orographic clouds', *Journal of Atmospheric Sciences* 67, 3286–3302 (2010).

[143] V. E. Zuev, A. A. Zemlyanov, Yu. D. Kopytin, and A. V. Kuzikovskii, *High Power Laser Radiation in Atmospheric Aerosols*, VE Reidel Publ, Dordrecht (1985).

[144] H. S. Kwok, T. M. Rossi, W. S. Lau, and D. T. Shaw. 'Enhanced transmission in CO_2-laser–aerosol interactions', *Optics Letters* 13, 192–195 (1988).

[145] V. K. Pustovalov and I. A. Khorunzhii. 'Thermal and optical processes in shattering water aerosol droplets by intense optical radiation', *International journal of Heat and Mass Transfer* 35, 583–589 (1992).

[146] A. A. Zemlyanov, Y. E. Geints, A. M. Kabanov, and R. L. Armstrong. 'Investigation of laser-induced destruction of droplets by acoustic methods', *Applied Optics* 35, 6062–6067 (1996).

[147] R. G. Pinnick, A. Biswas, R. L. Armstrong, S. G. Jennings, J. D. Pendleton, and G. Fernandez. 'Micron-sized droplets irradiated with a pulsed CO_2 laser: Measurement of explosion and breakdown thresholds', *Applied Optics* 29, 918–925 (1990).

[148] P. I. Singh and C. P. Knight. 'Pulsed laser-induced shattering of water drops', *AIAA Journal* 18, 96–100 (1988).

[149] P. Kafalas and A. Ferdinand. 'Fog droplet vaporization and fragmentation by a 10.6-μm laser pulse', *Applied Optics* 12, 29–33 (1973).

[150] F. Courvoisier, V. Boutou, J. Kasparian, *et al.* 'Ultraintense light filaments transmitted through clouds', *Applied Physics Letters* 83, 213–215 (2003).

[151] G. Méjean, J. Kasparian, E. Salmon, *et al.* 'Multifilamentation transmission through fog', *Physical Review E* 72, 026611 (2005).
[152] S. Skupin, L. Bergé, U. Peschel, and F. Lederer. 'Interaction of femtosecond light filaments with obscurants in aerosols', *Physical Review Letters* 93, 023901 (2004).
[153] C. Favre, V. Boutou, S. C. Hill, *et al.* 'White-light nanosource with directional emission', *Physical Review Letters* 89, 035002 (2002).
[154] C. Jeon, D. Harper, K. Lim, *et al.* 'Interaction of a single laser filament with a single water droplet', *Journal of Optics* 17, 055502 (2015).
[155] F. Courvoisier, V. Boutou, C. Favre, S. C. Hill, and J. P. Wolf. 'Plasma formation dynamics within a water microdroplet on femtosecond time scales', *Optics Letters* 28, 206–208 (2003).
[156] A. Lindinger, J. Hagen, L. Socaciu, T. Bernhardt, L. Woeste, and T. Leisner. 'Time-resolved explosion dynamics of H_2O droplets induced by femtosecond laser pulses', *Applied Optics* 43, 5263 (2004).
[157] P. Stockel, I. M. Weidinger, H. Baumgärtel, and T. Leisner. 'Rates of Homogeneous Ice Nucleation in Levitated H_2O and D_2O Droplets', *Journal of Physical Chemistry A* 109, 2540–2546 (2005).
[158] K. N. Liou. 'Influence of cirrus clouds on weather and climate processes: A global perspective', *Monthly Weather Review* 114, 1167–1199 (1986).
[159] M. Matthews, F. Pomel, C. Wender, *et al.* 'Laser vaporization of cirrus-like ice particles with secondary ice multiplication', *Science Advances* 2, e1501912 (2016).
[160] H. Rabeony and P. Mirabel. 'Experimental study of vapor nucleation on ions', *Journal of Physical Chemistry* 91, 1815–1818 (1987).
[161] J. Wolk and R. Strey. 'Homogeneous nucleation of H_2O and D_2O in comparison: The isotope effect', *Journal of Physical Chemistry B* 105, 11683–11701 (2001).
[162] T. Leisner, D. Duft, O. Mohler, *et al.* 'Laser-induced plasma cloud interaction and ice multiplication under cirrus cloud conditions', *Proceedings of the National Academy of Sciences* 110, 10106–10110 (2013).
[163] N. Jhajj, J. K. Wahlstrand, and H. M. Milchberg. 'Optical mode structure of the air waveguide', *Optics Letters* 39, 6312–6315 (2014).
[164] L. De la Cruz, E. Schubert, D. Mongin, *et al.* 'High repetition rate ultrashort laser cuts a path through fog', *Applied Physics Letters* 109, 251105 (2016).
[165] J. R. Vaill, D. A. Tidman, T. D. Wilkerson, and D. W. Koopman. 'Propagation of high-voltage streamers along laser-induced ionization trails', *Applied Physics Letters* 17, 20 (1970).
[166] D. W. Koopman and T. D. Wilkerson. 'Channeling of an ionizing electrical streamer by a laser beam', *Journal of Applied Physics* 42, 1883–1886 (1971).
[167] K. A. Saum and D. W. Koopman. 'Discharges guided by laser-induced rarefaction channels', *Physics of Fluids* 15, 2077 (1972).
[168] T. Shindo, Y. Aihara, M. Miki, *et al.* 'Model experiments of laser-triggered lightning', *IEEE Transactions on Power Delivery* 8, 311–317 (1993).

[169] T. Shindo, M. Miki, Y. Aihara, and A. Wada. 'Laser-guided discharges in long gaps', *IEEE Transactions on Power Delivery* 8, 2016–2022 (1993).

[170] M. Miki, Y. Aihara, and T. Shindo. 'Development of long gap discharges guided by a pulsed CO_2 laser', *Journal of Physics D* 26, 1244–1252 (1993).

[171] M. Miki and A. Wada. 'Guiding of electrical discharges under atmospheric air by ultraviolet laser-produced plasma channel', *Journal of Applied Physics* 80, 3208–3214 (1996).

[172] M. Miki, T. Shindo, and Y. Aihara. 'Mechanisms of guiding ability of CO_2 laser-produced plasmas on pulsed discharges', *Journal of Physics D* 29, 1984–1996 (1996).

[173] S. Uchida, Y. Shimada, H. Yasuda, *et al.* 'Laser-triggered lightning in field experiments', *Journal of Optical Technology* 66, 199–202 (1999).

[174] D. Wang, Z. I. Kawasaki, K. Matsuura, *et al.* 'A preliminary study on laser-triggered lightning', *Journal of Geophysical Research D* 99, 16907–16912 (1994).

[175] D. Wang, T. Ushio, Z. I. Kawasaki, *et al.* 'A possible way to trigger lightning using a laser', *Journal of Atmospheric and Terrestrial Physics* 57, 459–466 (1995).

[176] X. M. Zhao, J. C. Diels, C. Y. Wang, and J. M. Elizondo. 'Femtosecond ultraviolet laser pulse induced lightning discharges in gases', *IEEE Journal of Quantum Electronics* 31, 599–612 (1995).

[177] P. Rambo, J. Schwarz, and J. C. Diels. 'High-voltage electrical discharges induced by an ultrashort-pulse UV laser system', *Journal of Optics A* 3, 146–158 (2001).

[178] H. Pepin, D. Comtois, F. Vidal, *et al.* 'High-voltage electrical discharges induced by an ultrashort-pulse UV laser system', *Physics of Plasmas* 8, 2532–2539 (2001).

[179] D. Comtois, C. Y. Chien, A. Desparois, *et al.* 'Triggering and guiding leader discharges using a plasma channel created by an ultrashort laser pulse', *Applied Physics Letters* 76, 819–821 (2000).

[180] B. La Fontaine, D. Comtois, C. Y. Chien, *et al.* 'Guiding large-scale spark discharges with ultrashort pulse laser filaments', *Journal of Applied Physics* 88, 610–615 (2000).

[181] A. Bondiou and I. Gallimberti. 'Theoretical modelling of the development of the positive spark in long gaps', *Journal of Physics D* 27, 1252 (1994).

[182] M. Rodriguez, R. Sauerbrey, H. Wille, *et al.* 'Triggering and guiding megavolt discharges by use of laser-induced ionized filaments', *Optics Letters* 27, 772–774 (2002).

[183] R. Ackermann, G. Mechain, G. Mejean, *et al.* 'Influence of negative leader propagation on the triggering and guiding of high voltage discharges by laser filaments', *Applied Physics B* 82, 561–566 (2006).

[184] B. Forestier, A. Houard, I. Revel, *et al.* 'Triggering, guiding and deviation of long air spark discharges with femtosecond laser filament', *AIP Advances* 2, 012151 (2012).

[185] D. Comtois, H. Pepin, F. Vidal, *et al.* 'Triggering and guiding of an upward positive leader from a ground rod with an ultrashort laser pulse. I. Experimental results', *IEEE Transactions on Plasma Science* 31, 377–386 (2003).

[186] D. Comtois, H. Pepin, F. Vidal, *et al.* 'Triggering and guiding of an upward positive leader from a ground rod with an ultrashort laser pulse. II. Modeling', *IEEE Transactions on Plasma Science* 31, 387–395 (2003).

[187] M. Henriksson, J. F. Daigle, F. Theberge, M. Chateauneuf, and J. Dubois. 'Laser guiding of Tesla coil high voltage discharges', *Optics Express* 20, 12721–12728 (2012).

[188] Y. Brelet, A. Houard, L. Arantchouk, *et al.* 'Tesla coil discharges guided by femtosecond laser filaments in air', *Applied Physics Letters* 100, 181112 (2012).

[189] J. F. Daigle, F. Theberge, P. Lassonde, *et al.* 'Dynamics of laser-guided alternating current high voltage discharges', *Applied Physics Letters* 103 (18), 184101 (2013).

[190] L. Arantchouk, G. Point, Y. Brelet, *et al.* 'Large scale tesla coil guided discharges initiated by femtosecond laser filamentation in air', *Journal of Applied Physics* 116, 013303 (2014).

[191] T. Fujii, M. Miki, N. Goto, *et al.* 'Leader effects on femtosecond-laser-filament-triggered discharges', *Physics of Plasmas* 15, 013107 (2008).

[192] E. Schubert, D. Mongin, J. Kasparian, and J. P. Wolf. 'Remote electrical arc suppression by laser filamentation', *Optics Express* 23, 28640–28648 (2015).

[193] G. J. Fishman, P. N. Bhat, R. Mallozzi, *et al.* 'Discovery of intense gamma-ray flashes of atmospheric origin', *Science* 264, 1313–1316 (1994).

[194] J. R. Dwyer, M. A. Uman, H. K. Rassoul, *et al.* 'Energetic radiation produced during rocket-triggered lightning', *Science* 299, 694–697 (2003).

[195] H. Tsuchiya, T. Enoto, S. Yamada, *et al.* 'Detection of high-energy gamma rays from winter thunderclouds', *Physical Review Letters* 99, 165002 (2007).

[196] K. Sugiyama, T. Fujii, M. Miki, *et al.* 'Submicrosecond laser-filament-assisted corona bursts near a high-voltage electrode', *Physics of Plasmas* 17, 043108 (2010).

[197] A. Sasaki, Y. Kishimoto, E. Takahashi, S. Kato, T. Fujii, and S. Kanazawa. 'Percolation simulation of laser-guided electrical discharges', *Physical Review Letters* 105, 075004 (2010).

[198] S. Eto, A. Zhidkov, Y. Oishi, M. Miki, and T. Fujii. 'Quenching electron runaway in positive high-voltage-impulse discharges in air by laser filaments', *Optics Letters* 37, 1130–1132 (2012).

[199] A. V. Gurevich, G. M. Milikh, and R. A. Roussel-Dupre. 'Runaway electron mechanism of air breakdown and preconditioning during a thunderstorm', *Physics Letters A* 165, 463–468 (1992).

[200] A. V. Gurevich and K. R. Zybin. 'Runaway breakdown and the mysteries of lightning', *Physics Today* 58, 37–43 (2005).

[201] J. R. Dwyer. 'The initiation of lightning by runaway air breakdown', *Geophysical Research Letters* 32, L20808 (2005).
[202] S. Tzortzakis, B. Prade, M. Franco, and A. Mysyrowicz. 'Time-evolution of the plasma channel at the trail of a self-guided IR femtosecond laser pulse in air', *Optics Communications* 181, 123 (2000).
[203] S. Bodrov, V. Bukin, M. Tsarev, *et al.* 'Plasma filament investigation by transverse optical interferometry and terahertz scattering', *Optics Express* 19, 6829 (2011).
[204] E. Schubert, J. G. Brisset, M. Matthews, A. Courjaud, J. Kasparian, and J. P. Wolf. 'Optimal laser-pulse energy partitioning for air ionization', *Physical Review A* 94, 033824 (2016).
[205] F. Paschen. 'On the potential difference required for spark initiation in air, hydrogen, and carbon dioxide at different pressures', *Annalen der Physik* 273, 69 (1889).
[206] Z. Z. Ji, J. Zhu, Z. Wang, *et al.* 'Low resistance and long lifetime plasma channel generated by filamentation of femtosecond laser pulses in air', *Plasma Science and Technology* 12, 295–299 (2010).
[207] X. L. Liu, X. Lu, J. L. Ma, *et al.* 'Long lifetime air plasma channel generated by femtosecond laser pulse sequence', *Optics Express* 20, 5968–5973 (2012).
[208] K. M. Guo, J. Q. Lin, Z. Q. Hao, *et al.* 'Triggering and guiding high-voltage discharge in air by single and multiple femtosecond filaments', *Optics Letters* 37, 259–261 (2012).
[209] Z. Zhang, X. Lu, W. X. Liang, *et al.* 'Triggering and guiding HV discharge in air by filamentation of single and dual fs pulses', *Optics Express* 17, 3461–3468 (2009).
[210] X. Lu, S. Y. Chen, J. L. Ma, *et al.* 'Quasi-steady-state air plasma channel produced by a femtosecond laser pulse sequence', *Scientific Reports* 5, 15515 (2015).
[211] T. Produit, P. Walch, G. Schimmel, *et al.* 'HV discharges triggered by dual- and triple-frequency laser filaments', *Optics Express* 27, 11339 (2019).
[212] P. Panagiotopoulos, P. Whalen, M. Kolesik, and J. V. Moloney. 'Super high power mid-infrared femtosecond light bullet', *Nature Photonics* 9, 543 (2015).
[213] J. C. Diels and X. M. Zhao. 'Discharge of lightning with ultrashort laser pulses', US Patent 5, 175, 664 (1992).
[214] P. Rambo, J. Biegert, V. Kubecek, *et al.* 'Laboratory tests of laser-induced lightning discharge', *Journal of Optical Technology* 66, 194–198 (1999).
[215] J. Papeer, M. Botton, D. Gordon, P. Sprangle, A. Zigler, and Z. Henis. 'Extended lifetime of high density plasma filament generated by a dual femtosecond–nanosecond laser pulse in air', *New Journal of Physics* 16, 123046 (2014).
[216] J. Papeer, M. Botton, D. Gordon, P. Sprangle, A. Zigler, and Z. Henis. 'Corrigendum: Extended lifetime of high density plasma filament

generated by a dual femto–nanosecond laser pulse (2014 *New J. Phys.* **16** 123046)', *New Journal of Physics* 17, 089501.

[217] Z. Q. Hao, J. Zhang, Y. T. Li, *et al.* 'Prolongation of the fluorescence lifetime of plasma channels in air induced by femtosecond laser pulses', *Applied Physics B* 80, 627–630 (2005).

[218] B. Zhou, S. Akturk, B. Prade, *et al.* 'Revival of femtosecond laser plasma filaments in air by a nanosecond laser', *Optics Express* 17, 11450–11456 (2009).

[219] M. Scheller, N. Born, W. B. Cheng, and P. Polynkin. 'Channeling the electrical breakdown of air by optically heated plasma filaments', *Optica* 1, 125–128 (2014).

[220] P. Polynkin and J. V. Moloney. 'Optical breakdown of air triggered by femtosecond laser filaments', *Applied Physics Letters* 99, 151103 (2011).

[221] M. Clerici, Y. Hu, P. Lassonde, *et al.* 'Laser-assisted guiding of electric discharges around objects', *Science Advances* 1, e1400111 (2015).

[222] G. Méjean, R. Ackermann, J. Kasparian, *et al.* 'Improved laser triggering and guiding of megavolt discharges with dual fs-ns pulses', *Applied Physics Letters* 88, 021101 (2006).

[223] A. A. Ionin, S. I. Kudryashov, A. O. Levchenko, *et al.* 'Triggering and guiding electric discharge by a train of ultraviolet picosecond pulses combined with a long ultraviolet pulse', *Applied Physics Letters* 100, 104105 (2012).

[224] E. W. Rosenthal, N. Jhajj, I. Larkin, S. Zahedpour, J. K. Wahlstrand, and H. M. Milchberg. 'Energy deposition of single femtosecond filaments in the atmosphere', *Optics Letters* 41, 3905–3911 (2016).

[225] G. Point, E. Thouin, A. Mysyrowicz, and A. Houard. 'Energy deposition from focused terawatt laser pulses in air undergoing multifilamentation', *Optics Express* 24, 6271–6282 (2016).

[226] A. Houard, V. Jukna, G. Point, *et al.* 'Study of filamentation with a high power high repetition rate ps laser at 1.03 μm', *Optics Express* 24, 7437–7448 (2016).

[227] L. Arantchouk, B. Honnorat, E. Thouin, G. Point, A. Mysyrowicz, and A. Houard. 'Prolongation of the lifetime of guided discharges triggered in atmospheric air by femtosecond laser filaments up to 130 μs', *Applied Physics Letters* 108, 173501 (2016).

[228] F. Theberge, J. F. Daigle, J. C. Kieffer, F. Vidal, and M. Chateauneuf. 'Laser-guided energetic discharges over large air gaps by electric-field enhanced plasma filaments', *Scientific Reports* 7, 40063 (2017).

[229] J. Kasparian, R. Ackermann, Y. B. Andre, *et al.* 'Electric events synchronized with laser filaments in thunderclouds', *Optics Express* 16, 5757–5763 (2008).

[230] W. Rison, R. J. Thomas, P. R. Krehbiel, T. Hamlin, and J. Harlin. 'A GPS-based three-dimensional lightning mapping system: Initial observations in central New Mexico', *Geophysical Research Letters* 26, 3573–3576 (1999).

[231] C. Herkommer, P. Krötz, S. Klingebiel, *et al.* 'Towards a Joule-class ultrafast thin-disk based amplifier at kilohertz repetition rate', *Conference on Lasers and Electro-Optics OSA Technical Digest*, paper SM4E.3, (2019).
[232] T. Produit, P. Walch, C. Herkommer, *et al.* 'The Laser Lightning Rod Project', *European Physical Journal – Applied Physics*, 93, 10504 (2021).
[233] S. Zahedpour, J. Wahlstrand, and H. Milchberg. 'Quantum control of molecular gas hydrodynamics', *Physical Review Letters* 112, 143601 (2014).
[234] F. Calegari, C. Vozzi, S. Gasilov, *et al.* 'Rotational Raman effects in the wake of optical filamentation', *Physical Review Letters* 100, 123006 (2008).
[235] M. Renard, E. Hertz, S. Gurin, H. R. Jauslin, B. Lavorel, and O. Faucher. 'Control of field-free molecular alignment by phase-shaped laser pulses', *Physical Review A* 72, 025401 (2005).
[236] S. Varma, Y. H. Chen, J. P. Palastro, *et al.* 'Molecular quantum wake-induced pulse shaping and extension of femtosecond air filaments', *Physical Review A* 86, 023850 (2012).
[237] N. Jhajj, Y.-H. Cheng, J. K. Wahlstrand, and H. M. Milchberg. 'Optical beam dynamics in a gas repetitively heated by femtosecond filaments', *Optics Express* 21, 28980 (2013).
[238] N. Berti, P. Béjot, J.-P. Wolf, and O. Faucher. 'Molecular alignment and filamentation: Comparison between weak-and strong-field models', *Physical Review A* 90, 053851 (2014).
[239] A. Rouzee, E. Hertz, B. Lavorel, and O. Faucher. 'Towards the adaptive optimization of field-free molecular alignment', *Physical Review A* 41, 074002 (2008).
[240] J. Huang, C. Wu, N. Xu, Q. Liang, Z. Wu, H. Yang, and Q. Gong. 'Field-induced alignment of oxygen and nitrogen by intense femtosecond laser pulses', *Journal of Physical Chemistry A* 110, 10179–10184 (2006).
[241] Y.-H. Chen, S. Varma, A. York, and H. M. Milchberg. 'Single-shot, space-and time-resolved measurement of rotational wavepacket revivals in H_2, D_2, N_2, O_2, and N_2O', *Optics Express* 15, 11341–11357 (2007).
[242] M. Schroeder, I. Larkin, T. Produit, E. Rosenthal, H. Milchberg, and J. P. Wolf. 'Molecular quantum wakes for clearing fog', *Optics Express* 28, 11463 (2020).
[243] E. W. Rosenthal, I. Larkin, A. Goffin, *et al.* 'Dynamics of the femtosecond laser-triggered spark gap', *Optics Express* 28, 24599 (2020).
[244] D. Mongin, V. Shumakova, S. Ališauskas, *et al.* 'Conductivity and discharge guiding properties of mid-IR laser filaments', *Applied Physics B: Lasers and Optics* 122, 267 (2016).
[245] A. V. Mitrofanov, A. A. Voronin, D. A. Sidorov-Biryukov, *et al.* 'Mid-infrared laser filaments in the atmosphere', *Scientific Reports* 5, 8368 (2015).
[246] H. Liang, D. L. Weerawarne, P. Krogen, *et al.* 'Mid-infrared laser filaments in air at a kilohertz repetition rate', *Optica* 3, 678 (2016).

[247] D. Kartashov, S. Ališauskas, A. Pugžlys, *et al.* 'Mid-infrared laser filamentation in molecular gases', *Optics Letters* 38, 3194 (2013).
[248] S. Tochitsky, E. Welch, M. Polyanskiy, *et al.* 'Megafilament in air formed by self-guided terawatt long-wavelength infrared laser', *Nature Photonics* 13, 41–46 (2019).

Chapter 13

Long-range applications of ultrafast laser filaments

Martin Richardson,[1] Haley Kerrigan,[1] Danielle Reyes,[1] Jessica Peña,[1] Nathan Bodnar,[1] Shermineh Fairchild,[1,2] and Robert Bernath[1,3]

13.1 Introduction

The early observation that stabilized filament formation in the atmosphere [1–3] could potentially project ultrafast laser pulses with high intensities, and therefore high-energy fluences over long distances without the use of classical light focusing soon stimulated speculation of the possible applications of this unique form of high-power light propagation. Could one guide high-voltage discharges over long distances [4]? Could free-space filament propagation replace guided-wire triggering of lightning [5], following on previous attempts with CO_2 lasers [6]? A high-profile experiment currently underway on the Mount Säntis in the Swiss Alps is the most ambitious test so far in this direction [7]. The ability to project not only the intense ultrashort laser radiation intrinsic to filaments but also the nonlinear white light associated with the filaments makes possible such applications as white light Lidar [8–10] and white light spectroscopy [11,12]. The capability to project high intensities to long distances without large focusing optics opens many application pathways for stand-off materials identification using laser-induced breakdown spectroscopy (LIBS) [13–15], such as for substances in heritage sites [15] and gas pollutants in the atmosphere [16]. However, since the individual pulse energy is clamped, so too is the signature plasma emission, which, being isotropic, poses signal-to-noise constraints on spectral detection. One of the most notable achievements was the use of LIBS to explore the prospects of life on Mars, and its rock formations with the French–US Curiosity project [17]. Filament-induced Raman spectroscopy [18,19] is another approach but usually suffers from the same limitation. Filamentation studies have

[1]Laser Plasma Laboratory and Center for Directed Energy, College of Optics and Photonics, University of Central Florida, Orlando, USA
[2]NanoSpective Inc., Orlando, USA
[3]Townes Institute Science and Technology Experimentation Facility, TISTEF, University of Central Florida, Orlando, USA

opened new avenues in THz spectroscopy [20,21], especially through so-called two-color filamentation [22,23]. In recent times, the interaction of filaments with clouds has been explored. These studies include propagation through aerosol clouds [24,25], interaction studies with individual aerosols [25], and the creation of cloud condensation [26]. Researchers have also been exploiting the possibility of creating spatially structured filaments [27,28]. More recently, such structures have been demonstrated to propagate to ranges of 1 km [29]. These structures potentially allow for the guiding of microwaves and other forms of electromagnetic radiation [30].

This brief explanation of some of the potential applications of filamented laser beams in no way summarizes the large body of work performed by many authors. Even so, perhaps it paints an image of the richness of this field and, as experimentation and understanding of filamentation continues to progress, of future applications.

13.2 Fundamental properties of NIR filamentation in air

Before delving into the many long-range applications enabled by filaments, it is necessary to first revisit a discussion of the multifaceted properties of filaments that enable these applications. To begin, the nonlinear processes (i.e., Kerr self-focusing, plasma defocusing) that give rise to a laser filament achieve an active stabilization that confines the energy in the pulse to a narrow core diameter (~100 μm) and surrounding energy reservoir (~1 mm) over extended distances. This property, referred to as self-channeling, enables the delivery of high intensities (~10^{13} W/cm^2) to long ranges [29,31,32] without the need for external focusing mechanisms. The narrow diameter of the core, which is smaller than the inner scale of turbulence, also makes filaments resilient in the presence of atmospheric turbulence [33]. In addition, the propagation of filaments involves a dynamic energy exchange between the filament core and reservoir, whereby energy in the core diffracts into the reservoir while energy from the reservoir refuels the core. This dynamic spatial replenishment picture [34] of filamentation explains the unique ability of filaments to self-heal after an obstruction perturbs a portion of the filament profile [35–37]. This feature is advantageous to the long-range delivery of filaments through the atmosphere, where aerosols and other media may intercept the beam path. Another property of note is the production of a spectral supercontinuum that results primarily from the self-phase modulation of the pulse [38]. A discussion of fundamental filament properties would not be complete without mentioning the channel of low-density plasma (~10^{16} cm^{-3}) that forms in the wake of the pulse [39–42]. This plasma not only plays a critical role in the balancing act that enables self-channeling and other beneficial filament properties, but it also enables unique applications at distance that will be described later in this text. (Note that this discussion pertains primarily to NIR filaments; though the concepts may be extended to other wavelengths, the values and certain details provided will not apply.)

The remainder of this section will provide additional fundamental background knowledge required to understand the necessary preconditions for producing and delivering filaments through the atmosphere. The two primary considerations when determining if the pulse parameters are sufficient to filament are the pulse power and the focusing conditions (i.e., the numerical aperture, or NA, if applicable). Both of these will be discussed separately later.

An ultrashort pulse must have an initial peak power exceeding a value known as the critical power in order to overcome diffractive effects and self-focus, given by

$$P_{crit} = \frac{\alpha\lambda_0^2}{8\pi n_0 n_2},$$

where α is a parameter associated with beam shape, λ_0 is the central wavelength of the laser, n_0 is the linear refractive index, and n_2 is the nonlinear refractive index [38,43]. For a CW Gaussian beam with λ_0=800 nm in air, the theoretical threshold for catastrophic self-focusing is 3.2 GW. In practice, this value for ultrashort pulses may be much higher (e.g., P_{crit}=10 GW for τ=42 fs), owing to a decrease in the molecular contribution to the Kerr effect producing a lower effective n_2 and higher effective P_{crit} [44]. Though the critical power theoretically describes the condition for focusing of the beam to a singular point; in reality, other effects saturate the pulse self-focusing [45]. In the case of NIR filaments, multiphoton ionization plays this critical role.

While P_{crit} is an important parameter in filament discussions, this threshold alone is insufficient to describe the power necessary for a stable filament to form. The value required for filamentation is approximated as $P_{fil} = (\pi^2/4)P_{crit} \approx 2.5P_{crit}$ [46]. Additional power beyond $2.5P_{crit}$ may be necessary depending on the effective coupling of beam power into the filament, which is affected by the diameter and convergence of the initial profile (ideal for $w < 1.5$ mm and NA < 0.002) [38]. It is important to understand that once a pulse exceeds the filamentation threshold power and collapses into a filament, it will exhibit all the typical clamped properties of a filament, despite variations in input power above P_{fil} (but below $2P_{fil}$). An increase in initial pulse power will not lead to an increased coupling of energy into the filament nor to higher intensities or plasma densities. The additional energy in the pulse may either diffract out and be lost or (if the power exceeds $2P_{fil}$) may restructure into additional filaments. This second case, known as multiple filamentation (or multi-filamentation), describes the breakup of the beam due to the redistribution in power across inhomogeneities in the profile that produces localized spikes in power exceeding the threshold and yielding distinct filaments. The number of filaments that will be formed given an initial pulse power may be estimated as $N \approx P_{in}/P_{fil}$ [47]. Regardless of the initial power, the amount of energy that couples into a single filament is limited to ~1 mJ (demonstrated for a pulse with τ=400 fs, w=4 mm, and E=5–30 mJ) [1]. Based on the previous discussion, it is evident that the necessary power for filamentation in practice may vary depending on a large number of factors, including pulse duration, beam diameter and focusing (if applicable), and atmospheric conditions (or any alteration to the effective value of n_2).

In addition to the pulse power, thought must be given to the initial pulse-focusing conditions. Many studies, even those pertaining to long-range propagation through the atmosphere, are conducted in-lab and require the use of external focusing elements to assist the filament collapse. The convergence of the beam plays a critical role in the development of the pulse into a filament. There are three primary contributions to the wavefront curvature in this case: geometric focusing, Kerr focusing, and plasma defocusing. The NA of the beam, defined as the ratio between the beam $1/e^2$ half-width and the focal length of the assisting optic ($NA=w_0/f$), proves to be a useful parameter for delineating the effects of focusing on filament formation. Namely, a transition NA exists, NA_T, at which all three of these contributions to the wavefront are equal. Above this NA, the focusing can be said to be tight, and geometric focusing dominates. Below this NA, the focusing is effectively loose, and nonlinear effects dominate. The appropriate value of NA_T for a particular set of experimental conditions may be numerically simulated as outlined in [48]. This transition threshold varies with pulse duration and power. For a pulse with $\tau=50$ fs and $E=0.8$ mJ, NA_T is 4.25×10^{-3}, which for a 4.25-mm beam corresponds to an $f\sim1$ m optic. For larger initial beam diameters, experiments must be designed to accommodate longer focal lengths and propagation paths in order to fall in the loose-focusing, or low NA, regime, where nonlinear effects dominate the propagation.

Many beam properties are affected by the focusing regime. For example, in the nonlinear regime, spectral broadening occurs toward the blue and the red (i.e., the behavior typically associated with a filament), while in the linear regime, it occurs primarily toward the visible. Broadening is at its minimum at the transition NA [48]. In addition, the plasma shape and density are dramatically different between the two regimes. In the nonlinear regime, the plasma forms as an elongated channel with well-documented values of peak density (~10^{16} cm^{-3}) and decay time (~1 ns), and these values are clamped in relation to the core intensity, such that an increase in input energy does not enhance these quantities. By contrast, the plasma in the linear regime forms as a bubble with densities potentially exceeding 10^{16} cm^{-3} by a few orders of magnitude, depending on the input pulse energy as core intensity rises [41]. Another important property to consider is the temporal profile of the pulse. The characteristic filamentation phenomenon of pulse-splitting is seen to occur only in the nonlinear regime [48]. The many differences in properties between the regimes highlight the importance of appropriate selection of NA, in order for the results of an experiment to apply to long-range propagation of filaments.

13.3 Advanced forms of filament propagation through the atmosphere

While the balancing act that leads to stable filament formation is responsible for the many beneficial properties of these structures, it simultaneously constrains the beam parameters. For example, though the extended plasma column has potential use toward electric discharge guiding, this application would profit from higher electron densities or longer plasma lifetimes, neither of which can be increased in a standard filament. Tuning of the various filament properties that are typically clamped has been explored through the exploitation of nonlinear beam interaction, both in space and time. Advanced forms of filament propagation have been achieved through the formation of spatial and temporal filament structures (Figure 13.1).

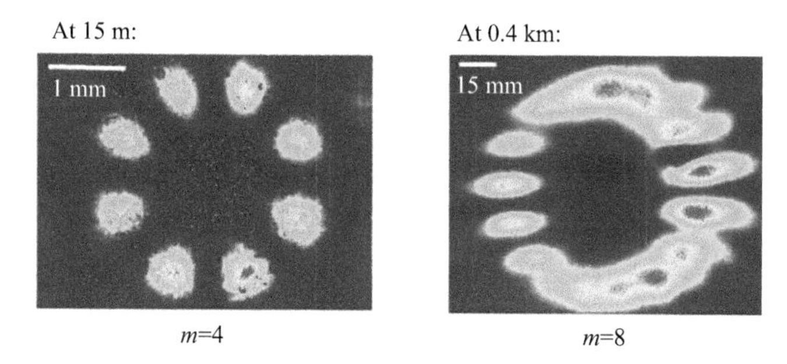

Figure 13.1 Engineered filament arrays produced using phase plates showing clusters of filaments propagating over extended distance [29]

This concept of engineering filaments into structured spatial arrays has existed for some time, primarily motivated by the potential use of these arrays to guide microwaves [30,49]. Interesting structures have been imagined and implemented, including a transient photonic lattice [30] and helical filaments [28]. However, recent studies have indicated that the spatial structuring of filaments can be used strategically to enhance and, to a degree, tune the filament properties [42,50–53]. It has been well documented that two filaments that are in phase with one another will attract and even fuse, while filaments that are out of phase will repel, when they are within a certain proximity [54–57]. This communication between filaments exists when the neighboring energy reservoirs overlap. One recent study demonstrated that two nonlinear beams with powers below the critical threshold separated by less than 330 μm would merge to form a single filament with properties equivalent to a filament formed by a single beam with equivalent total energy [53]. That is to say, an array of subcritical beams could theoretically be co-propagated collinearly or at a shallow angle with an identical array of subcritical beams to produce a filament structure along propagation, while reducing risk of damage to optical components. Another recent study noted the three-time enhancement of the plasma density and the extension of plasma lifetime of a filament formed by the merging of two independently filamenting beams when initially separated by 180 μm [42]. Despite the total power exceeding $2P_{fil}$, only a single filament was formed. In addition, more energy was coupled into that filament than would have been possible with a single beam containing the same total power. Therefore, structures could be envisioned that take an advantage of this interaction effect by controlling the relative position of filaments within the array to tune the electron density above the typical clamped value.

The notion of temporal engineering of filaments is slightly more recent in its conception. This effort entails the temporal concatenation of successive filaments, referred to here as stitching. Achieving stitching requires inter-pulse separations that are shorter than the filament plasma lifetime and requires that successive filaments propagate coaxially, so that the same plasma channel is revived and sustained by this temporal stacking. Several attempts have been made with varying levels of success [58–61]. Some studies involved inter-pulse separations that were

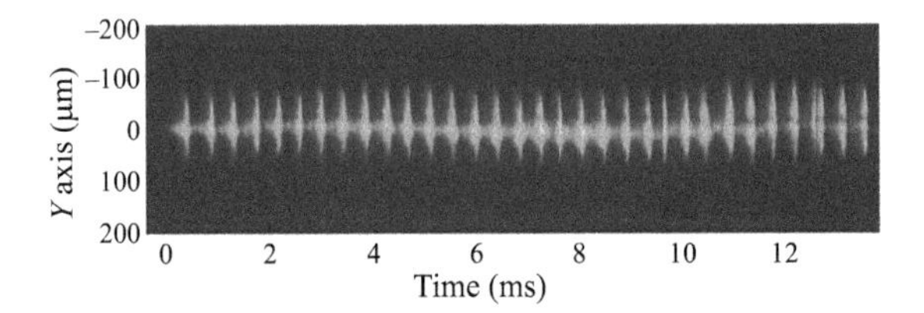

Figure 13.2 Demonstration of filament stitching using streak camera images placed end-to-end, each with durations of 2 ns, showing a single axis of the beam profile vs. time for the full duration of the filament burst, [62]

greater than the plasma lifetime, while others were performed in the linear regime and therefore do not apply to long-range propagation of filaments, and still others failed to verify spatial overlap of consecutive pulses (Figure 13.2). The first successful demonstration of stitching was reported in 2019 [62]. In this case, a 32-pulse burst of filaments separated in time by 415 ps each was observed to deliver all 32 filaments to the same point at distance within 13$\pm$12 μm. In addition, for an 8-pulse burst with inter-pulse separation of 830 ps, the plasma was confirmed to persist for the full duration of the burst, here 6 ns. This new paradigm of filament engineering has many advantages toward existing applications, while also opening the door to new and innovative applications.

To conclude this section, consider now the marriage of these two concepts: spatial and temporal engineering. With sufficient energy in a single laser shot, a structure containing hundreds of filaments could be envisioned, with each filament positioned precisely in both space and time, optimized for a particular application. This premise makes the most efficient use of energy in a single firing of the laser. More of the energy in the pulse is coupled into filaments and delivered to distance in a manner that is effective for application.

13.4 Ablation of materials with filaments

Filamentation eliminates the need for large telescopes to produce high laser intensities over long distances, which is advantageous to remote laser–material interactions. Long-range material modifications are possible with laser filaments, which maintain a high intensity central core over many times the Rayleigh range as a result of nonlinear processes. The intensity of the core, which is approximately 10^{13}–10^{14} W/cm^2 for filaments formed by laser pulses centered around 800 nm in air, is more than sufficient for the removal of material and the generation of plasma on the surfaces of target substrates with a single pulse. The ablation of a variety of target materials has been demonstrated with filaments at meaningful distances [14,32,63–65]. The longest range interaction experiments with filaments were reported using the Terawatt & Terahertz mobile laser facility [32], in which damage on germanium at the kilometer range was produced by 600 pulses of 200 mJ at

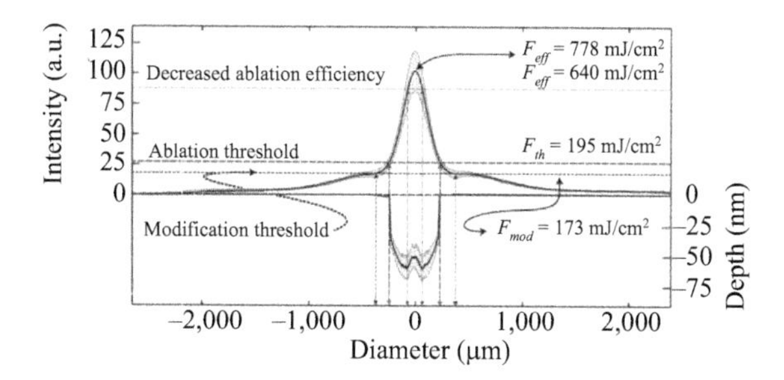

Figure 13.3 Comparison between relative filament fluence profile (top) and single-shot GaAs ablation crater (bottom). Shaded regions represent the standard deviation of multiple single-shot measurements [72]

10 Hz. A prominent application of remote filament ablation is target identification with LIBS [14,63,66], or in this case filament-induced breakdown spectroscopy, which uses the light emitted from plasma generated during ablation by high-intensity laser pulses to determine the elemental constituents of a target. For solid targets, the signal strength is closely tied to the amount of ablated materials.

13.4.1 Filament ablation dynamics

The modification and ejection of materials in laser ablation involves a number of processes that depend strongly on the properties of the laser light as well as the target material. Filamentation in air typically occurs for ultrashort pulses with pulse durations ranging from tens of femtoseconds to a few picoseconds. In general, filament ablation involves the same ablation mechanisms as non-filament femtosecond pulses. This process, sometimes referred to as "cold ablation," is dominated by nonthermal mechanisms as a result of the pulse duration being shorter than the electron cooling time [67,68]. In other words, laser energy is absorbed faster than the electrons can redistribute energy into the bulk of the material. Ultrafast pulses can produce craters with highly controllable dimensions because heat does not propagate beyond the laser–target interaction zone, making them extremely useful for applications such as micromachining [69]. This precision distinguishes ultrashort pulse ablation from ablation with longer pulses, such as nanosecond or millisecond duration pulses, which induce longer thermal ablation mechanisms [68,70]. When the pulse duration is longer than the electron cooling time, heat is distributed into the bulk of the target beyond the region of irradiance, causing significant modifications to the material beyond the targeted interaction area [67,68]. High-intensity nanosecond pulses deposit significant energy into the plasma generated above the target surface during illumination, which can reduce ablation efficiency compared to femtosecond pulses [71]. This process, known as plasma shielding, can be advantageous for applications such as LIBS that benefit from hotter, sustained plasmas. Another vital feature that distinguishes femtosecond ablation is the occurrence of self-focusing and filamentation.

The unique properties of filaments affect the ablation in several ways. The Townesian intensity profile is a feature that is reflected in the shape of the ablation crater. The majority of material removal is confined to the region irradiated by the high-intensity core, while the lower intensity peripheral field does no not have sufficient intensity for ablation. The light surrounding the core can modify the surface it irradiates but does not meaningfully contribute to the overall ablation [72]. The effect of the filament intensity distribution on ablation is illustrated in Figure 13.3.

The structure of the filament affects the ablation rate. Intensity clamping, which distributes laser energy in excess of that required for filamentation into the lower intensity peripheral field, produces consistent ablation rates [64,65]. That is, shot-to-shot fluctuations in the pulse energy have less effect on the resulting ablation, as the extra energy is contained in the lower intensity peripheral field, which does not produce significant ablation. This repeatability is advantageous for certain applications, such as the calibration of LIBS detectors. It should be noted that while ablation rates are consistent shot-to-shot, the location of the filament on the target can vary significantly as a result of changes in the atmosphere [64]. The effects of turbulence hinder the overall impact of the interaction when the filament does not strike the same spot on the target. Additionally, intensity clamping also means that high ablation rates are difficult to achieve by only raising the pulse energy. Reduced ablation rate due to intensity clamping is a well-studied phenomenon [64,65,73,74]. Several methods have been proposed to improve the rates of filament ablation to further long-range filament applications.

13.4.2 Augmented filament ablation

Improvements to filament ablation are possible by supplementing the filament with additional energy from a separate source to increase the intensity on the target despite intensity clamping. This technique is most commonly implemented in the form of a superimposed collinear nanosecond pulse. Notable improvements to the ablation rate and plasma brightness yielded have been demonstrated with combinations of high-intensity femtosecond and nanosecond pulses [75–79]. In these experiments, the resulting ablation is significantly dependent on the temporal delay between the two pulses. The changing state and temperature of the target surface after incidence of the first pulse has a considerable effect on the absorption of the second pulse and ensuing ablation dynamics. For example, Kerrigan *et al.* observed enhanced single-shot ablation of gallium arsenide when a femtosecond pulse in the filamentation intensity regime was followed by a lower intensity 1-μm nanosecond pulse for delays ranging from tens of nanoseconds to a few hundred nanoseconds. For longer delays in this regime, the residual thermal effects from the femtosecond pulse interaction amplified the thermal ablation processes induced by the nanosecond pulse [78]. Shorter delays up to approximately 50 ns involved interactions between the nanosecond pulse and the plasma generated by the femtosecond pulse. In this regime, improved ablation corresponded to enhanced absorption of the 1 μm pulse by the plasma and the generation of a supersonic shockwave on the expanding front of the plasma [79]. In experiments with different target materials and laser pulse parameters, the optimal temporal delay for enhanced ablation varies. Similar principles of enhancement can be applied to temporally and spatially structure arrays of filaments. This framework for enhancing ablation, involving a single source, simplifies implementation for long range experiments and is discussed in the following section.

13.4.3 Ablation with multiple filaments

Ultrashort pulses with energies many times the threshold for filamentation produce multiple filaments, where the formation of each plasma core depends on hot spots in the intensity profile and air density fluctuations in the atmosphere. While this effect limits the peak intensity of the filament core, it can also produce conditions that enhance the material removal. One way that co-propagating filaments can generate more ablation is through the production of adjacent shockwaves. In this scenario, the interaction of the shockwaves detonated by each filament core leads to damage, delamination, and ablation in the region between the craters formed by each pulse [80]. This effect has also been shown to support gains in signal strength in LIBS experiments with increases in pulse energy into the multifilament regime [81]. The ablation dynamics of adjacent filaments can be fully utilized with the propagation of controlled filament arrays. The implementation of adaptive optics capable of compensating for fluctuations in the beam intensity profile and atmospheric turbulence may provide further control of the spatial distribution of filaments at range.

Significant improvements to the ablation can also be obtained through the engineering of the temporal distribution of filament intensity. It is well known that the efficiency of ultrashort pulse ablation can be improved by employing two or more temporally separated pulses compared to a single pulse of equivalent energy. Enhanced ablation efficiency (Figure 13.4) has recently been demonstrated with bursts of up to 32 filaments with temporal pulse separations <1 ns [82]. These pulses, which propagated with a temporally concatenated plasma [62], had extremely precise overlap on the target at distances up to 30 m. In conjunction with improved rates of material removal, improvements to the plasma brightness and acoustic shockwave strength were also observed. Filament bursts present a promising means of enhancing ablation at long ranges. Future advances in high-energy ultrashort pulse lasers could lead to the marriage of temporal and spatial filament control technologies and greatly extend the capabilities of long-range filament ablation.

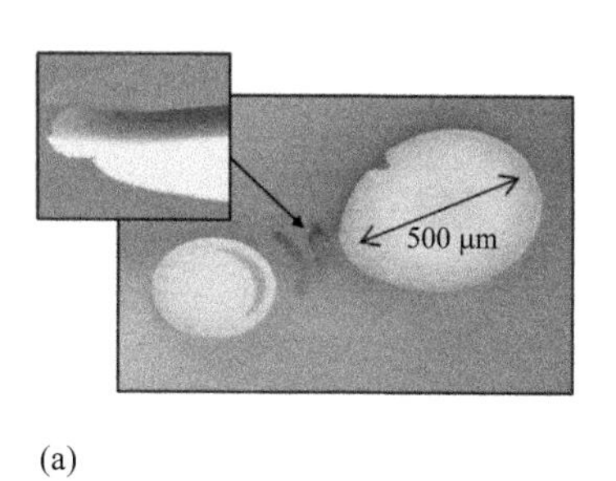

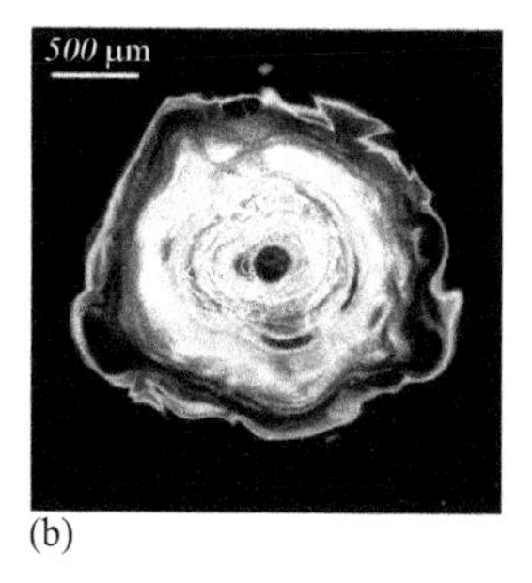

Figure 13.4 Ablation enhancements by (a) adjacent filaments observed by Bernath et al. [80] and (b) a 500-mJ burst of 32 filaments on coated silicon observed by Kerrigan [82]

13.5 Radio-frequency emission generated by filament plasmas

The irradiation of solid matter with ultrafast pulses has been shown to generate RF radiation both in vacuum [84,85] and in air [86]. When matter is irradiated with low intensity pulses, the irradiated surface will undergo melting and vaporization consistent with conventional heating. However, for femtosecond pulses possessing an irradiance of 10^{13}–10^{14} W/cm^2, as in filaments, the associated electric field is (0.9–2.7) $\times$ 10^8 V/cm, which falls along the threshold of dielectric breakdown [87]. The high irradiance present in the core of the filament will convert the material surface directly from a solid to a plasma on a sub-picosecond timescale [88]. The plasma and electron density resulting from this process is far in excess of the 10^{16} cm^{-3} electron density obtained within the filament core, enabling the generation of transient currents far greater than those found within the filament core and consequently a far stronger radio frequency emission. Experimental results indicate that filament interaction with a target does indeed produce a strong transient current and a localized dipole RF emission (Figure 13.5). This RF generation is typically broadband, although it has been shown that the target materials will often generate a unique spectral emission in the RF regime [58,89].

13.5.1 Methodology to measure up to 40 GHz of filament-induced radio-frequency

Given the duration of the laser pulses used to drive the plasma responsible for RF radiation, and the temporary duration of the resulting plasma, the produced RF radiation is expected to be short lived and broadband. For this reason, single-shot, broadband measurements are required to evaluate the RF radiation. To carry out such measurements, a single-shot oscilloscope could be coupled to a custom-built heterodyne receiver, to extend the acquisition range. Such a measurement was previously performed [58,89–92] using a heterodyne receiver that had a 4 GHz channel bandwidth to match both the bandwidth and time response of the oscilloscope and enable a 1–40 GHz spectral range. A pair of polarization sensitive, broadband horn antennas, namely the 1–18 GHz Sunol Sciences DRH-118 and the

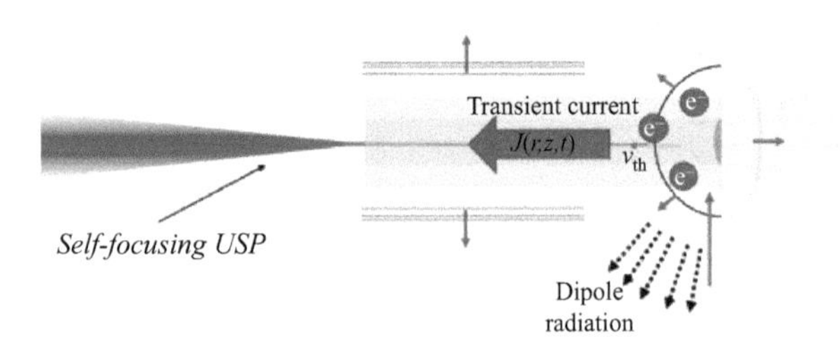

Figure 13.5 Cartoon illustrating the remote USP generation of localized radiation

18–40 GHz Q-Par Angus QPS180K, were used to couple RF radiation generated at the target. The schematic in Figure 13.6 can be referenced for additional details on this RF measurement methodology. This technique has been used to study the filament-material interaction in both the single pulse and burst pulse regimes.

13.5.2 High-frequency experiments in the single-filament regime

In the single pulse regime, initial experiments [89,91,92] were carried out by irradiating a copper target with a filament produced by a 50 fs, 12 mJ pulse. Filament collapse was assisted by a concave mirror located 9 m from the target. Laser filamentation began 3 m from the mirror, and the pulse propagated over the remaining 6 m as a filament before irradiating the target surface.

A time-domain image and corresponding compiled spectrum is shown in Figure 13.7. The time-domain plot is characteristic of all RF signals above the detection threshold of the measurement system. All such pulses were sub-nanosecond in duration and limited to the bandwidth of the instrumentation. The measured spectra corresponding to the RF pulses are relatively uniform and extend across all measured frequencies, suggesting that the true spectrum extends beyond the 21–35-GHz frequency range evaluated in this experiment.

Rudimentary radiation pattern measurements were carried out by evaluating the filament induced 17 RF radiation at three separate angles relative to the surface normal within the horizontal plane. For 18 each angle, measurements were carried out with the antenna both horizontally and vertically 19 aligned and using both horizontal and vertically polarized laser light.

The RF radiation was measured using the QPS180K horn antenna, which was aligned to and located 10 cm from the irradiated surface. The antenna was oriented at angles of 10°, 45°, and 90° relative to the surface normal in the horizontal plane and placed directly above the target, maintaining its 10 cm distance from the irradiated surface, to obtain angularly resolved RF measurements. Again, for each orientation, measurements were taken using both antenna polarizations, obtained by 90° rotations of the antenna, and for both laser polarizations, in order to determine laser and field polarization dependence of the RF radiation.

For horizontal polarizations, clear angular dependence was observed, with maximum radiated power obtained at 90° relative to the surface normal. For vertical polarizations, the total field irradiance was both independent of angle and weaker than the field irradiance obtained for horizontal polarization at any angle. This leads to the conclusion that the RF obtained from copper is horizontally polarized, with an antenna pattern that peaks at 90° relative to the surface normal and approaches the background value observed for vertical polarizations as the angle relative to the surface normal approaches zero. This polarization arrangement is consistent with a dipole radiation source aligned to the target normal.

13.5.3 The characteristics of ultrafast burst mode generated RF

Frequency-selective RF generation is achievable with recent advances in filamenting laser pulse format [83,93], i.e. the temporal stitching described in section

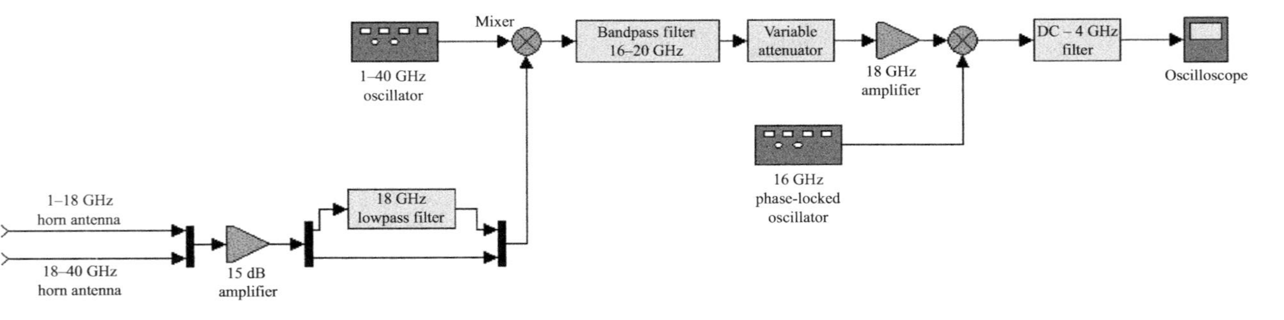

Figure 13.6 Heterodyne signal-processing system used to evaluate RF radiation

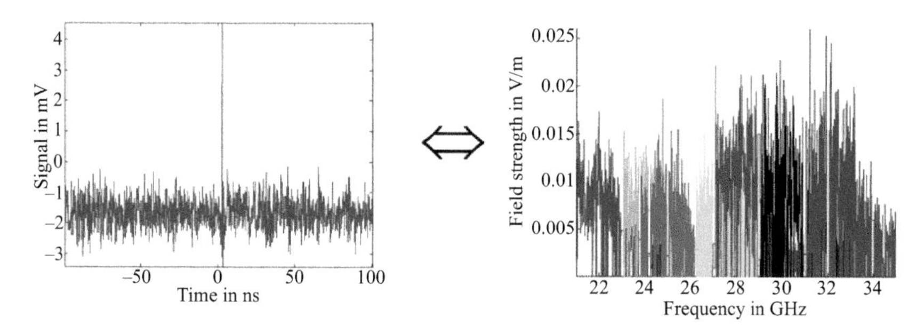

Figure 13.7 Sample time-domain measurement of filament-induced RF radiation and corresponding spectra

3.0. In this mode of operation, a train of equally spaced pulses narrows the bandwidth of the RF signal around a chosen frequency, which is determined by the temporal separation of the pulses. Higher harmonics of the selected tones are also produced, thereby generating a frequency comb at the target. This method therefore enables the production of multiple distinct frequencies, including frequencies well above the fundamental set frequency. Multi-GHz tones can be produced using so-called burst mode filamentation, with frequency tuning achieved by modifying the fundamental pulse separation.

The filament burst mode pulse format demonstrated using the Multi-Terawatt Filamentation Laser (MTFL) at the University of Central Florida (UCF) was able to produce several unique RF signatures from the interaction of the filaments with an aluminum target (Figure 13.8), by adjusting the interpulse separations from 2 ns to 1 ns then to 0.5 ns.

13.6 Acoustic shock generation by filaments

Supersonic shockwaves are the product of a large impulse of energy, such as the energy deposited by a high-intensity laser pulse in matter. A supersonic shockwave is created when energy is transferred to the surrounding medium at a rate that exceeds the speed of sound in that medium. When a large amount of laser energy is absorbed by a material over a short amount of time, the rapid heating and expansion of the absorbing matter sends a compression wave into the surrounding material. Laser-generated micro-explosions are widely employed in materials processing and medical applications. More recently, laser shockwaves have been proposed as a tool for interacting with distant objects and environments. Filamentation extends the applications of shockwaves to multi-kilometer ranges.

13.6.1 Filament-generated shockwaves

During filamentation of ultrashort pulses, interactions between the self-focused pulse and the propagation medium result in the absorption of high-intensity laser

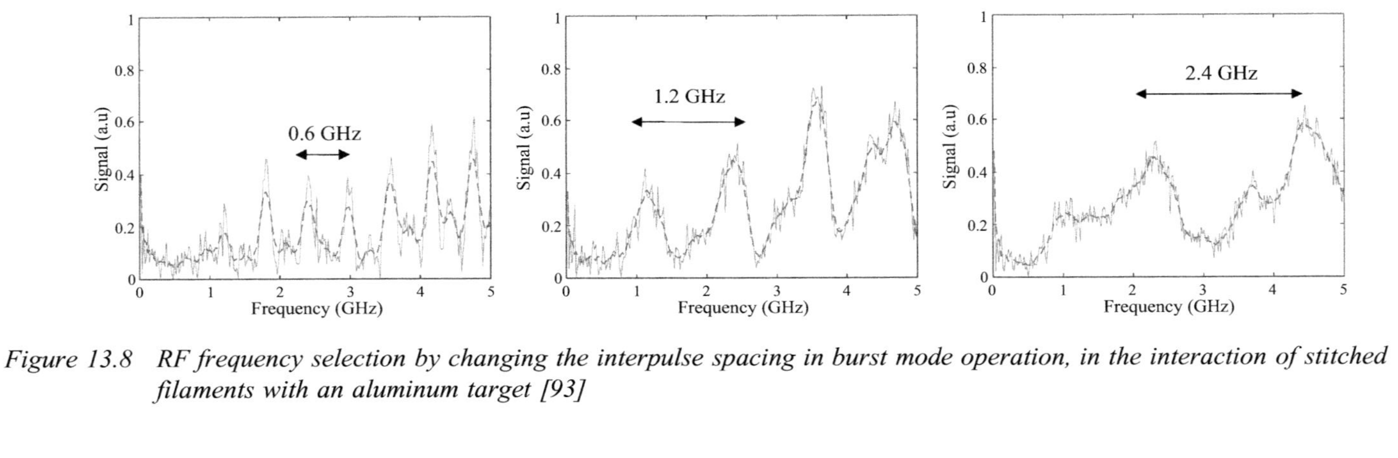

Figure 13.8 RF frequency selection by changing the interpulse spacing in burst mode operation, in the interaction of stitched filaments with an aluminum target [93]

Figure 13.9 Shadowgraph image showing transverse disturbance in air created by a bundle of filaments produced by a femtosecond pulse in the multifilamentation regime [58]

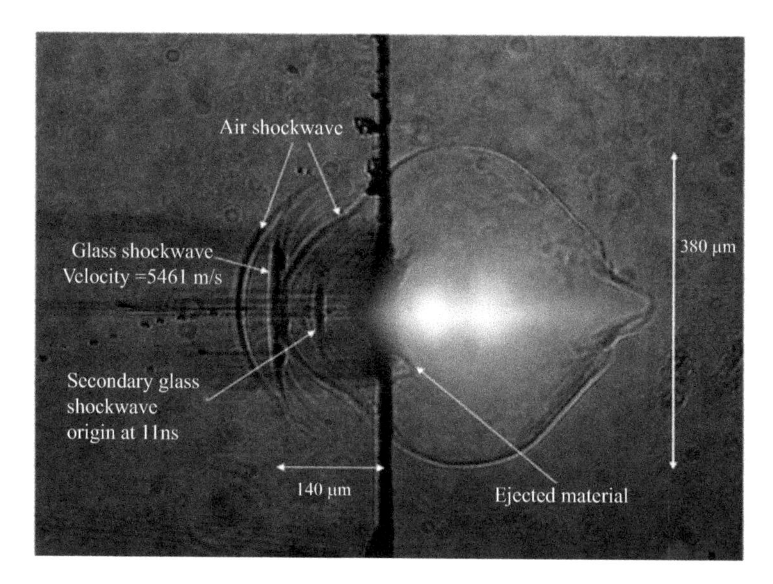

Figure 13.10 Multiple shockwaves observed 25 ns after a filament (8 mJ, 120 fs, 800 nm in air) irradiated the edge of glass microscope slide of 150-μm thickness [82]

light. The resulting localized heating by the high-intensity filament core causes the rapid expansion of the heated material, which produces a disturbance in the density of the surrounding matter. In air, the result is a very weak acoustic wave. In filamentation experiments, the amplitude of the shockwaves acoustic signal provides insight on the density of the plasma channel [89]. In Figure 13.9, shockwaves originating from the plasma channels of a bundle of filaments are visualized by deformations in the density of the surrounding air environment [58].

Stronger shockwaves are produced when filaments interact with solid targets. The effects of the interaction of a filament with a glass microscope slide are shown in

Figure 13.10, which was captured precisely 25 ns after the interaction using an ultrafast probe [58]. The fronts of the expanding supersonic shockwaves are identified by the compressed region of air ahead of the shockwave front, which is visible as a dark line in the image. The weak acoustic wave produced by the air plasma channel is faintly visible on the right-hand side of the image, closest to the laser source. To the right of the glass–air interface, a supersonic shockwave has expanded in the air to a diameter of 380 μm. The protrusion of the shockwave front in the path of the laser beam is a result of reduced air density in the filament plasma channel. Within the air, shockwave plasma can be identified by its bright, broad-spectrum radiation that has saturated the image. The ejection of molten material can also be observed. In the bulk of the substrate, multiple shockwaves are created. The main shockwave, generated by the initial expansion of the irradiated glass, was determined to have a velocity of 5,461 m/s, which exceeds the material acoustic velocity of 5,425 m/s. A secondary glass shockwave appears to have formed at a later time, most likely as a consequence of conserving momentum from material expulsion. Additional shockwaves as well as effects resulting from the interactions of multiple shockwaves have also been observed. The exact properties of the shockwaves depend on the characteristics of the filament, such as wavelength and pulse duration, the thermal properties of the propagation medium, and the shape and nature of the target.

13.6.2 Shockwave measurement and analysis

During filament interactions with matter, the absorbed laser energy is converted to many forms, including heat, phase changes of target, the ejection of plasma and other forms of material, and acoustic shockwave formation. Determining the energy of the shockwave is thus important to understanding shockwave formation and its overall role in filament interactions. The mathematical description of a strong explosion was first solved by Sedov in 1946 [90] and Taylor in 1950 [91], who related the energy of the strong explosion to its rate of shockwave expansion and the properties of the propagation medium. In the Sedov–Taylor theory [92], the energy E is related to the radius of a spherical shockwave R as a function of time t through the following equation:

$$R = \xi \left(\frac{E}{\rho} \right)^{1/5} t^{2/5},$$

where ξ is a constant and ρ is the density of the surrounding medium (equal to ~1 and 1.2 kg/m^3 for air in standard atmosphere conditions). The Sedov–Taylor theory can also be used to determine other parameters of the shockwave, such as temperature and pressure.

With the Sedov–Taylor theory, the detonation energies of laser-generated shockwaves can be extracted from measurements of its radial expansion. Techniques for detecting the position of the shockwave rely on the index change due to the compression of shockwave propagation medium ahead of the shockwave front. These techniques include Schlieren photography, shadowgraphy, and interferometry. The previous two figures were captured with shadowgraphy with an ultrafast probe that enables measurements to be made with sub-nanosecond

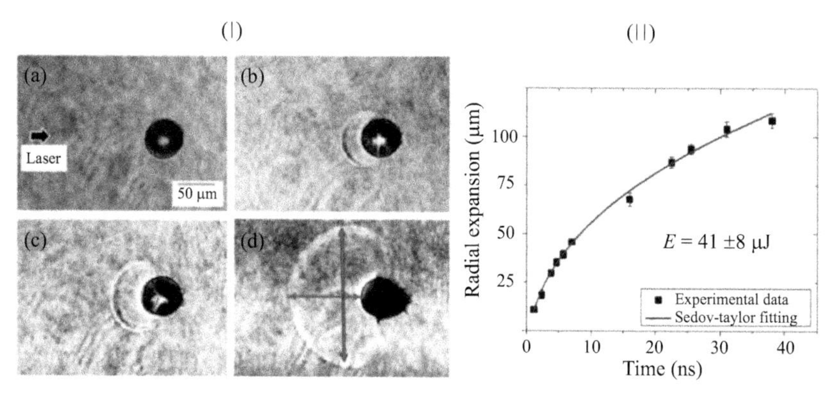

Figure 13.11 *Ultrafast shadowgraph images (I) and the Sedov–Taylor analysis (II) of the shockwave generated by the interaction of a filament (2.2 mJ, centered at 800 nm in air) with a water microdroplet in air by Jeon et al. [25]. The displayed shadowgraphs were captured at delays of (a) 0, (b) 2.7, (c) 5, (d) 21.5 ns after the interaction. The red and blue arrows (Id) specify the width and the radial expansion of the shockwave used in the analysis*

temporal resolution. These techniques provide valuable information on the laser–matter interactions and the strength of laser-generated shockwaves.

An example of ultrafast shadowgraphy and Sedov–Taylor analysis has been applied to the interaction of a filament with a water microdroplet (Figure 13.11) [25]. This analysis revealed that approximately 75% of the energy in the fraction of the filament core that directly impacted the droplet was absorbed. In this scenario, the remaining energy in the pulse is sufficient for filamentation, so the filament reforms after the interaction. This analysis provides insight on the propagation of filaments through aerosols. The shockwaves produced by filaments impacting water droplets have been proposed as a means of clearing paths through aerosols in the atmosphere [93,94]. This cloud cutting technique has the potential to allow for better transmission of a subsequent laser through the cloud without experiencing wavefront distorting effects typically associated with propagation through a cloud [95].

13.6.3 Enhanced shockwave generation

The shockwave energy depends on the fluence absorbed by the target. Just as intensity clamping limits the ablation by filaments, the strength of filament-generated shockwaves is also constrained by the energy confined in the core. To advance the capabilities of remote shockwave effects, several methods for improving the strength of filament-generated shockwaves have been proposed.

Figure 13.12, illustrated by Bernath *et al.* [80], illustrates the effects of different pulse configuration on the shockwaves generated in the bulk of a solid target. The same mechanisms of enhancement can be applied to the shockwaves generated in the air background. Compared to a single filament (a), collinear pulses, in the

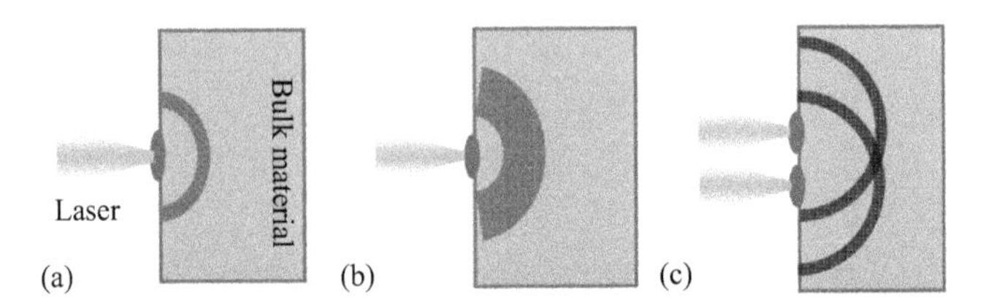

Figure 13.12 The shockwave front generated by generated by a single ultrafast pulse (a) compared to enhanced shockwaves produced by a collinear burst of pulses (b) and by spatially separated filaments (c), illustrated by Bernath et al. [80]

form of two or more pulses superimposed on the target with some temporal delay (b) or spatially separated by a small distance (c), yield stronger shock effects. Several experimental investigations have measured stronger shockwaves utilizing the configuration in (b). In this regime, more of secondary pulse energy is converted into acoustic energy through the improved absorption of the modified target surface or by reenergizing the plasma created by the previous pulse. In the scenario presented in (c), the interaction of adjacently produced shockwaves leads to enhanced effects [80,96].

Figure 13.13 shows two experiments that have yielded stronger shockwaves by superimposing a filament with secondary radiation. In an experiment by Kerrigan *et al.* (I) [79], a filament was superimposed with a delayed 1-μm nanosecond pulse. The Sedov–Taylor analysis of shockwaves produced by the individual (Ia) and combined (Ib) pulses on a gallium arsenide target revealed the energy of the shockwave generated by the nanosecond pulse could be increased by more than 3× through optimization of the delay and laser–plasma interactions. This effect has also been demonstrated with a combination of two temporally delayed filaments in an experiment by Peña *et al.* [97]. The results from this experiment imply that droplet clearing can be improved with the addition of a delayed secondary filament. Further improvements to the shockwave energy and impact can be achieved with the inclusion of more pulses in the form of a burst of filaments with optimized inter-pulse delays.

13.7 Propagation at high altitudes

Some applications may require filament propagation to or at high altitudes where the air pressure is a fraction of that at sea level. Since the process of filamentation relies on the nonlinear properties of the medium the pulse is traveling through, the reduced air density at high altitudes, which changes the nonlinear properties of the air, in turn changes the preconditions for filamentation and the resulting filament properties compared to those at sea level [98,99,100].

Experimentally generating filaments at low pressures in laboratory conditions proves challenging due to the constraints on the pulse preconditions required to ensure that the pulse propagates with critical power in the nonlinear regime.

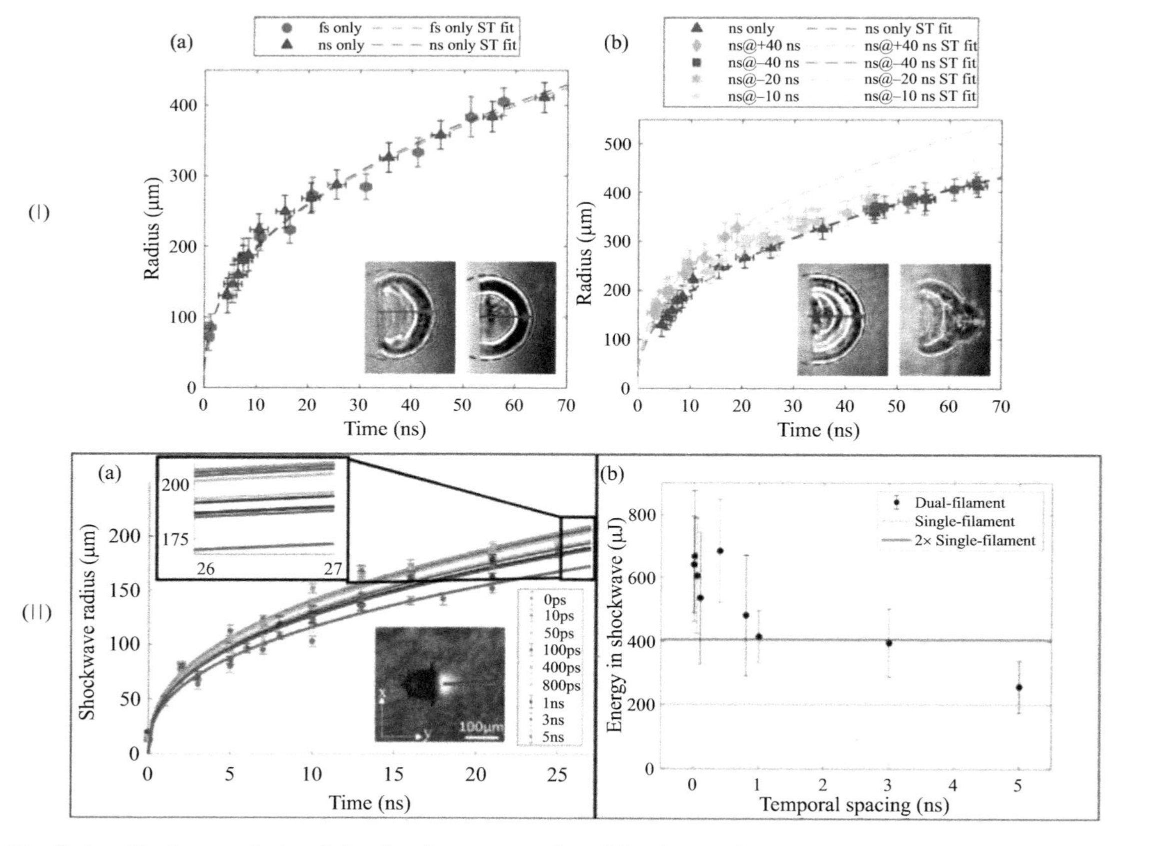

Figure 13.13 The Sedov–Taylor analysis of the shockwaves produced by the combination of a 800-nm filament and 1-μm nanosecond pulse on gallium arsenide by Kerrigan et al. [79] (I) and two 800-nm filaments with a water microdroplet by Peña et al. [97] (II) with varying inter-pulse delays

(Section 2.0 may be referenced for a discussion of the importance of meeting the necessary critical power and NA conditions to generate a filament in air.) Filament-like plasma has been observed for pressures as low as 0.002 atm, but this plasma was generated in the highly linear regime for the given pressure [106,107]. When considering a change in altitude, the scaling of the nonlinear refractive index of air with pressure, [105], impacts the calculation of both P_{crit} and NA_T. The critical power is found to be inversely proportional to pressure, indicating that more energy per pulse is required to generate a filament at lower pressures [107]. In determing the transition NA, both the Kerr self-focusing and plasma defocusing effects are found to scale with pressure, while the geometric sag contribution remains pressure-independent. Ultimately, the calculation yields that NA_T is smaller at lower pressure [107]. Practically, this means that the beam size must be smaller or the external focus must be longer to achieve nonlinear propagation at low pressures. The figure below depicts pressure scaling of both P_{crit} and NA_t based on the model described above (Figure 13.14).

Properly scaling filament preconditions with pressure is required to accurately design simulations and experiments at low pressures. Several parameters from standard filament propagation models are scaled with pressure to simulate low pressure propagation [107]. These studies can observe propagation at a constant pressure [38,107] or through a pressure gradient, such as starting in vacuum and propagating towards sea level conditions to simulate space to ground lasing.

While scaling the pulse preconditions with pressure ensures that filamentation at high altitude conditions, the filament properties are also impacted by the decreased pressure and therefore decreased air density. Studies, both experimental and theoretical, that scaled the critical power with the decreasing pressure found that the filament length and diameter increased as the pressure decreased [100,102,103]. Experimental data taken without scaling the critical power or in the

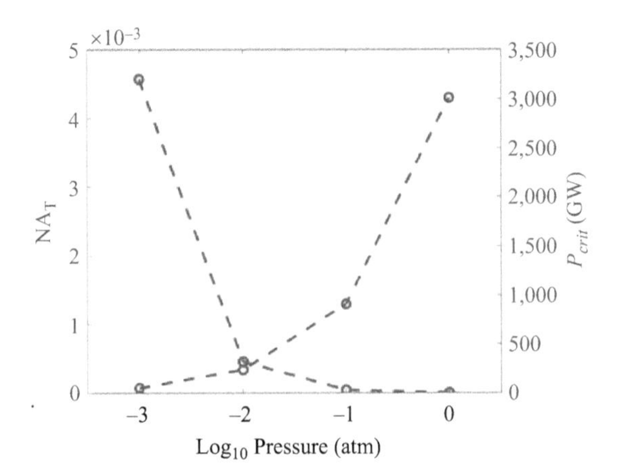

Figure 13.14 Plot showing the scaling of P_{crit} and NA_T as the air pressure decreases from 1 to 0.001 atm [100]

linear regime of propagation saw the opposite effect, or a shortening of the plasma and a shift downstream in the filament start location [37,102,104]. The filament plasma has been theoretically and experimentally predicted to change as the pressure decreases [99,100,104]. The plasma electron density is expected to decrease as pressure decreases [100]. Experimental studies have shown that the plasma decay rate decreases (i.e., the plasma is longer lived) at low pressures [99]. At sea level, white light generation and spectral broadening toward the red and blue are indicative of filamentation and nonlinear regime propagation. This spectral broadening has been theoretically and experimentally explored at low pressures, and studies indicate the extent of the broadening is highly pressure dependent [98,105]. The only filament property that is anticipated to be unaffected by altitude change is the clamped intensity of the filament [98,100,106]. The larger filament diameter and the increase in the energy coupled into the filament (from the increase in critical power) balance to maintain a clamped intensity of $\sim 10^{13}$ W/cm^2 even as the pressure decreases and other filament properties change. These changes in filament properties at high altitude may impact and even benefit filament applications like those described in the previous sections. Applying burst mode to high-altitude scenarios would allow for even more energy to be coupled into the burst due to the higher powers of the individual filaments within the burst. Ablation and material interactions in both burst mode and single pulse ought to be explored for low-pressure scenarios. Recent advances in understanding fundamental filament propagation at high altitudes will lead to effective implementation of filament applications in these conditions.

13.8 The future of filament propagation through the atmosphere

In this report, some of the applications of filamentation explored in recent times have been summarized and placed in the context of filamentation studies performed over the past 25 years. Nonetheless, no pretense is made of this review being comprehensive of all current applications of filamentation, let alone those of the future. Most reviewers would be unwise to do so.

This being said, some predictors are offered of where future applications will have impact.

Most filamentation studies have been made with Ti:sapphire CPA laser systems. Even recent studies at longer wavelength [107] utilize these same systems. This technology will never be amenable for compact, rugged systems that can be deployed on mobile platforms or in adverse environments. New technologies based on fiber laser and solid-state hybrid designs will accomplish this goal. The Laser Plasma Laboratory (LPL) at UCF has already developed many of the important elements of such a system, including all-fiber mode-locked oscillators [108], pulsed fiber amplifiers, and high-power broadband dispersion optics. The real challenges to this technology are in the thermal management of the pump and laser systems to permit high repetition rates at high powers.

Few comprehensive application, studies have so far been demonstrated in the field. This will change dramatically over the next few years because funding agencies are becoming more focused on commercial or governmental utility, and also because long-distance laser ranges such as UCF's TISTEF [29] are equipped for high-intensity laser propagation studies and satisfy compliance regulations for radiation and security.

Lastly, a general trend that is evident from today's research programs is that as new applications develop to higher levels of technical readiness, more R&D funds become available that further accelerates the evolution and progress of new ideas in the field.

Acknowledgments

The authors gratefully acknowledge conversations with Professors Jean-Claude Diels, Andre Mysyrowicz, Arnaud Couairon, and others in the filamentation community, the technical support of new graduate students entering our laboratory, Owen Thome, LaShae Smith, and contractual support from Army Research Office, Air Force Office of Scientific Research, the High Energy Laser Joint Technology Office, (HE-JTO), and the State of Florida.

References

[1] A. Braun, G. Korn, X. Liu, D. Du, J. Squier, and G. Mourou, Self-channeling of high-peak-power femtosecond laser pulses in air. *Opt. Lett.* 20, 73–75 (1995).

[2] E. T. J. Nibbering, P. F. Curley, G. Grillon, *et al.*, Conical emission from self-guided femtosecond pulses in air. *Opt. Lett.* 21, 62–64 (1996).

[3] A. Brodeur, F. A. Ilkov, and S. L. Chin, Beam filamentation and the white light continuum divergence. *Opt. Commun.* 129, 193–198 (1996).

[4] X. M. Zhao, J. C. Diels, C. Y. Wang, and J. M. Elizondo, Femtosecond ultraviolet laser pulse induced lightning discharges in gases. *IEEE J. Quantum Electron.* 31, 599–612 (1995).

[5] J. R. Dwyer, H. K. Rassoul, M. Al-Dayeh *et al.*, A ground level gamma-ray burst observed in association with rocket-triggered lightning. *Geophys. Res. Lett.* 31 (2004).

[6] D. W. Koopman and T. D. Wilkerson, Channeling of an ionizing electrical streamer by a laser beam. *J. Appl. Phys.* 42, 1883–1886 (1971).

[7] A. Thoss, in *Laser Focus World*. (2021).

[8] A. Iwasaki, N. Aközbeki, B. Ferlandi, *et al.*, A LIDAR technique to measure the filament length generated by a high-peak power femtosecond laser pulse in air. *Appl. Phys. B* 76, 231–236 (2003).

[9] M. Petrarca, S. Heninand, N. Berti, *et al.*, White-light femtosecond Lidar at 100 TW power level. *Appl. Phys. B* 114, 319–325 (2014).

[10] S. Rostami, M. Chini, K. Lim, *et al.*, Dramatic enhancement of supercontinuum generation in elliptically-polarized laser filaments. *Sci. Rep.* 6, 20363 (2016).

[11] S. Mitryukovskiy, Yi Liu, P. Ding, A. Houard, A. Couairon, and A. Mysyrowicz, Plasma luminescence from femtosecond filaments in air: Evidence for impact excitation with circularly polarized light pulses. *Phys. Rev. Lett.* 114, 063003 (2015).

[12] Q. Luo, S. A. Hosseini, B. Ferland, and S. L. Chin, Backward time-resolved spectroscopy from filament induced by ultrafast intense laser pulses. *Opt. Commun.* 233, 411–416 (2004).

[13] H. L. Xu, J. Bernhardt, P. Mathieu, G. Roy, and S. L. Chin, Understanding the advantage of remote femtosecond laser-induced breakdown spectroscopy of metallic targets. *J. Appl. Phys.* 101, 033124 (2007).

[14] K. Stelmaszczyk, P. Rohwetter, G. Méjean, *et al.*, Long-distance remote laser-induced breakdown spectroscopy using filamentation in air. *Appl. Phys. Lett.* 85, 3977–3979 (2004).

[15] S. Tzortzakis, D. Anglos, and D. Gray, Ultraviolet laser filaments for remote laser-induced breakdown spectroscopy (LIBS) analysis: Applications in cultural heritage monitoring. *Opt. Lett.* 31, 1139–1141 (2006).

[16] Q. Luo, H. Xu, S. Hosseini *et al.*, Remote sensing of pollutants using femtosecond laser pulse fluorescence spectroscopy. *Appl. Phys. B* 82, 105–109 (2006).

[17] N. Melikechi, R. Wiens, H. Newsom, and S. Maurice, Zapping Mars: Using lasers to determine the chemistry of the red planet. *Opt. Photonics News* 24, 26–33 (2013).

[18] D. Faccio, A. Couairon, and P. Di Trapani, Stimulated Raman X waves in ultrashort optical pulse filamentation. *Opt. Lett.* 32, 184–186 (2007).

[19] F. Calegari, C. Vozzi, M. Negro, S. D. Silvestri, and S. Stagira, Filamentation-assisted time-resolved Raman spectroscopy in molecular gases. *J. Mod. Opt.* 57, 967–976 (2010).

[20] D. J. Cook and R. M. Hochstrasser, Intense terahertz pulses by four-wave rectification in air. *Opt. Lett.* 25, 1210–1212 (2000).

[21] T. Bartel, P. Gaal, K. Reimann, M. Woerner, and T. Elsaesser, Generation of single-cycle THz transients with high electric-field amplitudes. *Opt. Lett.* 30, 2805–2807 (2005).

[22] M. Kress, T. Löffler, S. Eden, M. Thomson, and H. G. Roskos, Terahertz-pulse generation by photoionization of air with laser pulses composed of both fundamental and second-harmonic waves. *Opt. Lett.* 29, 1120–1122 (2004).

[23] X. Xie, J. Dai, and X. C. Zhang, Coherent control of THz wave generation in ambient air. *Phys. Rev. Lett.* 96, 075005 (2006).

[24] J. Kasparian and J. P. Wolf, Ultrafast laser spectroscopy and control of atmospheric aerosols. *Phys. Chem. Chem. Phys.* 14, 9291–9300 (2012).

[25] C. Jeon, D. Harper, K. Lim, *et al.*, Interaction of a single laser filament with a single water droplet. *J. Opt.* 17, 5 (2015).

[26] P. Rohwetter, J. Kasparian, K. Stelmaszczyk, *et al.*, Laser-induced water condensation in air. *Nat. Photonics* 4, 451–456 (2010).
[27] N. Barbieri, M. Weidman, G. Katona *et al.*, Double helical laser beams based on interfering first-order Bessel beams. *J. Opt. Soc. Am. A* 28, 1462–1469 (2011).
[28] N. Barbieri, Z. Hosseinimakarem, K. Lim, *et al.*, Helical filaments. *Appl. Phys. Lett.* 104, 261109 (2014).
[29] D. Thul, R. Bernath, N. Bodnar, *et al.*, The mobile ultrafast high energy laser facility – A new facility for high-intensity atmospheric laser propagation studies. *Opt. Lasers Eng.* 140, 106519 (2021).
[30] Z. A. Kudyshev, M. C. Richardson, and N. M. Litchinitser, Virtual hyperbolic metamaterials for manipulating radar signals in air. *Nat. Commun.* 4, 2557 (2013).
[31] G. Méchain, G. Méjean, R. Ackermann, *et al.*, Range of plasma filaments created in air by a multi-terawatt femtosecond laser. *Opt. Commun.* 247, 171–180 (2005).
[32] M. Durand, A. Houard, B. Prade, *et al.*, Kilometer range filamentation. *Opt. Express* 21, 26836–26845 (2013).
[33] G. DiComo, M. Helle, D. Kaganovich, A. Schmitt-Sody, J. Elle, and J. Peñano, Nonlinear self-channeling of high-power lasers through controlled atmospheric turbulence. *J. Opt. Soc. Am. B* 37, 797–803 (2020).
[34] M. Mlejnek, E. M. Wright, and J. V. Moloney, Dynamic spatial replenishment of femtosecond pulses propagating in air. *Opt. Lett.* 23, 382–384 (1998).
[35] M. Kolesik and J. V. Moloney, Self-healing femtosecond light filaments. *Opt. Lett.* 29, 590–592 (2004).
[36] S. Skupin, L. Bergé, U. Peschel, and F. Lederer, Interaction of femtosecond light filaments with obscurants in aerosols. *Phys. Rev. Lett.* 93, 023901 (2004).
[37] C. Jeon, PhD Dissertation, University of Central Florida, Orlando, FL (2016).
[38] A. Couairon and A. Mysyrowicz, Femtosecond filamentation in transparent media. *Phys. Rep. Rev. Sec. Phys. Lett.* 441, 47–189 (2007).
[39] J. Liu, Z. Duan, Z. Zheng, *et al.*, Time-resolved investigation of low-density plasma channels produced by a kilohertz femtosecond laser in air. *Phys. Rev. E* 72, 026412 (2005).
[40] Y. H. Chen, S. Varma, T. M. Antonsen, and H. M. Milchberg, Direct measurement of the electron density of extended femtosecond laser pulse-induced filaments. *Phys. Rev. Lett.* 105, 215005 (2010).
[41] D. Reyes, M. Baudelet, M. Richardson, and S. R. Fairchild, Transition from linear- to nonlinear-focusing regime of laser filament plasma dynamics. *J. Appl. Phys.* 124, (2018).
[42] D. Reyes, J. Pena, W. Walasik, N. Litchinitser, S.R. Fairchild, and M. Richardson, Filament conductivity enhancement through nonlinear beam interaction. *Opt. Express* 28, 26764–26773 (2020).
[43] G. Fibich and A. L. Gaeta, Critical power for self-focusing in bulk media and in hollow waveguides. *Opt. Lett.* 25, 335–337 (2000).

[44] W. Liu and S. L. Chin, Direct measurement of the critical power of femtosecond Ti:sapphire laser pulse in air. *Opt. Express* 13, 5750–5755 (2005).
[45] A. Couairon and L. Bergé, Infrared femtosecond light filaments in air: Simulations and experiments. *J. Opt. Soc. Am. B* 19, 1117–1131 (2002).
[46] A. Couairon and L. Bergé, Modeling the filamentation of ultra-short pulses in ionizing media. *Phys. Plasmas* 7, 193–209 (2000).
[47] F. Vidal and T. W. Johnston, Electromagnetic beam breakup: Multiple filaments, single beam equilibria, and radiation. *Phys. Rev. Lett.* 77, 1282–1285 (1996).
[48] K. Lim, M. Durand, M. Baudelet, and M. Richardson, Transition from linear-to nonlinear-focusing regime in filamentation. *Sci. Rep.* 4, 8 (2014).
[49] M. Châteauneuf, S. Payeur, J. Dubois, and J.-C. Kieffer, Microwave guiding in air by a cylindrical filament array waveguide. *Appl. Phys. Lett.* 92, 091104 (2008).
[50] B. Shim, in Frontiers in Optics 2008/Laser Science XXIV/Plasmonics and Metamaterials/Optical Fabrication and Testing. (Optical Society of America, Rochester, NY, 2008), pp. FTuV4.
[51] J. Ding, Z. Liu, S. Sun, Y. Shi, and B. Hu, Spatiotemporal pulse-splitting of a filamentary femtosecond laser pulse via multi-filament interaction. *Laser Phys.* 25, 105401 (2015).
[52] A. A. Kolomenskii, J. Strohabe, N. Kaya, G. Kaya, A. V. Sokolov, and H. A. Schuessler, White-light generation control with crossing beams of femtosecond laser pulses. *Opt. Express* 24, 282–293 (2016).
[53] S. R. Fairchild, W. Walasik, D. Kepler, *et al.*, Free-space nonlinear beam combining for high intensity projection. *Sci. Rep.* 7, 10147 (2017).
[54] A. A. Ishaaya, T. D. Grow, S. Ghosh, L. T. Vuong, and A. L. Gaeta, Self-focusing dynamics of coupled optical beams. *Phys. Rev. A* 75, 023813 (2007).
[55] Y. Y. Ma, X. Lu, T. T. Xi, Q. H. Gong, and J. Zhang, Filamentation of interacting femtosecond laser pulses in air. *Appl. Phys. B* 93, 463–468 (2008).
[56] Hua Cai, Jian Wu, Hao Li, Xueshi Bai, and Heping Zeng, Attraction and repulsion of parallel femtosecond filaments in air. *Phys. Rev. A* 80, 051802 (2009).
[57] A. C. Bernstein, M. McCormick, G. M. Dyer, J. C. Sanders, and T. Ditmire, Two-beam coupling between filament-forming beams in air. *Phys. Rev. Lett.* 102, 123902 (2009).
[58] R. Bernath, Dissertation, University of Central Florida, Orlando, FL (2006).
[59] J. Zhonggang, J. Zhu, Z. Wang, *et al.*, Low resistance and long lifetime plasma channel generated by filamentation of femtosecond laser pulses in air. *Plasma Sci. Technol.* 12, 295–299 (2010).
[60] Xiao-Long Liu, Xin Liu, Jing-Long Ma, *et al.*, Long lifetime air plasma channel generated by femtosecond laser pulse sequence. *Opt. Express* 20, 5968–5973 (2012).
[61] X. Lu, S. Y. Chen, J. L. Ma, *et al.*, Quasi-steady-state air plasma channel produced by a femtosecond laser pulse sequence. *Sci. Rep.* 5, 15515 (2015).

[62] D. Reyes, H. Kerrigan, J. Pena, *et al.*, Temporal stitching in burst-mode filamentation. *J. Opt. Soc. Am. B Opt. Phys.* 36, G52–G56 (2019).
[63] P. Rohwetter, K. Stelmaszczyk, L. Wöste, *et al.*, Filament-induced remote surface ablation for long range laser-induced breakdown spectroscopy operation. *Spectrochim. Acta, Part B At. Spectrosc.* 60, 1025–1033 (2005).
[64] M. Weidman, K. Lim, M. Ramme, M. Durand, M. Baudelet, and M. Richardson, Stand-off filament-induced ablation of gallium arsenide. *Appl. Phys. Lett.* 101, 034101 (2012).
[65] A. Valenzuela, C. Munson, A. Porwitzky, M. Weidman, and M. Richardson, Comparison between geometrically focused pulses versus filaments in femtosecond laser ablation of steel and titanium alloys. *Appl. Phys. B Lasers Opt.* 116, 485–491 (2014).
[66] M. Weidman, M. Ramme, B. Bousquet, *et al.*, Angular dependence of filament-induced plasma emission from a GaAs surface. *Opt. Lett.* 40, 4548–4551 (2015).
[67] B. N. Chichkov, C. Momma, S. Nolte, F. von Alvensleben, and A. Tunnermann, Femtosecond, picosecond and nanosecond laser ablation of solids. *Appl. Phys. A Mater. Sci. Process.* 63, 109–115 (1996).
[68] S. Nolte, C. Momma, H. Jacobs, *et al.*, Ablation of metals by ultrashort laser pulses. *J. Opt. Soc. Am. B Opt. Phys.* 14, 2716–2722 (1997).
[69] A. Zoubir, L. Shah, K. Richardson, and M. Richardson, Practical uses of femtosecond laser micro-materials processing. *Appl. Phys. A* 77, 311–315 (2003).
[70] L. Shah, J. Tawney, M. Richardson, and K. Richardson, Femtosecond laser deep hole drilling of silicate glasses in air. *Appl. Surf. Sci.* 183, 151–164 (2001).
[71] A. Semerok, C. Chaléard, V. Detalle, *et al.*, Experimental investigations of laser ablation efficiency of pure metals with femto, pico and nanosecond pulses. *Appl. Surf. Sci.* 138–139, 311–314 (1999).
[72] M. Weidman, University of Central Florida, Orlando, FL (2012).
[73] Z. J. Xu, W. Liu, N. Zhang, M. W. Wang, and X. N. Zhu, Effect of intensity clamping on laser ablation by intense femtosecond laser pulses. *Opt. Express* 16, 3604–3609 (2008).
[74] S. S. Harilal, J. Yeak, and M. C. Phillips, Plasma temperature clamping in filamentation laser induced breakdown spectroscopy. *Opt. Express* 23, 27113–27122 (2015).
[75] J. Scaffidi, W. Pearman, J. C. Carter, B. W. Colston, and S. M. Angel, Temporal dependence of the enhancement of material removal in femtosecond–nanosecond dual-pulse laser-induced breakdown spectroscopy. *Appl. Optics* 43, 6492–6499 (2004).
[76] F. Theberge and S. L. Chin, Enhanced ablation of silica by the superposition of femtosecond and nanosecond laser pulses. *Appl. Phys. A Mater. Sci. Process.* 80, 1505–1510 (2005).
[77] Cheng-Hsiang Lin, Zheng-Hua Rao, Lan Jiang, *et al.*, Investigations of femtosecond–nanosecond dual-beam laser ablation of dielectrics. *Opt. Lett.* 35, 2490–2492 (2010).

[78] H. Kerrigan, S. R. Fairchild, and M. Richardson, Nanosecond laser coupling for increased filament ablation. *Opt. Lett.* 44, 2594–2597 (2019).

[79] H. Kerrigan, M. Masnavi, R. Bernath, S. R. Fairchild, and M. Richardson, Laser-plasma coupling for enhanced ablation of GaAs with combined femtosecond and nanosecond pulses. *Opt. Express* 29, 18481–18494 (2021).

[80] R. Bernath, C. Brown, J. Aspiotis, M. Fisher, and M. Richardson, Shock-wave generation in transparent media from ultra-fast lasers. *Defense and Security Symposium* (SPIE, 2006), vol. 6219.

[81] P. J. Skrodzki, M. Burger, and I. Jovanovic, Transition of femtosecond-filament-solid interactions from single to multiple filament regime. *Sci. Rep.* 7, 8 (2017).

[82] H. Kerrigan, University of Central Florida, Orlando, FL (2021).

[83] F. S. Felber, Dipole radio-frequency power from laser plasmas with no dipole moment. *Appl. Phys. Lett.* 86, 231501 (2005).

[84] Z.-Y. Chen, J.-F. Li, J. Li, and Q.-X. Peng, Microwave radiation mechanism in a pulse-laser-irradiated Cu foil target revisited. *Phys. Scr.* 83, 055503 (2011).

[85] H. Hamster, A. Sullivan, S. Gordon, W. White, and R. W. Falcone, Subpicosecond, electromagnetic pulses from intense laser-plasma interaction. *Phys. Rev. Lett.* 71, 2725–2728 (1993).

[86] D. von der Linde and H. Schüler, Breakdown threshold and plasma formation in femtosecond laser–solid interaction. *J. Opt. Soc. Am. B* 13, 216–222 (1996).

[87] D. von der Linde, K. Sokolowski-Tinten, and J. Bialkowski, Laser–solid interaction in the femtosecond time regime. *Appl. Surf. Sci.* 109–110, 1–10 (1997).

[88] N. Barbieri, College of Sciences, University of Central Florida, Orlando, FL (2013).

[89] J. Yu, D. Mondelain, J. Kasparian, *et al.*, Sonographic probing of laser filaments in air. *Appl. Opt.* 42, 7117–7120 (2003).

[90] L. Sedov, Propagation of strong blast waves. *Prikl. Mat. Mekh.* 10, 241–255 (1946).

[91] G. Taylor, The formation of a blast wave by a very intense explosion. *Proc. R. Soc. London, Ser. A* 201, 175–186 (1950).

[92] Y. Zel'dovich and Y. Raizer, in *Physics of Shock Waves and High-Temperature Hydrodynamic Phenomena*. (Dover Publications, Mineola, NY, 2002).

[93] L. de la CruzE. Schubert, D. Mongin *et al.*, High repetition rate ultrashort laser cuts a path through fog *Appl. Phys. Lett.* 109, 4 (2016).

[94] G. Schimmel, T. Produit, D. Mongin, J. Kasparian, and J. P. Wolf, Free space laser telecommunication through fog. *Optica* 5, 1338–1341, (2018).

[95] M. Richardson and R. Bernath, Presented at DEPS Ultrashort Pulse Laser Workshop, Boulder CO, Sept. 3–5, 2008).

[96] P. J. Skrodzki, M. Burger, and I. Jovanovic, Transition of femtosecond-filament-solid interactions from single to multiple filament regime. *Sci. Rep.* 7, 12740 (2017).

[97] J. Peña, H. Kerrigan, and M. Richardson, Shockwave enhancement from temporally separated filaments interacting with a water droplet *J. Opt. Soc. Am. B*, (2021).

[98] I. Dicaire, V. Jukna, C. Praz, C. Milián, L. Summerer, and A. Couairon, Spaceborne laser filamentation for atmospheric remote sensing. *Laser Photonics Rev.* 10, 481–493 (2016).

[99] N. L. Aleksandrov, S. B. Bodrov, M. V. Tsarev *et al.*, Decay of femtosecond laser-induced plasma filaments in air, nitrogen, and argon for atmospheric and subatmospheric pressures. *Phys. Rev. E* 94, 013204 (2016).

[100] D. Reyes, College of Sciences, University of Central Florida, Orlando, FL (2020).

[101] Á. Börzsönyi, Z. Heiner, A. P. Kovács, M. P. Kalashnikov, and K. Osvay, Measurement of pressure dependent nonlinear refractive index of inert gases. *Opt. Express* 18, 25847–25854 (2010).

[102] S. A. Hosseini, O. Kosareva, N. Panov, *et al.*, Femtosecond laser filament in different air pressures simulating vertical propagation up to 10 km. *Laser Phys. Lett.* 9, 868–874 (2012).

[103] A. Couairon, J. Biegert, C. P. Hauri, U. Keller, and A. Mysyrowicz, Femtosecond filamentation in air at low pressures: Part I: Theory and numerical simulations. *Opt. Commun.* 259, 265–273 (2006).

[104] G. Méchain, M. Francoand, A. Couairon, T. Olivier, B. Prade, and A. Mysyrowicz, Femtosecond filamentation in air at low pressures. Part II: Laboratory experiments. *Opt. Commun.* 261, 322–326 (2006).

[105] C. Jeon, S. Rostami L. Shah, M. Baudelet, and MM. Richardson, in *Propagation Through and Characterization of Atmospheric and Oceanic Phenomena*. (Optical Society of America, Washington, DC, 2016), pp. Tu2A.3.

[106] X. Qi, C. Ma, and W. Lin, Pressure effects on the femtosecond laser filamentation. *Opt. Commun.* 358, 126–131 (2016).

[107] T. Popmintchev, Ming-Chang Chen, D. Popmintchev, *et al.*, Bright coherent ultrahigh harmonics in the keV x-ray regime from mid-infrared femtosecond lasers. *Science* 336(6086), 1287–1291 (2012).

[108] N. Bodnar *et al.* (to be published).

Index